W9-CHI-052

Practical Machinery Management for Process Plants
Volume 4

Major Process Equipment Maintenance and Repair

Gulf Publishing Company
Book Division
Houston, London, Paris, Tokyo

Practical Machinery Management for Process Plants
Volume 4

Major Process Equipment Maintenance and Repair

Heinz P. Bloch
Fred K. Geitner

Dedicated with gratitude to those who taught us, who inspired us, and who gave us their support and encouragement.

Practical Machinery Management for Process Plants
Volume 4

Major Process Equipment Maintenance and Repair

Library of Congress Cataloging in Publication Data

Bloch, Heinz P., 1933–

Major process equipment maintenance and repair.

(Practical machinery management for process plants; v. 4)
Bibliography: p.
Includes index.
1. Chemical plants—Equipment and supplies—Maintenance and repair. I. Geitner, Fred K.
II. Title. III. Series.

TP157.B56 1985 660.2′8′0288 84-15782

ISBN 0-87201-454-1
Series ISBN 0-87201-675-7

First Printing, April 1985
Second Printing, May 1989

Note: The reader is reminded that many of the techniques and procedures described herein are of a general nature and may have to be modified or adapted to be directly applicable to the specific machinery in his plant. In case of conflict, observe the manufacturer's instructions or ask the manufacturer to assist in resolving any differences.

Contents

Foreword

The readers of the four volumes on "Machinery Management" can be divided, in my opinion, into three categories:

- those who can say: That's exactly what happened to me back in 19-- !
- those who can say: Why didn't I know this back in 19-- ?!
- those who can say: I hope I'll remember all this when I am in charge!

In other words, those with a lot, a little, and no experience stand to benefit from studying these four volumes. Maybe some of the people with a lot of experience could find other ways to solve a particular case, but even they cannot match the knowledge and experience that the authors amassed in these books.

In the past, many a good Machinery Manager was "made" through many years of experience, and also through many costly mistakes. These "experts" passed on their experience to the people they worked with, but seldom could experience gained in one particular location prepare someone for the multitude of things that can go wrong. It is because of this that the authors must be commended for their effort to disseminate not only their experience, but also the lessons they learned from many other experts.

Volume 4 complements the first three books by focusing on major equipment installation and repair—foundations, pumps, blowers, turbines, electric motors, and lubrication and storage. These four volumes contain a wealth of information on machinery found in most petrochemical plants, and in their quest for perfection, three principal groups will benefit from this text: Those who design machinery, those who maintain machinery, and those who operate machinery.

As a manufacturer of machinery, I realize that only knowledgeable people can fully utilize our efforts to make the best machines, to give guidelines on how to optimally maintain these machines, and finally how to best operate these machines. Used in conjunction with the preceding three volumes or used alone, this book will make the reader a knowledgeable person.

Michael M. Calistrat
Baltimore, MD

Acknowledgments

"Our best thoughts come from others".
- Ralph Waldo Emerson

As in the preceding volumes of our four-part series of books on process machinery management we had to depend on the expert input of many individuals and companies to compile this material. We extend our thanks and appreciation to these able collaborators and contributors to Volume 4:

Union Pump Company, Canada (Horizontal Centrifugal Pumps)
Terry L. Henshaw (Reciprocating Pumps)
Pacific Pumps—Dresser (Centrifugal Pumps)
Goulds Pumps, Inc. (Centrifugal Pumps)
Byron Jackson Pump Division (Vertical Pumps)
John W. Dufour (Machinery Installation Guidelines)
Perry C. Monroe (Rotating Equipment Checklists)
J. V. Picknell and Nash Engineering Co. (Liquid Ring Vacuum Pumps)
SIHI (Liquid Ring Vacuum Pumps)
Henry Y. Hung and M-D Pneumatics, Inc. (Positive Displacement Rotary Blowers)
Canadian Blower—Canada Pumps Ltd. (Large Fan Blowers)
Mixing Equipment Co., Inc. (Mixers and Agitators)
Cooper Energy Services (Reciprocating Compressors and Gas Engines)
James R. Partridge (Lufkin Industries—Power Transmission Gears)
T. B. Woods Company (V-Belt Drives)
Westinghouse Canada, Inc. (Special Purpose Steam Turbines)
S. W. Mazlack (Break-In of Steam Turbine Carbon Ring Seals)
Elliott Company (General Purpose Steam Turbines)
Rotoflow Corporation (Turboexpanders)
Brian Turner (On-Stream Cleaning of Turbomachinery)
D. H. Jacobson and Westinghouse Canada, Inc. (Gas Turbines)
R. S. Adamski and Woodward Governor Co. (Hydraulic Governors)
Delta Enterprises (Sarnia) Ltd. (Electric Motors and Apparatus)
Bob Vigna/Ashland Oil (Electric Motor Repairs)
Electrical Apparatus Service Association (Standards for the Electrical Apparatus Sales and Service Industry)
A. M. Clapp (Lubrication Concepts, Training, Application Methods)
P. E. Knoeller—Exxon Company USA (Oil Mist, Greases)
Bently-Nevada (Vibration Measurement)
J. S. Mitchell and J. L. Frarey (Machinery Condition Monitoring)

Again, we thank our experienced colleague and friend Uri Sela for unselfishly giving of his personal time to review and improve our work. Bill Clark, Sig Zierau, Greg Piegari and especially Art Parente deserve our gratitude for manuscript screening and support in securing Exxon Chemical Company approval to publish. As always, we are indebted to our editor, Brad Sagstetter, for his help in getting it all together.

Heinz P. Bloch
Fred K. Geitner

Part I

Installation and Repair of Major Process Equipment

Chapter 1

Installation, Maintenance, and Repair of Horizontal Pumps

The most common centrifugal pump in the petrochemical industry is the horizontal single stage process pump. This pump has many different external designs. Perhaps the most common is the end suction top discharge design shown in Figure 1-1.

There are many features about this pump that make it adaptable for most applications. Designs can be small and inexpensive, or they can be designs that meet API 610* standards as well as AVS** with ANSI*** specifications. The top centerline discharge provides excellent stability when subjected to piping stresses and high temperatures. Larger pump models incorporate a double volute internal passageway that helps to balance radial loading on the impeller. This pump design has a vertical radial split casing with centerline supports and an overhung impeller mounted on a shaft supported by bearings. By changing impeller designs, this pump can be adapted to all kinds of product applications from light hydrocarbons to slurries.

AVS** pumps differ from API* designs as follows: They are chemical process pumps designed in accordance with ANSI B73.1-1974 (horizontal end suction) and ANSI 72.2-1975 (vertical inline). AVS pumps (Figure 1-2) are generally supplied with open impellers.

Temperatures are usually limited to 300°F and pressures to 300 psi maximum, depending on the material and flange type. Capacity ranges from 0–5,000 gpm, and materials are mostly ductile iron cases and impellers. Often stainless steel is used together with 316 stainless steel shaft

* API = American Petroleum Institute
** AVS = American Voluntary Standard
*** ANSI = American National Standards Institute

Figure 1-1. Horizontal single stage process pump to API (American Petroleum Institute) Standard. (Courtesy Byron Jackson.)

Figure 1-2. Typical AVS horizontal process pump with foot mounted casing. (Courtesy Byron Jackson.)

sleeves. Pump suction and discharge will normally have 150 lb raised face flanges.

Mechanical seals provided in AVS pumps are normally unbalanced, single inside, but single outside seals are also quite common. Face materials are often carbon versus ceramic or tungsten carbide. Other materials can be substituted where applicable. Seal flush is usually configured as recirculation from pump discharge.

Motors: TEFC (totally enclosed fan cooled) 460 volt (560 in Canada), three phase at 60 Hertz are standard drivers for North American applications.

Base Plates: Normally fabricated from steel plate with smaller base plates cast. Pump and motor are mounted on the base plate and connected with a coupling. For maintenance and repair work the coupling will have to be removed and the pump internals can be removed from the pump case without disturbing the piping.

AVS vertical in-line pumps are made in three basic designs: Style "A" is identified by the rigid spacer coupling which connects the pump stub shaft to the motor shaft. This design allows pump mechanical seal and impeller to be removed without disturbing the motor or pump flanges. All radial and thrust loads are transferred to the motor bearings. This style of pump is shown in Figure 1-3.

Figure 1-3. Vertical inline centrifugal pump. Rigid coupling, impeller, stuffing box and mechanical seal can be removed without disturbing motor and piping. (Courtesy Union Pump (Canada) Ltd.)

Style "B" consists of an AVS horizontal pump mounted in vertical position with a special in-line casing and motor support. The motor is mounted on top of the support and is connected to the pump with a flexible coupling that allows pull out of pump without disturbing the piping. The advantage of this pump design is that radial and thrust conditions are controlled by the pump bearings rather than the motor bearings. Also some parts are interchangeable with horizontal models. Figure 1-4 shows this style of pump.

Style "C" is the close-coupled design. The motor shaft is extended, and the impeller and mechanical seal mounted on it (no separate pump shaft needed). One disadvantage with this is that if anything goes wrong with the seal or pump, it can also cause damage to the motor. This design

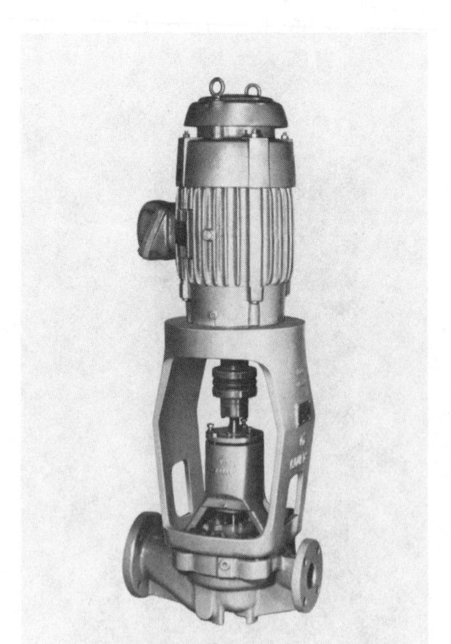

Figure 1-4. Vertical inline pump. Pump shaft is supported with its own independent bearings which also protect against radial thrust and shaft run-out. Source: Duriron Co.

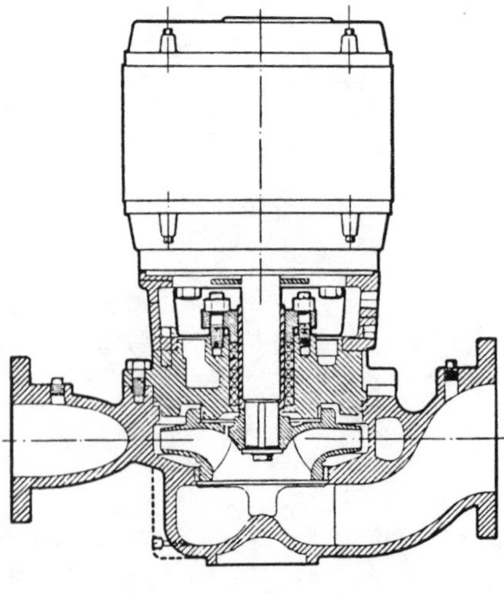

Figure 1-5. Vertical inline pump, close coupled design. Impeller and mechanical seal are mounted on motor shaft.

(see Figure 1-5) is being used less and less. API vertical in-line pumps will be discussed later.

API (American Petroleum Institute) style pumps are designed for petrochemical services in accordance with API Standard 610. API standards define minimum requirements for pumps in heavy duty hydrocarbon service.

API pumps are generally specified in steel or noncorrosive materials with 300 lb raised face (RF) flanges. The pump casings are available with centerline supports rather than foot supports to reduce alignment distortion at high temperatures. The pressure limitations are at approximately 700 PSI with a maximum temperature of 850°F. API pumps are required to have closed impellers with case and impeller wear rings.

The base plate and motor requirements will be the same as the AVS pumps, although the procurement of sturdier baseplates is advisable.

Principles of Installation of Pumps and Drivers*

The correct installation of pumps and drivers is an often overlooked requirement. Incorrect installation indirectly costs millions of dollars a year in increased maintenance and lost production due to premature equipment failure.

This segment of our text provides a set of guidelines that will result in a good pump installation. While centered around a typical single-stage, overhung, centrifugal pump, and a motor driver, it can easily be adapted to all types of machinery—the principles are the same. And, while these guidelines prescribe the minimum requirements to be performed by the installing agency, any specific instructions provided by the equipment manufacturer should also be observed. Conflicts should be resolved prior to installation. Refer also to checklists in Appendix 1B.

Preinstallation and Equipment Preservation Measures

When a pump is shipped from the manufacturer to the field, it normally is "suitably prepared" for short duration storage up to approximately six months. It is very important that the integrity of the equipment be maintained during the construction phase of the installation. Many pieces of equipment have been ruined before they had a chance to operate, because of mishandling in the field prior to unit start-up.

* Source: J. W. Dufour, Amoco Oil Company, Chicago, IL.

A good preinstallation program should accomplish the following:

1. Inspect all equipment upon arrival for any shipping damage.
2. Ensure good lifting practices are followed when transporting all equipment; a pressure gauge makes a vulnerable spot to place a lifting strap while off-loading a pump.
3. All nozzles and openings should be kept covered or plugged until piping is attached. Besides keeping out the elements, this will prevent foreign material such as welding rods, rags, waste paper, etc., from getting in the machine and causing damage. Disassembling a pump during a unit start-up to remove debris can be quite expensive.
4. If more than six months will pass before the equipment is expected to run, consideration should be given to respraying the pump internals with a suitable rust preventive. Better yet, the pump could be preserved by the application of oil mist. Ensure that whatever is used is compatible with any elastomers it may come in contact with, is easily removable, and is applied according to manufacturer's instructions.
5. Fill all oil-lubricated bearings with the proper lube oil as soon as possible. If the bearing is to be oil mist lubricated, consider attaching and using the oil mist generator during the construction phase of the project. If not, fill those bearings with oil also. Greased bearings present a different problem. All greased bearings should be repacked with the correct grease as soon as possible. Follow manufacturer's instructions; however, ensure that *all* the old grease is displaced by the new grease. Different greases have different additives that normally are not compatible with each other. Mixing two noncompatible greases will reduce the beneficial properties of either grease.
6. Coat all exposed machined surfaces with either a rust preventive or grease to protect them from the environment.
7. In order to prevent corrosion of the shaft sleeve, packed pumps received with the packing installed should have the packing completely removed immediately on arrival and the shaft sleeve and gland greased. Normally, packed pumps are shipped with an extra set of packing. Of course, just before starting, this packaging should be installed in the pump.
8. Likewise, steam turbines received with the carbon rings installed should have the rings removed immediately on arrival and the shaft greased. The rings should then be reinstalled just prior to start-up. Caution should be taken when removing and installing rings. Axial ring orientation and location as well as the direction of shaft rota-

tion is critical. Be sure to consult the manufacturer's instruction manual for details.

9. Pump mechanical seals are precision components and therefore require special handling during transport and installation. When moving pumps with seals, the pumps should be securely restrained to prevent excessive vibration and/or damage to the shaft and seal by dropping or bumping the shaft. When installing or lifting the pump, do not use the shaft as a leverage or lifting point.

   On new installations, if a mechanical seal is to fail, it normally will do so within the first few hours of operation. The primary causes can often be traced to improper installation of the seals, or mishandling of seals during pump installation.

Foundation And Anchor Bolts

The design of equipment foundations and the different characteristics of concrete and grouts are thoroughly discussed in Volume 3 of this series, so we will not go into great detail here. However, there are some general guidelines to follow that will ensure a good installation.

1. Assuming that the forms and steel reinforcing rods are all sized and placed according to approved drawings, the next most important step is the placement of the anchor bolts. Prior to the actual concrete pour the anchor bolts should be:

   a. Accurately set according to the foundation drawings and firmly secured to prevent shifting during the pouring process.
   b. Dimensionally checked (and rechecked) versus the foundation drawings for proper length, diameter, thread length, etc.
   c. Checked for proper projection; i.e., checked for correct elevations as referenced to an established benchmark. It can be very embarrassing to set a baseplate on a new foundation and find that the anchor bolts are not long enough to pass through the baseplate and the hold-down mats.
   d. Install metal or plastic anchor bolt sleeves. Remember, sleeves are not intended to encourage careless positioning of the anchor bolts. However, they will allow for slight errors in baseplate hole layouts and small shifting of the anchor bolt during the concrete pouring process.
   e. Ensure that the exposed threads are protected by coating with heavy grease or with paste wax. The exposed bolts should be covered with plastic wrap and the wrap firmly secured with wire.

2. After the pour, the surface of the foundation should be chipped to remove all laitance and defective or weak concrete. Normally, a chipping hammer should be used; sand blasting or using a needle gun is not effective. The amount of concrete removed should be such that the final baseplate or soleplate elevation allows for one to two inches of grout between the surface of the foundation and the lower baseplate flange or the underside of the soleplates. After chipping, the top surface should be reasonably level and free of all oil, grease, and loose particles.
3. Baseplates or soleplates should not be placed on the foundation until a minimum of ten days has elapsed after pouring the normal concrete. High early-strength concrete may be used in some specific applications but is not usually required. In any event, baseplates and soleplates should not be placed on foundations until the concrete has had time to dry and cure so that 85 percent of the shrinkage has taken place.
4. Protect the surface of the foundation according to the type of grout to be used. When using epoxy grout, the concrete surface must be dry at the time the grout is applied. When using cement-based grout, keep the foundation wet for the period of time recommended by the grout manufacturer prior to grouting.
5. If used, remove the tops of the plastic anchor bolt sleeves and ensure that the sleeves are free of foreign material.

Baseplate and Soleplate Preparation

1. While the practice varies from company to company, it is suggested that all equipment be removed from its baseplate or soleplate prior to grouting. This aids in leveling the plate and prevents unwanted distortion of the baseplate. The machinery can easily be reinstalled after the baseplate or soleplate has been grouted.
2. Normally, baseplates and soleplates are provided by the equipment supplier and manufactured in accordance with some company or industrial specification. The installing agency must inspect and verify that the baseplate or soleplate is in accordance with these specifications but, as a minimum, it should have the following:

    a. All baseplate and soleplate surfaces (except on mounting pads and in threaded holes but including the outside edges) that will be in contact with the grout should be coated with an inorganic zinc silicate or other primer compatible with the grout being used. Base metal, blistered, or rusted surfaces are unacceptable.

Note: Depending on the type of epoxy grout used, if the primer has been on the baseplate for an extended period, the surface may gloss over and thus prevent bonding. If this occurs, the baseplate must be stripped of all old primer by sand blasting to near white metal and recoated prior to grouting. Check the manufacturer's instructions carefully to determine if this is a potential problem.

b. Ensure all baseplates are provided with at least one grouting opening in each bulkhead section and/or each 12 sq ft of base area as a minimum. Vent holes should be provided at the corners of each bulkhead compartment. These will ensure that the grout will flow from the pour hole to the extremities of each compartment and that no voids are created by trapped air.
c. The corners of all baseplates and soleplates should be rounded to at least a 20-in. radius. As the grout cures, there will always be some shrinkage. Rounding the corners prevents stress concentrations in the grout that would eventually cause cracking.

3. Before setting the baseplate, ensure that all surfaces to be in contact with the grout are free of oil, grease, and rust.
4. Position the baseplate or soleplate on the prepared foundation, supporting it on leveling screws, rectangular leveling shims, or wedges having a small taper. These support pieces should be placed next to each foundation bolt to prevent distortion. Cover all leveling screws with grease or a heavy paste wax to prevent the grout from adhering. If using an epoxy grout, wax, mask off, or grease all areas that require protection from grout splatter.
5. Use a precision level and level the baseplate or soleplate side-to-side, end-to-end, and diagonally to within .002 in. per ft. Remember, it is mandatory that the machined mounting surfaces be flat and parallel. Corresponding mounting surfaces in the same place should be within .002 in. parallel overall. This mounting surface tolerance must be maintained after all anchor bolts have been adequately tightened. This will prevent overstressing and distortion of the equipment and/or base once the machinery is remounted.
6. After leveling, check that all support wedges or shims are in contact with the foundation and plate, then tighten the foundation bolts evenly but not too tight, and recheck level.
7. Check the elevation of the machined mounting surfaces of the baseplate or soleplates. Remember, the proper elevation should allow for a minimum of 1/8-in. shim thickness under the equipment. If everything checks out properly, the baseplate is ready for grouting.

Grouting Overview*

Epoxy Grout

1. Timing and proper mixing are the keys to successful grouting. The grout supplier's instructions must be followed implicitly. Before mixing the various components, everything else should be ready—surfaces cleaned and dry, forms completed and sealed, pushing tools, rags, cleaning solvents available, and adequate manpower allocated.

   Note: In general, epoxy grout is flammable, toxic, poisonous, and corrosive. Therefore, material should be kept away from open flame, high heat sources, or sparks. It should be mixed in a well-ventilated area. Workmen should wear eye protection, gloves, and protective clothing at all times during mixing and placing of grout and aggregate.

2. Grout forms should be built of materials of adequate strength and should be securely anchored and shored to withstand the pressure of the grout under working conditions.

   Epoxy grout forms must be coated with a paste wax, e.g., colored floor type wax, on areas that will come in contact with the grout to keep them from becoming bonded to the grout. All wax should be removed from the concrete, baseplate, or soleplate before grouting. To permit easy clean-up, wax or cover all surfaces where grout may splash.

   Forms should be liquid tight because epoxy grout will flow through even the smallest opening. Any open spaces or cracks in forms or at the joint between forms and the foundation should be sealed off using rags, cotton, foam rubber, caulking compound, etc.

   Because of epoxy grout's higher compressive and tensile strengths and its readiness to bond to metals, the top of the grout outside the baseplate or soleplate should be brought up along the side of the baseplate or soleplate to give some protection against lateral movement. The top of the grout on baseplates with flange-type support should be at the top of the flange. The top of the grout on

---

* Refer to Volume 3 for a detailed discussion of grouting procedures.

baseplates with solid sides and soleplates should be 1 in. above the bottom of the baseplate or the underside of the soleplate. The outside top edge of the grout should be chamfered at 45°.

3. Foundation anchor bolt sleeve should be filled with a nonbonding, pliable material such as asphalt or silicone rubber molding compound to prevent a water pocket around the anchor bolt.

4. A split hose or duct tape can be used around the exposed threads of anchor bolts to prevent direct contact between the epoxy grout and anchor bolts.

5. The foundation should be protected from the rain since it is important that the foundation be clean and dry at the time of grouting. Normal grouting temperature should be between 40° and 90°F. Due to the accelerated rate of curing at high temperature, the shading of the foundation from summer sunlight for at least 24 hours before and 48 hours after grouting may be required. In hot summer weather, it is preferable to place the grout during the afternoon, so that the initial cure will occur during the cooler evening hours.

   In cold weather, the grouting materials (including the aggregate) should be stored at a temperature above 70°F for 24 hours prior to mixing. When the temperature is below 65°F, the grout manufacturer should be consulted before mixing and placing the grout.

   However, for best results in cold weather, fabricate a temporary shelter around the baseplate or soleplate to be grouted and prewarm the baseplate or soleplate and foundation. When prewarming the installation, use convection-type space heating equipment and be careful not to overheat localized areas. Do not use radiant heating or open steam. Radiant heating warms the grout upper surface more than the lower surface. The grout surface therefore cures in a thermally expanded state, and after dissipation of the heat, produces stresses that tend to make the grout "curl-up," resulting in cracks in the concrete at the foundation corners just below the grout line.

6. Epoxy grout has a limited shelf life. Check the grout manufacturer's instructions prior to use.

7. Epoxy grout has a limited pot life. Check the grout manufacturer's instructions prior to use.

8. Epoxy grout should have a consistency very similar to that of a hydraulic cement slurry but with self-leveling flow characteristics. Epoxy grouts can generally be handled with the same methods and tools used with flow grade, sand-cement grouts. Epoxy grout can be manually mixed in a wheelbarrow using a mortar mixing hoe or a small cement or mortar mixer. Over-mixing and/or violent mixing whips air into the grout, and results in a weaker grout.

9. The actual placing of the epoxy grout can be accomplished by several means. Some companies prefer to force the grout into place, while others use their ingenuity and place the grout by various devices. Epoxy grout is very viscous; however, it will flow and seek its own level given time and an ambient temperature above 35°F. Generally, it is best to start placing the grout at one end of the baseplate or soleplate and work toward the other end in such a manner as to force the air out from beneath the baseplate or soleplate to eliminate voids as the grout moves along. A floating trowel is very helpful in forcing grout underneath by simply applying pressure on top. Plywood strips, sheet metal strips, wires, and push rods may also be used to place the grout completely under the baseplate or soleplate, but care should be exercised to prevent working air into the grout.

   Note: Check the forms frequently for leaks. Leaks do not self-seal. If not stopped, they will cause voids.

10. Epoxy grout curing rate depends on the temperature and pour thickness. Lower ambient temperatures and very thin layers of grout require longer curing time.

    Forms may be removed when the epoxy grout is adequately cured. This generally occurs in approximately 12 to 24 hours at 75°F or when the surface becomes firm and not tacky to the touch. When an accelerator is used, follow the manufacturer's instructions to determine the typical curing times required.

11. After the grout has cured, the baseplate and soleplate should be checked for complete grouting by tapping the baseplate or soleplate with a steel bar. If grouting voids are found based on a "hollow" sound, holes can be drilled in the baseplate or soleplate deck at each end of the voids and the voids filled with epoxy grout without aggregate; one hole should be used for the grout and the other hole as an air vent. A grease gun is normally used to force the

grout into voids. When pressure injection is used, install dial indicators on the baseplate or soleplate deck to confirm that epoxy placement is being accomplished without lifting the baseplate or soleplate deck.

12. The leveling shims or wedges used to level the baseplate or soleplate can be left in place after grouting. If for some reason they are removed after the grout has cured, the resulting voids should be filled with epoxy grout without aggregate.

    If leveling screws are used, they should be removed after the grout has cured to allow the full equipment weight to be distributed evenly over the grouted area.

    The foundation anchor bolts can now be retightened and the pump and driver installed. The pump and driver are now ready for alignment.

Cement-Based Grout

1. The grout manufacturer's requirements and instructions should be strictly followed.

2. For dry packing of cement-based grout, refer to Volume 3, Chapter 3 of this series.

3. Grout forms should be built of materials of adequate strength and should be securely anchored and shored to withstand the pressure of grout under working conditions. Forms should be tight against all surfaces, and joints be sealed with tape.

   Grout forms must be coated with form oil on areas that will come in contact with the grout to keep them from becoming bonded to the grout. All oil should be removed from the concrete, baseplate, or soleplate before grouting. To permit easy cleanup, cover all surfaces where grout may splash.

4. Prior to placing the grout, the top surface of the concrete foundation should be saturated with water for the time period recommended by the grout manufacturer. Excess surface water and water in the foundation bolt holes should be removed just prior to placing the grout.

5. Foundation anchor bolt sleeves should be filled with a nonbonding, pliable material such as asphalt or silicone rubber molding compound to prevent a water pocket around the anchor bolt.

6. A split hose or duct tape may be placed around the exposed threads of anchor bolts to prevent direct contact between the grout and anchor bolts.

7. The temperature of the baseplate or soleplate, supporting concrete foundation, and the grout should be maintained between 40 and 90°F during grouting and for a minimum of 24 hours thereafter.

8. Grout should be mixed with only water to produce the desired consistency according to the procedures recommended by the manufacturer.

   Caution: Check the quality of the water being used; ensure that it is oil free.

9. The placement of the grout should be rapid and continuous so as to avoid cold joints under the baseplate or soleplate. Generally, it is best to start placing the grout at one end of the baseplate or soleplate and work toward the other end to force the air from beneath the baseplate or soleplate to eliminate voids as the grout moves along. A floating trowel is very helpful in forcing grout underneath by simply applying pressure on top. Plywood strips, sheet metal strips, wires, and push rods may also be used to place the grout completely under the baseplate or the soleplate, but care should be exercised in order to prevent working air into the grout.

10. Grout should be cut back to the bottom outer edge of the baseplate or soleplate and tapered to the existing concrete. The top of the grout on baseplates with flange-type support should be at the top of the flange. The top of the grout on baseplates with solid sides and soleplates should be 1 in. above the bottom of the baseplate or the underside of the soleplate. The outside top edges of the grout should be chamfered at 45°.

11. After the grout has reached an initial set (the grout can be cut with a steel trowel and will stand up without support), it should be trimmed back to the level indicated on the drawings.

12. Grout should be cured according to the manufacturer's specifications and recommendations.

13. After the grout has cured, the baseplate or soleplate should be checked for complete grouting by tapping the baseplate or soleplate with a steel bar. If grouting voids are found based on a "hollow" sound, holes should be drilled in the baseplate or soleplate deck at each end of the voids and the voids filled with epoxy grout without aggregate; one hole should be used for the grout and the other hole as an air vent. A grease gun is normally used to force the grout into voids. When pressure injection is used, install dial indicators on the baseplate or soleplate deck to confirm that epoxy placement is being accomplished without lifting the baseplate or soleplate deck.

14. Forms should remain in place for a minimum of 24 hours except where form removal is needed to trim back grout.

15. The leveling shims or wedges used to level the baseplate or soleplate may be left in place after grouting. If for some reason they are removed after the grout has cured, the resulting voids should be filled with grout without aggregate.

    If leveling screws are used, they should be removed after the grout has cured to allow the full equipment weight to be distributed evenly over the grouted area. The holes should be caulked with putty.

    The foundation anchor bolts can now be retightened and the pump and driver installed. The pump and driver are now ready for alignment.

Machinery Alignment

Chapter 5, Volume 3 of this series deals extensively with equipment alignment, and it is not our intent to duplicate the efforts of others in this area. However, here are several general steps to follow which will result in a well-aligned, trouble-free machine:

1. The owner should insist that the installing agency use the reverse indicator method of alignment, or the laser alignment method, whenever the separation between shaft ends is larger than 50 percent of the diameter at which the dial indicators contact the coupling rim. The advantages of using this system far outweigh the arguments for rim-and-face and other mechanical alignment methods. If you or your contractor are unfamiliar with reverse indica-

tor alignment, get a book and learn, or purchase one of the small hand-held calculators now available that are based on this system.

3. Measure and adjust the distance between the driver and the pump shaft ends (D.B.S.E.). This distance should be in accordance with the pump layout drawing and within the tolerance provided by the coupling manufacturer or guideline value of Chapter 5, Volume 3 of this series. The D.B.S.E. should be set with the pump and driver shafts pulled toward each other for turbine drives and motor drives with antifriction bearings. For motor drives with sleeve bearings, the D.B.S.E. should be set with the motor shaft at its magnetic center.

3. Measure and adjust the distance between the driver and the pump shaft ends (D.B.S.E.). This distance should be in accordance with the pump layout drawing and within the tolerance provided by the coupling manufacturer. The D.B.S.E. should be set with the pump and driver shafts pulled toward each other for turbine drives and motor drives with antifriction bearings. For motor drives with sleeve bearings, the D.B.S.E. should be set with the motor shaft at its magnetic center.

4. Both the driver and pump shafts should be checked for mechanical runout using a dial indicator. Mechanical runout should not exceed .002-in. total indicator reading (TIR).

5. Driver-to-pump and pump-to-driver alignment targets should be provided to the installation contractor before starting the alignment. These targets should include allowances for thermal growth of hot pumps and steam turbines. If the actual cold targets are not provided, the amount of vertical growth due to temperature may be estimated using the following formula:

$$\text{Vertical growth} = \frac{(\text{Oper.temp.} + \text{Amb.temp.}\ ^\circ\text{F})}{2}(6.0 \times 10^{-6}\ \text{in./in.} - {}^\circ\text{F})(\text{Ht.in.})$$

6. Initial alignment should be made at ambient temperature and "pipefree"—no pipe forces or weight on the equipment (pump and turbine flanges should be unbolted).

7. To eliminate adding mechanical runout to alignment error, indicator readings should be taken by turning the pump and driver shafts together.

8. Tighten the equipment hold-down bolts and recheck the alignment; adjust as necessary.

   Check for a "soft foot" by loosening each hold-down bolt in turn while measuring with a dial indicator movement between machine foot and soleplate or baseplate. If movement exceeds approximately .001-in. (.025 millimeters) at any foot, shim changes should be made to eliminate the "soft foot" and alignment rechecked before proceeding.

9. The alignment should be checked and recorded after both bolt-up of all piping to the pump and bolt-up of all piping to the driver. No significant strain should be present as indicated by any change in the pump-driver alignment. A change in alignment of more than .001-in. from the pipe-free condition should be investigated and the piping strain corrected. When "cold set" has been included in the piping design, the final alignment must be checked and the tolerances met after the system has reached normal temperature.

10. Pumps operating at over 300°F and all steam turbine drivers should be hot aligned; that is thermally cycled to normal operating temperatures and realigned, if at all possible, while hot. This will ensure that the alignment tolerances are still being met under operating conditions.

11. After the final alignment has been approved, the support pads for the pumps and drivers should be drilled at two locations and tapered dowel pins with threaded ends to facilitate removal should be installed. Unless specifically located by the equipment manufacturer, the dowel pins should be placed near the thrust bearing end of the equipment.

Pre-Operational Checks

The following is a checklist of items that should be looked at before actually starting up the equipment. The list is by no means complete for all types of machinery but should act as a mind jogger *before* initial startup. Refer also to Appendix 1B.

1. For pumps with double mechanical seals or packing with external gland oil, the gland oil supply piping should be cleaned by oil or solvent flushing prior to connecting to the pump. The oil system should be flushed at the design flow rate at a temperature of about

160°F (or lower, as component design dictates) using the system pump and lint-free cloth filter bags at all seal or packing inlet connections; oil should be circulated for a minimum of four hours. The filter bags should be examined and cleaned at approximately 30 minute intervals. Flushing should be continued until there is no evidence of particle pickup for two consecutive 30 minute periods.

2. For pumps with tandem mechanical seals, the overhead reservoir and all flush oil supply piping should be thoroughly cleaned by oil flush prior to connecting to the pump.

3. It is important for good seal or packing performance that dirt and/or foreign debris not be introduced into the seal or packing cavity.

   Note: Pumps with double mechanical seals should not be flushed out, steamed out, pressure tested, or operated without the seal gland oil system in operation at the specified pressure level. The gland oil system pressure should be 10–15 psi higher than the pump side pressure on the inner seal for all operating conditions. This will prevent inadvertent blowing open of the seal with pumped product that could cause seal failure or contamination of the flush system.

4. The flushing and steaming of pumps with single or tandem mechanical seals should be held to a minimum period of time. This will minimize the amount of debris entering the seal cavity, and prevent the destruction of the static seal elements by overheating. Static seal elements should not be heated above the temperatures listed in Chapter 10 of Volume 3.

5. All cooling water piping on pumps and turbines should be flushed and connected prior to operation.

   Prior to operation, any lube oil in bearing housings and constant level oilers should be drained and new, clean lube oil added, and the proper oil level established. Verify that the oil level in the bearing housing is correct and that the constant level oiler is functioning properly. The owner should verify that the installing agency is using the proper grade of lube oil.

6. On equipment to be lubricated by an oil mist system: install all local oil mist piping or tubing with a slope toward the equipment without any sags or low spots in the piping or tubing runs; install

transparent sight bottle on bottom of bearing housing. Prior to operating the equipment, verify that oil mist flows to each bearing.

Note: Unless you prelubricate with a high-viscosity lube oil, the oil mist system must have been in operation for a minimum of 12 hours prior to attempting to run any equipment.

7. Motor drivers should be power-rotated to check for proper direction of rotation prior to coupling to the driven equipment.

8. Turbine overspeed trip setting and governor operation must be checked prior to coupling to the driven equipment.

9. Gear type couplings should be packed with the proper grease and the pump and driver coupled up. Recheck the coupling float and verify that it is within the coupling manufacturer's tolerances.

10. The coupling guard should be installed prior to rotating any shaft under power.

11. If a separate lube oil system is provided, the system should be cleaned and flushed and all alarms and shutdowns set and tested prior to operation of the equipment.

A Final Note

Many things can influence the operation and reliability of pumps. One area often overlooked is the initial installation. A poor installation may cause premature failure due to misalignment, excessive piping strain, improper lubrication, etc.

It is relatively easy and inexpensive to eliminate one major source of pump failure: install it right the first time. An ounce of prevention is worth a pound of cure.

Pump Preparation for Start-Up

After the pump has been installed and coupling alignment completed, the appropriate checklist in Appendix 1B may be consulted and these steps should be followed for a successful start-up:

1. Pump and driver should be checked for sufficient and proper lubrication.
2. Driver should be checked for correct rotation.
3. Pump suction valve should be fully opened. (Check pump and piping for leaks.)

4. Pump case should be vented. (Open vent at top of pump casing until all air is expelled from casing.)
5. If product is hot, ample time should be allowed for pump case to heat up. (Pump case and rotating assembly could distort from uneven heat transfer.)
6. Before starting, rotate pump shaft by hand. (Should be free, no rubbing.)
7. Crack open discharge valve—don't fully open. (A centrifugal pump uses less horsepower at start-up with the discharge valve nearly closed; also this practice will prevent initial cavitation.)
8. *Start Pump,* watch discharge pressure gauge, and as soon as pump pressure stabilizes, open discharge valve slowly. Watch discharge gauge; discharge pressure will fall off for a few turns of the valve until existing head conditions are met. Once pressure stabilizes, you can fully open the discharge valve.
   Important! Never allow pump to run too long with discharge valve closed.

The Pump in Operation

1. During operation, a centrifugal pump requires occasional inspection (Data sheets in Appendix 1B may prove helpful).
2. Make sure that there is flow as the discharge valve is opened by watching for a drop in discharge pressure.
3. Watch for fluctuations in suction and discharge pressure to make sure the pump does not cavitate.
4. After the pump has run for a few minutes, the operator should touch the pump and motor bearings to determine if they are overheating.

   Note: The Operator always touches the motor with the back of the hand so that in case of shock the hand can be pulled away.

5. The mechanical seals should be checked for leakage particularly during the first hours of operation. A minor leak through the seal usually stops after a short time, but if it continues, the pump should be stopped and the seal fixed.
6. When operating the pump at a discharge pressure below the rated point, the motor should be watched carefully. The discharge valve should be throttled to build up head to a safe point. Should the low head condition persist, the pump should be shut down. Centrifugal pumps should not be operated at greatly reduced capacity or with the discharge valve pinched because the energy required to drive the pump is converted into heat and the temperature of the liquid may reach the boiling point. Furthermore, many pumps are subject to flow instability at low flows.

Shutting Down the Pump

The discharge valve on a centrifugal pump should be partially closed before the driver is stopped in order to prevent reverse flow. Usually, there is a check valve in the discharge line to prevent such reverse flow.

Diagnosing Pump and Seal Problems in the Field

Severe operating conditions in most refineries and chemical plants subject process pumps to high temperatures, abrasion, corrosion and premature bearing and mechanical seal failures.

Damage to the pump can occur not only inside the mechanical surfaces, but on the outside as well. Surrounding atmospheric conditions can also shorten the life of any pump, especially in corrosive environments. The life expectancy of pumps and mechanical seals in this type of environment is very dependent on proper maintenance procedures.

Many mechanical seal failures have been the result of wear or deterioration of pump bearings or internal pump components. Troubleshooting pump and mechanical seal difficulties should begin at the pump while it is installed and running. Maintenance and operating personnel need to determine first if a process deficiency might be causing pump or mechanical seal problems. The investigation should involve a thorough study of pump hydraulics to determine if the pump is performing per design. Accurate suction and discharge pressure readings need to be taken. The pump should also be checked for excessive vibration, shaft deflection, noisy bearings, and excessive temperature. If pump hydraulics appear to be normal, but the pump is noisy and vibrating, it's quite possible that the pump could be misaligned, or the coupling could be faulty, or possibly the pump and/or motor bearings are defective. By using a vibration analyzer and monitoring the frequency of the vibration, one can determine the probable source, and the problem can be eliminated. If the pump bearings have been subjected to severe vibration, the pump will have to be removed to the shop for repairs, and if the mechanical seal is leaking it will also need replacing.

A more thorough coverage of this subject can be found in Volume 2 of this series, *Machinery Failure Analysis and Troubleshooting.*

Pump Preventive Maintenance

Earlier we had attempted to define the components of machinery maintenance strategy. We believe that preventive maintenance activities around process pumps have to be shared by vigilant operators and maintenance personnel. Table 1-1 is presented as a guide for this task.

Table 1-1
Recommended Preventive Maintenance Checks
Centrifugal Pumps and Drivers

Intervals	Routine
Daily	—Check pump for noisy bearings & cavitation noise.
Daily	—Check bearing oil for water, discoloration & contamination.
Daily	—Feel all bearings for temperature.
Monthly	—Add oil if required.
Monthly	—Clean oiler bulbs & level windows as required.
Fall & Summer	—Do seasonal oil change-out if required by Lube Guide.
Daily	—Inspect bearings & oil rings through filling ports. Wipe bearing covers clean.
Monthly	—Ascertain that oil level is correct distance from shaft center line. Adjust oiler as required.
Daily	—Check for oil leaks at gaskets, plugs & fittings.
½ Year	—Machines not running—standby service:—Overfill bearing housing to bottom of shaft & rotate several turns by hand to coat shaft & bearings with oil. Drain back down to re-establish proper level.
Daily	—Self flushed pumps—hand check flush line temperature to determine flow through line. External flushed—pumps—determine if flow indicator & needle valve adjustment is O.K.
Daily	—Determine if mechanical seal condition is normal.
Daily	—Check any water cooling for effective operation. Hand test temperature differential across coolers, jackets & exchangers. Disassemble & clean out as required.
Fall	—Where cooling water is decommissioned, ensure that no water remains in jackets, coolers or piping.
Fall	—Inspect for damaged or missing insulation.
Daily	—Check for operation of heat tracing.
Yearly	—Thoroughly inspect disc coupling for signs of wear & cracks in laminations. Tighten bolts.
Yearly	—Dial indicator check coupling alignment in coupled condition. Use special coupling indicator clamps where possible. Ensure that thermal growth allowance is correct.
Yearly	—With indicator clamped to coupling, depress & lift on each coupling and note dial indicator change. Determine if deflection is normal for this machine.
Yearly	—Dial indicator check axial float of pump & driver shafts in similar manner.
½ Year	—Apply light coat of Rust Ban to exposed machined surfaces to prevent rust & corrosion.
Monthly	—Clean out debris from bearing brackets. Drain hole must be open.
Daily	—Determine if steam leakage at packing & valves is normal.
Daily	—Check for leaks at pressure casing & gaskets. Determine if steam traps are operating properly—no continuous blow & no water in casing or drain lines.
½ Year	—Clean & oil governor linkage & valve stems.
Yearly	—Remove turbine sentinel valve. Shop test & adjust to proper setting.
Yearly	—Inspect trip valve & throttle valve stems & linkages for wear. Check overspread mechanism for wear. (Turbine not running.)
Yearly	—Remove mechanical governor cover & inspect flyball seats, spring, bearing & plunger for wear.
Yearly	—Uncouple from pump & overspeed turbine. Ensure that trip valve will stop turbine with steam supply valve (throttle valve) fully open. Compare tripping speed with previous record. Adjust trip mechanism & repeat if necessary. Follow manufacturer's instructions when making adjustments.
Yearly	—Where process permits, test run turbine coupled to pump. When not possible, run uncoupled. With tachometer—verify proper governor operation & control. Determine if hand (booster) valves are completely closed when not required to carry load. This influences steam economy.
½ Year	—Exercise overspeed trip & valve steam linkage on turbines not running.
Yearly	—Change oil in hydraulic governors.
Monthly	—Determine if hydraulic governor heater is working.
Monthly	—Check for proper oil level & leaks at hydraulic governor. Check for oil leaks at lines, fittings & power piston.
Monthly	—Replace guards (repair if required).
Monthly	—Determine if pump unit requires general cleaning by others.

Pump Repair

Field Checks Before Removal

Most pump repairs in the petrochemical environment are breakdown repairs as a consequence of component failure. Typical failure causes are:

- Leaking shaft seal
- Reduced pumping rate
- Pump binding or stuck
- Failed bearings
- Excessive vibration
- Leaking casing

The following steps should be taken before the removal of the pump:

1. Check with operator as to perceived failure cause.
2. Run the pump where possible and attempt to diagnose failure by:

   - Looking
   - Listening
   - Feeling
   - Smelling
   - Measuring bearing temperatures
   - Measuring power
   - Analyzing vibration
   - Measuring flow and pressures

Field Checks During Removal

If diagnosis shows the pump has to be removed, a sequence of field checks will still be appropriate:

1. Check coupling for wear or lack of grease.
2. Visually check oil and oil level.
3. Remove pump, check body gaskets, seats.
4. Visually check impeller and casing wear rings. Also check impeller vs. casing wear ring clearance, check impeller, volutes and balance holes for plugging.
5. Check flush lines and quench lines for internal corrosion or plugging.
6. Visually check condition of gauges, etc.
7. Remove pump to shop for repair.

If failed bearings are suspected in pump or motor:

- Check radial clearance and end float in motor.
- Run motor and check for abnormal noise, vibration.
- If motor is bad, remove and repair.

Diagnosing Pump and Seal Problems in the Shop

While the pump is being repaired it is advisable to carefully examine every component. A recommended procedure is to match mark all parts prior to disassembly and to make the following checks while dismantling the pump:

1. Visually check impeller and nut for wear, erosion, corrosion and other deterioration.
2. Remove seal flange nuts and check seal tension.
3. Record impeller position in relation to pump frame.
4. Remove impeller nut and impeller.
5. Inspect wear rings inboard, if any.
6. Check and record throttle bushing clearance.
7. Check body gasket faces.
8. Remove stuffing box body from pump frame.
9. Check stuffing box gasket face, bore, and pilots.
10. Remove and inspect all shaft keys.
11. Remove sleeve, seal, sleeve gasket and sleeve flange. If necessary, determine the cause of seal failure and inspect condition of parts.
12. Check pump bearings for roughness. Record shaft end float, check shaft for wear, erosion, corrosion and straightness.
13. Excessive shaft axial end play:

    Excessive shaft movement can result in pitting, fretting, or wear at points of contact in shaft packing and mechanical seal areas. It can cause over or under-loading on springs resulting in high wear rates and leakage. It can also cause excessive strain and wear on pump bearings. Defective bearings in turn can cause excessive shaft end play.

    To check for this condition a dial indicator should be installed so that its stem bears against the shoulder on the shaft (Figure 1-6).

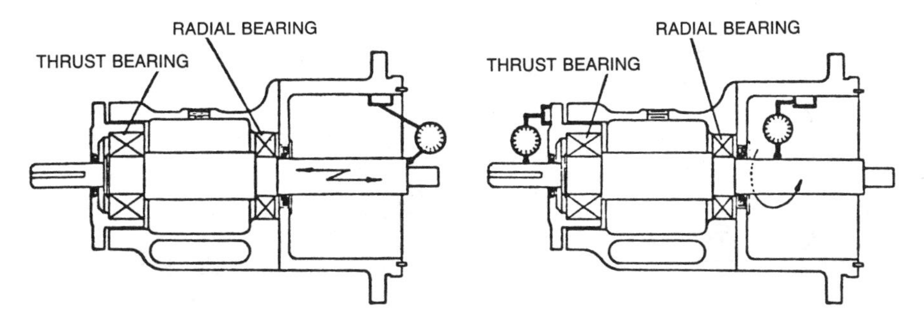

Figure 1-6. Checking for end play.

Figure 1-7. Checking for bent shaft.

A soft hammer should be used to lightly tap the shaft on one end and then the other. Total indicated end play should be between .001 in. and .004 in. for proper assembly.

14. Bent shaft:

When a pump shaft is bent or out of alignment, bearing life, seal life, and performance are impaired. Bent shafts also cause vibration and coupling failures. To check for this condition, install a dial indicator to the pump housing and adjust so that the stem bears on shaft outside diameter. Rotate shaft and check for run-out. If run-out is greater than .002 in. the shaft should be straightened (Figure 1-7).

The shaft should be checked in several different locations.

15. Check all pilot fits for concentricity. Also check for excessive shaft radial movement:

Excessive radial shaft movement allows shaft and seal to whip, deflect, and vibrate. This type of movement is caused by improper bearing fit in pump bearing housings or possibly an undersized shaft. If the bearing bore is oversized, determine if it was caused by corrosion, wear or improper machining. To check for this condition, a dial indicator should be placed on the shaft OD as close to the bearings as possible. The shaft should be lifted, or light pressure applied to shaft. If the total movement exceeds .003 in. maximum, bearings and bearing fits should be checked and necessary repairs made (Figure 1-8).

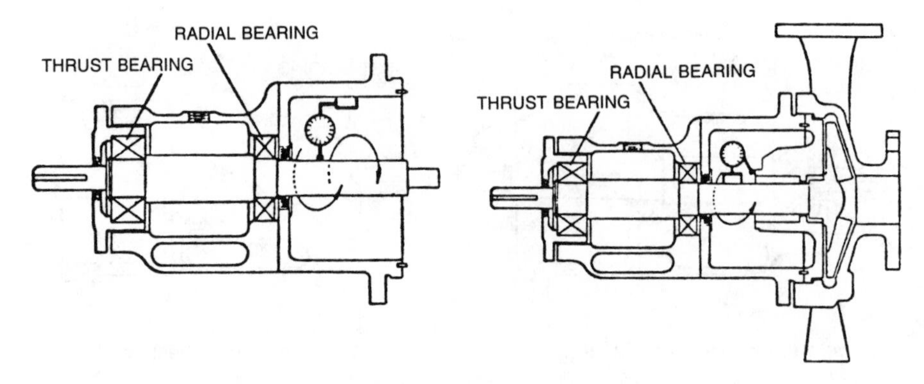

Figure 1-8. Checking for whip or deflection.

Figure 1-9. Checking for stuffing box squareness.

16. Stuffing box squareness:

    If the face of the pump stuffing box is not perpendicular to the shaft axis, the mechanical seal gland will tilt when installed. This may cause the seal to wobble and could lead to seal failure.
    To check for this condition, clamp a dial indicator to the shaft with the stem against the face of the stuffing box, after the cover has been bolted in place. Total indicator measurement should not exceed .002 in. If face measurement should exceed this tolerance, the cover should be placed in a lathe and machined square. Stuffing box faces should always be checked for pitting, nicks, burrs, and possible erosion before installing the seal (Figure 1-9).

17. Check for bore concentricity:

    The concentricity of a stuffing box bore and shaft can be difficult to measure because of rust or corrosion due to leaking gaskets. Concentricity is critical and may have to be reestablished by welding and remachining. On large double-ended pumps where there is a large separation between stuffing boxes it is very important that the concentricity be held to design tolerances.

    To check for concentricity, attach a dial indicator to the shaft and sweep as shown in Figure 1-10.

    Stuffing boxes should be concentric to the shaft axis within .005 in. total indicator reading. If readings are in excess of this, the pump may have to be realigned and redowelled.

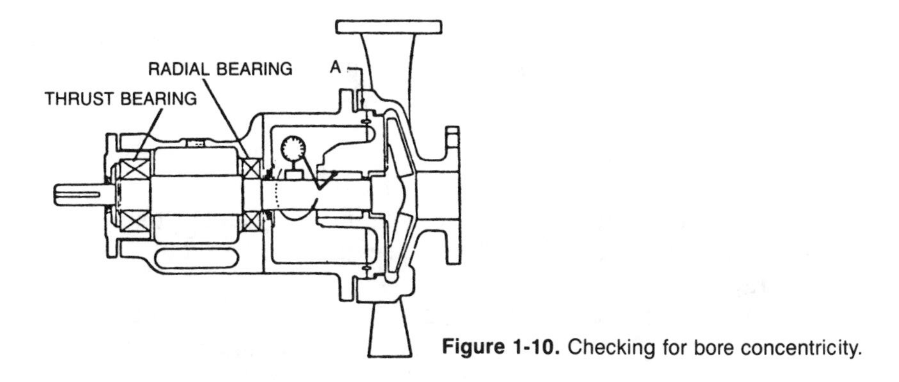

Figure 1-10. Checking for bore concentricity.

18. If bearings are found to be rough or the end float is excessive:
    - Remove pump shaft and bearing from housing.
    - Remove bearings from shaft.
    - Check shaft fits, coupling, bearings.
    - Check shaft straightness and polish lightly.
    - Clean and check bearing fits in housing.
    - Repair or replace all faulty and worn parts prior to reassembly.

Detailed Inspection Procedures

There are several basic rules that should be observed when inspecting and repairing process pumps. Some of these are:

1. Have a good understanding what clearances and fits should be met.
2. Record all data and measurements on suitable inspection forms. (See Appendixes A and B at the end of this Chapter.) Record all unusual deterioration found while dismantling the pump.
3. Use new gaskets and O-rings when reassembling the pump.
4. Keep the work place clean.

Inspection of Parts

Shafts

1. Check for straightness: Runout is not to exceed .002 in. Bearing seats must be in good condition.
2. Inspect threads, keyways, and shoulders on shaft. Repair if damaged.

3. Measure and record all shaft fits. Undersized or damaged fits should be repaired by the procedures outlined in Volume 3 of this series.

Case End Wall and Cover

1. Measure and record all fits between pump casing and mating parts.
2. Remove all plugs and fittings to inspect threads. Reinstall all plugs and fittings.
3. Inspect and indicate mounting pads to ensure they are flat and parallel with pump centerline. Machine, if out of alignment.

Bearing Housing and Bearings

1. Observe good anti-friction bearing mounting procedures (see Volume 3 for details).
2. Ball bearings: Replace if worn, loose, or rough and noisy when rotated. If dirty, clean with solvent, dry and coat with a good lubricant. New bearings should not be unwrapped until ready for use. Whenever in doubt about the condition of a bearing, scrap it. But if the bearing is still relatively new, and feels and looks good, don't discard it.
3. Sleeve bearings: Check surfaces of bearing and shaft for imperfection, babbitt build-up, and hot spots. Small imperfections do not harm the bearing. A typical diametral clearance is .0015 in. per in. of shaft diameter. For proper operation, clearances should never exceed .003 in. per in. of shaft diameter on typical pumps.

Mechanical Seals

Refer to Chapter 8 in Volume 3 for maintenance and repair of mechanical seals.

Impellers

1. Replace if excessively worn or corroded. The impeller should have been statically and dynamically balanced at the factory, and static and dynamic balance must be maintained for proper operation of your equipment.
2. Inspect and measure impeller bore and if worn or deteriorated, machine true. Recondition the shaft to fit revised impeller bore size. Refer to Volume 3 for guidance.
3. Measure outside diameter of impeller wear rings and record size. Refer to Table 1-2 for diametral clearances.

Table 1-2
Required Diametral Clearances—Process Pumps Wear Rings*

Wear Ring Diameter	Diametral Clearance Under 500°F	Diametral Clearance Over 500°F
3½ in. through 5 in.	.016	.018
5 in. through 6 in.	.017	.019
6 in. through 7 in.	.018	.020
7 in. through 8 in.	.019	.021
8 in. through 9 in.	.020	.022
9 in. through 10 in.	.021	.023
10 in. through 11 in.	.022	.024
11 in. and over	.023	.025

* *An additional diametral clearance of .005 in. is provided if both wear rings are made of austenitic stainless steel, Monel or other materials with high galling tendencies.*

Casing and impeller wear rings are provided at both sides of the impeller on API-type pumps. These rings allow a small clearance to be maintained between the rotating impeller and stationary casing rings. For proper hydraulic performance these clearances should approximate the experience values indicated in Table 1-2. Rings should be replaced when clearances have increased to a point where hydraulic requirements cannot be met or where inefficient operation would prove wasteful. For API values refer to Table 1-3.

Why do wear ring clearances deserve our attention? The following section will provide the answer.

Keep Pumps Operating Efficiently**

In centrifugal pumps, it is essential to pump operability and hydraulic performance that excessive internal leakage (or recirculation) be prevented. This is accomplished by establishing and maintaining close running clearances between stationary and rotating wear rings which restrict fluid flow to seal between the inlet and outlet of each impeller and between stationary and rotating interstage bushings. These bushings effect sealing between the stages of a multistage pump. Certain types of pumps contain hydraulic thrust balancing devices, another source of internal pump leakage.

** From "Keep Pumps Operating Efficiently," by J. Lightle and J. Hohman, Dresser Industries, Pacific Pump Division, in *Hydrocarbon Processing*, Sept. 1979. By permission of Dresser Industries, Pacific Pump Division.

As the close clearances become larger through wear, corrosion, erosion or perhaps questionable maintenance practices, internal leakage rates increase. The increased leakage must be pumped and repumped continuously by the impeller, requiring additional input horsepower.

The amount of added power to continuously recirculate excessive internal leakage is a function of the pump specific speed*. In low specific speed pumps (low capacity-high head) excessive running clearances result in larger percentage changes in power requirements than occur in high specific speed pumps (high capacity-low head). This is reflected in the empirical data plotted in Figure 1-11.

* For an explanation of pump specific speed refer to Figure 1-13.

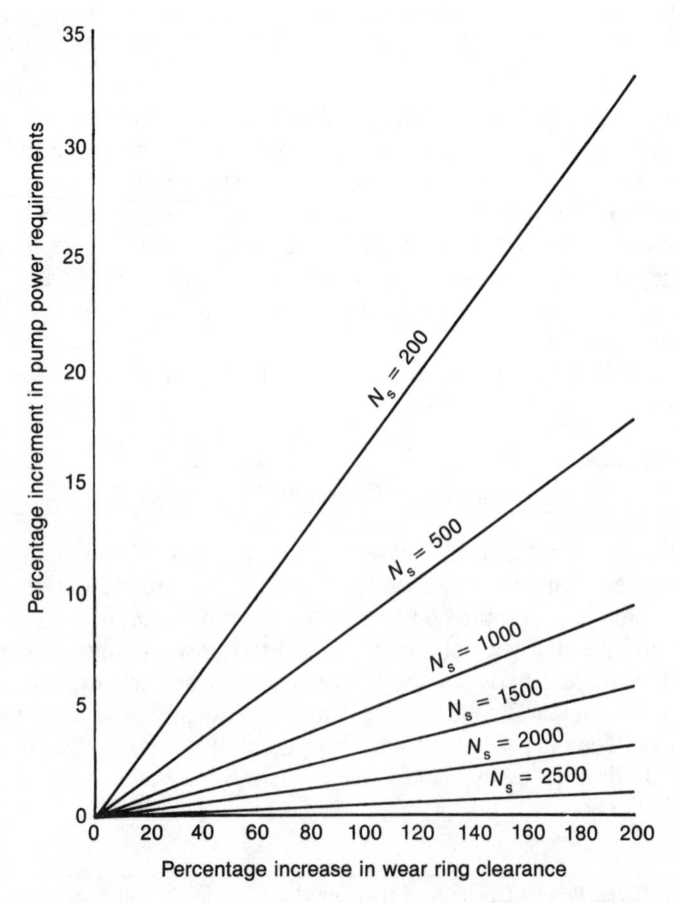

Figure 1-11. Added power resulting from excessive wear ring clearance for different specific speeds.

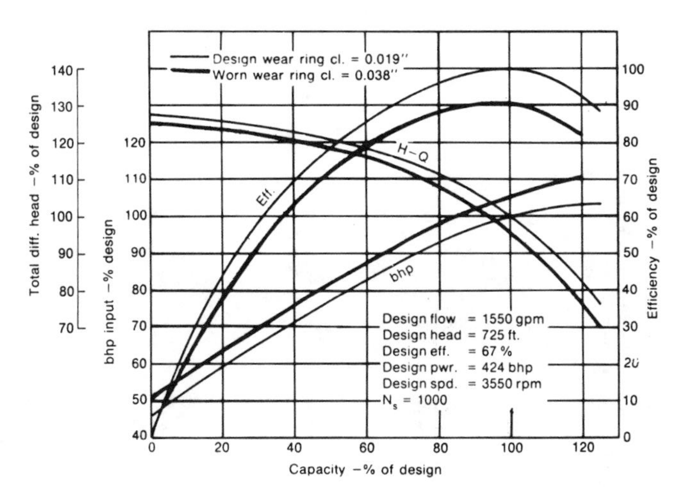

Figure 1-12. Pump performance curves.

The data in Figure 1-11 are somewhat misleading since it may be easy to conclude that high specific speed pumps do not cause excessive costs resulting from worn clearances. Beware, however, that small percentage changes of large horsepowers result in large annual costs. Also, as noted in the following example, mechanical operation may be adversely affected by excessive clearances in pumps of various specific speed ranges.

A typical example: Consider a single stage, overhung process pump—one designed to produce a total head of 725 ft at 1,550 gpm when operating at 3,550 rpm. Such a unit can be considered a typical process pump. Figure 1-12 shows the characteristic performance curves for an example pump; all scales are shown as a percentage of the design conditions. The solid curves indicate performance of the pump in new condition.

At the design operating capacity, the unit is 67 percent efficient, requiring 424 bhp* input horsepower (assuming the pumpage has a specific gravity of 1.0).

Referring to the specific speed nomogram (Figure 1-13), it is determined that our example pump has a specific speed of 1,000.

Now, going back to Figure 1-11, we see that if the wear rings have worn to the point where running clearances have doubled (increased by 100 percent), a pump having a specific speed of 1,000 will suffer an in-

---

* Brake horsepower

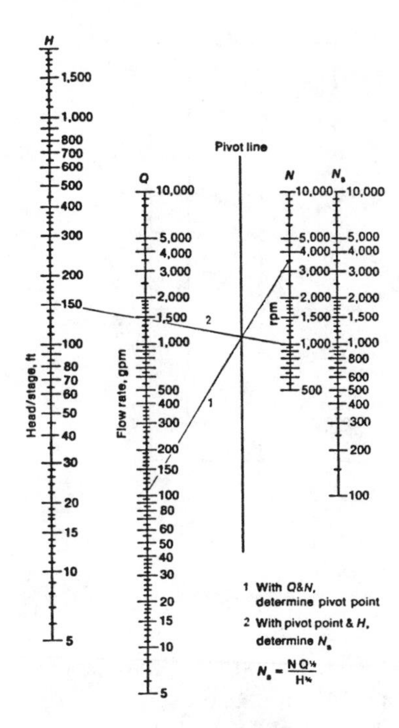

Figure 1-13. Specific speed nomogram.

crease in required horsepower input of approximately 4.8 percent; in our example, this amounts to approximately 20 brake horsepower. The .038 in. wear performance curve on Figure 1-12 shows the worn-condition performance characteristics of the example pump.

Figure 1-14 shows the annual power cost this extra 20 brake horsepower will represent to you, based on 300 days per year operation.

If your power cost is 6¢/kWh, your annual power cost resulting from internal wear in this pump would be $6,440. If yours is a "typical" 100,000 bbl/day refinery using 25,000 pump horsepower, an overall increase of 5 percent in your pump horsepower requirements could represent additional costs of $400,000 per year.

Maintenance practices. Normal operational wear is not the only cause of excessive part clearances in pumps, nor are wasted dollars and fuel the only adverse effects.

Intentional opening up of wear ring or other wearing part clearances is used by some maintenance people to solve certain pump operating problems. Unfortunately, such practices sometimes appear to be effective—over the short run. Over a period of time, however, such practices can create other problems. The resulting increased internal leakages within the pump (and the accompanying increased power required to pump the additional flow) seem to many to be a small price to pay, if in fact such criteria are considered at all. But, from a purely mechanical standpoint, the stability of the rotor is perhaps safeguarded only as long as normal running clearances are maintained. Typical consequences of liberally open clearances are likely to include excessive vibration, overheating and ultimately pump or driver bearing failure, shaft breakage, driver overloading, and possible total pump destruction. Ultimate maintenance costs can be very high and unit operation can be compromised through premature and repeated outages.

If two or more pumps are designed for parallel operation and share total capacity, then unequal running clearances can cause unequal load sharing by the pumps. One or more of the units can be forced to operate at significantly more or less than its design flow rate. Efficiency falls off and brake horsepower requirements increase even beyond those caused by excessive running clearances.

Running clearances. Greater than normal wear ring clearances at the impeller inlet eye increase the flow rate through the impeller (not out the discharge nozzle of the pump), increase the effective inlet fluid tempera-

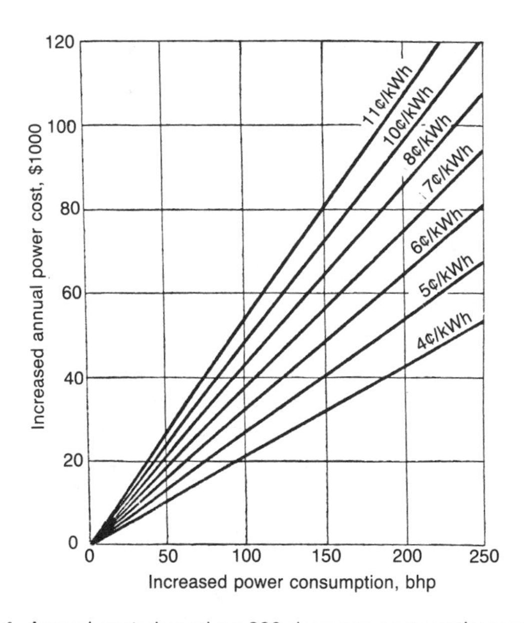

Figure 1-14. Annual costs based on 300 days per year continuous operation.

Table 1-3
Minimum Running Clearances

Diameter of Rotating Member at Clearance (inches)	Minimum Diametral Clearance (inches)
<2	0.010
2.000–2.499	0.011
2.500–2.999	0.012
3.000–3.499	0.014
3.500–4.999	0.016
5.000–5.999	0.017

ture in the impeller eye, and can introduce undesirable flow patterns in the impeller inlet. On pump installations having marginal NPSH* available, these effects can result in noise, vibration and physical damage normally associated with cavitation.

Original design wear ring clearances can be obtained from the pump manufacturer, and usually agree with the diametral clearances specified by API Standard 610, Section 19. The API recommended clearances are shown in Table 1-3 and are comparable to the experience values shown in Table 1-2.

The proper running clearances have been established based on operating economy consistent with good pump reliability.

Over a period of time, the design clearances will change due to wear, corrosion or perhaps as a result of operational problems such as overheating, thermal shock or problems with bearings or shaft sealing systems.

The best way to maintain minimum operating and maintenance costs resulting from increased wear part clearances is to establish a standard practice of measuring the running clearances whenever a pump is disassembled for any reason. On certain multistage pumps having thrust balancing devices, monitoring external balance leak-off flow can be used as a means of gauging wear of internal parts. Increased balancing flows are a direct indication of wear. In addition to measuring running clearances during normal downtimes, wear parts should be checked for eccentricity, out of roundness, and signs of excessive corrosion or erosion.

For best operation, obtain needed replacement parts from a reputable source. This will ensure correct materials, proper heat treatment to maintain correct hardness and wear properties, and proper manufacturing tol-

* Net Positive Suction Head

erances. It is especially important that special types of wear rings be obtained from a knowledgeable producer; typical special parts include serrated, stepped, or reverse-threaded wear rings, bushings, and thrust balancing devices.

It is always incumbent on process plant management, purchasing, operators and maintenance people to make a conscious effort to minimize operating costs. One very effective way to accomplish this goal is to be alert to the adverse effects of excessive wear part clearances in your pumps.

Calculating the Cost of Your Excess Clearances

Use Figures 1-11, 1-13, and 1-14 to calculate cost of excess wear ring clearances for the pumps in your plant. Since horsepower losses from excessive wear ring clearances vary widely by pump type and index, specific speed (N_S) can be used to simplify these calculations. Use Figure 1-13 and the instructions in step 1 to determine the specific speed for your pumps.

Step 1—the nomogram, Figure 1-13, solves the N_S equation:

$$N_S = NQ^{1/2}/H^{3/4}$$

Where:

N_S = pump specific speed,
N = pump speed in rpm,
Q = pump capacity in gpm,
H = total head per stage in ft.

To determine N_S from this nomogram, draw a line connecting Q and N for your pump. Note the intersection of this line with the pivot line. Then draw a line connecting this pivot point (the pivot line intersection of the QN-line) and H; extend the line to the N_S scale and read N_S for the pump.

Step 2—Use Figure 1-11 to determine the percent increase in pump bhp for the percentage increase in wear ring clearance and N of your pump.

Step 3—Determine the normal power requirement of your pump from the manufacturer's performance curve, or calculate the design horsepower as follows:

$$\text{bhp} = (QH/3960e)\ (\text{s.g.})$$

Where:

e = pump efficiency,
s.g. = specific gravity of the pumpage,
Q & H = as defined above.

Multiplying the normal pump power requirement by the percentage determined in step 2 will yield the estimated wasted bhp resulting from excessive wear ring clearance.

Step 4—Figure 1-14 can now be used to estimate the increased annual cost in your plant based on your location's cost of electricity.

Pump Assembly Procedures*

Horizontal Process Pump Disassembly (Figure 1-15)

Dismantling Rotating Element. The back pull-out design of the pump shown in Figure 1-15 allows the complete rotating assembly to be removed without disturbing the suction and discharge piping or the driver. Disconnect all auxiliary piping and drain the oil from the bearing housing. After disconnecting the spacer type coupling (see separate instructions), remove casing stud nuts. Screw bolts into the tapped holes in the casing cover (02) and tighten these jack bolts evenly to facilitate removal of the rotating assembly. The complete rotating assembly can now be moved to a clean area for further dismantling.

Dismantling Casing Cover with Mechanical Seal

After the complete rotating element has been taken to a clean work area, the unit can be fully dismantled by following these instructions and referring frequently to the sectional drawing:

1. Remove impeller nut (21-1) {L.H. Threads}.
2. With a suitable puller, remove impeller (05).
3. Remove impeller key (11-1).
4. Unbolt seal gland (07-2) from casing cover and slide back against deflector.

* Courtesy Union® Pump (Canada) Ltd. Note that these procedures are typical and may have to be modified to suit different pump models.

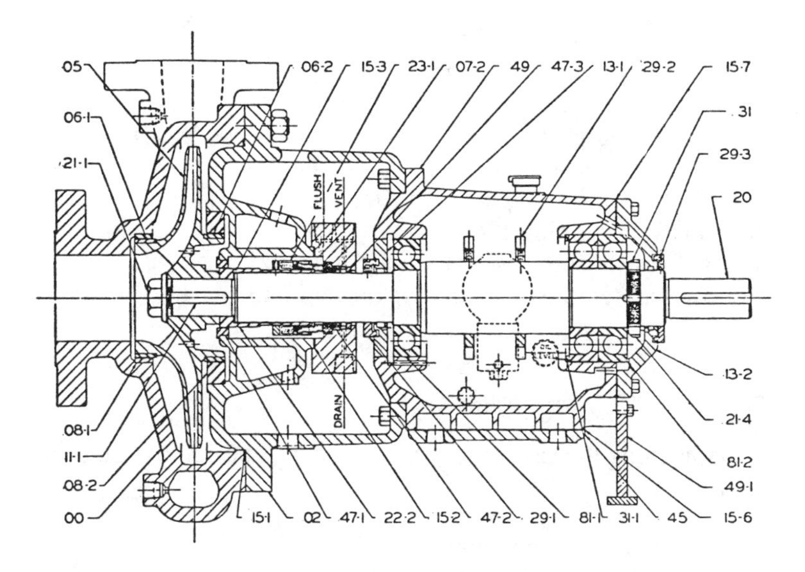

PART NO.	DESCRIPTION	PART NO.	DESCRIPTION
00	CASING	21-1	IMPELLER NUT
02	CASING COVER	21-4	BEARING LOCKNUT
05	IMPELLER	22-2	SEAL SLEEVE
06-1	IMPELLER RING-FRONT	23-1	MECHANICAL SEAL
06-2	IMPELLER RING-BACK	29-1	DEFLECTOR-INBOARD
07-2	SEAL GLAND	29-2	OIL FLINGER
08-1	CASE RING-FRONT	29-3	DEFLECTOR-OUTBOARD
08-2	CASE RING-BACK	31	BRG. LOCKWASHER
11-1	IMPELLER KEY	31-1	BEARING BACKUP RING
13-1	BEARING END CAP-INBOARD	45	COOLER COVER
13-2	BEARING END CAP-OUTBOARD	47-1	THROAT BUSHING
15-1	CASING GASKET	47-2	SEAL RETAINER
15-2	GLAND GASKET	47-3	THROTTLE BUSHING
15-3	SLEEVE GASKET	49	BRG. BRACKET
15-6	COOLER COVER GASKET	49-1	BRG. BRACKET SUPPORT
15-7	ARTUS PLASTIC SHIMS	81-1	RADIAL BEARING
20	SHAFT	81-2	THRUST BEARING

Figure 1-15. Cross section and parts list of a horizontal overhung, single stage process pump. (Courtesy Union Pump (Canada) Ltd.)

5. Unscrew cap screws holding bearing housing (49) to casing cover (02).
6. Pull bearing housing (49) from casing cover (02). Be careful not to damage the mechanical seal.
7. Loosen set screws holding seal rotating member to sleeve.
8. Pull seal sleeve (22-2) and the mechanical seal rotating element off shaft.
9. Remove mechanical seal from seal sleeve. *Note:* Mechanical seals have lapped sealing faces. Handle with care, keep wrapped in clean cloth and avoid contacting seal faces.

10. Remove sleeve gasket (15-3).
11. Slide seal gland (07-2) from shaft (20).
12. Pull stationary mechanical seal from gland.
13. Remove gland gasket (15-2) and mechanical seal O-ring.
14. Press throttle bushing (47-3) from gland.
15. Grind off weld between cover (02) and case ring (08-2), and remove case wear ring (08-2).
16. Press throat bushing (47-1) from cover (02).
17. Follow separate instructions for dismantling bearing housing.

Dismantling Bearing Housing

After dismantling the casing cover, the bearing housing can be dismantled (Figure 1-15).

1. Remove pump half coupling.
2. Remove deflectors (29-1 and 29-3).
3. Remove outboard end cap (13-2).
4. Slide shaft assembly out of bearing housing.
5. Remove ball bearing locknut and lockwasher (21-4 and 31).
6. Remove thrust bearing (81-2) and radial bearing (81-1).
7. Remove oil flinger (29-2).
8. "Tap-out" inboard bearing end cap (13-1).

Dismantling of Between Bearings Process Pump

General: The process pump shown in Figure 1-16* allows for complete change-out of bearings and mechanical seals without the necessity of disassembling the impeller or casing. To provide clear understanding, the disassembly and reassembly procedures have been broken down into specific sections:

- Bearing housings—ball bearing construction
- Bearing housings—sleeve bearing construction
- Stuffing boxes with various sealing arrangements
- Complete disassembly of pump rotating element

After the pump has been shut down and the motor secured in the off position, drain pump casing of all liquid. Drain oil from bearing housing and disconnect water and flush piping where necessary.

---

* Courtesy Union® Pump (Canada) Ltd. Note that these procedures are typical and may have to be modified to suit different pump models.

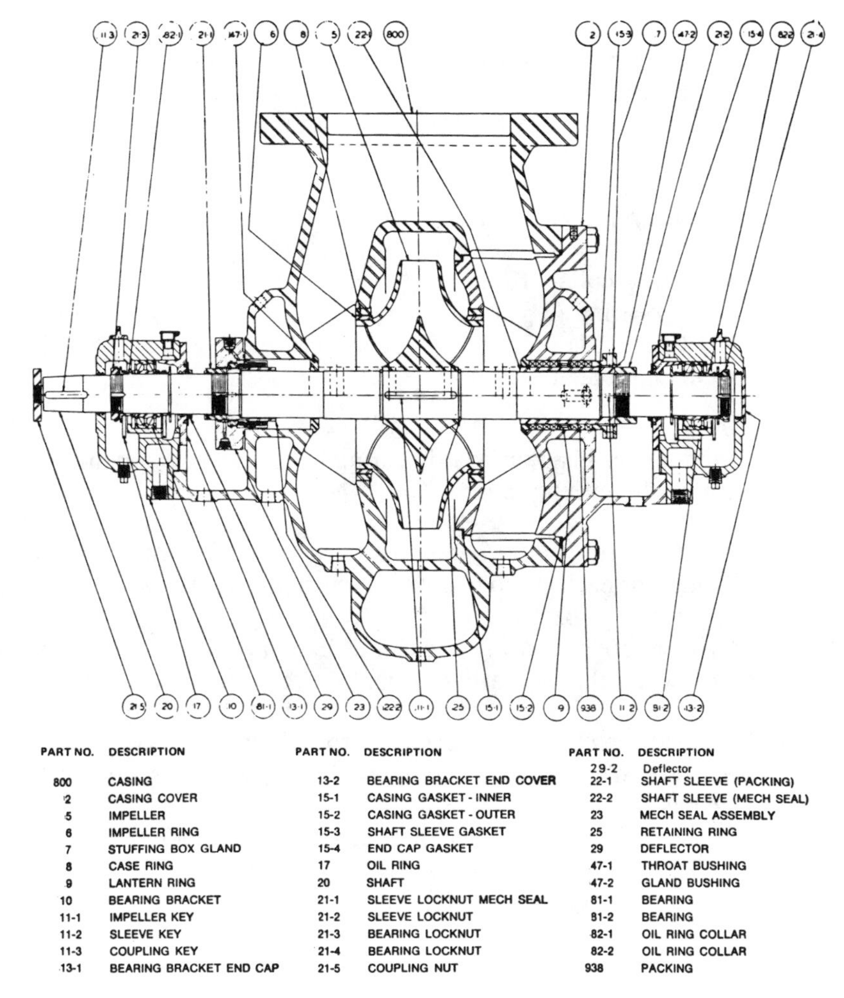

PART NO.	DESCRIPTION	PART NO.	DESCRIPTION	PART NO.	DESCRIPTION
				29-2	Deflector
800	CASING	13-2	BEARING BRACKET END COVER	22-1	SHAFT SLEEVE (PACKING)
2	CASING COVER	15-1	CASING GASKET - INNER	22-2	SHAFT SLEEVE (MECH SEAL)
5	IMPELLER	15-2	CASING GASKET - OUTER	23	MECH SEAL ASSEMBLY
6	IMPELLER RING	15-3	SHAFT SLEEVE GASKET	25	RETAINING RING
7	STUFFING BOX GLAND	15-4	END CAP GASKET	29	DEFLECTOR
8	CASE RING	17	OIL RING	47-1	THROAT BUSHING
9	LANTERN RING	20	SHAFT	47-2	GLAND BUSHING
10	BEARING BRACKET	21-1	SLEEVE LOCKNUT MECH SEAL	81-1	BEARING
11-1	IMPELLER KEY	21-2	SLEEVE LOCKNUT	81-2	BEARING
11-2	SLEEVE KEY	21-3	BEARING LOCKNUT	82-1	OIL RING COLLAR
11-3	COUPLING KEY	21-4	BEARING LOCKNUT	82-2	OIL RING COLLAR
13-1	BEARING BRACKET END CAP	21-5	COUPLING NUT	938	PACKING

Figure 1-16. Cross-section and parts list of a horizontal, double suction impeller, beam type process pump. (Courtesy Union Pump (Canada) Ltd.)

Ball Bearing Disassembly and Reassembly—(Figure 1-16)

1. Remove coupling spacer and coupling nut (21-5).
2. Tap pump half coupling off its tapered seat.
3. Unbolt bearing housing (10).
4. Remove cap screws between bearing housing (10) and end cap (13-1) and loosen socket head set screws in deflectors (29).

5. Slide deflector and end cap against sleeve nut.
6. Lift oil ring over oil ring collar to clear housing.
7. Slide bearing housing off bearings.
8. Remove in sequence the following parts:

   - Oil ring (17)
   - Bearing lock nut (21-4)
   - Bearing lock washer (31)
   - Oil ring collar (82-2)
   - Ball bearing (81-1)
   - Oil ring (17)
   - Oil ring collar (82-2)

Note: Both inboard and outboard bearing housing and components are identical, with the exception of:

a. Inner oil ring collar at outboard bearing is machined to special width for each pump to obtain correct setting of shaft in relation to stuffing box face.
b. The outboard bearing housing utilizes a bearing spacer (25) to allow the thrust bearing to position the rotating element.

9. Slide end cap (13-1) and deflector from shaft. After inspection has been carried out to verify component integrity, reassembly can be made as follows:
   a. Slide deflectors (29) on shaft against sleeve lock nut.
   b. Slide end cap (13-1) over shaft against deflectors.
   c. Install in sequence these parts:

      - Oil ring collars (82-2) (special at thrust bearing)
      - Oil ring (17)
      - Oil ring collar (82-2)
      - Bearing lock washer (31)
      - Bearing lock nut (21-4)
      - Oil ring (17)

   d. Place 1/16 in. thick gasket (15-4) over inboard end cap and push inboard bearing housing over bearing (*watch oil rings*); insert dowels and bolt into place.
   e. Bring inboard end cap forward and tighten cap screws.
   f. Without placing end cap gaskets, slide outboard housing over bearing, bring end cap forward and tighten cap screws between end cap and housing finger tight.

g. Measure gap between housing and end cap and remove bearing housing from bearings.
h. Insert gaskets with a thickness of .003 in. greater than measured gap and push bearing housing over bearings, insert dowels and bolt into place.
i. Bring outboard end cap forward and tighten cap screws.
j. Check shaft assembly for end play. End play should be .002 in.
k. Slide deflectors into place and lock to shaft with socket head set screws.
l. Rotate shaft to check that it is free to rotate and does not bind.
m. Reinstall coupling. Make sure coupling is firmly seated on shaft, and coupling key does not interfere with proper mounting.
n. Fill housing with proper grade of oil.
o. Check alignment of unit and lubricate coupling (lubricated type only).

Bearing Housing—Sleeve Bearing Construction

This arrangement consists of ring-oiled sleeve-type babbitt-lined radial bearings combined with an angular contact ball thrust bearing. The thrust bearing is relieved in the bore of the housing to assure freedom from interference with the radial alignment, and is oil-ring lubricated. Only portions of Figure 1-16 apply.

1. Remove coupling spacer and coupling nut (21-5).
2. Tap pump half coupling off its taper seat on shaft.
3. Loosen socket head set screws in deflectors and slide deflector back against sleeve seat.
4. Remove top halves of bearing housings and sleeve bearings.
5. Remove in sequence:

   a. End cap (13-1)
   b. Oil ring (17)
   c. Thrust bearing lock nut (21-4)
   d. Thrust bearing lock washer
   e. Oil ring collar (82-2)
   f. Thrust bearing and bearing mounting sleeve
   g. Thrust bearing spacer

6. Remove lower halves of sleeve bearing and unbolt bearing housings.
7. Slide deflectors off shaft.

After inspections have been carried out and proper clearances verified, these data should be logged for future reference. When all parts are cleaned and corrected as necessary, reassembly can be made as follows:

1. Slide deflectors (29) on shaft against sleeve lock nut.
2. Bring bearing housing lower halves up, install dowels and bolt into place.
3. Place oil rings over shaft into correct housing pocket.
4. Oil shaft and sleeve bearings and "roll" lower half bearings into housing.
5. Push thrust bearing spacer onto shaft. *Note:* This spacer is machined to special width for each pump to obtain correct setting of shaft in relation to stuffing box face.
6. Assemble thrust bearings into bearing mounting sleeve and install as a unit on shaft (duplex bearings are mounted "back-to-back").
7. Assemble in sequence the following parts:

    a. Oil ring collar
    b. Bearing lock washer
    c. Bearing lock nut (21-4)
    d. Oil ring (17)
    e. Top half of sleeve bearings
    f. Top half of bearing housings

8. Without gaskets, push thrust bearing end cap into position and tighten cap screws between housing and cap finger tight.
9. Measure gap between housing and end cap and remove bearing end cap.
10. Place gaskets with a thickness of .003 in. greater than the measured gap over end cap and install cap. Draw cap screw down tight.
11. Check shaft assembly for end play. End play should be .001 in. to .003 in.
12. Slide deflectors into place and lock to shaft with socket head set screws.
13. Rotate shaft and check that it is free to rotate and does not bind.
14. Reinstall coupling. Make sure coupling is firmly seated on shaft and coupling does not interfere with proper mounting.
15. Fill housing with proper grade of oil.
16. Check alignment of unit and lubricate coupling (lubricated type only).

Stuffing Box With Mechanical Seal

Most pump stuffing boxes are designed to accommodate different types and makes of mechanical seals. The modified cartridge design allows easy installation and additionally provides protection against sleeve movement on high suction pressure application (Figures 1-15 and 1-16).

After the bearings have been disassembled, the mechanical seals can be removed as follows:

1. Unbolt seal gland and slide off shaft.
2. Loosen socket head set screws in sleeve seat and remove nut with spanner wrench.
3. A 10-24 tapped hole is provided in key to assist in removing sleeve key.
4. Pull sleeve and mechanical seal rotating assembly off shaft.
5. Remove mechanical seal from seal sleeve. Mechanical seals have lapped faces—handle with care; keep wrapped in clean cloth and avoid contacting seal faces.
6. Remove sleeve gasket (15-3).
7. Pull stationary mechanical seal from gland (07).
8. Remove gland gasket and mechanical seal O-ring.
9. Press throttle bushing from seal gland.

After inspection has been carried out and all parts are cleaned and corrected as necessary, the mechanical seal can be reinstalled.

1. Press throttle bushing into seal gland.
2. Insert stationary seal face into gland, being careful not to damage O-ring.
3. Slide sleeve gasket against shaft shoulder.
4. Install rotating assembly of mechanical seal on shaft sleeve. For special seals follow instructions given separately; for standard seals use the following procedure:

    a. Wipe both lapped sealing faces with a clean, soft cloth.
    b. Apply some oil to sleeve and slide rotating member onto sleeve. Be careful *not* to put oil on lapped seal faces.
    c. Position rotating member on sleeve as indicated on detailed seal drawing to be supplied by seal manufacturer.
    d. Tighten set screws which hold rotating member to sleeve.

5. Oil pump shaft and slide sleeve and sleeve assembly against shaft shoulder.

6. Insert sleeve key, tighten sleeve nut with spanner wrench and lock sleeve nut in place with socket head set screw.
7. Bolt seal gland into place.
8. Reassemble bearing housing.
9. Verify that shaft is free to rotate and does not bind.
10. Follow detailed start-up instructions.

Casing Disassembly

The pump shown in Figure 1-16 is built in two basic casing arrangements:*

1. Top Suction: Top discharge, centerline supported for hot applications.
2. Side Suction: Side discharge, foot mounted for general applications.

Both arrangements incorporate heavy supports to carry pipe strain and protect the casing from distortion. To reduce the radial load on the impeller and to obtain added strength, all sizes have *double volute casings. Confined spiral-wound* gaskets are provided between casing and casing cover. These gaskets not only ensure positive sealing to the atmosphere, but also positively seal the discharge from the suction passages to eliminate casing wash-out at this point. Metal-to-metal contact between casing and cover assures positive alignment and eliminates the need for feeler gauges or other checking devices.

If it becomes necessary to disassemble the complete rotating element, the following procedure can be followed:

1. Dismantle inboard bearing housing
2. Dismantle inboard stuffing box
3. Unbolt casing cover and with help from jack bolts remove remaining rotating assembly. *Note:* This assembly is very heavy and mechanical lifting devices are necessary. The eye bolt is not at the center of gravity and care must be exercised to balance the total unbalanced weight.
4. While taking the assembly to a different work area, support the coupling end of the shaft at all times.
5. Clamp the cover flange in a vise and again support the free end of the shaft.
6. Disconnect the thrust bearing housing and the stuffing box.

---

* Union® Pump Class "HOL"

7. Slide shaft and impeller assembly from casing cover.
8. Remove impeller retaining rings and press impeller from shaft.
9. Case wear rings, throat bushings and impeller wear rings are tack-welded in place. Grind tack-weld off and remove rings and bushings.

After inspection of parts has been carried out and relevant dimensions recorded for future reference, reassembly can be carried out following these procedures:

1. Press case wear rings, throat bushings and impeller wear rings into place and tack weld in three places.
2. Install one impeller retaining ring and impeller key, and press impeller on shaft against ring.
3. Install second impeller retaining ring.
4. Return all parts to casing and with coupling end first place shaft and impeller assembly into casing.
5. Install gaskets, inner gasket into case and outer gasket on cover, and hold in place with heavy grease.
6. Slide casing cover over shaft (watch bearing surfaces and threads) and bolt into place.
7. Reassemble stuffing box.
8. Reassemble bearing housing.

Reassembling Casing Cover with Mechanical Seal

After inspection has been carried out as outlined in the inspection section, and all parts are cleaned and corrected as necessary, the casing cover can be reassembled by following the instructions given below and by frequently referring to the appropriate sectional drawing, Figure 1-15 or Figure 1-16.

1. Press throat bushing (47-1) into casing cover (02).
2. Press case ring (08-2) into casing cover (02) and tack weld in three places.
3. Press throttle bushing (47-3) into seal gland.
4. Insert stationary seal face into seal gland (07-1), being careful not to damage O-ring or seal face.
5. Slide gland over shaft against deflector and place sleeve gasket (15-3) on shaft.
6. Install rotating assembly of mechanical seal on shaft sleeve. For special seals follow instructions given separately; for standard seals use the following procedure:

a. Wipe the lapped sealing faces of rotating and stationary elements perfectly clean with a soft cloth.
b. Oil rotating member lightly and slide on sleeve, taking care *not* to get oil on the seal faces.
c. Position rotating member on sleeve as indicated on detailed seal drawing.
d. Lightly tighten set screws which hold rotating member to sleeve (see Tables 1-4 and 1-5).

7. Oil pump shaft and slide sleeve and seal assembly over shaft.
8. Insert impeller key (11-1).
9. Press sleeve against shaft shoulder and firmly tighten set screw holding seal rotating member.
10. Insert gasket (15-2) into groove on cover (02).
11. Slide casing cover (02) over pump shaft (20) and seal, insert and tighten cap screws between housing (49) and cover. Check location of seal flush connection.
12. Bring seal gland forward and start gland nuts.
13. Squirt a few drops of light oil into the flush connection.
14. Check shaft and make sure it is free to rotate.
15. Draw gland nuts up evenly until metal to metal contact is realized between gland and cover.
16. Push impeller onto shaft and draw up impeller nut (21-1) (L.H. Threads).
17. Return complete back pull-out assembly to pump.
18. Slide casing gasket (15-1) over cover.
19. Slide rotating element into casing and tighten casing stud nuts evenly.
20. Check shaft that it is free to rotate and does not bind.
21. Follow start-up instructions.

Reassembling Bearing Housing

After inspection has been carried out and all parts are cleaned and corrected as necessary, the bearing housing can be reassembled by following these instructions and by frequently referring to the sectional drawing in Figure 1-15:

1. Press inboard bearing end cap (13-1) into bearing housing (49) until face is flush with housing. Oil return hole must be properly located.
2. Assemble oil flinger (29-2) on shaft (20) against the shoulder and lock in place with two socket head set screws.

Table 1-4
Pump shaft diameters

TYPE OF PUMP	SHAFT DIAMETER	
	At Coupling	At Stuffing Box
Packed	1.2500 / 1.2495	1.502 / 1.498
Mechanical Seal	1.2500 / 1.2495	1.502 / 1.498

Table 1-5
Permissible shaft run-out per ft. of diameter

MAX RUN OUT WITH RESPECT TO SHAFT	MAX FACE RUN OUT	MAX SHAFT RUN OUT	SHAFT END PLAY
004 TIR	004 TIR	002 TIR	NOT TO EXCEED 003

3. Slide thrust bearing (81-2) on shaft (20) as far as possible by hand. Oil bearing seat on shaft. Place pipe or sleeve over shaft, being sure it rests against inner race only. Tap sleeve evenly until bearing is seated firmly against shaft shoulder.
4. Assemble thrust bearing lock washer (31) and lock nut (21-4).
5. Slide radial bearing (81-1) on shaft (20) as far as possible by hand. Oil bearing seat on shaft. Place pipe or sleeve over shaft, being sure it rests against inner race only. Tap sleeve evenly until bearing is firmly seated against shaft shoulder.
6. Install shaft (20) and bearing subassembly into bearing housing. Due to bearing and housing tolerances, it may be necessary to lightly tap shaft until the thrust bearing is seated against bearing housing.
7. Install bearing end cap O-ring or gaskets.
8. Assemble outboard bearing end cap (13-2).
9. Assemble inboard (29-1) and outboard (29-2) deflectors on shaft.
10. Install pump half coupling.

Appendix 1-A

Gall Resistance of Metals

	Cast Iron	3% Ni-Cast Iron	Ni-Resist (Type 1, 2)	Ductile Iron	Ductile Ni-Resist	"S" Monel	"K" Monel[1]	"B" Monel	"H" Monel	Duranickel	"G" Nickel[2]	"S" Nickel[2]	Inconel	"S" Inconel[3]	400 Stainless (Soft)	400 Stainless (Hard)	300 Stainless Steel	SAE 1000 to 6000 (Soft)	SAE 1000 to 6000 (Hard)	Bronze (Leaded)[5]	Ni-Vee Bronze "A"[4]	Ni-Vee Bronze "B"	Ni-Vee Bronze "D"	Ni-Al Bronze[6]	Hastelloy "A"—"B"	Hastelloy "C"	Hastelloy "D"	Nitrided Steel	Chrome Plate[7]	Stellite
Cast Iron	S	S	S	S	S	S	S	S	S	S	S	S	S	S	S	S	S	S	S	S	S	S	S	S	S	S	S	S	S	S
3% Ni-Cast Iron	S	S	S	S	S	S	S	S	S	S	S	S	S	S	S	S	S	S	S	S	S	S	S	S	S	S	S	S	S	S
Ni-Resist (Type 1, 2)	S	S	S	S	S	S	S	S	S	S	S	S	S	S	S	S	S	S	S	S	S	S	S	S	S	S	S	S	S	S
Ductile Iron	S	S	S	S	S	S	S	S	S	S	S	S	S	S	S	S	S	S	S	S	S	S	S	S	S	S	S	S	S	S
Ductile Ni-Resist	S	S	S	S	S	S	S	S	S	S	S	S	S	S	S	S	S	S	S	S	S	S	S	S	S	S	S	S	S	S
"S" Monel	S	S	S	S	S	S	F	F	S	S	S	S	F+	S	F	S	F	F	S	S	F	S	S	F	F+	S	S	S	S	S
"K" Monel[1]	S	S	S	S	S	F	F	N	F	F	S	S	N	S	F	F	N	N	F	S	F	S	S	F−	F	F	S	S	S	S
"B" Monel	S	S	S	S	S	F	N	N	F	N	F	F	N	S	N	F	N	N	F	S	F	S	S	F−	N	F	S	S	F	S
"H" Monel	S	S	S	S	S	S	F	F	F	F	F	F	N	S	N	F	N	N	F	S	F	S	S	F−	N	F	S	S	S	S
Duranickel	S	S	S	S	S	S	F	N	F	F	S	S	N	S	N	F	N	N	F	S	F	S	S	F−	F	F	S	S	S	S
"G" Nickel	S	S	S	S	S	S	S	F	F	S	S	F	F	S	F	S	F	F	S	S	S	S	S	S	S	S	S	S	S	S
"S" Nickel	S	S	S	S	S	S	F	F	F	S	F	S	F	S	F	S	F	F	S	S	S	S	S	S	S	S	S	S	S	S
Inconel	S	S	S	S	S	F+	N	N	N	N	F	F	N	F	N	F	N	N	F	S	F	S	S	F−	N	F	S	S	S	S
"S" Inconel[3]	S	S	S	S	S	S	S	S	S	S	S	S	F	S	F	S	F	F	S	S	S	S	S	S	S	S	S	S	S	S
400 Series Stainless Steel (Soft)	S	S	S	S	S	F	F	N	N	N	F	F	N	F	N	F	F	N	F	S	F	S	S	F−	N	F	S	F	F	S
400 Series Stainless Steel (Hard)	S	S	S	S	S	S	F	F	F	F	S	S	F	S	F	S	F	S	S	S	S	S	S	F+	F	S	S	S	S	S
300 Series Stainless Steel	S	S	S	S	S	F	N	N	N	N	F	F	N	F	F	F	N	N	F	S	F	S	S	F−	N	F	S	S	S	S
SAE 1000 to 6000 Steel (Soft)	S	S	S	S	S	F	N	N	N	N	F	F	N	F	N	S	N	N	S	S	S	S	S	F+	N	F	S	S	S	S
SAE 1000 to 6000 Steel (Hard)	S	S	S	S	S	S	F	F	F	F	S	S	F	S	F	S	F	S	S	S	S	S	S	S	F	S	S	S	S	S
Bronze (Leaded)[5]	S	S	S	S	S	S	S	S	S	S	S	S	S	S	S	S	S	S	S	S	S	S	S	S	S	S	S	S	S	S
Ni-Vee Bronze "A"[4]	S	S	S	S	S	F	F	F	F	F	S	S	F	S	F	S	F	S	S	S	F	S	S	F	F	S	S	F	S	S
Ni-Vee Bronze "B"	S	S	S	S	S	S	S	S	S	S	S	S	S	S	S	S	S	S	S	S	F	S	S	F	S	S	S	S	S	S
Ni-Vee Bronze "D"	S	S	S	S	S	S	S	S	S	S	S	S	S	S	S	S	S	S	S	S	S	S	S	S	S	S	S	S	S	S
Ni-Al Bronze[6]	S	S	S	S	S	F	F−	F−	F−	F−	S	S	F−	S	F−	F+	F−	F+	S	S	F	F+	S	F	F−	F+	S	F	S	S
Hastelloy "A"—"B"	S	S	S	S	S	F+	F	N	N	F	S	S	N	S	N	F	N	N	F	S	F	S	S	F−	N	F	S	S	S	S
Hastelloy "C"	S	S	S	S	S	S	F	F	F	F	S	S	F	S	F	S	F	F	S	S	S	S	S	F+	F	F	S	S	S	S
Hastelloy "D"	S	S	S	S	S	S	S	S	S	S	S	S	S	S	S	S	S	S	S	S	S	S	S	S	S	S	S	S	S	S
Nitrided Steel	S	S	S	S	S	S	S	S	S	S	S	S	S	S	F	S	S	S	S	S	F	S	S	F	S	S	S	S	S	S
Chrome Plate[7]	S	S	S	S	S	S	S	S	S	S	S	S	S	S	S	S	S	S	S	S	S	S	S	S	S	S	S	S	?[7]	S
Stellite	S	S	S	S	S	S	S	S	S	S	S	S	S	S	S	S	S	S	S	S	S	S	S	S	S	S	S	S	S	S

Degree of Resistance: S — Satisfactory; F — Fair; N — Little or None

Appendix 1-B

Checklist for Rotating Equipment

548-7120

DETAILED CHECKLIST FOR ROTATING EQUIPMENT:
COUPLINGS

MACHINE TAG NO.

DATE/BY

1. KEYS FILL KEYWAY AND DO NOT PROJECT BEYOND SHAFT END. ______

2. HUBS POSITIONED ON SHAFT PROPERLY. ______

3. SPACING BETWEEN HUBS ________ INCHES. ______

4. COUPLING SPACER IN PLACE, BOLTS NOT BOUND. ______

DATE		FILE REFERENCE
3/20/84	MACHINERY RELIABILITY PROGRAM	PAGE 1 OF 1

548-7120

DETAILED CHECKLIST FOR ROTATING EQUIPMENT:
COUPLING GUARD

MACHINE TAG NO.

		DATE/BY
1.	MATERIAL IS 12 GAUGE GALVANIZED STEEL.	______
2.	HINGED INSPECTION DOOR PROVIDED	______
3.	GUARD DOES NOT TOUCH SHAFTS OR HOUSINGS.	______
4.	GUARD ENCLOSES COUPLING TO PRECLUDE CONTACT BY PERSONNEL.	______
5.	GUARD DOES NOT DEFLECT WHEN CONTACTED BY PERSONNEL.	______
6.	GAPS BETWEEN GUARD AND ACCESSIBLE ROTATING PARTS NEVER LARGER THAN 1/2 INCH.	______

DATE		FILE REFERENCE
3/20/84	MACHINERY RELIABILITY PROGRAM	PAGE 1 OF 1

PUMP DATA SHEET
VERTICAL

ITEM NO.__________ YARD NO.__________ UNIT__________

MANUFACTURER __________ MODEL NO. __________ S/N __________

IMPELLER SIZE __________ IMPELLER SETTING __________ PUMP CURVE NO. __________

MECHANICAL SEAL ☐ PACKING ☐ DWG. NO.__________
MANUFACTURER __________ SIZE/TYPE __________ SEAL POT FLUID __________

COUPLING: GEAR ☐ DIAPHRAGM ☐ DRY SHIM PACK ☐ DWG. NO.__________
MANUFACTURER __________ SIZE/TYPE __________

PUMP LUBRICATION:
OIL MIST DRY SUMP ☐ RING OIL ☐ PRESSURE ☐ FLINGER ☐ GREASED ☐
LUBE OIL TYPE __________

MOTOR LUBRICATION:
OIL MIST DRY SUMP ☐ RING OIL ☐ PRESSURE ☐ FLINGER ☐ GREASED ☐
LUBE OIL TYPE __________

OPERATING CONDITIONS:
PUMPING TEMP. °F __________ SUCTION PRESSURE (PSIG) __________ DISCHARGE PRESSURE (PSIG) __________

MOTOR DATA:
MANUFACTURER __________ MODEL NO. __________ S/N __________ YARD NO. __________
HP __________ RPM __________ VOLTAGE __________ PHASE __________ FLA __________
MOTOR ROTATION (LOOKING TOWARD PUMP) CLOCKWISE ☐ COUNTERCLOCKWISE ☐
MOTOR DOWELLED: YES ☐ NO ☐ SIZE __________ (SHOW LOCATION ON DWG. BELOW)

FINAL ALIGNMENT:

T FACE B

T RIM B

SUCTION

DISCHARGE

POSITION 1 SHAFT LEVEL __________ (IN./FT.)

POSITION 2 SHAFT LEVEL __________ (IN./FT.)

VIBRATION DATA:

PUMP				MOTOR			
I.B. BEARING HORIZ.	MILS ____	IN/SEC ____		MILS ____	IN/SEC ____		
I.B. BEARING VERT.	MILS ____	IN/SEC ____		MILS ____	IN/SEC ____		
O.B. BEARING HORIZ.	MILS ____	IN/SEC ____		MILS ____	IN/SEC ____		
O.B. BEARING VERT.	MILS ____	IN/SEC ____		MILS ____	IN/SEC ____		

ORIGINAL PURCHASE ORDER NO. __________

SIGNATURE __________ DATE __________

UNIT __________

PUMP DATA SHEET
HORIZONTAL

ITEM NO.________ YARD NO.________ UNIT________

MANUFACTURER ________ MODEL NO. ________ S/N ________

IMPELLER SIZE ________ PUMP CURVE NO. ________

MECHANICAL SEAL ☐ PACKING ☐ DWG. NO. ________
MANUFACTURER ________ SIZE/TYPE ________ SEAL POT FLUID ________

COUPLING: GEAR ☐ DIAPHRAGM ☐ DRY SHIM PACK ☐ DWG. NO. ________
MANUFACTURER ________ SIZE/TYPE ________

PUMP LUBRICATION:
OIL MIST DRY SUMP ☐ RING OIL ☐ PRESSURE ☐ FLINGER ☐ GREASED ☐
LUBE OIL TYPE ________

MOTOR LUBRICATION:
OIL MIST DRY SUMP ☐ RING OIL ☐ PRESSURE ☐ FLINGER ☐ GREASED ☐
LUBE OIL TYPE ________

OPERATING CONDITIONS:
PUMPING TEMP. °F ________ SUCTION PRESSURE (PSIG) ________ DISCHARGE PRESSURE (PSIG) ________

MOTOR DATA:
MANUFACTURER ________ MODEL NO. ________ S/N ________ YARD NO. ________
HP ________ RPM ________ VOLTAGE ________ PHASE ________ FLA ________
MOTOR ROTATION (LOOKING TOWARD PUMP) CLOCKWISE ☐ COUNTERCLOCKWISE ☐

BASE LEVEL AFTER GROUTING (INCHES/FEET) AXIALLY ________ RADIALLY ________

CALCULATED TEMPERATURE RISE: PUMP ________ MOTOR ________

PUMP DOWELLED: YES ☐ NO ☐ SIZE ________ LOCATION ________

MOTOR DOWELLED: YES ☐ NO ☐ SIZE ________ LOCATION ________

FINAL ALIGNMENT:

T FACE B   T RIM B

VIBRATION DATA:

	PUMP		MOTOR	
I.B. BEARING HORIZ.	MILS ____	IN/SEC ____	MILS ____	IN/SEC ____
I.B. BEARING VERT.	MILS ____	IN/SEC ____	MILS ____	IN/SEC ____
O.B. BEARING HORIZ.	MILS ____	IN/SEC ____	MILS ____	IN/SEC ____
O.B. BEARING VERT.	MILS ____	IN/SEC ____	MILS ____	IN/SEC ____

ORIGINAL PURCHASE ORDER NO. ________

SIGNATURE ________ DATE ________

UNIT ________

548-7120

DETAILED CHECKLIST FOR ROTATING EQUIPMENT:
CENTRIFUGAL PUMP RUN-IN

MACHINE TAG NO.

		DATE/BY
1.	MOTOR IS LARGE ENOUGH TO DRIVE PUMP WITH RUN-IN LIQUID.	______
2.	TEMPORARY AMPMETER IS INSTALLED. MOTOR FULL LOAD AMPS. ______	______
3.	PRESSURE GAUGES ON BOTH SIDES OF SUCTION STRAINER.	______
4.	WRITTEN DETAILED RUN-IN PROCEDURE FOR PUMP ON HAND.	______
5.	PERSONNEL UNDERSTAND PROCEDURE.	______
6.	PUMP VENTED AND PURGED WITH N_2 (IF REQUIRED).	______
7.	SUCTION VALVE FULLY OPENED.	______
8.	MINIMUM FLOW BY-PASS VALVE FULLY OPENED.	______
9.	DISCHARGE VALVE OPENED APPROXIMATELY 20 PERCENT	______
10.	PUMP FULLY PRIMED.	______
11.	COOLING WATER TURNED ON.	______
12.	SEAL POT(S) HAVE OIL AT PROPER LEVEL.	______
13.	NO UNUSUAL NOISES OR VIBRATION AT START-UP.	______
14.	SUCTION PRESSURE REMAINS CONSTANT.	______
15.	POSITIVE DISCHARGE PRESSURE.	______
16.	SUCTION PRESSURE ______ PSIG APPROX.	______
17.	DISCHARGE PRESSURE ______ PSIG APPROX.	______
18.	MOTOR AMPS ______ APPROX.	______
19.	PHASE AMPERAGE READINGS ARE EQUAL.	______

DATE		FILE REFERENCE
3/20/84	MACHINERY RELIABILITY PROGRAM	PAGE 1 OF 2

548-7120

DETAILED CHECKLIST FOR ROTATING EQUIPMENT:
CENTRIFUGAL PUMP RUN-IN

MACHINE TAG NO.

DATE/BY

20. ABSENCE OF LEAKS ON CONNECTIONS AND PIPE FLANGES. ______

21. OBSERVED DATA TAKEN DURING RUN-IN. ______

	BEG.	1 HR.	2 HR.	3 HR.	4 HR.	
LOAD CURRENT	____	____	____	____	____	AMP.
DISP. (P-P) MOTOR, IN/OUT	____	____	____	____	____	MIL.
VEL. MOTOR, IN/OUT	____	____	____	____	____	IN./SEC.
DISP. (P-P), IN/OUT	____	____	____	____	____	MIL.
VEL. PUMP, IN/OUT	____	____	____	____	____	IN./SEC.
NOISE	____	____	____	____	____	DBA

22. FOUR HOUR RUN-IN COMPLETED. ______

23. SHUT DOWN PER RUN-IN PROCEDURE. ______

DATE		FILE REFERENCE
3/20/84	MACHINERY RELIABILITY PROGRAM	PAGE 2 OF 2

Chapter 2

Installation, Maintenance, and Repair of Vertical Pumps*

Types of Vertical Pumps for Process Plants

The types of vertical pumps most commonly used in process plants include multistage process and condensate pumps, single stage inline process pumps, cooling water and cooling tower circulating pumps, and deepwell pumps. The multistage process and condensate pumps as illustrated in Figure 2-1 utilize standardized diffuser bowl assemblies installed in an outer barrel that is under suction pressure. This type of pump is often used where the available NPSH is not sufficient to accommodate a horizontal pump, or where space is at a premium. The first stage impeller is located below foundation level, thus providing additional NPSH. Reference 1 classifies these as vertical canned pumps of the double-casing type. Double casing refers to the type of construction in which the pressure casing is separate and distinct from the pumping element it contains. The single stage inline process pump shown in Figure 2-2 is simple, and pipe strain has minimal impact on rotating element alignment. This type of pump is selected when low initial cost of the pump, foundation, and piping is especially significant, and sufficient NPSH is available. Cooling water and cooling tower circulating pumps as shown in Figure 2-3 take suction from an open pit, sump, lake, river, or ocean. They provide cooling water to process units and to steam generating plant condensers. A typical deepwell pump is shown in Figure 2-4. As the name implies, it pumps water from deep wells and is used to supply makeup and utility water to process plants in areas where the wa-

---

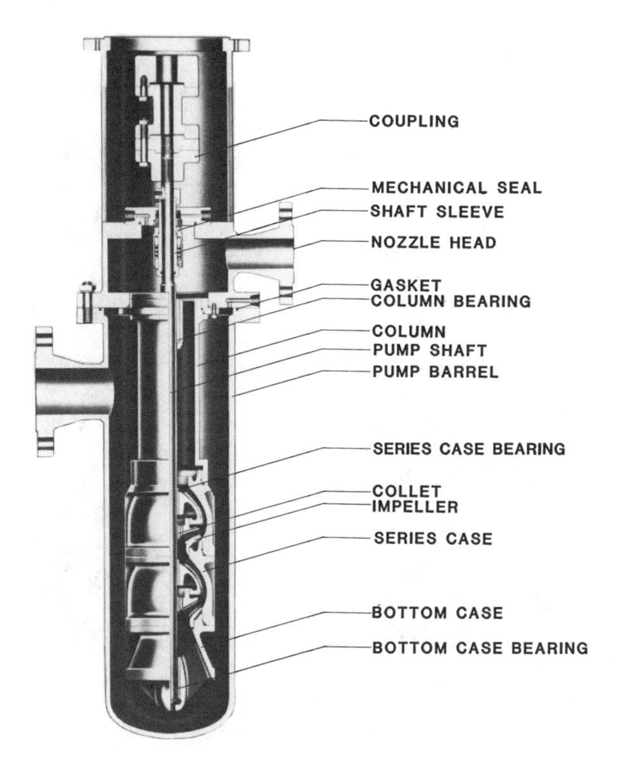

Figure 2-1. Multistage process or condensate pump. (Courtesy Byron Jackson Pump Division, Borg-Warner Industrial Products, Inc.)

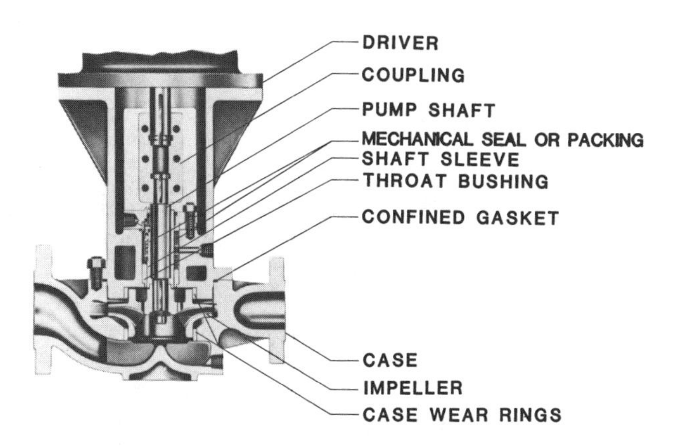

Figure 2-2. Inline process pump. (Courtesy Byron Jackson Pump Division, Borg-Warner Industrial Products, Inc.)

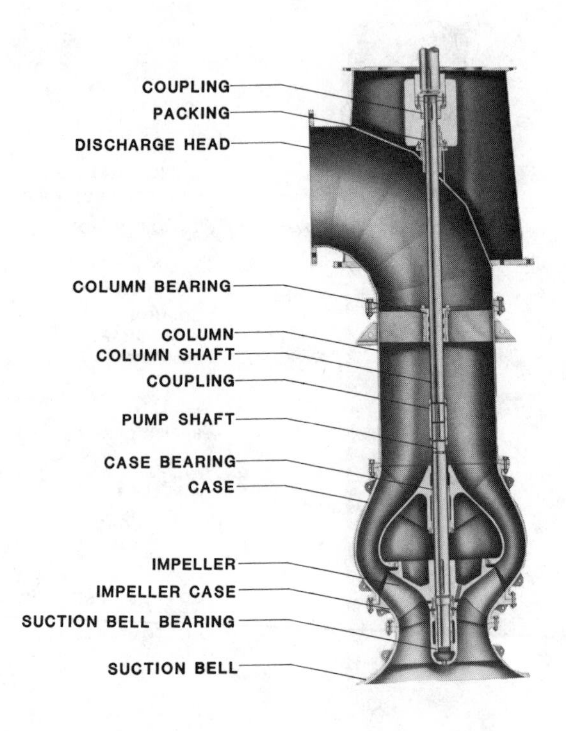

Figure 2-3. Vertical circulating pump. (Courtesy Byron Jackson Pump Division, Borg-Warner Industrial Products, Inc.)

ter table and water supply permits. Submersible pumps as illustrated in Figure 2-5 supply makeup and utility water from wells that are either too deep for reliable operation of deepwell pumps with line shafts or in locations that for environmental reasons require underground discharge installations. They also are used where motor noise would be a problem, and in locations subject to flooding.

Types of Drivers

Most vertical pumps in process plants are driven by vertical electric motors. The process pumps often have spares, driven by vertical steam turbines. Pumps driven through right angle gears by horizontal electric motors, steam turbines, or internal combustion engines are less common.

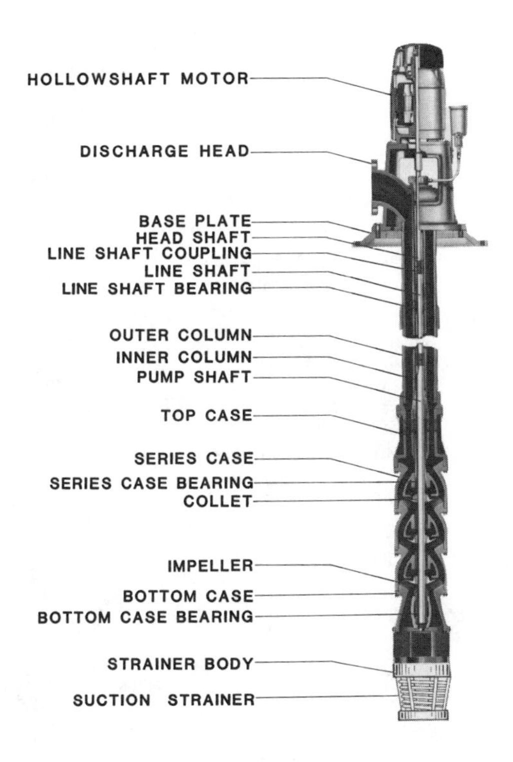

Figure 2-4. Deepwell pump. (Courtesy Byron Jackson Pump Division, Borg-Warner Industrial Products, Inc.)

Solid Shaft Drivers

Solid shaft drivers are used to drive pumps with relatively short shafts, less than 30 to 50 ft long. They therefore are used to drive almost all process pumps and circulators. They provide more positive shaft alignment which is especially important when the pumps have mechanical seals rather than packing. However, when a solid shaft driver is used, the axial adjustment of the pump rotor is limited by the height of the adjusting plate in the three-piece or four-piece coupling. The maximum axial adjustment is generally 1/2 to 3/4 in. Thrust bearings normally will carry either downthrust or upthrust. The construction of a solid shaft motor is shown in Figure 2-6.

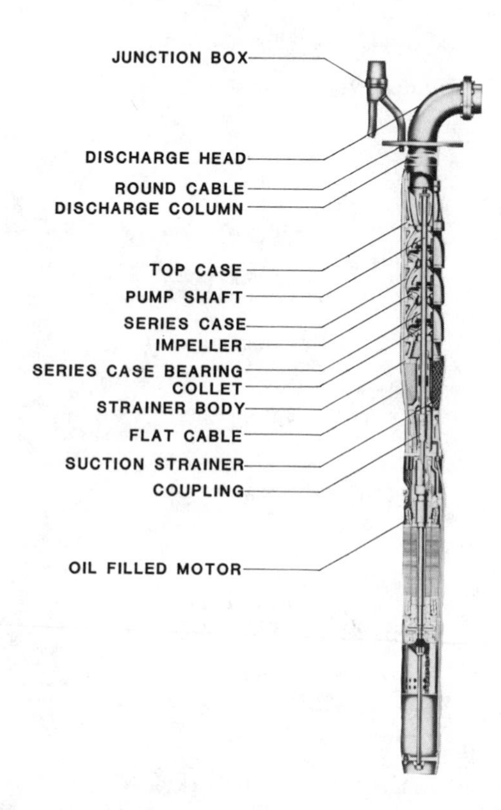

Figure 2-5. Submersible pump and motor. (Courtesy Byron Jackson Pump Division, Borg-Warner Industrial Products, Inc.)

Hollow Shaft Drivers

Hollow shaft drivers provide a means to adjust the axial position of the pump rotor for distances of several inches. They were developed many years ago to fill the need for large adjustments created by the demand for deepwell pumps of ever increasing settings, and are now standard for almost all deepwell pumps. Hollow shaft drivers normally are designed to carry downthrust only and special provisions must be made if upthrust capability is required. The construction of a hollow shaft motor is shown in Figure 2-7.

Deepwell Pump Shaft Adjustment

Details of the upper portion of a hollow shaft motor are shown in Figure 2-8. To adjust the axial position of a deepwell pump rotor:

1. Tighten the adjusting nut until the shaft starts to turn. This is an indication that the shaft has stretched to compensate for the weight of the rotor and that the impellers have lifted clear of the bowls.

2. Engage the locking arm or otherwise hold the self release coupling against rotation. As the coupling is keyed to the shaft, the shaft is also held against rotation.

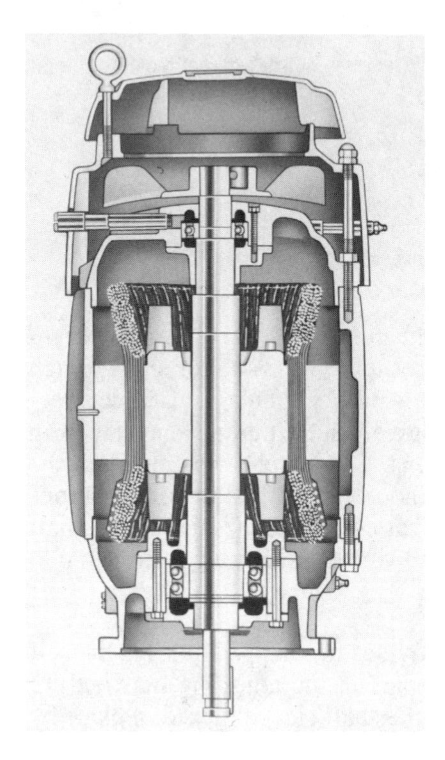

Figure 2-6. Totally enclosed fan-cooled vertical solid shaft normal thrust motor. (Courtesy U.S. Electrical Motors Division, Emerson Electric Company.)

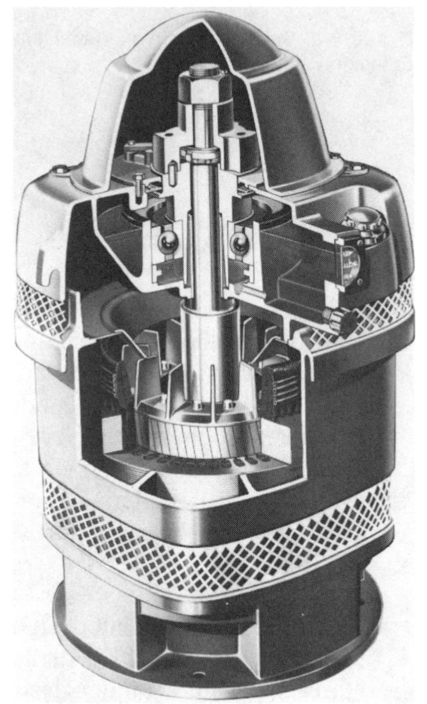

Figure 2-7. Weather-protected type I (WPI) vertical hollow shaft high thrust motor. (Courtesy U.S. Electrical Motors Division, Emerson Electric Company.)

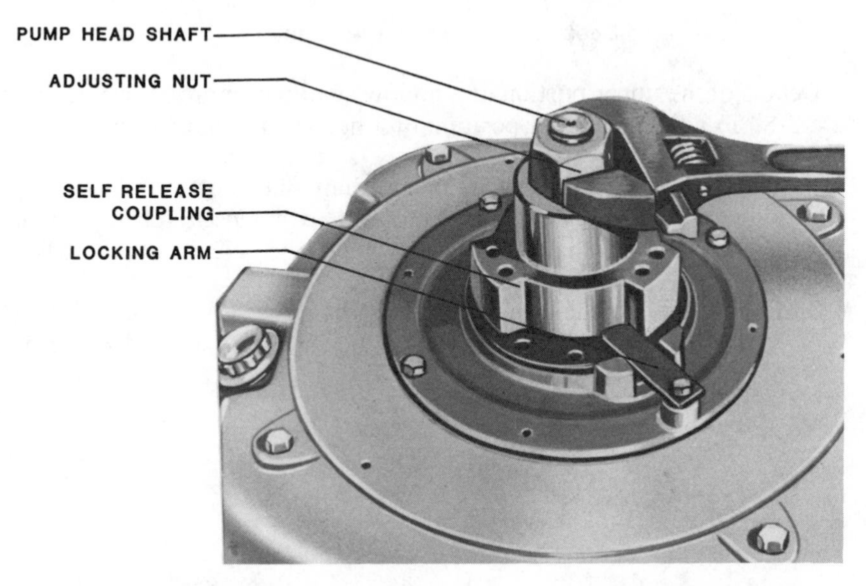

Figure 2-8. Deepwell pump shaft adjustment. (Courtesy U.S. Electrical Motors Division, Emerson Electric Company.)

3. Tighten the adjusting nut to lift the pump rotor the distance recommended by the pump manufacturer to compensate for shaft stretch caused by pump thrust and to allow proper running clearance. Recommended rotor lift is generally given on the pump nameplate or in the instruction manual. For pumps with semi-open impellers the running clearance is only a few thousandths of an inch. For pumps with closed impellers a typical running clearance is 1/4 in. Shaft stretch caused by pump thrust will vary considerably from pump to pump.

4. A typical adjusting nut design provides four tapped holes in the self release coupling, three drilled holes in the adjusting nut, and 12 threads per in. With this design, the shaft can be locked in place by installation of the locking screw in any 1/12 turn position, and the shaft setting will be within 1/144 or .007 in.

5. Disengage the locking arm.

An alternate method to set semi-open impellers is:

1. Raise the rotor as high as it will go, then lower it two turns on the adjusting nut. While raising the rotor, do not use excessive force. Especially if the impellers are collet mounted, be sure that they are not pushed downward. Just raise the rotor until a very slight drag is detected when the shaft is rotated.

2. Lock the adjusting nut, start the pump, record the ammeter reading with the pump running at or near shutoff or at the minimum flow allowed by the system.

3. Check to be sure that the ammeter reading is less than the amps on the motor nameplate.

4. Lower the rotor 1/12 turn, lock the adjusting nut, start the pump, record the ammeter reading with the pump running at or near shut-off or at the minimum flow allowed by the system.

5. Repeat Step 4 until the ammeter reading increases sharply because the impellers are rubbing on the bowls.

6. Raise the rotor 1/12 turn to its final position and lock the adjusting nut.

Hollow Shaft Driver Reverse Protection Clutch

If the driver is accidentally run with reverse rotation, the deepwell pump shaft couplings may unscrew and cause damage. Figure 2-9 shows a reverse protection clutch which automatically disengages the pump shaft from the driver if this occurs. In some shallow settings, particularly with closed impellers, the pump may develop a momentary upthrust during start-up. In these installations, it is necessary to install hold-down bolts in the clutch and to check that the driver thrust bearing is locked against upthrust.

Hollow Shaft Driver Nonreverse Ratchet

If the foot valve or check valve fails in a deepwell pump installation, when the driver is de-energized the water will run back down the discharge column, and the pump will act as an unloaded turbine and achieve a reverse rotation that can reach rather high speed. The torque does not reverse, so the threaded couplings do not unscrew as a result. Although

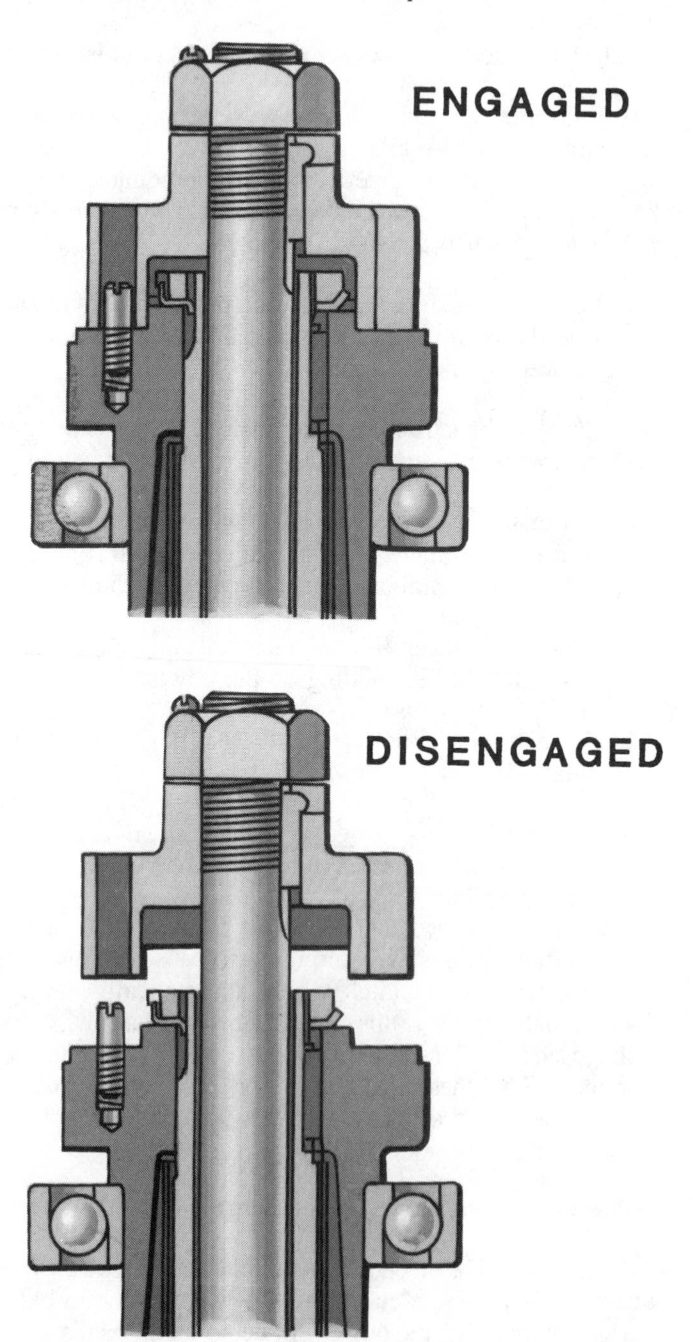

Figure 2-9. Reverse protection clutch. (Courtesy U.S. Electrical Motors Division, Emerson Electric Company.)

the reverse speed may be higher than normal pump rpm, it seldom causes damage unless water-lubricated column bearings are run dry, or an attempt is made to restart the driver while the pump is running backward. Hollow shaft drivers are often provided with nonreverse ratchets so that reverse rotation does not occur. Alternatively, some systems have time-delay relays to prevent premature restart. For deepwell pumps with settings of 500 ft or more, the long shaft may act as a torsional spring causing the nonreverse ratchet to be subject to torque reversals. This problem is controlled by providing a ratchet which is designed so that the movement caused by torque reversal is only 3 to 5°

Driver Alignment

Practically all vertical pumps now used in process plants within the United States utilize the driver thrust bearing to carry the combined thrust load of the driver and pump. The driver also provides radial alignment for the upper portion of the pump shaft. Reliability of the driver bearings is a prerequisite to the reliability of the pump. Accurate radial and angular alignment between the driver and the pump is therefore essential. Driver shaft runout and concentricity with the mounting fit must be checked prior to assembly or reassembly of the driver on the pump. Referring to Figure 2-10, the following steps are recommended:

1. Thoroughly clean the driver shaft and mounting face.
2. Attach a dial indicator to the shaft and rotate on the driver rabbet fit and mounting face.

   a. The concentricity of the rabbet fit must be within .002 in. total indicator reading (T.I.R.) per ft of rabbet fit diameter.
   b. The mounting face must be perpendicular to the shaft within .002 in. T.I.R. per ft of rabbet fit diameter.

3. Mount the dial indicator on the driver housing to check the shaft runout and end float.

   a. The shaft runout must not exceed .002 in. T.I.R. or .001 in. T.I.R. per in. of shaft diameter, whichever is greater.
   b. The squareness of the split ring groove to the shaft centerline must be within .002 in. T.I.R.
   c. Shaft end float must not exceed .010 in. T.I.R., and .005 in. T.I.R. is preferred if the pump has a mechanical seal.

These requirements are more stringent than NEMA standards but are essential for solid shaft drivers, particularly for pumps operating at 3600

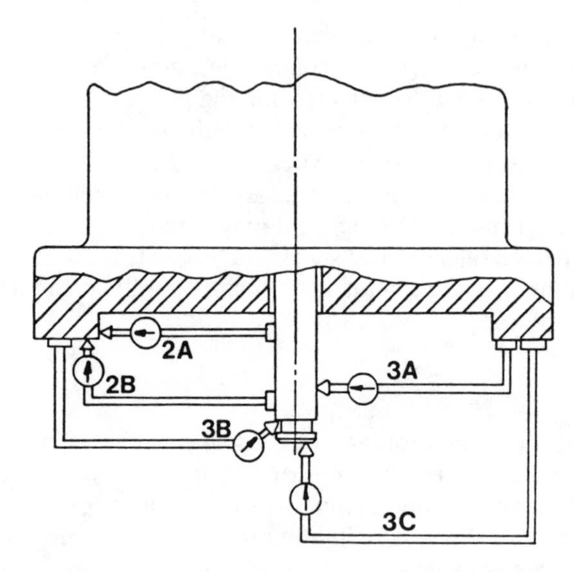

Figure 2-10. Checking solid shaft driver run-outs and concentricities.

rpm or pumps with mechanical seals. NEMA tolerances for concentricities and runouts, however, are quite adequate for deepwell turbine pumps utilizing hollow shaft motors or gears and operating at 1,800 rpm or less. less.

Some pump manufacturers do not use the driver rabbet fit for alignment but rather align the driver shaft to the pump stuffing box, tighten the driver mounting bolts and dowel the driver to the pump. This procedure eliminates the need for concentricity between the driver shaft and rabbet fit. It can also be utilized as a compensating procedure if an otherwise acceptable driver has a rabbet fit which is not concentric with the shaft.

1. The driver rabbet fit or the driver mounting flange of the pump is machined to allow clearance for radial movement of the driver.
2. If the driver is too large to be moved radially with a soft hammer, four jacking bolts should be installed.
3. The driver is then mounted, but the mounting bolts are left loose and the driver is aligned radially using a dial indicator mounted on the shaft and sweeping the stuffing box bore. This alignment should

be within .0005 in. per in. of stuffing box bore for pumps with mechanical seals and .001 in. per in. of stuffing box bore for packed pumps. If the coupling has an adjusting plate such as the four-piece coupling shown in Figure 2-11, it can be unscrewed until it engages the driver half coupling or spacer to verify alignment between the pump shaft and the driver shaft.

4. The driver bolts are then tightened and two dowels installed.
5. After the coupling is completely assembled, check the runout of the pump shaft or shaft sleeve, measured by a dial indicator immediately above the stuffing box or mechanical seal. Paragraph 2.8.2.7 of Reference 1 requires that this runout not exceed .002 in. T.I.R. for new pumps operating above 1400 rpm, or .004 in. T.I.R. for new pumps operating below 1,400 rpm.

Maintenance and Repair of Driver Bearings

Except for drivers with grease lubricated ball bearings, dry sump oil mist lubrication for driver antifriction bearings is recommended. Concurrent elimination of cooling water is generally feasible. If a self-contained lube oil reservoir is required, lube oil purification is recommended. These topics are detailed in Chapter 7 of Volume 1 of this series. The precautions required for proper assembly and removal of ball bearings are also covered in Chapter 7 of that volume.

Maintenance and Repair of Submersible Motors

Submersible pumping units are generally not pulled from the well until the pump performance has deteriorated by about 5 percent head loss, or the motor has burned out, or "megging" of the motor windings shows a low resistance to ground. Repair of a damaged motor should only be done by the manufacturer or by an authorized repair shop. Some manufacturers stock rebuilt motors for immediate shipment so that outage time is minimized. Instructions for the preparation of used motors for shipment must be followed with great care so that further damage is avoided. Detailed information on the maintenance of a particular type of submersible pumping unit is contained in Reference 2.

Maintenance and Repair of Driver to Pump Couplings

Two piece or three piece couplings are used to connect solid shaft drivers to pumps with packed stuffing boxes and closed impellers. Three-piece couplings have threaded adjusting plates for adjusting axial impeller clearances in vertical pumps with semi-open impellers. Four piece

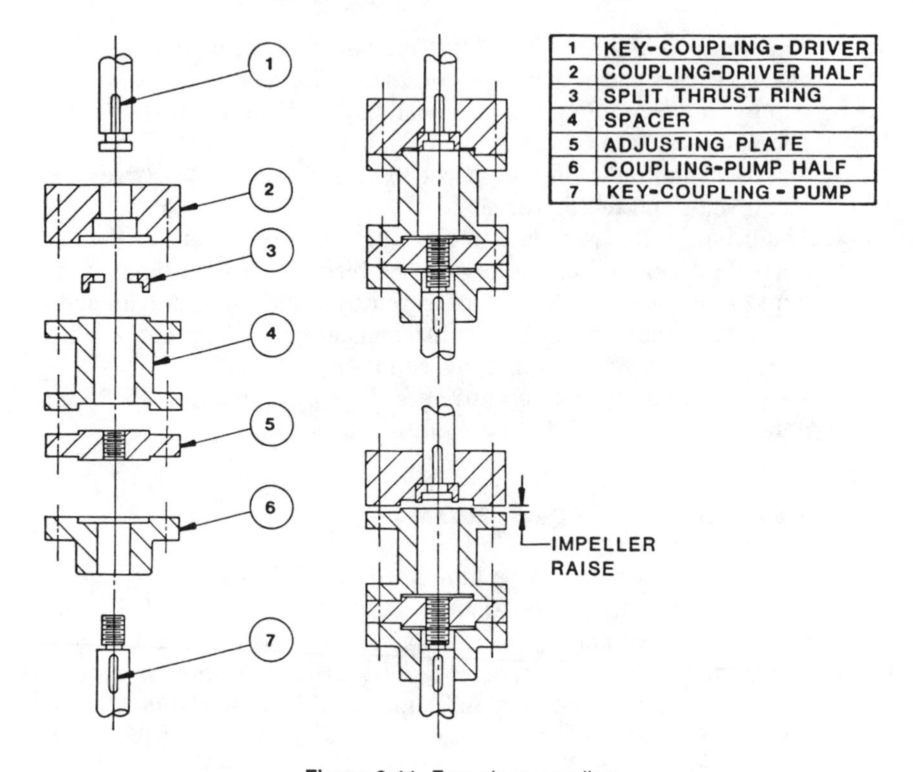

Figure 2-11. Four-piece coupling.

couplings as shown in Figure 2-11 are adjustable axially and have spacers so that mechanical seals may be removed without removing the drivers.

Maintenance of radial and angular alignment of all rigid couplings is extremely important. They do not normally wear but are subject to corrosion and handling damage. Except for very large couplings it is generally most economical to replace a coupling if there is time to obtain a new coupling from the pump manufacturer. Components of the two, three and four-piece couplings can be repaired to restore original concentricities and the squareness of flange faces.

1. Rabbet fits may be built up by metalizing or by careful peening with a ball peen hammer to provide stock for machining. Welding distorts the part and should be avoided.

2. Coupling hubs and spacers should be mounted on a mandrel to assure that the coupling is running absolutely true. Skim cuts are used

to restore the flatness and squareness of flange faces and the concentricities of rabbet fits. Only skim cuts should be used and concentricities and tolerances must be maintained.

3. Coupling bores may be restored by chrome plating followed by an internal grinding operation. However, this is generally an expensive procedure.

4. Both the driver half coupling and the pump half coupling should be installed with snug sliding fits.

5. Large couplings or couplings operating at 3,600 RPM and above should be placed on a mandrel and dynamically balanced, using procedures similar to those used for the dynamic balancing of impellers.

Maintenance and Repair of Pump Shaft Couplings

Both threaded and keyed shaft couplings are found in circulating, deepwell, and submersible pumps. Shaft couplings are relatively inexpensive. They should be stocked, and replaced if corrosion or damage makes them unserviceable.

Auxiliary Piping Requirements

Special attention to auxiliary piping requirements can reduce maintenance and repair costs, especially those related to barrels, seals, and packing. This is particularly true if the pumped liquid is very hot, very cold, volatile, or when the pump suction is under a vacuum or very high pressure. Reference 3 covers auxiliary piping requirements for these cases.

Maintenance and Repair of Packed Stuffing Boxes

Almost all process pumps are equipped with mechanical seals. Packed stuffing boxes are generally found only in deepwell pumps and circulating pumps handling cold clean water or with a cold clean water injection below the packing.

1. Soft braided fabric packing impregnated with lubricant is satisfactory for low pressure applications. Lubricated synthetic packing is advisable for higher pressures.

2. Preformed packing rings promote a quicker and better packing job. However, if the replacement packing is in a continuous coil it must

be cut into rings before installing. Tightly wrap one end of the packing around the shaft or a mandrel which is the same diameter as the shaft. Wrap one coil and mark it with a sharp knife. Leave a gap of 1/16 to 1/8 in. and mark the ends parallel. After cutting on the marks, a length of packing may be used as a template for cutting all the other rings.

3. Remove all old packing, using flexible packing hooks. Be sure to remove the old packing below the lantern ring.

4. Check the shaft or sleeve for nicks and scratches, remove any that are present, and then clean carefully. Clean the bore of the stuffing box. Check the lantern ring to make sure that the channels and holes are not plugged.

5. Coat at least the outside diameter of each packing ring with grease or oil.

6. Place the first ring around the shaft and press it evenly to the bottom of the stuffing box. Seat it firmly against the face of the throat bushing. Rotate the shaft by hand until it turns freely. This helps provide initial running clearance. Install succeeding rings in the same manner, with each joint 90° clockwise from the preceding one. Each ring must be seated firmly as it is installed to avoid overloading the rings next to the gland which do most of the sealing. Whenever possible, use split bushings as shown in Figure 2-12 to seat each ring separately. They prevent cocking and compress each ring evenly in the stuffing box. Wood, babbitt or brass bushings may be used. When all packing rings are installed, the shaft should turn freely without binding.

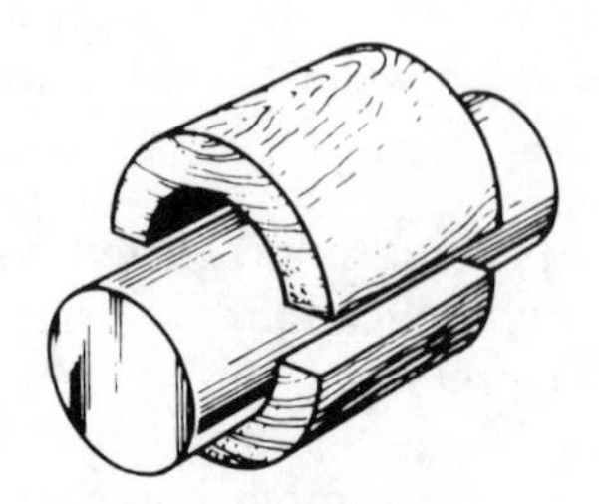

Figure 2-12. Use split bushings to seat each ring of packing separately.

7. Install the packing gland and draw the gland nuts hand tight. Do not compress the packing.

8. When a pump is first operated with new packing, it should be loose enough to allow a small stream of liquid to flow from the gland. There should be no leakage around the OD of the packing. Do not operate the pump with packing too tightly compressed, and do not make final adjustment until the pump has been operated for several hours. Then tighten the gland nuts alternately, very slowly and evenly, until the leakage rate is reduced to a steady drip. Gland nuts should be tightened $^{1}/_{6}$ turn (one flat) at a time and the packing leakage observed for 10 minutes before tightening another $^{1}/_{6}$ turn.

9. Check the gland temperature and loosen the gland nuts to increase the leakage rate if overheating of the packing is observed. The gland adjustment should be checked periodically and readjusted as needed.

Note: Refer also to the packing instructions in Chapter 8 of Volume 2.

Maintenance and Repair of Pump Bearings

Table 2-1 shows nominal diametral clearances for vertical pump bearings. Bearings should always be made from nongalling materials or the surfaces should be nitrided or overlayed with Stellite or Colmonoy so that they are nongalling.

Bowl bearings are product lubricated and normally have sufficient differential pressure across them to provide ample flow for lubrication and

Table 2-1
Nominal Diametral Clearances for Vertical Pump Bearings

Standard Shaft Size	Diametral Bearing Clearance
$^{3}/_{4}$	.006
1	.006
$1^{3}/_{16}$	.008
$1^{7}/_{16}$	.008
$1^{9}/_{16}$	.009
$1^{11}/_{16}$	.009
$1^{15}/_{16}$	.010
$2^{3}/_{16}$	.011
$2^{7}/_{16}$ and larger	.012

(Dimensions are in Inches)

cooling. The load-carrying ability of bowl bearings is enhanced by the Lomakin effect which is a function of the magnitude of the differential pressure across the bearing and is explained in detail in Reference 4. Unless the product being pumped contains foreign material, these bearings are best designed without spiral or axial grooves. Grooves interfere both with the Lomakin effect and with the hydrodynamic load carrying ability of the bearings.

Product Lubricated Column Bearings should have spiral or axial grooves to provide proper cooling because they generally are not subject to differential pressure.

Grease Lubricated Column Bearings should have axial grooves.

Oil Lubricated Column Bearings should have spiral grooves to provide a downward pumping action.

Prelubrication of Column Bearings that are installed above the water level in a deepwell or circulating pump is necessary. It can be achieved by oil or grease lube, by fresh water injection from a source other than the pump discharge, or by prelube piping around the check valve in the discharge piping of a deepwell pump.

The bottom case or suction bell bearings in deepwell turbine pumps and circulating water pumps are usually lubricated by a heavy non-watersoluble grease which when installed in a properly designed bearing housing with a large grease reservoir will last until the pump needs to be dismantled for other reasons (Figure 2-13). These bearings should be inspected and repacked during routine maintenance. The bearing grooves and grease reservoir should be filled completely and the grease plug reinstalled so that the bearing is dead-ended and the grease will not wash out.

Tension nut bearings should have oil or grease lubrication or injection of pumped product that has been passed through a filter or a cyclone separator. Cyclone separators are inexpensive and reliable, provided that no large solid particles are in the product, there is a large density difference between the liquid and the solids, and the orifice does not erode out.

Avoidance of Wear Rings

Semi-open impeller designs should be utilized whenever possible so that wear rings are avoided. If wear ring repair costs are excessive, complete bowl assemblies of closed-impeller construction may be replaced by semi-open impeller bowl assemblies after design review by the pump manufacturer. This review should include determination that the driver thrust bearing is adequate and that the coupling can be replaced with one that is axially adjustable. The initial efficiency of semi-open impeller

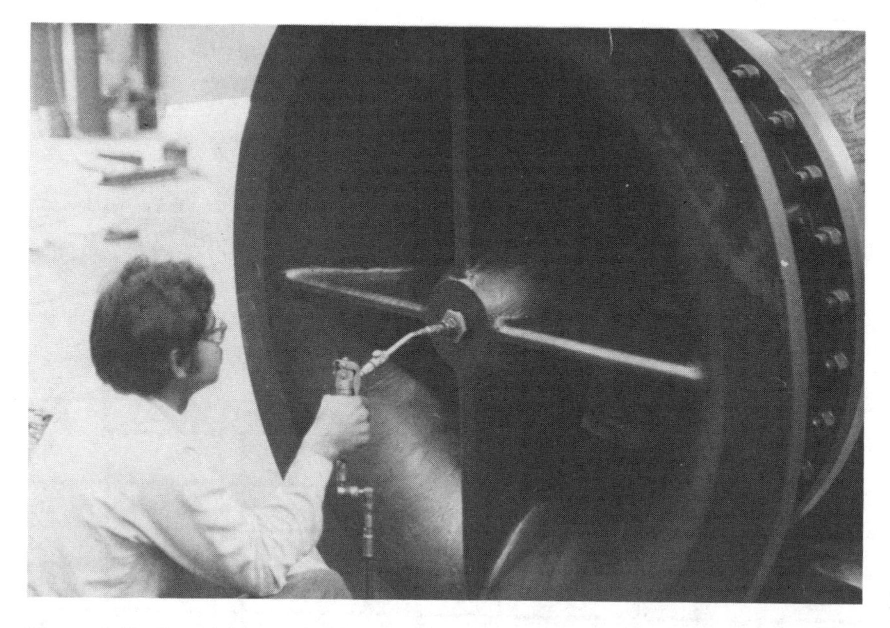

Figure 2-13. Greasing a suction bell bearing. (Courtesy Byron Jackson Pump Division, Borg-Warner Industrial Products, Inc.)

units can be up to two points higher because of reduced disk friction and more accurate and better finished water passages. The high efficiency can be sustained by periodic axial adjustment of the rotor without dismantling the pump. Closed impellers are necessary, however, when pumping hot liquids, because of the axial thermal expansion of the pump, and generally when the pumps are more than 100 ft in length.

Maintenance and Repair of Wear Rings

Although most users specify wear rings on newly installed pumps, this is not economically justified as long as the pumps are designed for future installation of wear rings. When they can be utilized, deepwell turbine pump components are economical and readily available both as complete bowl assemblies and as spare parts. Leading manufacturers will furnish machining dimensions, in a format such as Figure 2-14. When these pumps require service the impeller wear surfaces can be turned to a predetermined diameter and the bowls machined to receive wear rings of matching dimensions. Original clearance is restored by installing only one wear ring per stage. These machining and assembly operations are

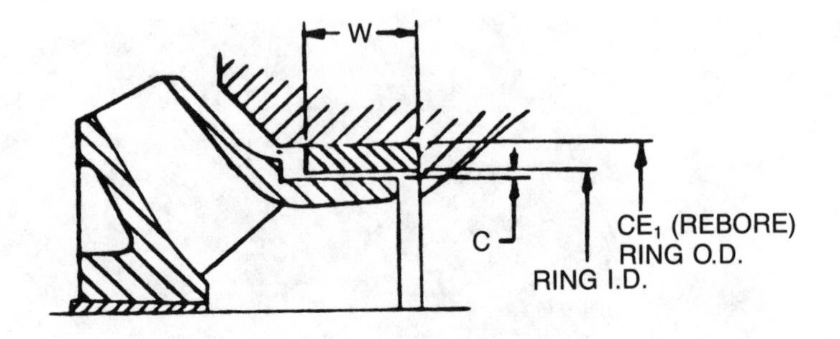

SIZE	TYPE	CE_1 +.001	WEAR RING O.D. -.001	WEAR RING I.D. +.001	WEAR RING W	C
6	MQ	3.562	3.565	3.000	.656	.016
6	HQ	4.250	4.253	3.750	.843	.015
7	MQ	4.125	4.128	3.562	.968	.016
8	MQ	4.875	4.878	4.250	.968	.016
8	HQ	5.875	5.878	5.250	1.062	.016
10	MQ	5.750	5.753	5.000	1.125	.016
10	HQ	7.375	7.378	6.625	1.125	.016
11	MQ	6.500	6.503	5.750	1.250	.018
12	MQ	7.000	7.003	6.250	1.312	.018
12	HQ	7.687	7.690	6.875	1.062	.011
12	HQR	8.812	8.815	8.000	1.343	.018
13	MQ	7.687	7.690	7.000	1.375	.018
15	MQ	8.562	8.565	7.875	1.718	.018
15	HQ	9.625	9.628	8.875	1.500	.018
17	MQ	9.844	9.847	9.000	1.968	.021
17	HQ	11.281	11.284	10.500	1.750	.018
20	MQ	11.001	11.004	10.250	2.343	.021
20	HQ	13.250	13.253	12.375	1.875	.021
23	HO	15.125	15.128	14.250	2.125	.021

Figure 2-14. Wear ring repair dimensions.

no more costly than the removal of case and impeller wear rings and the installation of new ones. Significant savings can be achieved by reduction of spare parts inventories.

As is the case with bowl bearings, the wear rings have differential pressure across them and will act to some degree as bearings provided that they are not grooved. Grooved wear rings should only be used in installations where the product contains abrasive material. We acknowledge, however, that some users report acceptable results from the use of grooved wear rings in nonabrasive services as well.

Replacement Impellers

Comparison testing has shown that replacement impellers which are not furnished by the pump manufacturer can be very expensive in the long run. In addition to potential shortening of pump life between repairs, inadequate NPSH performance, improper curve shape and invalidation of pump warranties, the cost of additional power can be significant. In one well documented case (Reference 5) the power requirement increased from 134.7 horsepower to 139.8 horsepower resulting in an estimated additional electric power cost of $3,000 per year. Although the cost of purchasing the replacement impeller from the pump manufacturer may be greater, economic analysis will show that purchasing the impeller from another source is false economy. Some pump users have purchased impellers from others because delivery time from the pump manufacturer was too long. The problem of delivery time can be avoided by anticipation of the requirement and timely placement of the order, or by handling of the order on a rush basis by the manufacturer.

Impeller Dynamic Balance

The dynamic balance must be restored prior to reinstallation if any work has been done on an impeller or if it shows appreciable wear. Impellers which are out of balance can cause vibration and rapid wear of bearings and wear rings.

Vertical pump impellers normally are balanced individually to quality grade G-6.3 as indicated on the nomograms Figure 2-15 for impellers which weigh 50 kilograms (110 pounds) or less and Figure 2-16 for impellers which weigh more than 50 kilograms. By alignment of the pump operating speed in rpm with the impeller weight in kilograms, the tolerance or allowable unbalance in gram-millimeters is given by the center scale on each nomogram. The weight in kilograms equals 0.454 times weight in lbs, i.e., a 100 lb impeller weighs 45.4 kilograms. To convert the allowance in gram-millimeters to inch-ounces, multiply the value in gram-millimeters by .00139 to get the value in inch-ounces, i.e., an al-

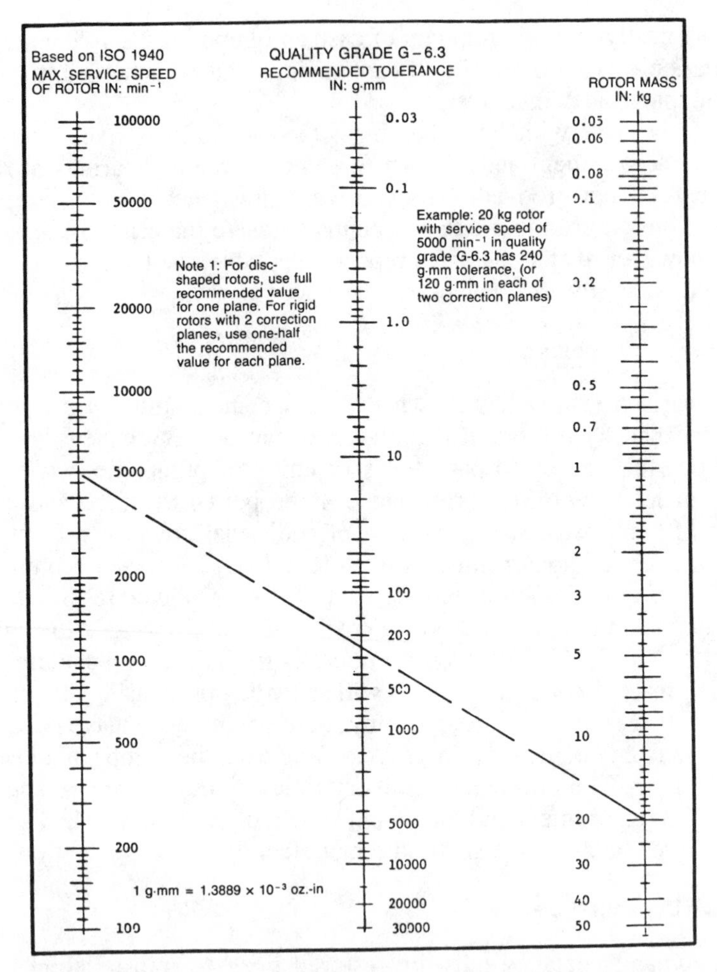

Figure 2-15. Balance tolerance diagram for rotor mass up to 50 kilograms.

lowance of 100 gram-millimeters equals an allowance of .139 inch-ounces.

1. The impeller is mounted on a mandrel for balancing.
2. If it has a keyway, a key or half key must be installed.
3. Correction of unbalance is achieved by removal of metal from the impeller shrouds, generally by grinding with a handheld portable grinder (Figures 2-17 and 2-18).
4. Metal should not be removed from the outer ½ in. of the shroud, and no more than ⅓ of the shroud thickness should be removed.

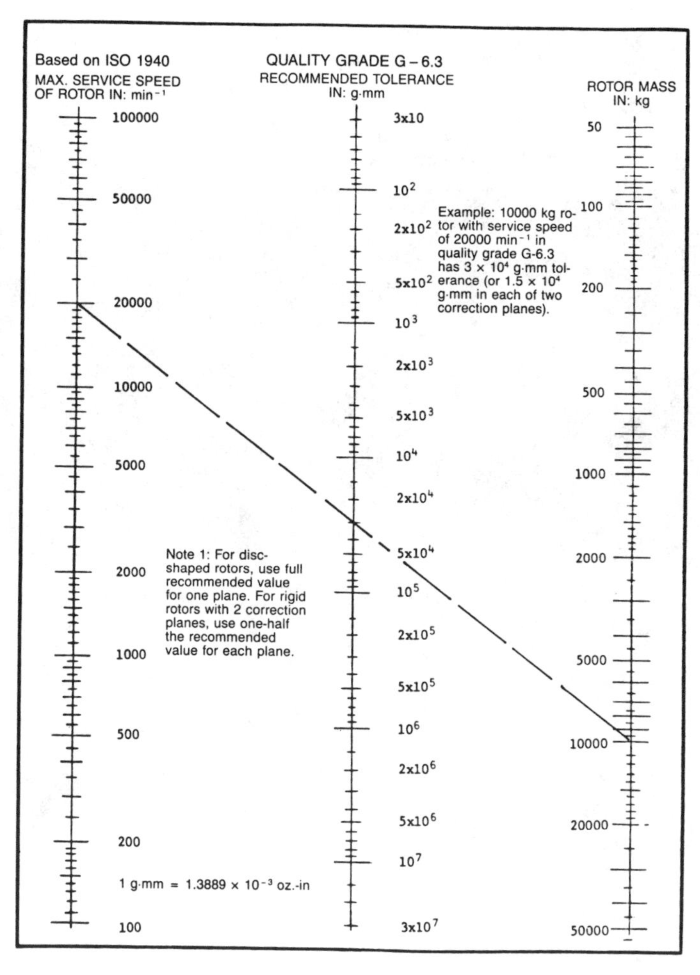

Figure 2-16. Balance tolerance nomogram for rotor mass above 50 kilograms.

Mounting of Impellers on the Shaft

The use of collets is a common and satisfactory method for mounting impellers on the shafts of small vertical turbine-type pumps, generally limited to pumps with bowl outside diameters less than 18 in. and shaft diameters less than 2³/₁₆ in. The collets are split along one side, have an inside diameter equal to the diameter of the shaft and are tapered on the outside to fit the tapered bore of the impeller. When clean, free from nicks and scratches and properly driven into the impeller bores, collets will lock the impellers in place axially and radially, and will transmit the

Figure 2-17. Impeller on dynamic balancing machine. (Courtesy Byron Jackson Pump Division, Borg-Warner Industrial Products, Inc.)

Figure 2-18. Correcting unbalance by removal of metal from the impeller shroud. (Courtesy Byron Jackson Pump Division, Borg-Warner Industrial Products, Inc.)

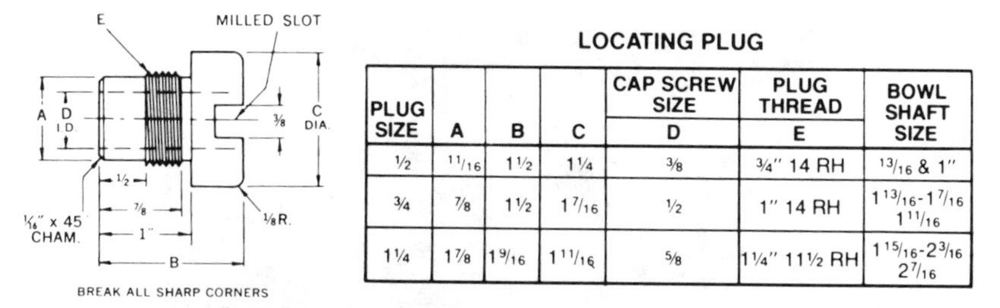

LOCATING PLUG

PLUG SIZE	A	B	C	CAP SCREW SIZE D	PLUG THREAD E	BOWL SHAFT SIZE
1/2	11/16	1 1/2	1 1/4	3/8	3/4" 14 RH	13/16 & 1"
3/4	7/8	1 1/2	1 7/16	1/2	1" 14 RH	1 13/16-1 7/16 1 11/16
1 1/4	1 7/8	1 9/16	1 11/16	5/8	1 1/4" 11 1/2 RH	1 15/16-2 3/16 2 7/16

Figure 2-19. Dimensions for shaft locating plug.

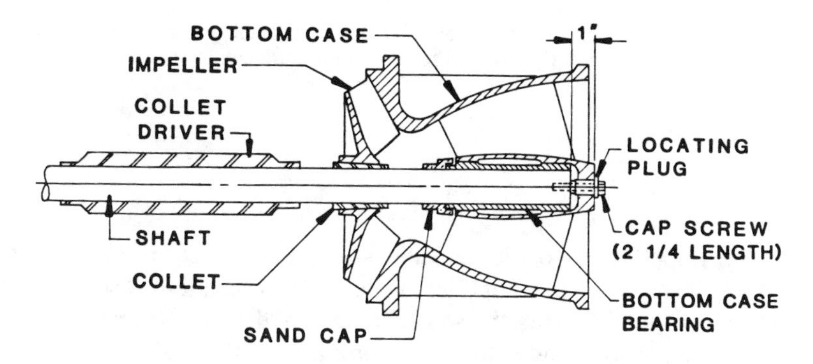

Figure 2-20. Shaft position for collet installation.

torque from the shaft to the impellers. They are quite satisfactory for small process pumps, but should not be used for temperatures above 160°F.

To drive the collets into the impeller bores properly, the shaft should be positioned firmly in the bottom case with a shaft locating plug and capscrew.

1. Figure 2-19 shows typical dimensions for the shaft locating plug.
2. Figure 2-20 shows the assembly of the shaft locating plug into the bottom case. The shaft is held in place by a 2¼ in. long capscrew installed in the tapped hole at the bottom of the shaft. Prior to assembly all parts including the collet driver must be inspected to be sure that they are clean and free from nicks, burrs or scratches.

Figure 2-21. Using a screwdriver to keep the collet from binding on the shaft. (Courtesy Byron Jackson Pump Division, Borg-Warner Industrial Products, Inc.)

3. The impeller is positioned against the bottom case. The collet is then inserted in the impeller bore. Collets usually close in on the bore and tend to grab the shaft, so a screwdriver is inserted to open the collet and allow it to slide along the shaft, as shown in Figure 2-21. Next, the collet driver is slipped over the shaft with the large diameter end facing the collet.
4. The collet is then driven firmly into the impeller bore utilizing the impact of the collet driver. Hold the impeller tightly against the case by hand while driving the collet. Let go of the collet driver before impact and allow its momentum to drive the collet (Figure 2-22).
5. Check the impeller to be sure that it is tight on the shaft.
6. The collet driver impact can loosen the 2¼ in. long capscrew. Retighten it after each impeller is mounted.
7. Additional series cases and impellers are installed until the bowl assembly is complete. As an added precaution the capscrew can be released and the shaft rotated after each stage is assembled to be sure that there is no binding.
8. The 2¼ in. long capscrew and the locating plug must be removed when the bowl assembly is complete.

Never use heat or a pin, dowel or setscrew to install or reposition a collet-mounted impeller.

An alternate method of securing the impeller to the shaft uses a key to transmit the torque and an axial locking device, such as a snap ring or split ring, to locate the impeller axially and transmit axial thrust from the impeller to the shaft. Split rings are superior. If snap rings are used, they should be confined so that impeller thrust does not make them tend to pop out of the snap ring groove. The fit between the impeller and the shaft should be a snug sliding fit.

For multi-stage pumps with semi-open impellers, correct sizing of split rings is essential to ensure maximum pump efficiency and trouble-free operation. Original split rings are individually sized at the factory during pump assembly to precisely locate the impellers. Used split rings, when reused, must be installed only with the same impellers, and in their original location and orientation. Always use new split rings whenever the pump shaft, cases or impellers are replaced. Machine new split rings to size, or verify the correct sizing of used split rings, as follows:

1. T-section split rings with thrust collars are shown in Figure 2-23.

2. When feasible, assemble the pump bowl assembly in a vertical position to assure the most accurate split ring sizing. During horizon-

Figure 2-22. Collet driver in position to drive a collet. (Courtesy Byron Jackson Pump Division, Borg-Warner Industrial Products, Inc.)

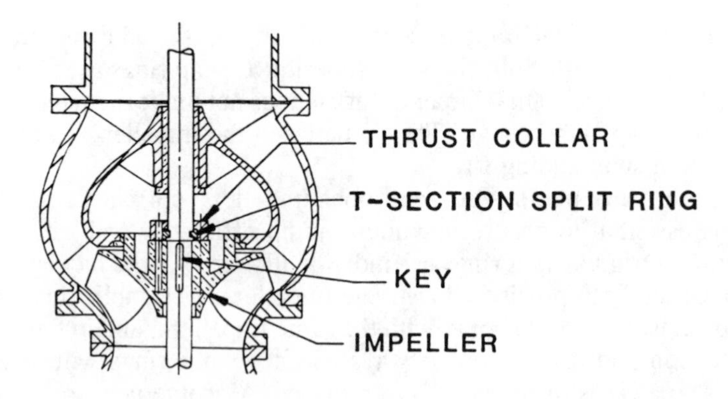

Figure 2-23. Impeller mounted with key and T-section split ring with thrust collar.

tal assembly, be sure to support pump shaft and cases adequately to prevent any sagging, which could lead to alignment errors.

3. With first-stage of pump assembled, install a locating plug and capscrew as shown in Figure 2-20 through the hole in the bottom of the suction bell or bottom case and into the pump shaft. Tighten the capscrew to locate the rotating element during assembly of the second stage. If the first-stage impeller is closed and the series impellers are semi-open, place a spacer, such as a piece of key stock, between the first-stage impeller and the impeller case or suction bell. This will raise the impeller the distance recommended by the manufacturer, generally 1/4 in. Then tighten the capscrew. (Be sure to remove these parts after the pump bowl assembly is completed.)

4. Slip the second-stage impeller down the pump shaft until it bottoms against the first stage case.

5. Accurately measure distance "B" from the face of the impeller hub to the nearest edge of the split ring groove, as shown in Figure 2-24a. This can be done by putting key stock in the groove and measuring the gap with a feeler gauge.

6. Machine one side of split ring, as shown in Figure 2-24b, to "B" + .002 in. This side will install toward the impeller. Mark the OD of the split ring in two places 180° apart so that both halves

will be marked after the ring is cut in half. This is to assure correct orientation at assembly.

7. Next, turn the split ring around and place it against the thrust collar with (backwards orientation), as shown in Figure 2-24c. Machine so that the protrusion of the split ring beyond the thrust collar is .002 to .004 in.

8. After machining to size, saw-cut the new split ring in half, with a saw-cut width of approximately $^{1}/_{32}$ in., and deburr.

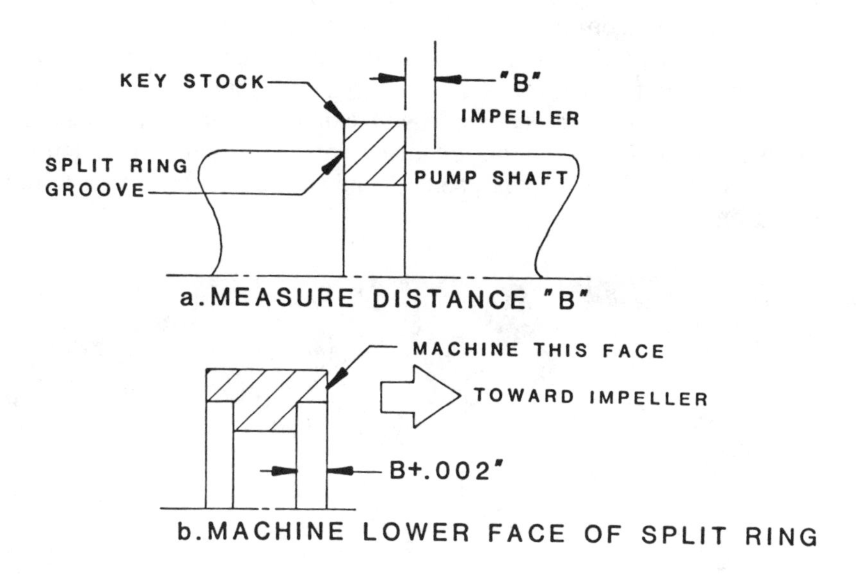

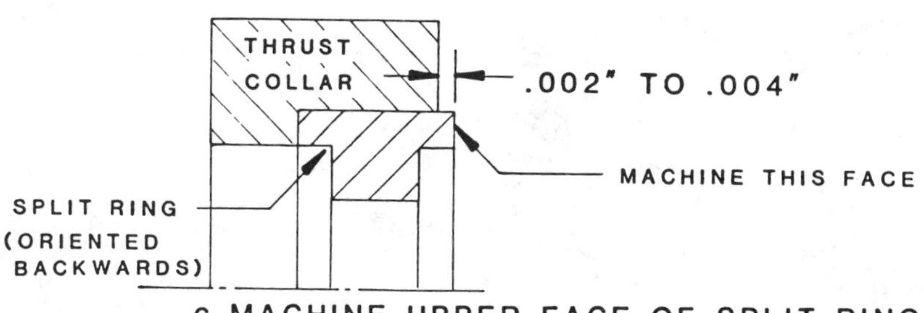

Figure 2-24. Sizing of T-section split ring.

9. Repeat Steps 4 through 8 to size the split rings for each succeeding stage.

10. When the pump bowl assembly has been completed, remove the capscrew and locating plug, and the spacer for the first-stage impeller, if used. *Note:* Some pump designs utilize shims to position the impellers so that remachining of the split rings is not necessary. Figures 2-25 through 2-27 show further details.

Maintenance and Repair of Shafts

Most vertical multistage pumps have wearing surfaces on the shaft under the bowl bearings, bottom bearings, column bearings, tension nut bearings and packing. If the impellers are collet mounted there is very little machining labor in the shaft, and worn shafts generally are replaced rather than repaired. Unless provided with shaft sleeves, worn line shafts are turned end for end to provide new journals when line shaft bearings are replaced. Large shafts or shafts with extensive machining can be built up using hard chrome plating or metalizing. A grinding operation then assures that the shaft is round, smooth and within tolerance prior to reinstallation.

Figure 2-25. Removing a series case to expose the top of an impeller. (Courtesy Byron Jackson Pump Division, Borg-Warner Industrial Products, Inc.)

Figure 2-26. Removing the thrust collar. (Courtesy Byron Jackson Pump Division, Borg-Warner Industrial Products, Inc.)

Figure 2-27. Removing the split ring. (Courtesy Byron Jackson Pump Division, Borg-Warner Industrial Products, Inc.)

Figure 2-28. Checking shaft straightness. (Courtesy Byron Jackson Pump Division, Borg-Warner Industrial Products, Inc.)

Figure 2-29. Straightening a shaft. (Courtesy Byron Jackson Pump Division, Borg-Warner Industrial Products, Inc.)

Figure 2-30. Minor cavitation damage in an impeller eye. (Courtesy Byron Jackson Pump Division, Borg-Warner Industrial Products, Inc.)

All shafts must be checked and straightened if necessary prior to reinstallation. Straightening is performed at ambient temperature using a hydraulic press or arbor press. The supports used to hold the shafting during the straightening operation are generally "V" blocks with brass or copper inserts, but some small shafts are straightened on the rollers used to measure runout.

Straightening Sequence

1. Rotate the shaft on rollers located near each end of the shaft and check the runout total indicator reading (T.I.R.) at several points along the shaft using a dial indicator. See Figure 2-28.
2. Move the shaft onto the "V" blocks and straighten by applying force with a hydraulic press or arbor press. See Figure 2-29.
3. Check T.I.R. Maximum acceptable T.I.R. is .0005 in. per ft of shaft length, but not more than .001 in. within any one ft of length.
4. If T.I.R. is acceptable, except for deepwell pump shafts, thermal treat in a furnace to relieve peak residual stresses in accordance with Table 2-2. Thermal treatment must be performed vertically with the shaft hanging so that gravity tends to keep it straight.
5. Check T.I.R. after thermal treatment. If T.I.R. is unacceptable repeat Steps 1 through 4 until T.I.R. is acceptable.

Table 2-2
Thermal Treatment of Shafting After Straightening

Material Type	Thermal Treatment Temperature	Time
A 479 Type 410 Class 2	1100 ± 25°F.	Hold for 1 hour per inch of diameter or 2 hours minimum.
A 276 Type 410 Condition H	1050 ± 25°F.	
A 434 Class BC (4140)	850 ± 25°F.	
A 582 Type 416	1100 ± 25°F.	
K 500 Monel	725 ± 25°F.	
A 479/182 Type XM 19	725 ± 25°F.	
A 276 Type 304/316	725 ± 25°F.	
A 564 Type 630 17-4 (H1150)	1100 ± 25°F.	

Turned ground and straightened vertical pump line shaft material is readily available. However, this shafting may not be straight when received due to handling and shipment and should be checked when received. The standard material is Type 416 stainless steel, which contains approximately 12 percent chromium, has good strength and adequate corrosion resistance for most applications.

Upgrading of Materials

A generally accepted guide for selection of materials used in centrifugal pumps for process services is contained in API Standard 610, and reproduced as Tables 2-3A and 2-3B. Minor damage (see Figure 2-30) is to be expected. However, if excessive corrosion, erosion, wear, or breakage is observed in a particular pump, the materials of construction should be reviewed and upgraded to these standards. If unacceptable maintenance costs exist even though the guidelines have been followed, materials should be further upgraded, substituting for example 12 percent chrome steel for low carbon steel or 18-8 stainless steel for 12 percent chrome steel to solve corrosion problems. Wear rings can be overlayed with Stellite or Colmonoy for better wear resistance. Shafts can be upgraded from Type 416 stainless steel to Type 440 heat treated stainless steel if more strength is required but corrosion resistance is satisfactory. If both increased strength and corrosion resistance are needed 17-4 PH,

(Text continued on page 96.)

Table 2-3A
Materials Guidelines from API Standard 610

Part	ASTM Material?	I-1 CI / CI	I-2 CI / BRZ	S-1 STL / CI	S-3 STL / NI-RESIST
Pressure casing	Yes	Cast iron	Cast iron	Carbon steel	Carbon steel
Inner case parts (bowls, diffusers, diaphragms)	No	Cast iron	Bronze	Cast iron	Ni-resist
Impeller	Yes	Cast iron	Bronze	Cast iron	Ni-resist
Case wearing rings	No	Cast iron	Bronze	Cast iron	Ni-resist
Impeller wearing rings	No	Cast iron	Bronze	Cast iron	Ni-resist
Shaft (Note 5)	Yes	Carbon steel	Carbon steel	Carbon steel	Carbon steel
Shaft sleeves, packed pumps	No	12% Chrome, hardened	Hard bronze	12% Chrome, hardened	12% Chrome, hardened
Shaft sleeves, mechanical seals	No	18-8 Stainless steel or 12% chrome	18-8 Stainless steel or 12% chrome	18-8 Stainless steel or 12% chrome	18-8 Stainless steel or 12% chrome
Throat bushings	No	Cast iron	Bronze	Cast iron	Ni-resist
Interstage sleeves	No	Cast iron	Bronze	Cast iron	Ni-resist
Interstage bushings	No	Cast iron	Bronze	Cast iron	Ni-resist
Lantern ring (if packed pump)	No	Cast iron	Cast iron or bronze	Cast iron	Cast iron
Gland with packing or plate retaining mechanical seal	Yes	Carbon steel	Carbon steel	Carbon steel (Note 6)	Carbon steel (Note 6)
Gland studs or bolts	Yes	Carbon steel	Carbon steel	AISI 4140 steel	AISI 4140 steel
Case studs and bearing housing	Yes	Carbon steel	Carbon steel	AISI 4140 steel	AISI 4140 steel
Case gasket	No	Asbestos composition	Asbestos composition	Asbestos composition	Asbestos composition

[a] The abbreviation above the diagonal line indicates case material; the abbreviation below the diagonal line indicates trim material. Abbreviations are as follows: CI = cast iron, BRZ = bronze, STL = steel, 12% CHR = 12% chrome, SS = stainless steel.

(Continued on next page)

Table 2-3A. Continued.

Part	ASTM Material?	Material Class and Material Class Abbreviations[a]			
		S-4	S-5	S-6	S-9
		STL / STL	STL / STL 12% CHR	STL / 12% CHR	STL / MONEL
Pressure casing	Yes	Carbon steel	Carbon steel	Carbon steel	Carbon steel
Inner case parts (bowls, diffusers, diaphragms)	No	Cast iron	Carbon steel	12% Chrome	Monel
Impeller	Yes	Carbon steel	Carbon steel	12% Chrome	Monel
Case wearing rings	No	Cast iron	12% Chrome, hardened	12% Chrome, hardened	Monel
Impeller wearing rings	No	Cast iron	12% Chrome, hardened	12% Chrome, hardened	Monel
Shaft (Note 5)	Yes	Carbon steel	AISI 4140 steel	AISI 4140 steel (Note 4)	K-Monel
Shaft sleeves, packed pumps	No	Tungsten carbide-3 over 12% chrome	Tungsten carbide-3 over 12% chrome	Tungsten carbide-3 over 12% chrome	K-Monel, hardened
Shaft sleeves, mechanical seals	No	18-8 Stainless steel or 12% chrome	18-8 Stainless steel or 12% chrome	18-8 Stainless steel or 12% chrome	K-Monel, hardened
Throat bushings	No	Cast iron	12% Chrome	12% Chrome	Monel
Interstage sleeves	No	Cast iron	12% Chrome, hardened	12% Chrome, hardened	K-Monel, hardened
Interstage bushings	No	Cast iron	12% Chrome, hardened	12% Chrome, hardened	K-Monel, hardened
Lantern ring (if packed pump)	No	Cast iron	Cast iron	Cast iron	Monel
Gland with packing or plate retaining mechanical seal	Yes	Carbon steel (Note 6)	Carbon steel (Note 6)	Carbon steel (Note 6)	Carbon steel (Note 6)
Gland studs or bolts	Yes	AISI 4140 steel	AISI 4140 steel	AISI 4140 steel	K-Monel, hardened
Case studs and bearing housing	Yes	AISI 4140 steel	AISI 4140 steel	AISI 4140 steel	K-Monel, hardened
Case gasket	No	18-8 Stainless steel, jacketed asbestos (Note 3)	18-8 Stainless steel, jacketed asbestos (Note 3)	18-8 Stainless steel, jacketed asbestos (Note 3)	Teflon (Note 7)

[a] The abbreviation above the diagonal line indicates case material; the abbreviation below the diagonal line indicates trim material. Abbreviations are as follows: CI = cast iron, BRZ = bronze, STL = steel, 12% CHR = 12% chrome, SS = stainless steel.

Table 2-3A. Continued.

Part	ASTM Material?	C-6 12% CHR / 12% CHR	D-6 5% CHR / 12% CHR	A-7 18-8 SS / 18-8 SS (Notes 2 & 5)	A-8 316 SS / 316 SS (Note 2)
Pressure casing	Yes	12% Chrome	5% Chrome	18-8 Stainless steel	316 Stainless steel
Inner case parts (bowls, diffusers, diaphragms)	No	12% Chrome	12% Chrome	18-8 Stainless steel	316 Stainless steel
Impeller	Yes	12% Chrome	12% Chrome	18-8 Stainless steel	316 Stainless steel
Case wearing rings	No	12% Chrome, hardened	12% Chrome, hardened	Hard-faced 18-8 stainless steel (Note 1)	Hard-faced 316 stainless steel (Note 1)
Impeller wearing rings	No	12% Chrome, hardened	12% Chrome, hardened	Hard-faced 18-8 stainless steel (Note 1)	Hard-faced 316 stainless steel (Note 1)
Shaft (Note 5)	Yes	12% Chrome	12% Chrome	18-8 Stainless steel	316 Stainless steel
Shaft sleeves, packed pumps	No	Tungsten carbide-3 over 12% chrome	Tungsten carbide-3 over 12% chrome	Hard-faced 18-8 stainless steel (Note 1)	Hard-faced 316 stainless steel (Note 1)
Shaft sleeves, mechanical seals	No	18-8 Stainless steel or 12% chrome	18-8 Stainless steel or 12% chrome	18-8 Stainless steel	316 Stainless steel
Throat bushings	No	12% Chrome	12% Chrome	18-8 Stainless steel	316 Stainless steel
Interstage sleeves	No	12% Chrome, hardened	12% Chrome, hardened	Hard-faced 18-8 stainless steel (Note 1)	Hard-faced 316 stainless steel (Note 1)
Interstage bushings	No	12% Chrome, hardened	12% Chrome, hardened	Hard-faced 18-8 stainless steel (Note 1)	316 Stainless steel
Lantern ring (if packed pump)	No	12% Chrome	12% Chrome	18-8 Stainless steel	316 Stainless steel
Gland with packing or plate retaining mechanical seal	Yes	12% Chrome (Note 6)	12% Chrome (Note 6)	18-8 Stainless steel, carbon or (Note 6)	316 Stainless steel, carbon or (Note 6)
Gland studs or bolts	Yes	AISI 4140 steel	AISI 4140 steel	18-8 Stainless steel	316 Stainless steel
Case studs and bearing housing	Yes	AISI 4140 steel	AISI 4140 steel	AISI 4140 steel	AISI 4140 steel
Case gasket	No	18-8 Stainless steel, jacketed asbestos (Note 3)	18-8 Stainless steel, jacketed asbestos (Note 3)	18-8 Stainless steel, jacketed asbestos (Note 3)	316 Stainless steel, jacketed asbestos (Note 3)

[a] The abbreviation above the diagonal line indicates case material; the abbreviation below the diagonal line indicates trim material. Abbreviations are as follows: CI = cast iron, BRZ = bronze, STL = steel, 12% CHR = 12% chrome, SS = stainless steel.

Table 2-3B
Material Specifications from API Standard 610

ASTM Material Specifications for Centrifugal Pump Parts

Material	Pressure-Containing Castings	Forgings	Bar Stock	Bolts and Studs
Cast iron	ASTM A 48	—	—	—
Carbon steel	ASTM A 216, Grade WCA or WCB	ASTM A 105 or A 576	ASTM A 576, Grade 1015	—
AISI 4140 steel	—	—	ASTM A 322, Grade 4140	ASTM A 193, Grade B7
5% Chrome steel	ASTM A 217, Grade C5	ASTM A 182, Grade F5	—	—
12% Chrome steel	ASTM A 743, Grade CA15 or CA6NM	ASTM A 182, Grade F6	ASTM A 276, Type 410 or 416	ASTM A 193, Grade B6
18-8 Stainless steel (Note 2)	ASTM A 743, Grade CF8	ASTM A 182	ASTM A 276	ASTM A 193
316 Stainless steel	ASTM A 743, Grade CF8M	ASTM A 182, Grade F316	ASTM A 276, Type 316	ASTM A 193, Grade B8M
Bronze	ASTM B 584, UNS-C 87200	—	ASTM B 139	ASTM B 124, Alloy 655

General Notes

1. Hard-facing material (stellite, Colmonoy, tungsten carbide, etc.) for Classes A-7 and A-8 shall be selected by the vendor unless specified by the purchaser.
2. 18Cr–8Ni includes Types 302, 303, 304, 316, 321, and 347. If a particular type is desired, the purchaser shall so specify.
3. If pumps with horizontally split cases are furnished, an asbestos composition case gasket is acceptable.
4. For Class S-6, the shaft shall be 12-percent chrome if the temperature exceeds 350 F (177 C).

Table 2-3B. Continued.

Miscellaneous Materials

Material	Typical Description
Ni-resist	Type 1, 2, or 3 as recommended by International Nickel Co. for service conditions
Precipitation-hardening stainless steel	Acceptable types include ARMCO 17-7 PH and 17-4 PH, U.S. Steel Stainless W, Allegheny Ludlum AM 350 and AM 355
Stellite	Overlay-weld deposit of $\frac{1}{32}$ inch minimum finished thickness of Haynes Stellite AWS, Class RCoCr-C, RCoCr-A, or equal; solid cast stellite No. 3 or equal may be substituted for an overlayed part
Colmonoy	Sprayed or fused deposit of 0.010 inch or gas-weld deposit of $\frac{1}{32}$ inch minimum finished thickness of Wall-Colmonoy AWS, Class RNiCr-C or equal
Carbon	Suitable mechanical carbon as recommended by the mechanical seal manufacturer for the service
Asbestos composition	Long fiber with synthetic rubber binder suitable for 750 F, or spiral-wound stainless steel and asbestos
Tetrafluoroethylene (TFE)	Teflon, Kel-F, or similar material
Tungsten carbide-1	Kennametal K-6 (cobalt binder) or equal (solid part, not overlay)
Tungsten carbide-2	Kennametal K-801 (nickel binder) or equal (solid part, not overlay)
Tungsten carbide-3	METCO 31C, WALLEX 55, or equal (sprayed overlay; minimum finished thickness of 0.03inch)
Fluoroelastomer	DuPont Viton or equal
Alloy 20	ASTM A 296, CN7M; Carpenter 20CB3; or equal
Nitrile (Buna-N)	B.F. Goodrich HYCAR or equal
FFKM elastomer	ASTM D 1418, FFKM elastomer; DuPont Kalrez; or equal
Silicon carbide	Carborundum KT or equal
Graphite foil	Union Carbide Graphoil or equal

XM-19 or K 500 Monel shafting can be substituted (Table 2-4). XM-19 offers a good combination of strength, corrosion resistance and cost. If collet-mounted impellers have proved to be unsatisfactory in a particular service, and a change to keyed impellers is made, it may be necessary to upgrade the shaft material to compensate for the stress raisers created by cutting keyways and split ring grooves in the shaft. For additional information on materials, refer also to Appendix 1A in Chapter 1 of this volume.

Graphalloy® Bearings

Product lubricated bearings in clean liquids can be upgraded to a material such as Graphalloy. Graphalloy is a solid carbon graphite, the pores of which have been impregnated with molten metal in a high heat and high pressure process. During the impregnation process the metal permeates the graphite in long continuous metal filaments. It is these filaments which give Graphalloy its ductility, high strength and good heat dissipating properties. The metal increases its strength and ductility and removes heat generated at the bearing surface. Graphalloy contains no oil and does not produce any toxic emission which may contaminate the pumped product. It is extremely durable in clean products. Graphalloy bearings operate well at temperatures from −50°F to +300°F in fresh or salt water, gasoline, jet fuels, solvents, bleaches, caustics, dyes, liquified gases, acids and most chemical process and transfer services. Babbitt grade GM105.3 is generally recommended for pump bearings. However, Graphalloy is not recommended for applications in products containing abrasive material. In products with abrasive material, extremely hard bearings such as tungsten carbide, boron carbide, or Tribaloy® can be utilized running on a hard journal. However, some of these materials are extremely brittle, and cannot be externally shrink fitted.

Table 2-4
Typical Mechanical Properties of 1 in. Diameter Shaft Stock

	Ultimate Tensile Strength	Yield Strength (.2%)	Charpy V Notch Ft/lbs	Hardness (Rockwell)
XM-19	150	125	90	RC 30
Monel K-500	165	105	37	RC 30
Type 316	85	35	110	RB 80
Type 304	85	35	110	RB 80
17-4PH Cond. H-1150	145	125	50	RC 33

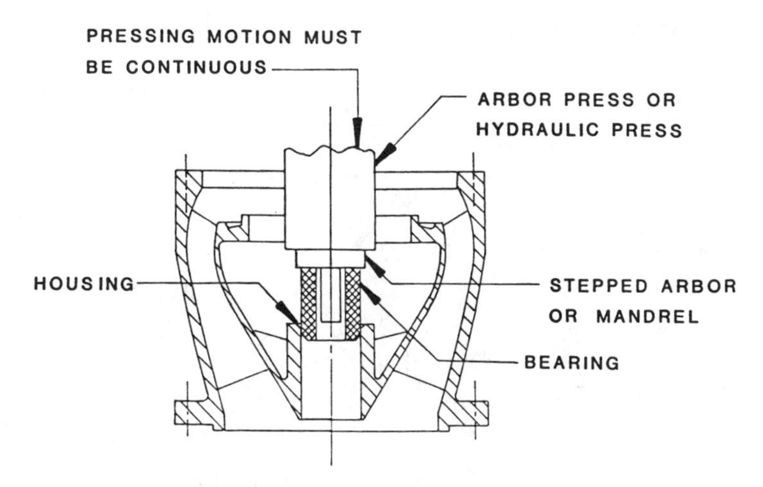

Figure 2-31. Pressing Graphalloy® bearing into housing using arbor press or hydraulic press.

Installing Graphalloy® Bearings

The preferred method for installing Graphalloy bearings utilizes an arbor press or hydraulic press as shown in Figure 2-31. The bearing or the bore into which the bearing is installed should have a 1/32 in. minimum × 45° chamfer to facilitate entry of the bearing. A stepped mandrel or arbor should be used to ensure that the bearing will be positioned straight with the hole before installation. The small outside diameter of the arbor should be 1/16 in. smaller than the inside diameter of the bearing, and the large outside diameter of the arbor should be larger than the outside diameter of the bearing. The pressing motion must be continuous with no interruption until the bushing is completely in place. As an alternative, the bearing may be pressed into the housing by the bolt-and-nut method; that is, with a plate against the upper end of the bushing as shown in Figure 2-32. The nut must be continuously drawn up.

Threaded Fasteners

Except in a passive environment such as refined petroleum products, studs and nuts are preferred to capscrews. The threaded fastener material should be resistant to any corrosion that may be present and should be upgraded if corrosion is observed.

All threads should be examined before reinstallation to ensure that no burrs, nicks or bad threads exist. Imperfect threaded fasteners must be replaced or the threads "chased" with a thread die. All threads both internal and external must be cleaned with solvent to remove all foreign matter including rust and old thread lubricant. Apply a uniform layer of

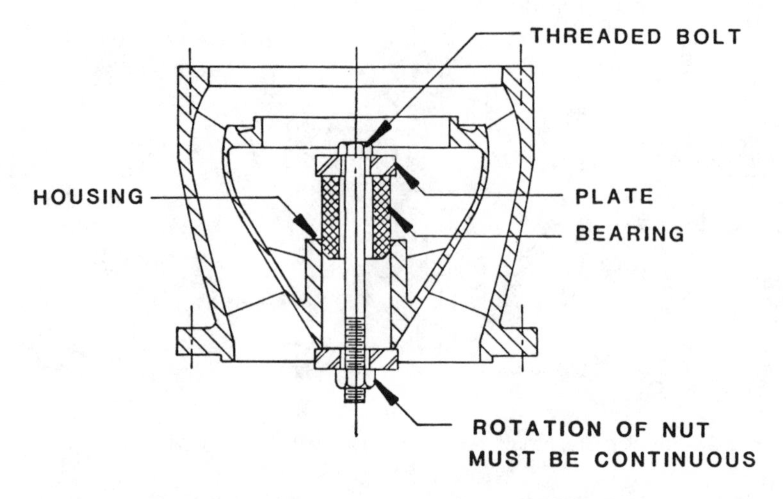

Figure 2-32. Bolt-and-nut method of pressing Graphalloy® bearing into housing.

thread lubricant Dag Dispersion® No. 156 or equal with a friction coefficient of 0.15 to all surfaces that experience relative motion, including threads, nuts, washers, and flanges in contact with nuts or washers. If these procedures are used the torque values indicated in Table 2-5 may be applied to the materials shown in Table 2-6.

Care of Large Threaded Joints

Threaded line shaft couplings are common in deepwell pumps and small circulators. The threads on the shaft are cut with great care to assure concentricity with the centerline of the shaft. Threads must be carefully inspected before assembly. Shaft ends must be wiped clean and inspected to be sure that they are free from foreign matter, burrs, nicks, and scratches. Thread compound should be used. The two shafts each are threaded approximately halfway into the coupling and the ends of the two shafts firmly butted together. It has been suggested that a Teflon® disk can be installed between shaft ends to compensate for some misalignment. This is a questionable practice. Line shaft couplings are designed so that a substantial amount of the torque is transmitted by friction through the shaft ends. Lubrication of the shaft ends by Teflon® can cause the coupling to break.

Pipe wrenches are normally used to assemble the line shaft. One wrench is applied to the shaft above the coupling and one on the coupling. Care must be exercised to avoid side strain on the shaft when locking the joint. Pipe wrenches must not be used on the shaft below the coupling as assem-

Table 2-5
Torque Values for Threaded Fasteners

Size	Torque (Ft-lbs) Category-I (1)	Category-II (2)
3/8-16UNC	16	8
7/16-14UNC	27	14
1/2-13UNC	40	20
9/16-12UNC	60	30
5/8-11UNC	80	40
3/4-10UNC	130	65
7/8-9UNC	210	105
1-8UNC	330	115
1-1/8-7UNC	520	260
1-1/8-8UN	470	240
1-1/4-7UNC	730	370
1-1/4-8UN	670	340
1-3/8-6UNC	970	490
1-3/8-8UN	910	460
1-1/2-6UNC	1170	590
1-1/2-8UN	1070	540
1-3/4-5UNC	2070	1040
1-3/4-8UN	2000	1000
2-4½ UNC	3000	1500
2-8UN	2930	1470

Notes: *(1) Based on approximately 40,000 psi prestress.*
(2) Based on approximately 20,000 psi prestress.

Table 2-6
Mechanical Properties of Common Threaded Fastener Materials

Category-I:

ASTM. No.	Common Name	Approximate Strength (KSI) Min. Yield	Tensile
A-193 GR B7	4140	105	125
A-193 GR. B6	410	85	110
A-193 GR. B16	Cr-Mo-V	105	125
A-193 GR. B5	501	80	100
A-325 TP. 1	1030	81	105
A-354 GR. BD	Alloy Steel	78	90
A-453 GR. 651 Cl. A	Stainless Steel	70	100

Category-II:

ASTM. No.	Common Name	Approximate Strength (KSI) Min. Yield	Tensile
A-576 GR. 1018	1018	32	58
A-193 GR. B8 Cl. 1	304	30	75
A-320 GR. B8 Cl. 1	304	30	75
A-479 GR. 302	302	30	75
A-479 GR. 304	304	30	75
A-479 GR. 316	316	30	75
A-479 GR. 410	410 (Annealed)	40	70
B-98 Alloy C 66100	Sil. Brz. (H02 Temper)	38	70
B-150 Alloy C 64200	Al. Brz.	45	85
B-164 Alloy N 04400	Monel (Annealed)	25	70

Note: Properties from ASTM 1982 and 1983 editions. Properties vary with fastener size and heat treatment.

bly of bearings, impellers, and collets will include sliding these parts down over this portion of the shaft. Strap wrenches are suitable for this purpose and are preferable. The use of pipe wrenches on shafting, although quite appropriate for servicing deepwell pumps, should be discouraged, because this practice easily could carry over to the servicing of process pumps where it is completely unacceptable. Threaded line shaft couplings always have left hand threads that tend to be tightened by the transmitted torque.

Threaded inner column generally has left hand threads because right hand threads would tend to loosen from the torque applied as the result of a bearing failure. The threads are internal and straight, not external and tapered like pipe threads. This is shown in Figure 2-4. The joints are sealed by the metal-to-metal contact between the smooth ends of the column. It therefore is necessary to avoid nicks, scratches, or burrs on the ends of the column. The threads should be coated with a thread lubricant or a thread sealant. Inner column may be tightened with pipe wrenches, strap wrenches, or chain tongs.

Threaded outer column and bowl assemblies generally are tightened with chain tongs. The threads are external, right hand and straight, not tapered. This is also shown in Figure 2-4. These joints also are sealed by metal-to-metal contact between the ends of the column. Column and bowl threads should be coated with a thread lubricant or a thread sealant.

Submersible pump column has tapered external pipe threads that must be coated with a thread sealant and tightened both to seal the joints and to keep them from loosening from the starting torque. The threaded bowl joints and the joint between the cast iron top case and the steel column should be cleaned with Loctite® Primer T and coated with Loctite® 277. The Loctite® 277 joints may require heat for disassembly.

Care of Flanged Pressure Joints

Flanged mating surfaces must be thoroughly cleaned prior to reassembly and sealing surfaces inspected to be sure that they are free from burrs, nicks, or scratches. Static seals such as O-rings, flat gaskets, and spiral-wound gaskets should not be reused but should be replaced after each dismantling of the pressure joint. Additional information on static seals is contained in Chapter 7 of Volume 1. After installation of the static seal, assemble the joint hand tight and check for uniform gap or metal-to-metal contact of the flange mating surfaces. Tighten bolts alternately in pairs 180° apart, then another pair 90° clockwise from the first pair, and so forth until the entire flange has been tightened. The bolt-tightening sequence for a 12 bolt flange is shown in Figure 2-33. Bolts should be tightened in three steps, 1/3 torque, 2/3 torque, and full torque.

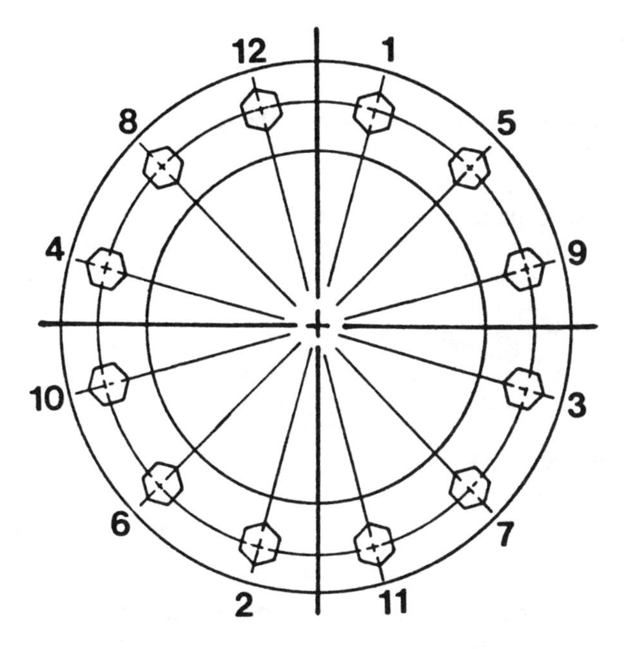

Figure 2-33. Bolt tightening sequence.

Handling

The crane, shackles, eye bolts, cable, and other lifting devices must be in good repair and capable of safely handling the weights. The crane should be positioned so that the hook is directly over the center of gravity of the piece to be lifted. Do not bump, push, or scrape components or assemblies. Motor, turbine, or gear lifting lugs or eye bolts are designed for lifting these components only. Do not lift the pump together with the motor, turbine, or gear. Use lifting lugs when provided. Otherwise, install eye bolts with washers and nuts in the flange bolt holes and attach lifting lines. Use care to avoid damage to the machined surfaces.

Tools and Work Area

Common millwright and pipe-fitter tools generally are sufficient to perform vertical pump maintenance. The work area should be clean and well lighted. Horizontal racks for dismantling and reassembly of vertical turbine type pumps are desirable to avoid handling damage, and to shorten turnaround time.

Personnel

Maintenance personnel should be familiar with the basic principles of vertical pumps and know how to handle the tools required for their maintenance and repair. Leading pump manufacturers have highly qualified service engineers who can visit periodically and assist in the training and updating of maintenance personnel with the latest techniques. Some manufacturers also have service training centers where maintenance personnel can receive up-to-date instruction on vertical pump maintenance.

Spare Parts

Economical repair and maintenance procedures require adequate availability of the necessary spare parts. Pump repair parts are classified according to anticipated use:

Level 1

Parts required for inspection of the pump during a scheduled shutdown. This includes gaskets, O-rings, bearings, and packing or mechanical seal.

Level 2

Parts replaced because of normal wear of the pump over a period of time. During normal operation, pump wearing clearances open up, causing less efficient operation and higher susceptibility to failure. In addition to this, wear rings and shaft sleeves are required.

Level 3

Parts required for overhaul in the event of a pump failure include all parts listed above plus a pump shaft and impellers. Stocking of a complete bowl assembly will minimize pump downtime and allow for an orderly and economical repair.

In process plants, Level 3 stocking of spare pump parts is justified for most vertical pumps. The general topic of spare parts is covered in Chapter 5 of Volume 1 of this series.

Service Records

The topic of Maintenance Records is covered in Chapter 6 of Volume 1 of this series.

Problem Solving

Detailed information relating to vertical pump problem solving is contained in Reference 6; for general troubleshooting matrixes consult Volume 2 of this series.

Installation and Operation

Installation guidelines may vary greatly for the various vertical pump types and the reader should consult the manufacturer for details. Refer to Appendix 1B, earlier in this Volume, for operating and documentation checklists.

References

1. "Centrifugal Pumps for General Refinery Services," API Standard 610, Sixth Edition, June 1981, American Petroleum Institute, 2101 L Street, Northwest, Washington, D.C. 20037
2. "Technical Manual for Installation, Operation and Maintenance of Type H Submersible Motor Pumping Unit," Manual No. 2565, July 1981, Byron Jackson Pump Division, Borg-Warner Industrial Products, Inc., P.O. Box 22634, Long Beach, California 90801
3. "The Vertical Process Pump, Don't Hold it Back," Bulletin 6000-A, Byron Jackson Pump Division, Borg-Warner Industrial Products, Inc., P.O. Box 22634, Long Beach, California 90801
4. Gopalakrishnan, S., Fehlau, R., and Lorett, J., "How to Calculate Critical Speed in Centrifugal Pumps," *Oil and Gas Journal,* December 7, 1981 (Reprints available from Byron Jackson Pump Division, Borg-Warner Industrial Products, Inc., P.O. Box 22634, Long Beach, California 90801)
5. "Identical Twins?," Section 0000, Page 0011.1, Issue A, Warren Pumps Division, Houdaille Industries, Inc., Warren, Massachusetts 01083
6. "Operation and Maintenance of Vertical Pumps, Problem Solving," Engineering Data No. 2-510, Byron Jackson Pump Division, Borg-Warner Industrial Products, Inc., P.O. Box 22634, Long Beach, California 90801

Chapter 3

Reciprocating and Liquid Ring Vacuum Pumps*

The justification for using a reciprocating pump in a petrochemical plant instead of a centrifugal or rotary pump must be cost—not just the initial cost but total cost, including costs for power and maintenance.

Some applications are inherently best suited for reciprocating units. Such services include high-pressure water cleaning (typically 20 gpm at 10,000 psig), glycol injection (typically 5 gpm at 1,000 psig), and ammonia charging (typically 40 gpm at 4,000 psig). Another application that practically mandates a reciprocating unit is abrasive and/or viscous slurries above about 500 psig. Examples of these services range from coal slurry to peanut butter.

The best feature of the power pump is its high efficiency. Overall efficiencies normally range from 85 to 94 percent. The losses of approximately 10 percent include all those due to belts, gears, bearings, packings, and valves.

Another characteristic of the reciprocating pump is that capacity is a function of speed, and is relatively independent of discharge pressure. Therefore, a constant-speed power pump that moves 100 gpm at 500 psig will handle very nearly 100 gpm at 3,000 psig.

The direct-acting pump has some of the same advantages as the power pump, plus others. These units are well suited for high-pressure low-flow applications. Discharge pressures normally range from 300 to 5,000 psig, but may exceed 10,000 psig. Capacity is proportional to speed from stall to maximum speed, regardless of the discharge pressure.

---

* From "Reciprocating Pumps," *Chemical Engineering*, Sept. 21, 1981 by T. L. Henshaw, Union Pump Co. Adapted by permission of the author.

Speed is controlled by throttling the motive fluid. The unit is normally self-priming—particularly the low clearance-volume type.

Direct-acting pumps are negligibly affected by hostile environments such as corrosive fumes, because of the absence of a bearing housing, crankcase, or oil reservoir (except for units requiring a lubricator). Some direct-acting pumps inadvertently inundated by flood-water have continued to operate without adverse effects. Direct-acting pumps are quiet, simple to maintain, and their low speeds and rugged construction lead to a very long life.

Both power and direct-acting pumps with special fittings and operating at low speeds have been successfully applied to abrasive-slurry services.

The low thermal efficiency of the direct-acting pump is sometimes used to advantage. When steam is the motive fluid, very little heat is lost from inlet to exhaust. The exhaust temperature is the same as that obtained by throttling. In those cases where high-pressure steam is throttled to a lower pressure for heating (such as for deaerating boiler feedwater), the steam can be used to drive a direct-acting pump, with the exhaust steam used for heating. In this circumstance, the drive end (piston rings, valves, etc.) is made to operate without lubrication, so that the exhaust steam will be oil-free.

Pump Classification

Reciprocating pumps are usually classified by their features:

- Drive end, i.e., power or direct-acting.
- Orientation of centerline of the pumping element, i.e., horizontal or vertical.
- Number of discharge strokes per cycle of each drive rod, i.e., single-acting or double-acting.
- Configuration of the pumping element, i.e., piston plunger or diaphragm.
- Number of drive rods, i.e., simplex, duplex, or multiplex.

Figure 3-1 illustrates this classification in chart form.

Figure 3-2 shows two examples of reciprocating pumps.

Cross-sectional drawings for power and direct-acting pumps are shown in Figures 3-3 and 3-4, respectively.

The size of a power pump is normally designated by listing first the diameter of the plunger (or the piston), and second the length of the stroke. In the United States, the units are inches. For example, a pump

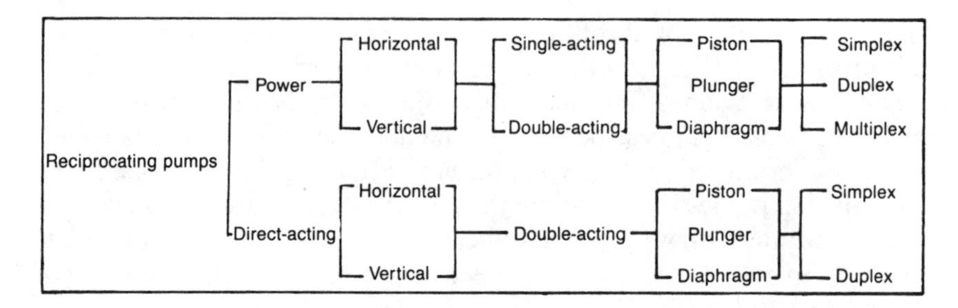

Figure 3-1. Classification of reciprocating action pumps.

Figure 3-2A. Horizontal, quintuplex, power pump.

Figure 3-2B. Direct-acting, duplex, double-acting piston pump. (Reciprocating pumps may be driven by electric power or a motive fluid.) Courtesy of Union Pump Company.

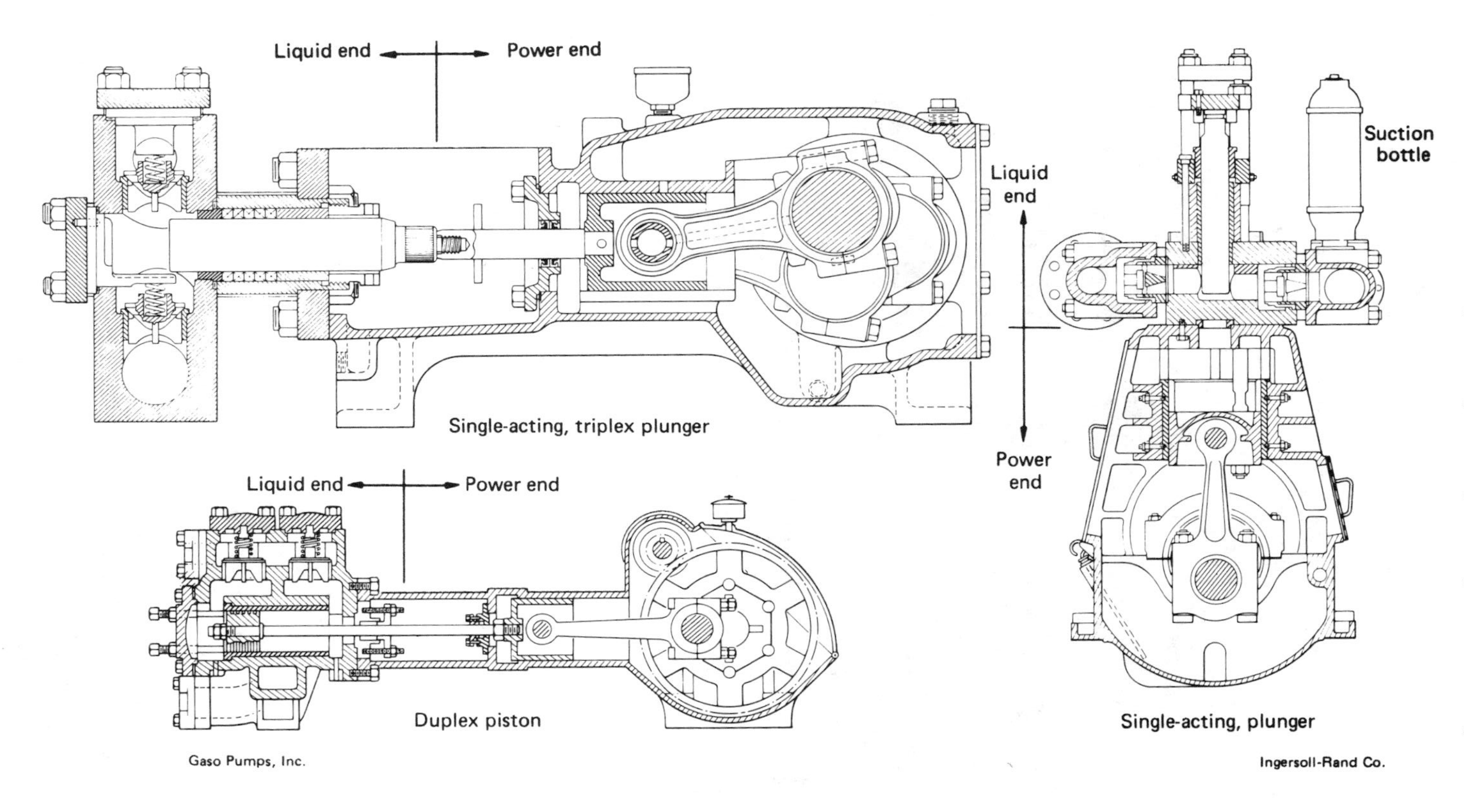

Figure 3-3. Power pump styles.

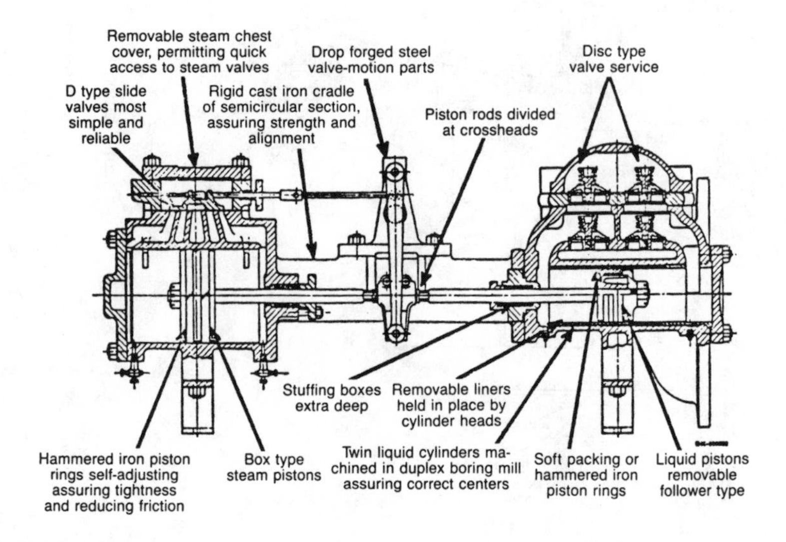

Figure 3-4. Typical action of a horizontal duplex pump.

designated as 2 × 3 has a plunger diameter of 2 in. and a stroke length of 3 in. For a direct-acting pump, the same convention is followed, except that the diameter of the drive piston precedes the liquid-end-element diameter. For example, a pump designated 6 × 4 × 6 has a drive-piston diameter of 6 in., a liquid-piston diameter of 4 in., and a stroke length of 6 in.

Liquid-End Components

All reciprocating pumps contain one or more pumping elements (pistons, plungers, or diaphragms) that reciprocate into and out of pumping chambers to produce the pumping action. Each chamber contains at least one suction and one discharge valve. The valves are simply check valves that are opened by the liquid differential pressure. Most valves are spring loaded.

The liquid end is that portion of the pump that does the pumping. Elements common to all reciprocating-pump liquid ends are the liquid cylinder, pumping element, and valves.

The liquid cylinder is the major pressure-retaining part of the liquid end, and forms the major portion of the pumping chamber. It usually contains or supports all other liquid-end components.

A piston ("a," Figure 3-5) is a flat cylindrical disk, mounted on a rod, and usually contains some type of sealing rings. A plunger ("b," Figure 3-5) is a smooth rod and, in its normal configuration, can only be single-acting. With a piston, the sealing elements move. With a plunger, they are stationary. A piston must seal against a cylinder or liner inside the pump. A plunger must seal only in the stuffing box, and touches only packing and possibly stuffing box bushings.

A piston pump is normally equipped with a replaceable liner (sleeve) that absorbs the wear from the piston rings. Because a plunger contacts only stuffing box components, plunger pumps do not require liners.

Sealing between the pumping chamber and atmosphere is accomplished in a stuffing box or packing box ("c," Figure 3-5). The stuffing box contains rings of packing that conform to and seal against the stuffing box ID and the rod.

If a lubricant, sealing liquid or flushing liquid is injected into the center of the packing, a lantern ring or seal cage is required. This ring provides an annular space between the packing rings so that the injected fluid can freely flow to the rod surface.

The valves in a reciprocating pump are opened by the liquid differential pressure, and allow flow in only one direction. They have a variety of shapes, including spheres, hemispheres, disk, and bevel seats (Figure 3-6).

Packing Maintenance

The biggest maintenance problem on most reciprocating pumps is packing. Although the life of standard packing in a power pump is about 2,500 hr, some installations with special stuffing box arrangements have experienced a life of more than 18,000 hr, at discharge pressures of up to 4,000 psig.

Short packing life can result from any of the following conditions:

1. Improper packing for the application.
2. Insufficient lubrication.
3. Misalignment of plunger (or rod) with stuffing box.
4. Worn plunger, rod, stuffing box bore or stuffing box bushings.
5. Packing gland too tight or too loose.
6. High speed or high pressure.
7. High or low temperature of pumpage.
8. Excessive friction (too much packing in box).

A

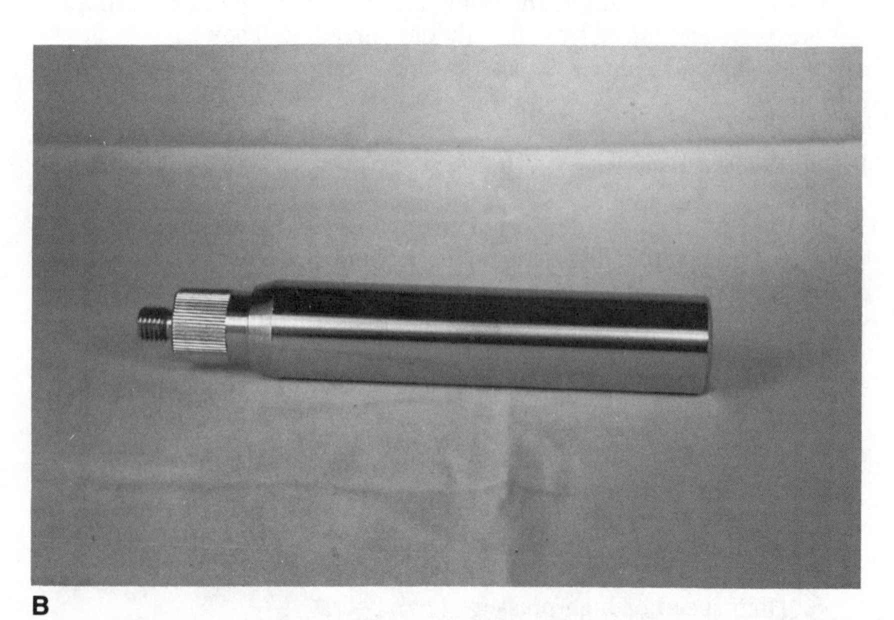

B

Figure 3-5. Steam piston mounted on rod (a). Plunger with hard metal coating (b). Cutaway stuffing box showing spring-loaded packing (c). (Courtesy of Union Pump Company.)

9. Packing running dry (pumping chamber gas-bound).
10. Shock conditions arising from entrained gas or cavitation, broken or faulty valve springs, or system problems.
11. Solids from the pumpage, environment or lubricant.
12. Improper packing installation or break-in (where required).
13. Icing caused by volatile liquids that refrigerate and form ice crystals upon leakage to atmosphere, or by pumping liquids at temperatures below 32°F.

As is evident from these conditions, short packing life can indicate problems elsewhere in the pump or system.

To achieve a low leakage rate, the clearance between the plunger (or rod) and packing must be essentially zero. This requires that the sealing rings be relatively soft and pliant. Because the packing is pliant, it tends to flow into the stuffing box clearances, especially between the plunger and follower bushing. If this bushing does not provide an effective barrier, the packing will extrude, and leakage will increase.

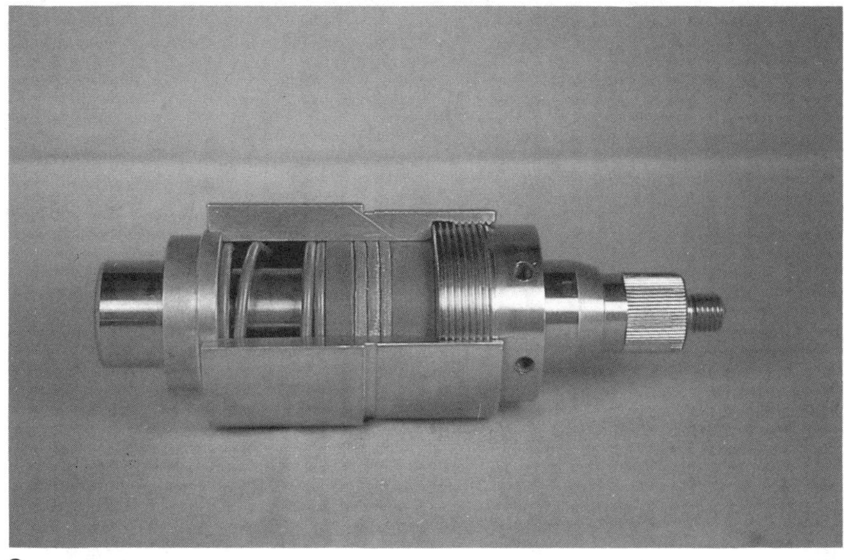

C

A

B

Figure 3-6. Wing-guided valve and seat pressed into cylinder (a). Disk-type valve and seat clamped to cylinder (b). Disk-valve assembly (c). (Courtesy of Union Pump Company.)

A set of square of V-type packing rings will experience a pressure gradient, during operation, as indicated in Figure 3-7. The last ring of packing adjacent to the gland-follower bushing will experience the largest axial loading of all rings—resulting in greater deformation, tighter sealing and, therefore, the largest pressure drop. Hence, the gap between the plunger and the follower bushing must be small enough to prevent packing extrusion. Most packing failures originate at this critical sealing point.

Because this last ring of packing is the most critical, does the most sealing, and generates the most friction, it requires more lubrication than do the others. In the nonlubricated arrangement (Figure 3-7), this ring must rely on the surface of the plunger to drag some of the pumpage back to it in order to provide cooling and lubrication. To maximize packing life in this situation, the overall stack height of the packing should not exceed the stroke length of the pump. Short packing life has resulted from nonlubricated operation of stuffing boxes equipped with lantern rings, especially on short-stroke pumps (approximately 2 in. stroke length). The lantern ring located in the center of the packing sometimes causes the overall stack height of the packing to exceed the stroke length.

C

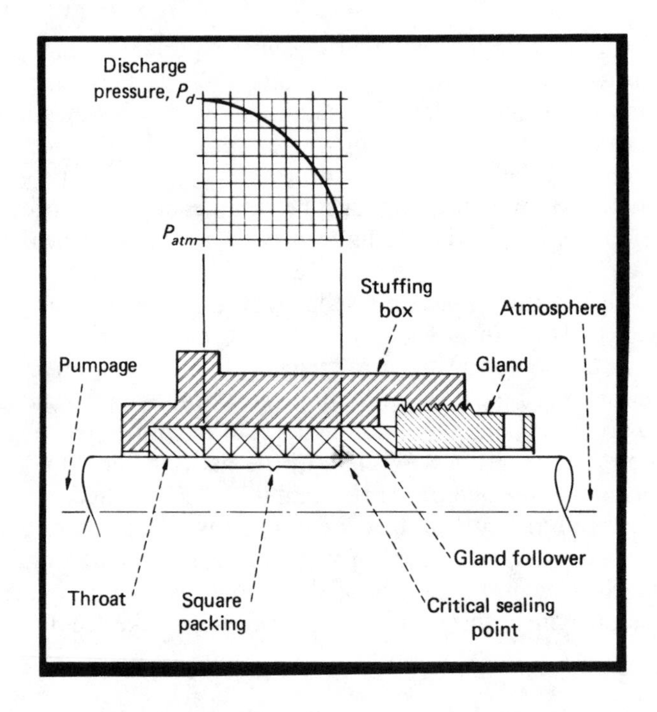

Figure 3-7. Pressure gradient across packing.

Because the last ring of packing requires more lubrication than do the others, lubrication of the packing from the atmospheric side is more effective than injection of oil into a lantern ring located in the center of the packing. Care must be exercised to get the lubricant onto the plunger surface and close enough to the last ring, so that the stroke of the plunger will carry the lubricant under the ring. If the lubricant drips onto the plunger aft of the gland, the plunger-stroke length may not be sufficient to carry the lubricant under the last ring of packing.

Because the last ring of packing deforms the most, it conforms to the irregularities in the bore of the stuffing box. Therefore, when the gland is tightened, most of the force is absorbed in the last ring, causing it to seal tighter in the box and on the plunger. Very little of the gland force gets transmitted to the inner rings of packing.

Hence, the bottom ring of packing must be firmly seated during installation, using a rod with a flat end or a stack of gland bushings. After the stuffing box has been completely assembled, with the plunger reinstalled, and before filling the liquid end with fluid, it is advisable to tighten the gland snugly by hand with the gland wrench. If allowed to sit with this imposed load, most packing will flow and conform to the stuffing box and plunger. It will often be found that after 10 minutes the gland can be further tightened. This process should be repeated two or three times, or until the gland cannot be further tightened. The gland should then be completely loosened, and the packing allowed to expand for 10 to 15 minutes. The gland should then be drawn up only finger tight (no wrench). Now, the block valves may be opened and liquid allowed into the pump.

Soaking the packing in oil prior to installation will enhance a proper break-in and increase packing life.

During the first few hours of pump operation following repacking, each stuffing box should be monitored for temperature. It is normal for some boxes to run warmer than others—as much as 50°F above the pumping temperature. Only if this exceeds the maximum temperature rating of the packing are steps required to reduce box temperature.

The best lubricant for most installations equipped with stuffing box lubricators has been found to be steam-cylinder oil. This oil is compounded with tallow, which gives it a tenacity for the plunger surface and makes it ideal for providing a lubricating wedge between the plunger and packing.

The concepts that a higher discharge pressure requires more rings of packing and that a larger number of rings lasts longer may have been true for long-stroke low-speed machines but has been disproven in some power-pump applications. Unless they are profusely lubricated, the larger number of rings create additional frictional heat and wipe lubricant from the plunger surface—thus depriving some rings of lubrication. On numerous salt-water injection pumps operating at pressures above 4,000 psig, packing life was reported to be only two weeks when twelve rings of packing were installed in each stuffing box. With three rings in each box, packing life was approximately six months.

Stuffing Boxes

Stuffing box designs—including the standard nonlubricated types and various lubrication and bleedoff schemes to minimize leakage and extend packing life—are shown in Figure 3-8.

The most significant advance in packing arrangements in recent years is the spring-loading of packing. Although the concept has been discussed in the literature for decades, and actually put into practice by one manufacturer for at least twenty years, only recently has this arrangement received general attention.

Spring loading is applied almost exclusively to V-ring (chevron) packing but also works well with square packing rings. The spring must always be located on the pressure side of the packing. Springs of various

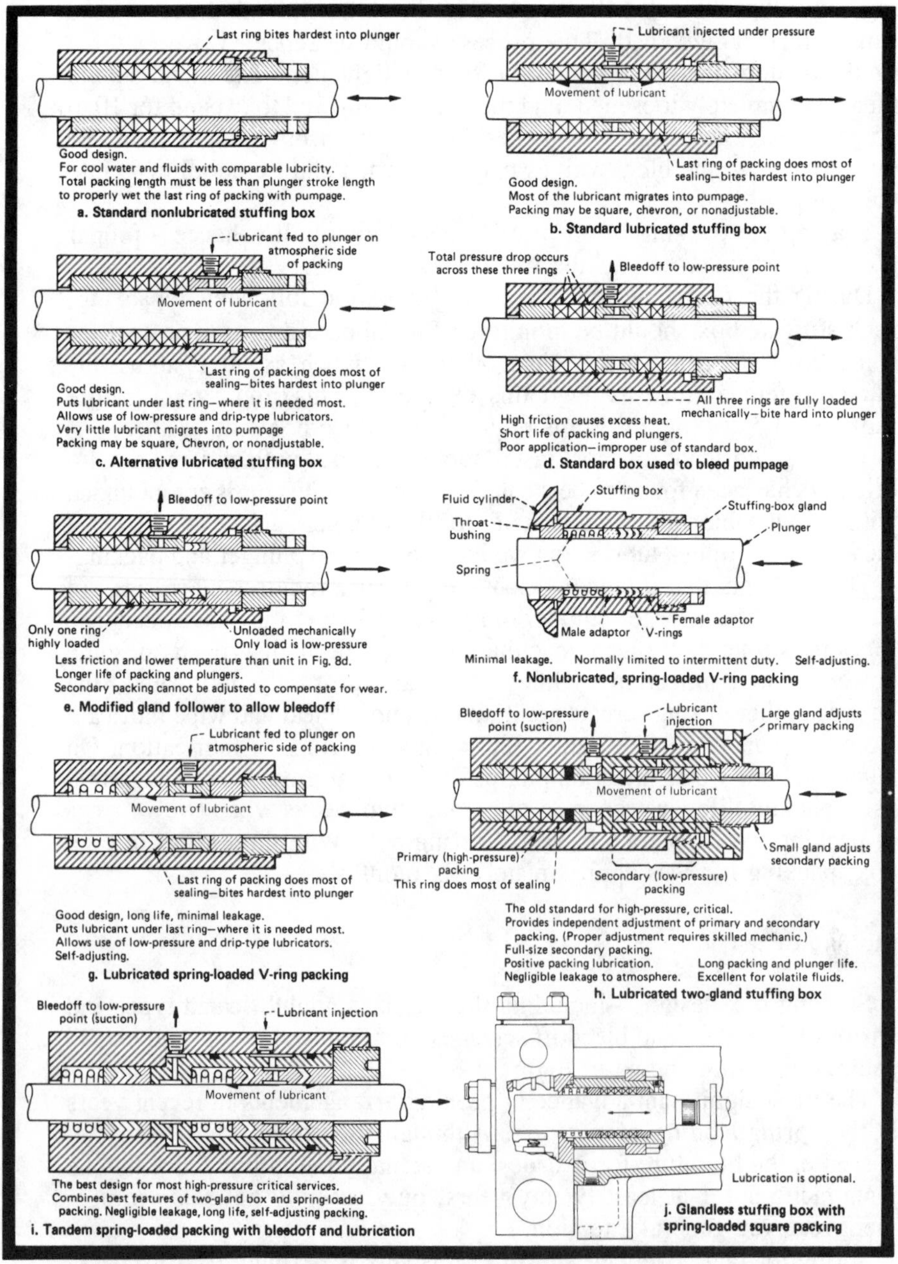

Figure 3-8. Stuffing box designs.

types can be used, including a single large coil, multiple coil, wave-washer, belleville, and a thick rubber washer.

The force required by the spring is small compared to the force imposed on the packing by the liquid. The major function of the spring is to provide a small preload to help set the packing, and to hold all bushings and packing rings in place during operation.

Spring loading of packings has many advantages. It:

- Requires no adjustment of the gland—the gland is tightened until it bottoms, then is locked. This removes one of the biggest variables in packing life—operator skill.
- Allows expansion—if the packing expands due to frictional heat during the initial break-in, the spring allows for the expansion.
- Takes up wear—as the packing wears, adjustment automatically occurs from inside the box. The problem of transmitting the force through the top packing ring during gland adjustment is eliminated.
- Provides a cavity—the spring cavity provides an annular space for the injection of a clean liquid for slurry applications.
- Eliminates the need for gland if pump design allows this—the stuffing box assembly (if a separate component) can be disassembled and reassembled on a workbench.

Disadvantages of spring-loaded packing are associated with the cavity created by the spring. Since this cavity communicates directly with the pumping chamber, the additional clearance volume can cause a reduction in volumetric efficiency if the pumpage is sufficiently compressible. This cavity also provides a place for vapors to accumulate. If the pump design does not provide for venting this space, a reduction in volumetric efficiency may occur.

Spring-loaded packing is the reciprocating pump's equivalent to the mechanical seal for rotating shafts. Leakage is low, life is extended, and adjustments are eliminated. Packing sets can be stacked in tandem (they must be independently supported) for a stepped pressure reduction, or to capture leakage from the primary packing that should not be allowed to escape to the environment.

Plunger Material

After the packing, the plunger is the component of a power pump that requires the most frequent replacement. The high speed of the plunger and the friction load of the packing tend to wear the plunger surface. For longer life, plungers are sometimes hardened. A more popular method is to apply a hard coating to the plunger surface. Such coatings are of chrome, various ceramics, nickel-based alloys, or cobalt-based alloys. Desired features for the coatings include hardness, smoothness, high bond strength, corrosion resistance, and low cost. No one coating optimizes all of these features.

The ceramic coatings are harder than the metals but are brittle, porous, and sometimes lower in bond strength. Porosity contributes to shorter packing life. Mixing of hard particles such as tungsten carbide into the less-hard nickel or cobalt alloys has resulted in longer plunger life at the expense of shorter packing life.

Drive-End Components

The drive end of a power pump is called a power end (see Figure 3-3). Its function is to convert rotating motion from a driver to reciprocating motion for the liquid end. The main component of the power end is the power frame, which supports all other power end parts and, usually, the liquid end. The second major item in the power end is the crankshaft (sometimes, a camshaft). The function of the crankshaft in a power pump is the same as a crankshaft in an internal-combustion engine, except that the flow of energy is opposite.

The main bearings support the shaft in the power frame. The connecting rod is driven by the throw of the crankshaft on one end, and drives a crosshead on the other. The crosshead moves in pure reciprocating motion, the crankshaft in pure rotating motion. The connecting rod is the link between the two.

Although similar in construction and motion to a piston in an internal-combustion engine, the crosshead is fastened to a rod called an "extension," "stub," or "pony" rod. The other end of this rod is fastened to the plunger or piston rod.

The function of the drive end (or steam end, or gas end) of a direct-acting pump is to convert the differential pressure of the motive fluid to reciprocating motion for the liquid end. The drive end is similar in construction to the liquid end, containing a double-acting piston and valving. The major difference is that the valve is mechanically actuated by a control system that senses the location of the drive piston, to cause the valve to reverse the flow of the motive fluid when the drive piston reaches the end of its stroke.

The main component of the drive end is the drive cylinder. This cylinder forms the major portion of the pressure boundary, and supports the other drive-end parts. Unlike the power-pump's power end, this cylinder does not support the liquid end.

Maintenance of Liquid Ring Vacuum Pumps*

Liquid ring vacuum pumps and compressors consist of a rotor with radial pumping vanes rotating in a casing, causing a ring of service liquid to form at the outer circumference of the rotor. Depending on the machine manufacturer, the rotor may be mounted with its rotational axis eccentric to the casing centerline, or concentric in a lobe-shaped casing. This allows the service liquid depth in the vanes to change depending on the rotational position of the vanes. This in turn causes a liquid piston effect with the cavities between successive vanes being filled and emptied of service liquid as the rotor turns. Porting in the casing is arranged so that the suction flow enters the rotor where the liquid ring depth is decreasing, and discharge occurs when the depth is nearing its maximum. The service liquid, vanes, and close-running clearances at the rotor ends serve to seal the compression "chambers." As the depth of the service liquid increases, the "chamber" volume decreases and compresses the gases. High pressure ratios are obtained by staging. Figures 3-9A and 3-9B illustrate the liquid ring principle.

Several clearances are critical to the successful operation of liquid ring machines. Rotor end clearances provide a leakage path from discharge

* Compiled by J. V. Picknell, Esso Chemical Canada, Ltd.

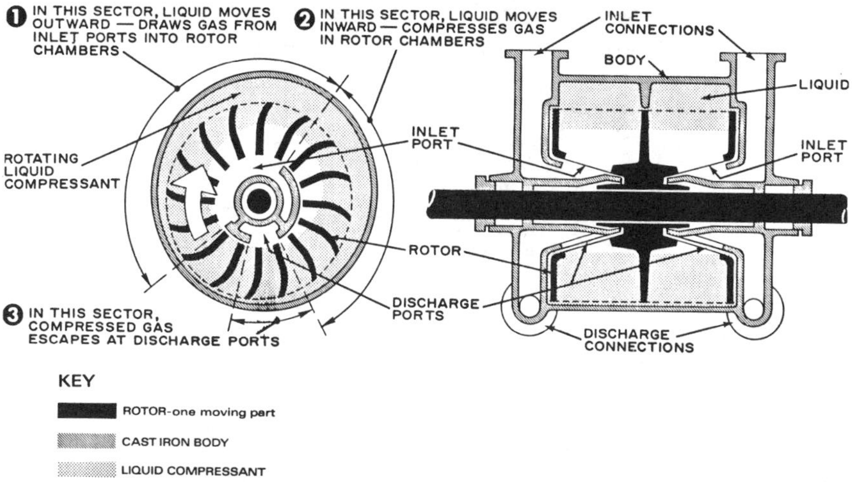

Figure 3-9A. Sectional and end view of a liquid ring vacuum pump. (Courtesy of Nash Engineering Company.)

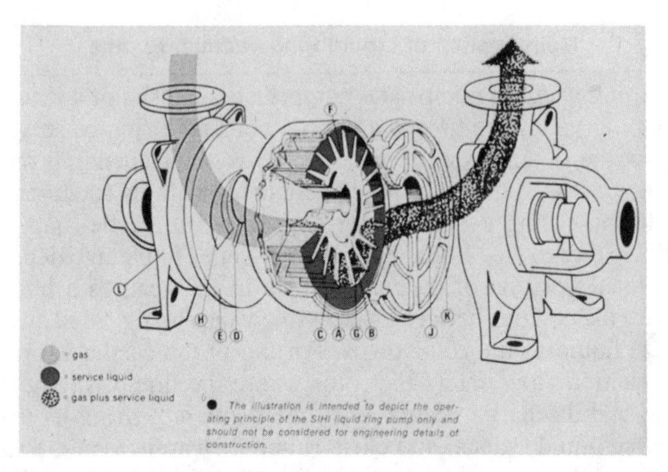

Figure 3-9B. Exploded view of a liquid ring vacuum pump with side ports. (Courtesy of SIHI.)

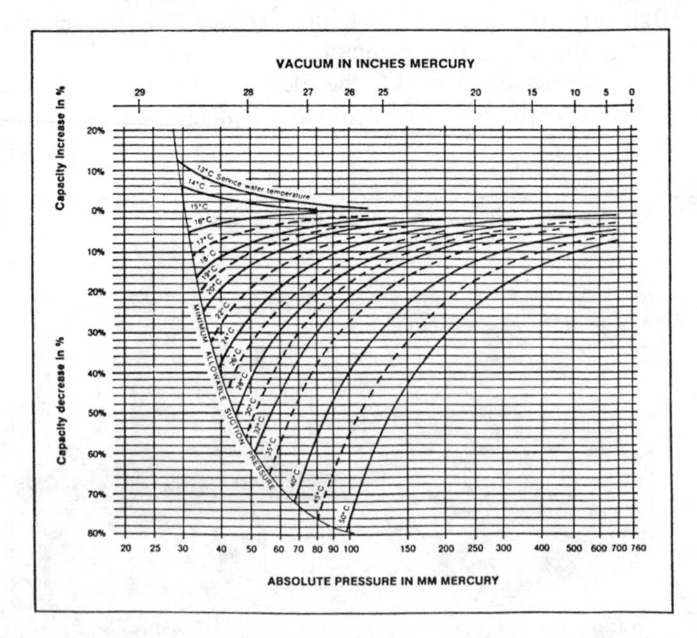

Figure 3-10. Performance reduction of a liquid ring vacuum pump due to service water temperature rise. (Courtesy of SIHI.)

back to suction where they are not filled with service liquid. Where suction and discharge ports are arranged in cones around the shaft, the clearances between the cones and the rotor are important for the same reason. The rotor tip clearance in the casing is less critical since this will be submerged in service liquid.

Also important is the service water temperature rise which can have an effect on the performance of liquid ring vacuum pumps. Figure 3-10 is typical of this interaction.

The importance of rotor end clearances in most designs necessitates particular care in rotor axial positioning while assembling, and attention to bearing fits so that the rotor cannot change position other than by thermal growth. This becomes more difficult, but equally important in multi-staged machines. During operation of the machines, bearing care is important to ensure that the rotor does not shift. Adequate lubrication and vibration, shock pulse and/or spike energy monitoring is important. Most liquid ring machines employ tight running clearances over fairly large areas (i.e., the entire rotor end area at both ends). Bearing failures resulting in axial or radial position changes can result in rubs and seizure of the machines. Assembly views reveal the close running tolerances employed (Figures 3-11 and 3-12).

Sealing of shafts is accomplished by either mechanical seals or by packing. Double and tandem seals may be used where process conditions

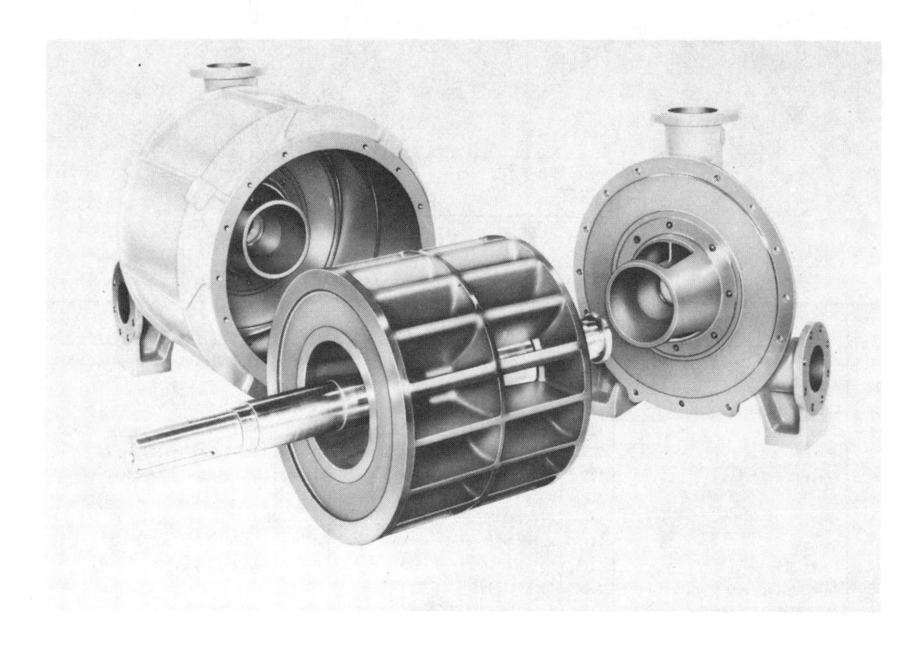

Figure 3-11. Disassembled view of liquid ring vacuum pump with side ports showing "sandwich" construction.

dictate it. Close attention to the maintenance of flushing and buffer fluid systems is necessary to ensure long life of the seals. Seal failure can result in significant reduction in vacuum pump capacity with no visible external leakage.

Performance of liquid ring machines can be significantly affected by the service liquid used. The volume of liquid employed alters the submergence of the rotor vane tips throughout the compression cycle and

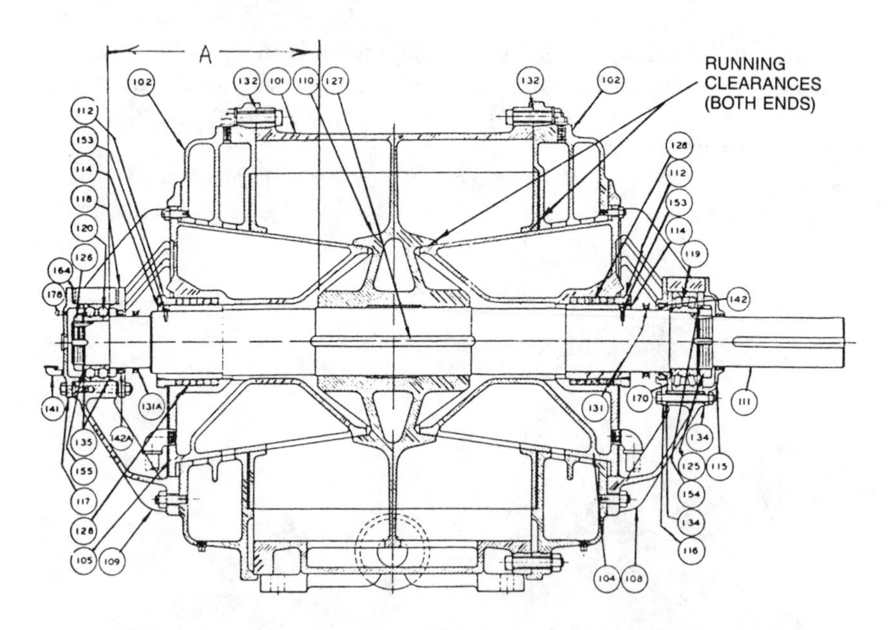

Part No.	Part Name	Part No.	Part Name	Part No.	Part Name
101	Body	117	Bearing Cap-Outer-Idle End	135	Gasket-Bearing Cap-Idle End
102	Head	118	Bearing Cap-Inner-Idle End	141	Oiler For Bearing
104	Cone-Drive End	119	Bearing-Drive End	142	Shaft Seal-Drive End
105	Cone-Idle End	120	Bearing-Idle End	142A	Shaft Seal-Idle End
108	Bracket-Drive End	125	Shaft Nut-Drive End	153	Lock Screw-Shaft Sleeve
109	Bracket-Idle End	126	Shaft Nut-Idle End	154	Lockwasher-Shaft Nut-Drive End
110	Rotor	127	Rotor Key	155	Lockwasher-Shaft Nut-Idle End
111	Shaft	128	Packing		
112	Gland	131	Water Slinger-Drive End	164	Set Of Shims
114	Shaft Sleeve	131A	Water Slinger-Idle End	170	Bearing Spacer
115	Bearing Cap-Outer-Drive End	132	Gasket Body		
116	Bearing Cap-Inner-Drive End	134	Gasket-Bearing Cap-Drive End	178	Vent Cap

Figure 3-12. Cross section and bill of material of a vacuum pump with cone type porting. (Courtesy of Nash Engineering Company)

will affect the volume of gas drawn into the machine and the compression ratio. Service liquid absorbs the heat of compression and must be cooled. This is accomplished by running the liquid through external coolers, or by make-up liquid, or by using a once-through system. Usually the service liquid circulating system employs a discharge vessel in which gas and liquid separation occurs. The level of liquid in this vessel is maintained at the level of the shaft centerline to ensure the correct amount of liquid is in the machine.

As mentioned, service liquid temperature also has a profound effect on the efficiency and capacity of liquid ring machines. As the temperature of the service liquid rises, so does its vapor pressure. This increases the partial pressure of the service liquid vapor in the machine and reduces the volume available for the process gas.

Final discharge pressure, where it can vary, can also affect overall performance. If the process gas contains a condensible vapor and the discharge pressure is high enough at compression temperatures to allow condensation, some liquid will condense. When this liquid leaks through running clearances back to suction, it can flash off and reduce inlet or suction capacity.

Starting of liquid ring machines must be done with the machine only half full of liquid. Failure to maintain the correct level for starting can result in either reduced capacity (level too low) or overloading (level too high). The latter is more serious as it can result in driver overload, belt wear, or coupling failure. These machines have only a limited capability to handle liquids in the process stream.

Large volumes of liquid in the process gas or vapor stream can overload the machine. The reader will appreciate that a volume of liquid greater than the volume between vanes at the discharge openings cannot be compressed. High vibration, overload, and machine failure can result. Particulate matter or solids in the process stream can be handled in small quantities. Solids can lodge between running faces and cause wear, and eventually open up clearances to reduce capacity. Large quantities of solids can plug up internal clearances and passages, reducing the capacity of the machine and possibly seizing it. Excess heat can be generated by the closing up of running clearances. This can cause excess thermal growth of the rotor and further wear.

Cavitation damage can often be found in the suction porting and on vacuum pump rotors. During operation this can be detected by the characteristic sound of "gravel on steel." Some vacuum pump systems employ an ejector in their suction lines to boost the suction pressure that the machine sees. Motive fluid for the ejector can be taken from the pump discharge. Where cavitation is found it may be worthwhile to consider the use of a suction line ejector, raising the suction pressure (where process conditions will allow), or using pump parts made of materials resistant to cavitation damage, such as high-chrome steels.

Maintenance of liquid ring machines is minimized because there are no wearing parts other than bearings and seals. The bulk of necessary maintenance efforts should be aimed at the service water, seal fluid, and pressure regulating systems. This necessitates careful monitoring by operating personnel. In addition to these it is necessary to monitor bearing health by vibration analysis, shock pulse, or spike energy methods.

When disassembly for repairs is necessary there are very important details to ensure that running clearances are reestablished correctly. In machines with cone type inlet and discharge porting, axial rotor position is critical, especially where running clearances are on tapered surfaces as in Figure 3-13. You can appreciate that an axial shift will close up the clearance on one end and open it up on the other. In Figure 3-13 the thrust end bearing is on the left hand side. Shimming is used at this end under the bearing cap to adjust the axial position. When machines of this design are disassembled, note the shim thickness removed. If the entire rotor assembly is being replaced, compare the distance marked "A" on the new and the old rotor and adjust the shim thickness accordingly.

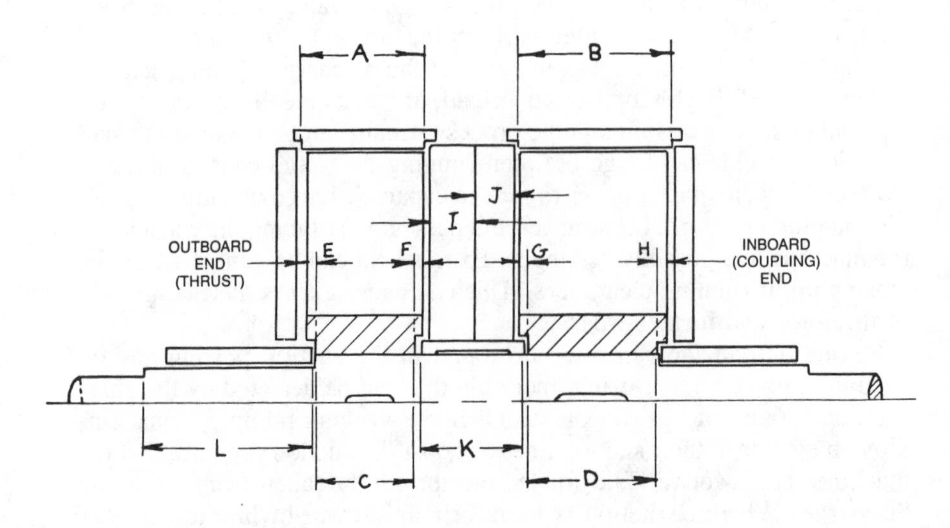

Figure 3-13. Clearance checks required for vacuum pump rotors.

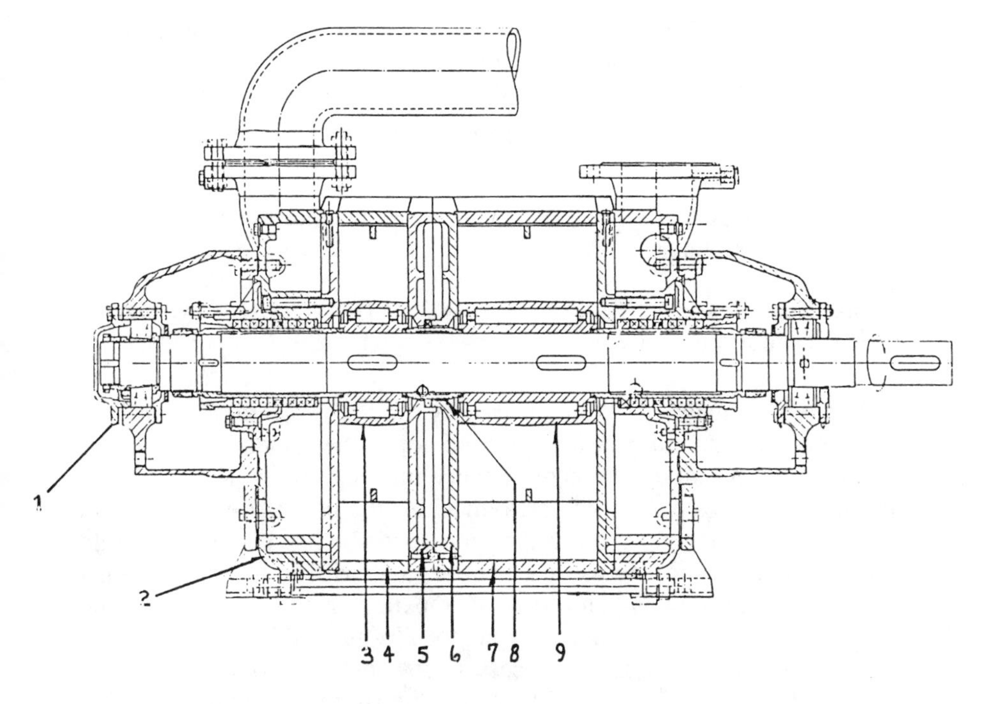

Figure 3-14. Cross section view of two-stage vacuum pump.

In "sandwich" construction machines such as in Figure 3-14, the correct clearances are more difficult to obtain. A two-stage machine with thrust end at the left is shown. Several measurements are required to ensure correct assembly (refer to Figures 3-13 and 3-14).

1. The length of the center bodies "A" and "B" (items 4, 7).
2. The length of the impeller hubs "C" and "D" (items 3, 9).
3. Check that the impeller ends are parallel and the tips .03 to .08 mm (.001 to .003 in.) narrower than the hubs.
4. The depths of the recesses at the impeller hubs "E," "F," "G," and "H."
5. Check that the surfaces of the intermediate plates (items 5, 6) are parallel and flat within .04 mm (.0015 in.) and record their thickness at the outer circumference "I," and "J."

6. The length of the distance bushing "K" (item 8) and ensure that its ends are parallel within .03 mm (.001 in.).
7. The distance from the inboard end of the sleeve to the first shaft step outboard of the sleeve at the thrust bearing end of the rotor "L."

The clearances that will result from assembly can now be calculated, and adjustments made:

Total End Clearance = (A + B + I + J) − (E + C + K + D + H)
Total Center Clearance = (K − F − G) − (I + J)

On a typical machine, these clearances should be .38 to .51 mm (.015 to .020 in.) per running surface pair. To adjust the center clearances it is necessary to machine the center distance bushing (alter "K"). To adjust the end clearances it is necessary to machine the impeller ends. If impeller ends are damaged by a rub they must also be machined. If impeller machining results in excess end clearances it will be necessary to machine the center bodies. The limit of impeller machining before center body machining is necessary is typically .76 mm (.030 in.). When one impeller is shortened it is also necessary to shorten the other impeller by an amount equal to the amount taken off the first impeller multiplied by the ratio of impeller lengths in order to maintain the original pressure ratio.

When impellers and center bodies are shortened, the internal volume of the machine, and hence capacity, is reduced. To restore the original capacity it will be necessary to purchase and install new impellers and center bodies.

The distance "L" is critical to ensure that the entire assembly is located at its correct axial position. Reassembly begins with the rotating element and intermediate plates. Start by assembling the thrust end shaft sleeve to the shaft with its locknut and washer, setting "L" to the original dimension. Stack up the rotor parts in order, including the intermediate plates and finishing with the other sleeve locknut. This is most easily accomplished with the rotor standing on its thrust end. Check the clearance of the intermediate plates to the impeller ends with feeler gauges. The body can now be assembled, again beginning at the thrust end. Start with the cover (item 2 in Figure 3-14). Use either O-rings or Permatex® at the body joints during this assembly. Insert the rotating assembly with the intermediate plates when the first center body is installed. Finish this step of the assembly with the other end cover and tie rods. With the body on its feet, level the assembly and tighten the tie rods evenly to the recommended torque values.

Finally, assemble the bearing brackets, bearings and seals or packing. Assemble the thrust end first. At the driven end the bearing must be centralized in its housing. Check the end float by loosening and tightening the thrust end bearing cover (item 1 in Figure 3-14). Measure axial movement with a dial indicator on the exposed shaft end. Adjust that bearing cover so that the rotor will be held in the middle of its axial float.

Chapter 4

Positive Displacement and Dynamic Blowers

Maintenance Instructions for Positive Displacement Rotary Blowers*

Process plants frequently employ rotary positive displacement blowers utilizing a 3-lobe rotor design. The rotors are driven by timing gears. Ball bearings at each end of the rotors provide support and control limited clearances. Internal parts are lubricated by splash oil lubrication. Lip seals are used to control process gas leakage. A cutaway picture of a typical blower is shown in Figure 4-1A.

Materials of Construction

Rotors with integral shafts, housing and end plates are made of ductile iron.

Disassembly (Figure 4-1B)

Disassembly procedures are generally simple and straightforward. Here is a typical sequence for the blower shown. It may have to be modified for different types or models:

1. Drain oil from blower.
2. Remove drive end cover (6) and flanged drive shaft (45) from drive gear.
3. Mark all parts with a center punch so they can be reassembled in same position.

* Courtesy of M-D Pneumatics Inc., compiled by H. Y. Hung, Esso Chemical Canada, Ltd., Sarnia, Ontario.

4. Remove nondrive end cover (7).
5. Remove flathead socket cap screw (29, 69), rotor shaft washer (25), and oil slinger (21) from nondrive end of each rotor.
6. Remove timing gears (8):
7. Place the blower on its side as shown in Figure 4-2A.
8. Remove the gear lock nuts (35) from shafts.
9. Rotate the gears to the position shown in Figure 4-2A—keyways in line and gear timing arrows matched. Mark gears with reference marks—five teeth below timing marks.
10. Turn the gears upward five teeth so the reference marks are matched as shown in Figure 4-2B. This gear position is necessary when pulling one gear first so rotors will clear and not jam.
11. Pull the driven gear first, using a gear puller. It is assembled in two parts—gear rim and hub. Do not disassemble. Do not interchange the dowel pins (58); they are select fitted.
12. Remove the drive gear. Keep the key (24) together with the gear.
13. Remove end plates (4 & 5):

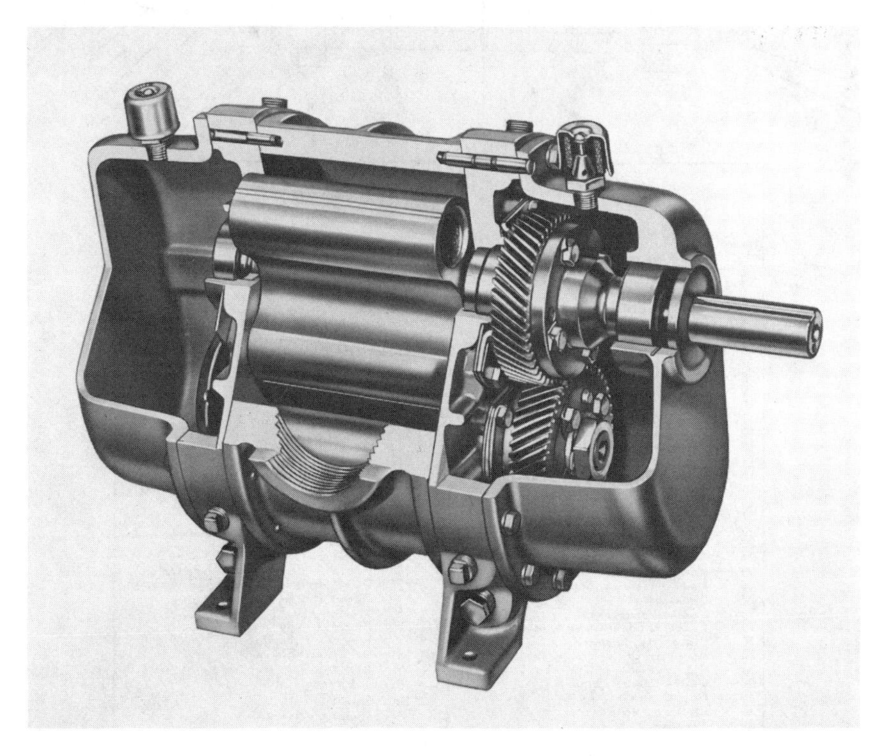

Figure 4-1A. Cutaway drawing of a three-lobe rotary displacement blower. (Courtesy MD Blowers.)

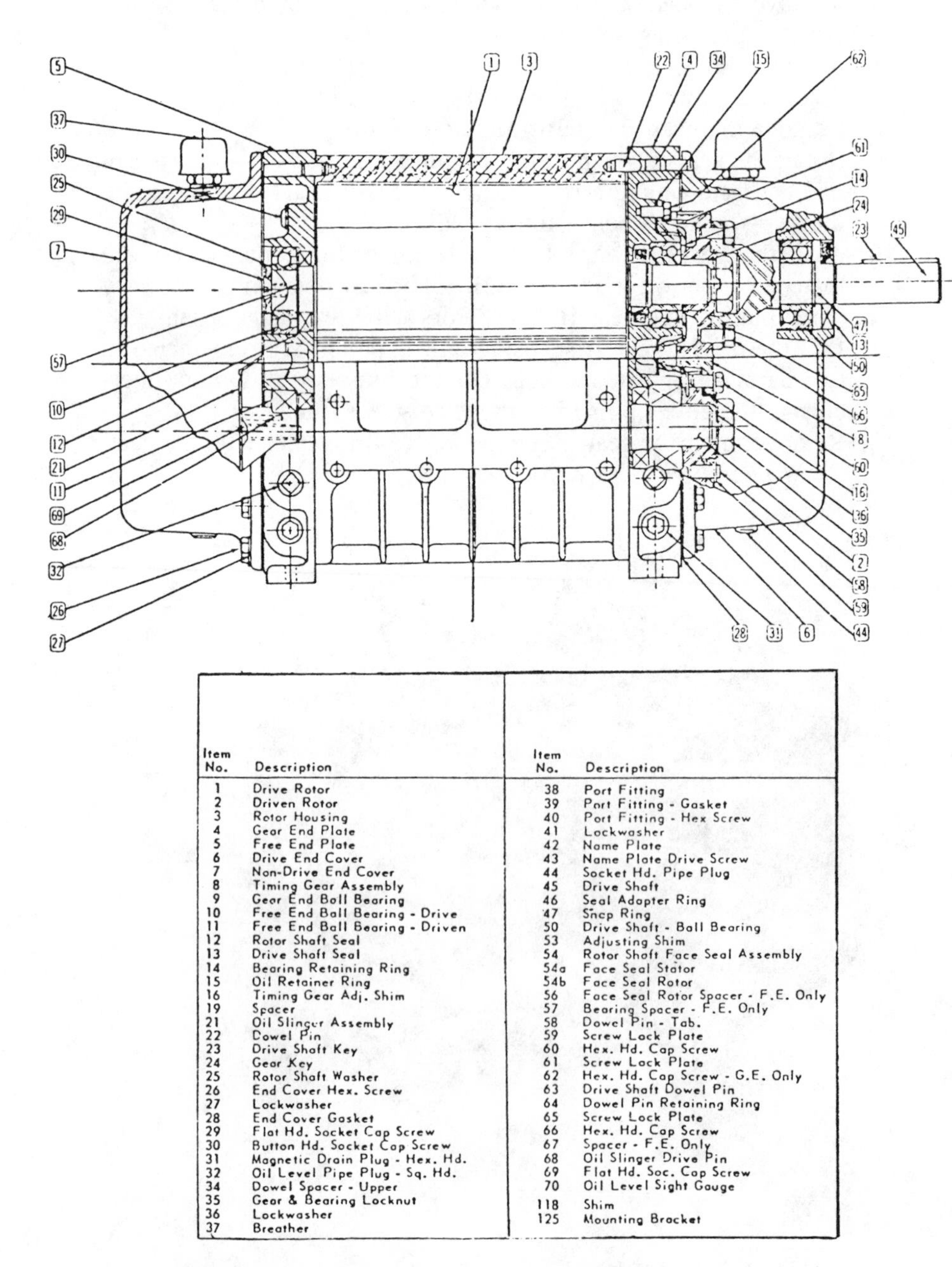

Item No.	Description	Item No.	Description
1	Drive Rotor	38	Port Fitting
2	Driven Rotor	39	Port Fitting - Gasket
3	Rotor Housing	40	Port Fitting - Hex Screw
4	Gear End Plate	41	Lockwasher
5	Free End Plate	42	Name Plate
6	Drive End Cover	43	Name Plate Drive Screw
7	Non-Drive End Cover	44	Socket Hd. Pipe Plug
8	Timing Gear Assembly	45	Drive Shaft
9	Gear End Ball Bearing	46	Seal Adapter Ring
10	Free End Ball Bearing - Drive	47	Snap Ring
11	Free End Ball Bearing - Driven	50	Drive Shaft - Ball Bearing
12	Rotor Shaft Seal	53	Adjusting Shim
13	Drive Shaft Seal	54	Rotor Shaft Face Seal Assembly
14	Bearing Retaining Ring	54a	Face Seal Stator
15	Oil Retainer Ring	54b	Face Seal Rotor
16	Timing Gear Adj. Shim	56	Face Seal Rotor Spacer - F.E. Only
19	Spacer	57	Bearing Spacer - F.E. Only
21	Oil Slinger Assembly	58	Dowel Pin - Tab.
22	Dowel Pin	59	Screw Lock Plate
23	Drive Shaft Key	60	Hex. Hd. Cap Screw
24	Gear Key	61	Screw Lock Plate
25	Rotor Shaft Washer	62	Hex. Hd. Cap Screw - G.E. Only
26	End Cover Hex. Screw	63	Drive Shaft Dowel Pin
27	Lockwasher	64	Dowel Pin Retaining Ring
28	End Cover Gasket	65	Screw Lock Plate
29	Flat Hd. Socket Cap Screw	66	Hex. Hd. Cap Screw
30	Button Hd. Socket Cap Screw	67	Spacer - F.E. Only
31	Magnetic Drain Plug - Hex. Hd.	68	Oil Slinger Drive Pin
32	Oil Level Pipe Plug - Sq. Hd.	69	Flat Hd. Soc. Cap Screw
34	Dowel Spacer - Upper	70	Oil Level Sight Gauge
35	Gear & Bearing Locknut	118	Shim
36	Lockwasher	125	Mounting Bracket
37	Breather		

Figure 4-1B. Cross-sectional drawing and bill of material of a three-lobe rotary displacement blower. (Courtesy MD Blowers.)

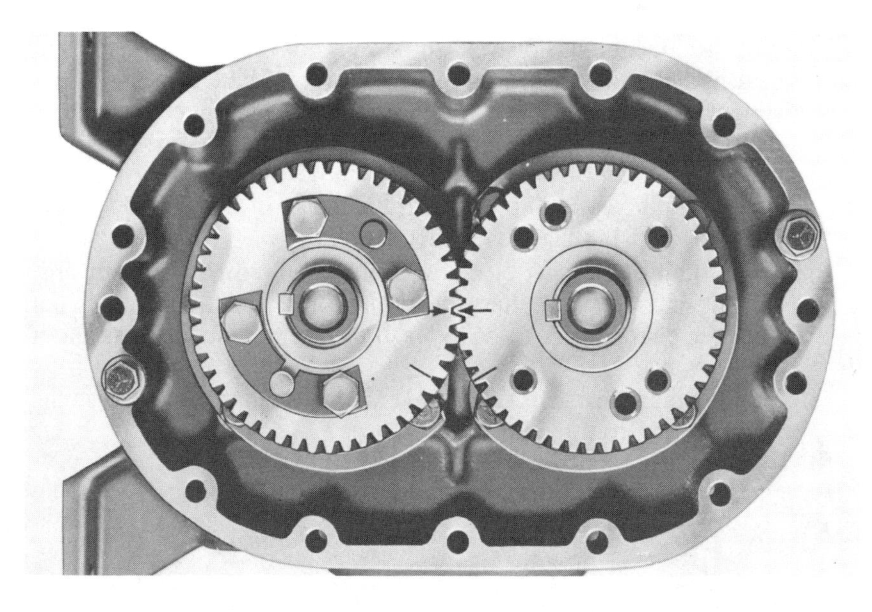

Figure 4-2A. Keyways in line and timing arrows matched. (Courtesy MD Blowers.)

Figure 4-2B. Timing arrows advanced five teeth—reference marks aligned. (Courtesy MD Blowers.)

14. Place support blocks on the bed of an arbor press. Set the blower, gear end pointing down, on the two blocks. Make sure the blocks support the rotor housing (3). Press the gear end plate (4) and rotors (1 & 2) out of the rotor housing (3) simultaneously.
15. Lift the housing off the rotors and remove the nondrive end plate (5) by tapping the end plate from the inside of the housing. Place the rotor housing back over the rotors.
16. Set the unit on support blocks with the gear end pointing upward. Do not extend blocks into the rotor bores. Press the rotors out from the gear end plate. Mark each rotor and note the position of the keyways.
17. Remove the bearing retainers (14) on drive end plate and push the bearings (9) from the end plates.
18. Remove lip seals from the end plates.
19. Clean all parts with clean solvent.

Assembly

Here is a typical assembly sequence and rule-of-thumb dimensions:

1. Install lip seals in end plates. Ensure the lip is towards the bearing side.
2. Install rotors in drive end plate.
3. Stand rotors on free end in arbor press. Keyways must be in line.
4. Place drive end plate (4) with lip seals installed, over rotors.
5. Press gear end ball bearings (9) on rotor shafts and into bearing bores in end plate.
6. Install gear end bearing retaining ring (14), and screws (62).
   Caution: The bearings used have flush ground faces and should be installed with manufacturer's bearing number toward the gear side. Do not use standard bearings which have not been flush ground within .001 in. tolerance.
7. Install timing gears.
8. Insert the gear keys in their proper location.
9. Install drive gear first. Press gear on drive rotor. To prevent jamming, timing marks must be arranged as shown on Figure 4-2B. Secure gears with lock plate (65) and hex screw (66). Check face runout not to exceed .001 in. TIR.
10. Check clearance between face of the drive end plate and the rotor lobe ends (.006 in. to .009 in.).
11. Install rotor and drive end plate assembly in rotor housing (3). Secure with 4 hex screws (26). Be sure dowel pins (63, 58) are in

housing. Apply a thin, even coat of silicone grease to the end of housing before assembly.

12. Install free end plate.
13. Secure nondrive end plate (5) to rotor housing (3) with hex screws (26). Be sure dowel pins are in housing.
14. Install spacer (57) and bearing (10) on each rotor shaft.
15. Place oil slinger (21) on lower bearing, install drive pin (68) and secure with flathead socket cap screw (69). Use rotor shaft washer (25) and flathead socket cap screw (29), to secure upper bearing (1) on rotor.
16. Check clearance between face of the nondrive end plate and the rotor lobe ends (.012 in. to .018 in.).
17. Check rotor tip to housing clearance.
    Inlet side: .0125 in. to .014 in.
    Discharge side: .0085 in. to .010 in.
18. Check interlobe clearance (See Figure 4-2C).
19. Adjust timing:
    The driven gear is made of two pieces. The outer gear shell is fastened to the inner hub with four cap screws (60) and is located with two dowel pins (58). A .030 in. thick shim (16), made up of .003 in. laminations, separates the hub and the shell. By removing or adding shim laminations, the gear shell is moved axially relative to the inner hub which is mounted on the rotor shaft. Being a helical gear, it rotates as it is moved in or out and the driven rotor turns with it, thus changing the clearance between rotor lobes (Figure 4-2D). Changing the shim thickness .003 in. will change the rotor lobe clearance .0015 in. When reassembling shell on hub, be sure that timing marks coincide as shown in Figure 4-2B.
20. Bolt nondrive end cover (7) to nondrive end plate (5). Be sure gasket (28) is placed between the cover and the end plate.
21. Mount bearing (50) and snap ring (47) on drive shaft (45) and fasten to drive gear (8), using four hex screws (66) and lock plate (65). Drive shaft shall not run out more than .001 in. TIR.
22. Assemble cover (6) to blower. Install lip seal (13) on cover.

Some useful information should be tabulated in the vendor's instruction manual:

1. Torque values for:

    - Gear and bearing locknut
    - Rotor and drive shaft flathead screw
    - Gear end ball bearing retainer screw

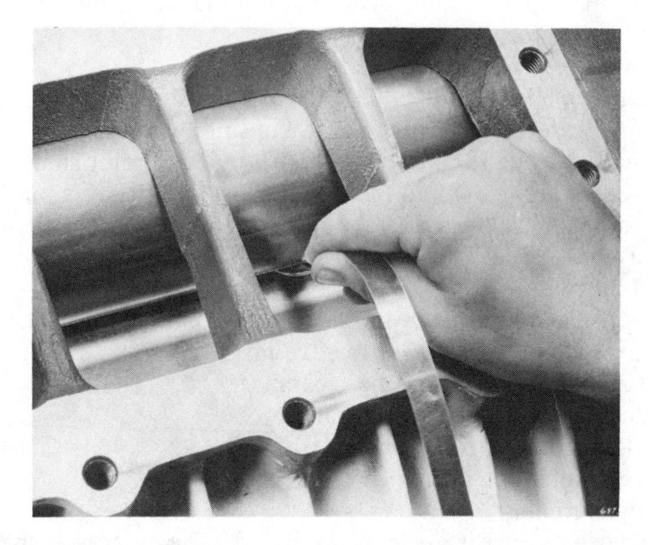

Figure 4-2C. Checking interlobe clearance. (Courtesy of MD Blowers.)

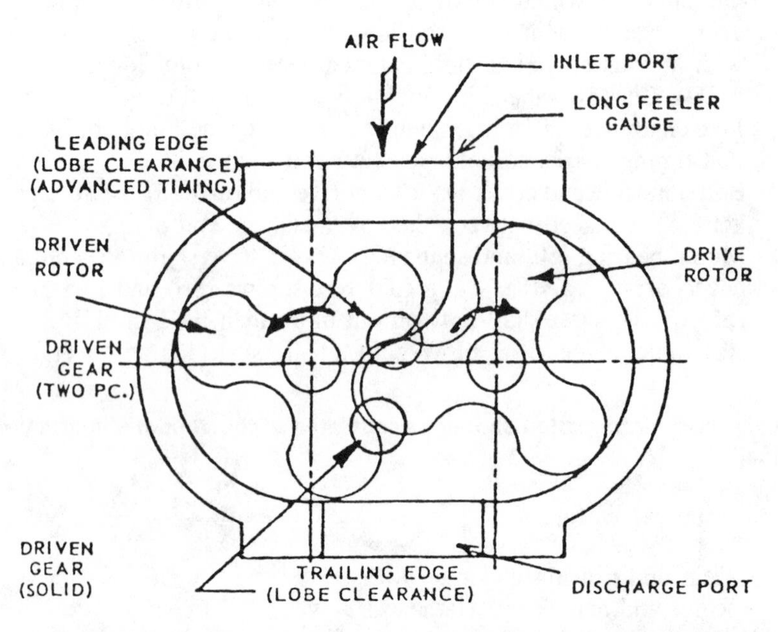

Figure 4-2D. Application of feeler gauge. (Courtesy of MD Blowers.)

Unless otherwise stated by the manufacturer, use the following values in ft-lbs:

- 1/4 in. cap screw 12
- 5/16 in. cap screw 24
- 3/8 in. cap screw 42
- 1/2 in. cap screw 90

2. Weight of blower
3. Oil capacity: drive end and nondrive end
4. Lubricating oil required
5. Oil change interval

Care and Maintenance of Fans*

Large Fan Blower Maintenance

The use of large fans, or blowers, in petrochemical process, power plant, and other industrial applications has experienced a tremendous growth in the past decades. Much of the growth has been brought about by the demand for cleaner air. Boilers, waste heat recovery equipment, furnaces and related systems have produced increasingly hotter, more corrosive and particulate matter-laden gases to be moved. With the handling of more demanding gas in greater volumes have come larger and faster fans, which must be considered an important part of the plant maintenance load.

Heavy duty fans dealt with in the following section are used in two major functions: Forced draft and induced draft. Forced draft fans are used to push air through a furnace, boiler, or process apparatus. Induced draft fans are exhaust fans which draw gases from the process—usually pollution control equipment. Both types of fans are usually found with one of four impeller arrangements, namely paddle wheel, radial tip, double inlet airfoil, and single inlet fan. Figure 4-3 shows an exploded view of the more common heavy duty single inlet fan. Figure 4-4 illustrates the less common vaneaxial fan arrangement.

Fan Component Nomenclature

Heavy duty fans have impeller diameters up to 166 in. and may weigh up to 10,000 lbs. Operating speeds range from 600 to 1,800 revolutions per minute depending upon the fan system and its application. Figure 4-5

---

* Courtesy of Canadian Blower/Canada Pumps Ltd., Kitchener, Ontario, Canada

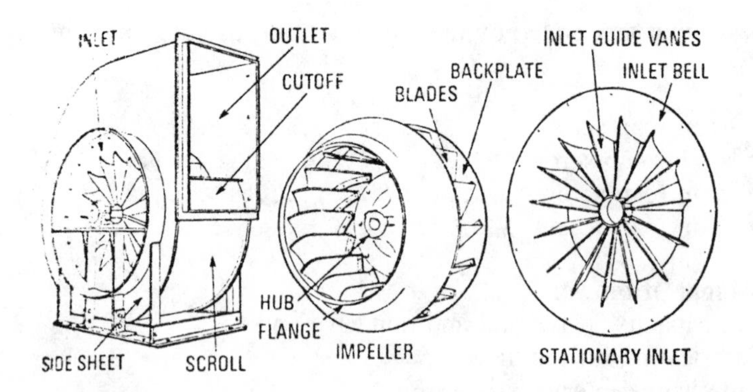

Figure 4-3. Exploded view of a centrifugal fan. (Courtesy Canadian Blower/Canada Pumps Ltd. (CB/C Ltd.))

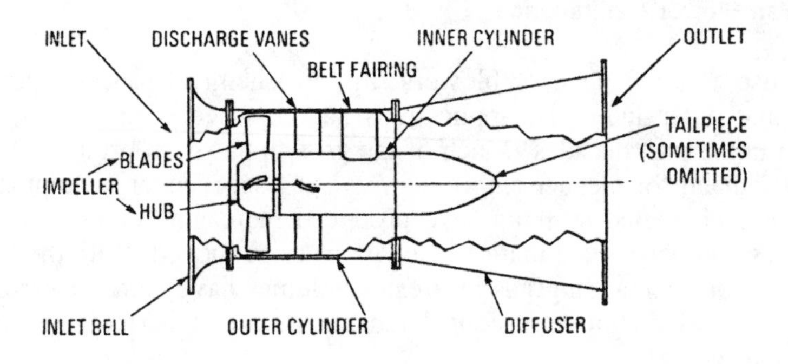

Figure 4-4. Cutaway view of a vaneaxial fan (courtesy CB/CP Ltd.).

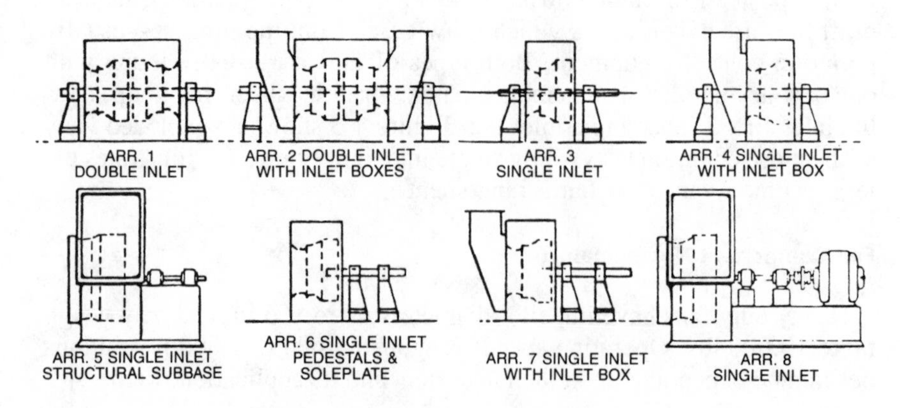

Figure 4-5. Standard fan arrangements.

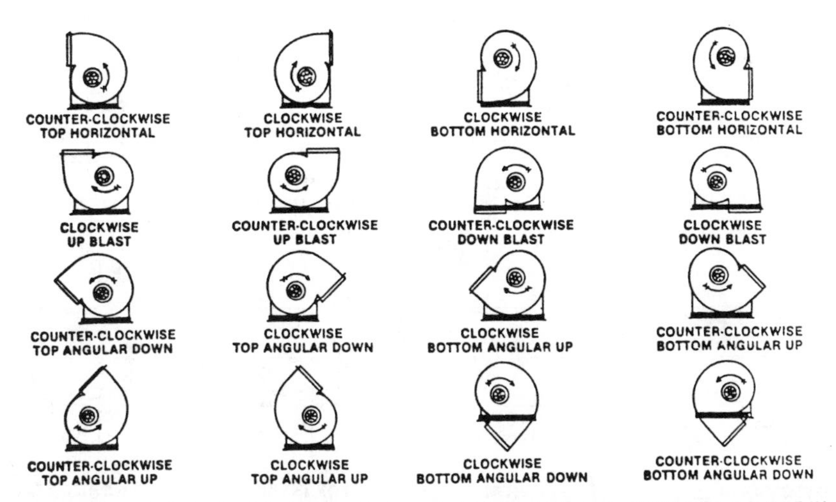

DIRECTION OF ROTATION IS DETERMINED FROM DRIVE SIDE FOR EITHER SINGLE OR DOUBLE INLET FANS. (THE DRIVING SIDE OF A SINGLE INLET FAN IS CONSIDERED TO BE THE SIDE OPPOSITE THE INLET REGARDLESS OF ACTUAL LOCATION OF THE DRIVE.) WHEN DUAL DRIVE IS USED THE DRIVE SIDE IS THE SIDE OF THE MAIN DRIVING UNIT.

Figure 4-6. Direction of fan rotation and discharge designation.

shows the terminology used by the fan industry to describe fan arrangements. Figure 4-6 illustrates the standard designation of rotation and discharge.

Housing and Inlet Boxes

Most heavy duty fan housings and inlet boxes are constructed of welded sheet metal. Units that must be shipped disassembled are first assembled at the manufacturing plant, match marked for identification and alignment and then disassembled into as large as possible sections so that erection at the job site is fast. Standard inlet box positions are shown in Figure 4-7.

Housings are designed for wheel installation or removal either through the fan inlet as in Arrangement 1 fans without boxes, or the housing is split. Flanges for bolting or welding are provided on the fan inlet or discharge.

Drains are either a pipe coupling or flanged. For pressures up to 18–20 in. water gauge, quick-opening access doors are used; at higher pressures, bolted access doors are used.

When fans are specified for abrasive conditions, side and scroll liners are provided (Figure 4-8). Liners are shipped already installed in the housing. These liners are replaceable.

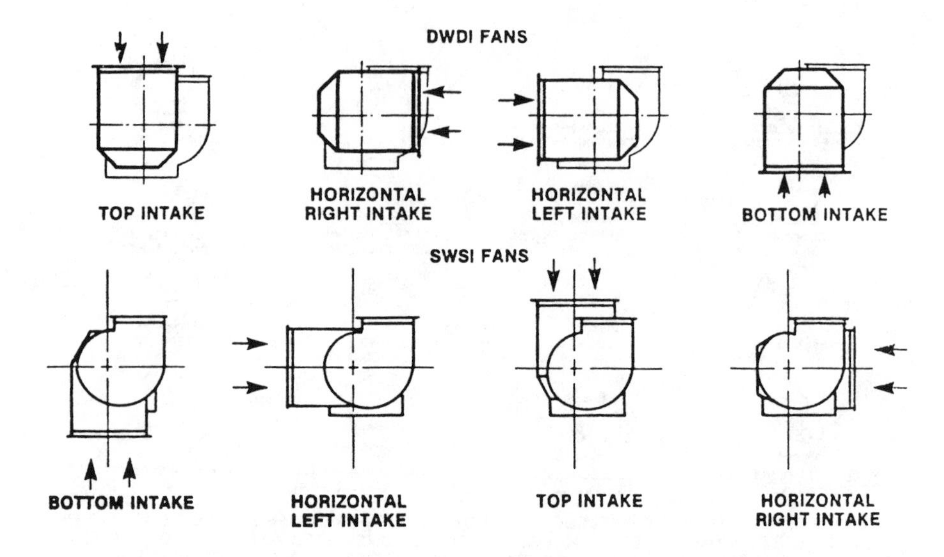

Figure 4-7. Standard inlet box positions. (Courtesy CB/CP Ltd.)

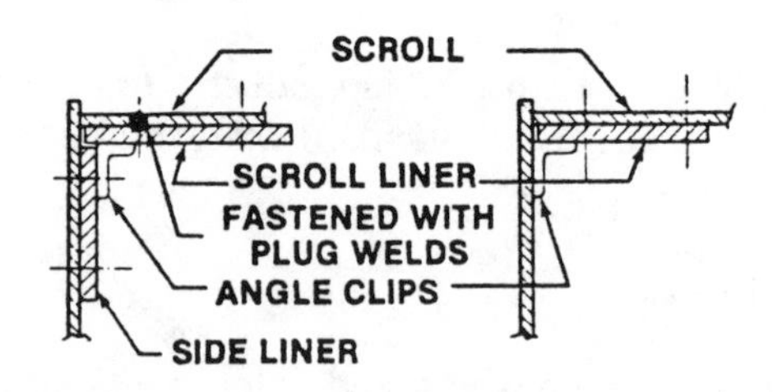

Figure 4-8. Side and scroll liners. (Courtesy CB/CP Ltd.)

Wheels

Heavy duty fan wheels are shipped as a single, assembled unit with all surfaces either painted or coated with rust preventive. Each wheel has been both statically and dynamically balanced. Standard wheel designs and direction of rotation are shown in Figure 4-9.

When specified for abrasive conditions, wear strips constructed of at least 1/4 in. floor plate are shipped as shown in Figure 4-10.

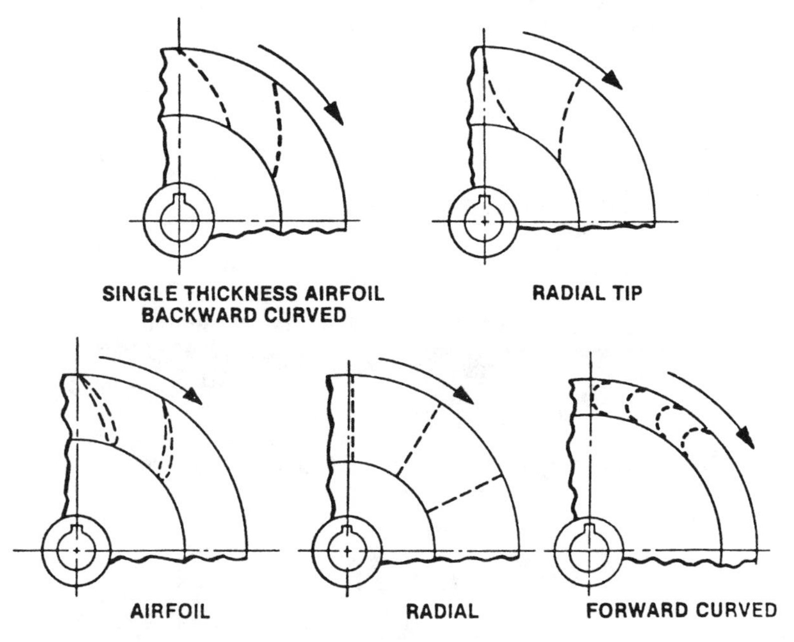

Figure 4-9. Wheel designs and rotations. (Courtesy CB/CP Ltd.)

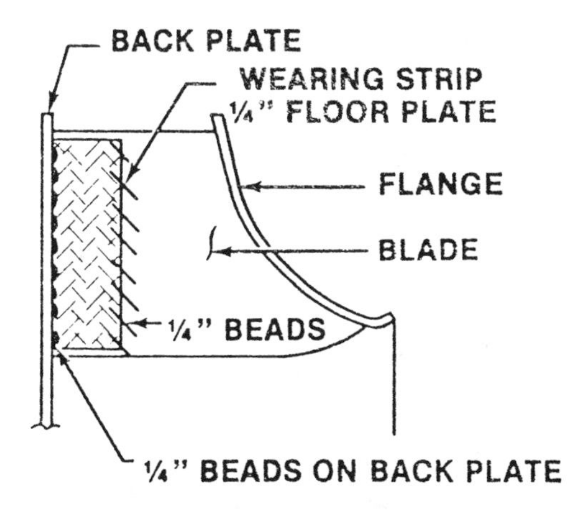

Figure 4-10. Wear strips. (Courtesy CB/CP Ltd.)

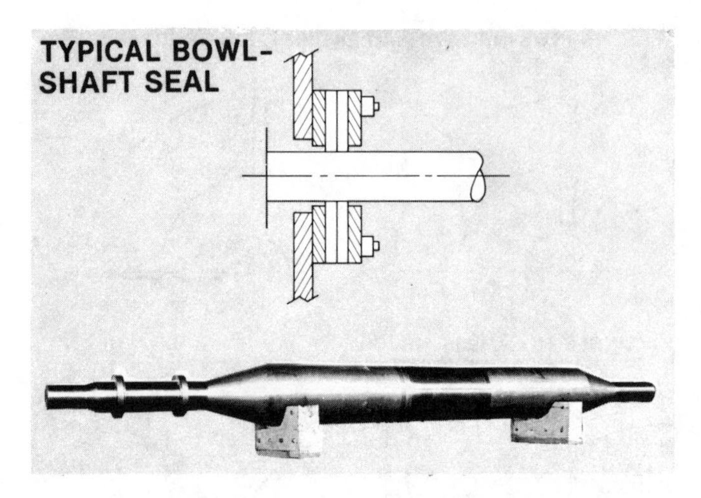

Figure 4-11. Shafts and seals. (Courtesy CB/CP Ltd.)

Shafts and Seals

Shafts for typical heavy duty fans up to about 8 in. diameter are hot rolled steel. Larger shafts are forgings normalized and tempered. Shafts are ground to close tolerances and fitted with keys. Thrust collars may be removable or an integral part of the shaft. Some larger fans for gas recirculation or high-pressure, forced draft service have a shrink fit on the shaft.

When required to minimize gas or air leakage around the shaft, the shaft seals are provided on fan housings and inlet boxes as shown in Figure 4-11. Depending upon how the fan is to be used, shaft seal types include mechanical, pressurized-air for gas recirculation and other specialized seals. If a unit is ordered with any seal type except the mechanical stuffing box style, special instructions are normally provided in the manufacturer's data package.

To reduce air or gas leakage through the fan housing shaft hole, the stuffing box has either three or four rows of packing and a grease-lubricated lantern ring. The stuffing box is packed and the lantern ring greased at the factory before shipment.

Bearings

Bearings are matched to the service conditions and application intended for the fan. Ball, roller, and sleeve bearings are used.

Variable Inlet Vanes (VIV's)

Variable inlet vanes regulate fan capacity using moveable vanes in the fan inlet. A common linkage joins the vanes together so they operate in unison when the control lever is moved. See Figure 4-12.

VIV's are completely assembled in the inlet bell and adjusted at the factory for operation.

To meet a range of conditions, fan manufacturers provide several VIV designs. In one design, the vanes are cantilevered with the operating mechanism inside the fan. They are easily serviced by removing the inlet bell that includes the VIV assembly. Each vane shaft is supported by two self-lubricating bushings separated by a steel sleeve. The area between the steel sleeve and vane shaft is packed with graphite grease to prevent the shaft from seizing. Service conditions for this design are up to 150°F

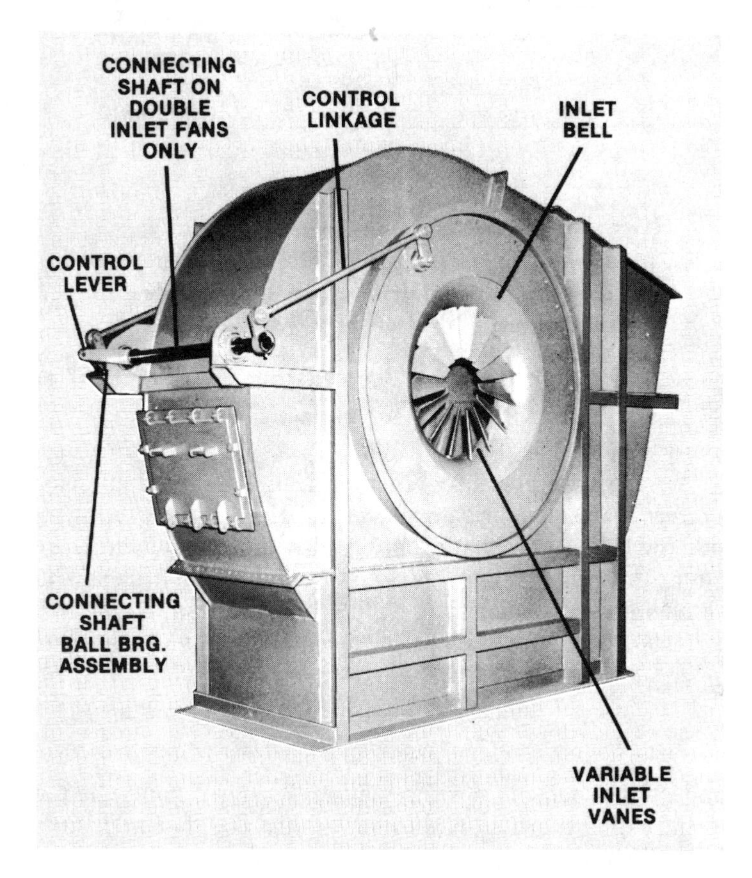

Figure 4-12. Common linkage variable inlet vanes (VIV). (Courtesy CB/CP Ltd.)

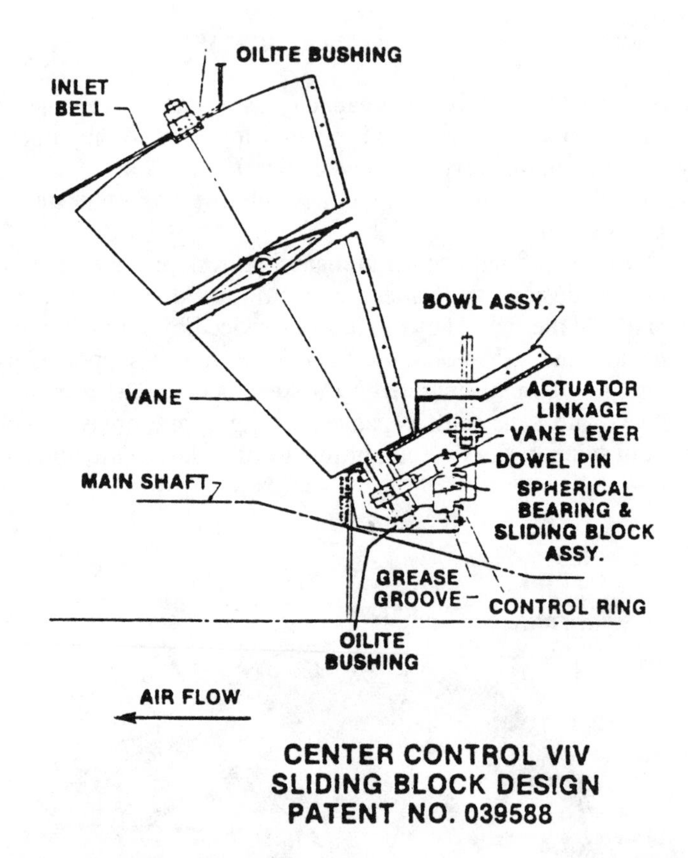

Figure 4-13. Center control variable inlet design vane. (Courtesy CB/CP Ltd.)

and 12 in. water gauge pressure. Another design uses prelubricated ball bearings rather than self-lubricated bushings. Here, the operating mechanism inside the bell is completely shrouded with a dust cover. In yet another design, the center control VIV is used on ID and clean, hot-gas fans. The operating mechanism, a sliding block design, is connected to the vanes outside the air flow next to the fan shaft as shown in Figure 4-13.

Turning Gear

For applications such as gas recirculation, fans may stand idle for long periods while exposed to temperatures as high as 800°F. Under such conditions, a thermal set can result in the shaft, causing extreme vibration upon startup.

To prevent these problems, one manufacturer, for instance, offers a turning gear option that attaches to the outboard end of the fan shaft. When the main drive shuts down, the turning gear engages and slowly rotates the wheel and shaft to distribute the heat evenly. When the unit is restarted, the turning gear automatically disengages.

Assembly and Installation of Fans

This section provides general procedures for installation of heavy-duty fans that are shipped disassembled. The procedure is specifically for a double inlet fan with inlet boxes and sleeve bearings on pedestals and soleplates. Installation of other fan types will vary somewhat. The assembly drawing furnished with the equipment should be checked. To ensure safety during installation, be sure to use only qualified personnel and proper equipment.

Parts Identification

Before starting to erect the fan, familiarize yourself with all the parts. Make use of the assembly drawing and read the installation manual. Each part of the assembly is marked with a factory identification number. Orders for multiple units also indicate a sequence number. For example, two identical fans could be identified as A1 for unit number one, A2 for unit number two, etc. Part must *not* be interchanged between fans.

Fan Shafts

Fan shafts assembled to the wheel at the factory but shipped separately should be stamped with a fit number in the keyway. The factory identification number should also be stamped on the end of the shaft.

Wheels

Wheels assembled to the fan shaft at the factory but shipped separately are generally stamped with a fit number on the wheel hub above the keyway. Double width fans have the keyways marked with the number one and two corresponding to the fan shaft. The fan identification number and fit numbers also are painted on the wheel flange except for unpainted or special material wheels which are usually tagged.

Housing and Boxes

The mating parts of all housings and inlet box sections are matchmarked with the fan identification number, a section number indicated by roman numerals and alignment arrows. All units are provided with a rotation arrow. Either a sticker or metal arrow attached to the housing indicates the direction of rotation.

Accessories

Stationary inlet vanes or variable inlet vanes (VIV's) that are shipped separately should have an arrow indicating rotation of the wheel relative to the vanes.

Note

Inlet vanes and VIV's must be installed correctly in relation to the wheel rotation or fan will not perform. Other items such as coupling guards, belt guards, crossover or connecting linkages for VIV's or inlet box dampers, etc. are tagged or marked with the fan identification number.

Check Foundations

Check that foundation bolts are located as called for on the assembly drawing and that the fan mounting holes will mate. For high temperature operation, provide for fan expansion in the foundation. Determine with a transit and mark the foundation centerline. This centerline is used as a reference throughout the erection period. Include the driver and determine shaft centerline height from the concrete or some other reference point in the foundation or a close-by fixed point.

Set Lower Fan Housing

Lift housing into place. If housing is split, move only the lower half into position. Do not lift the housing with a fork lift or car loader. When moving the housing, be sure to clear the foundation bolts. Lower the housing onto blocks so the base angles clear the foundation bolts. Next, align the base angle holes with the foundation bolts. Lower the housing carefully over the foundation bolts, taking care not to damage the threads. Under the housing put temporary shims appoximately as thick as the grout will be. Shims (about 4 in. wide) should be flat and flush with edges of the base angles. The weight of the unit should be carried by at least one shim on each side of every foundation bolt. Next, attach the inlet boxes to the housing. These may be bolted to the housing through a bar iron welded on the edge of the housing or fastened to locating clips on the housing side with a few bolts and then seal welded to the fan housing. If a special inlet box support leg is required, attach it now but do not pull it down tight to its foundation. During the final alignment procedure, this support leg will be securely fastened.

Set Bearing Base

Next, put the pedestals in place, maintaining approximate correct elevation to the center of the bearing. The held-fixed bearing should be set level. Use flat shims under the soleplates to level. Because the pedestals, tops of soleplates, and bearings are machined to exact tolerances, they will set in position at the correct level if the foundation was prepared to the design drawings. Because a foundation occasionally may vary slightly from the fan base or may settle, some fan erectors add 1/16 in. shims between the pedestal and bearing. These shims are added because dropping the level of the bearing is virtually impossible later if an error is found or a change in center height is required. The shims should have an adequate area to carry the weight of the parts and should be slotted to clear the bearing bolts for easy installation or removal.

Wheel and Shaft Installation

Remove the shaft from the box with a nylon or rope sling. *Do Not Lift by Bearing Journal Surfaces.* As shaft rests in saddles in the shipping box, slings can be placed underneath the shaft easily while the shaft is still in the box. Remove the shaft preservative with solvent and coat the journal area with clean oil using a clean applicator.

Note

Do *not* touch the cleaned journal surface with bare hands since perspiration can cause discoloration and pitting.

On double width wheels, clean the space between the hubs carefully and remove all foreign objects that could fall onto the shaft as the wheel is turned. Clean the bore of the wheel hubs with suitable solvent and lubricate to ease entrance of the shaft.

Turn set screws in to check that they are long enough to hold the wheel to the shaft. Now, turn the set screws out so the shaft will clear. If three set screw holes appear in the hub, use only two—one over the key, and the second leading the key in the direction of rotation. Be sure to support the shaft for lifting with two rope slings, one near the middle to carry the weight with the second sling balancing the shaft. Be sure that the rotation arrow on the housing corresponds to the rotation arrow on the wheel and that the shaft thrust collars are on the drive end. For proper wheel installation, compare blade shapes as shown in Figure 4-9 with wheel rotation. In the case of dual drive, the held-end bearing should be determined before the shaft is placed in the wheel. Place the shaft into the bore of the hub. After the shaft is partially into the wheel, set both ends of the shaft on supports, and turn the wheel to align the keyway. Alignment is simplified if the shaft keyway is on top during alignment. When aligned, push wheel to correct position. Select the key marked "1" or "2" in double wheel and, after coating the key with white lead and oil, tap into keyway as marked. Tighten the wheel set screws only after wheel and shaft are in place on the bearing and the inlet bell adjustment has been made.

Install Inlet Bells or VIV's

Place the inlet bells of fixed or variable vanes over the shaft, making sure that the rotation arrow mark on the inlet part corresponds to the housing and wheel marking. Take care not to damage the journal section of the shaft.

Prepare and Install Bearings

Remove the caps from the bearings and carefully clean the bearing surfaces with solvent. Apply a coating of clean oil with a clean applicator. Some installers pour oil over the surface so that contamination cannot occur. Cover immediately with a clean cloth to keep out contaminants. Inspect and clean the oil rings, then put the shaft seals and oil rings in a safe location so they will not be damaged before they are installed.

Bearing Installation

Depending upon the application, a variety of ball, roller, and sleeve bearings are used. Generalized information for each type follows:

1. If a bearing is disassembled, mark its position in relation to each part to avoid reassembly errors. Do not mix parts of one bearing with another.
2. Determine the type of pillow block and location of fixed bearing.
3. Check all nameplates on fan for any special instructions.
4. Mount bearings in position on the shaft per specific directions that apply to your type of bearing.
5. Clean the shaft and remove burrs or other irregularities. Be sure the bearing is not to be seated on worn flat sections.
6. After final alignment, tighten and dowel whenever possible.

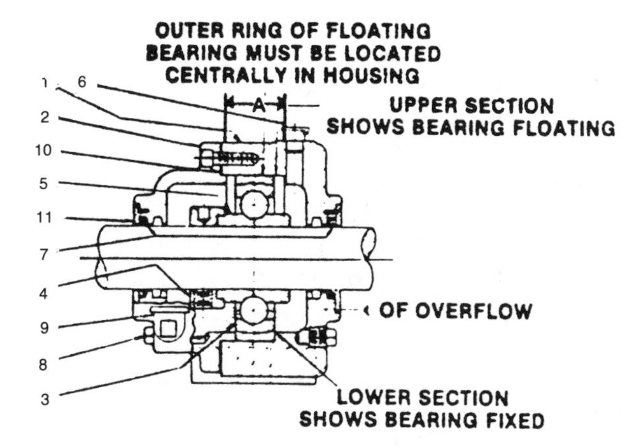

Figure 4-14. Pillow block assembly. (Courtesy CB/CP Ltd.)

Installing Fixed and Floating Pillow Blocks

Pillow blocks are often shipped with the bearings mounted in the housing but with the locking collars separate. To install pillow blocks of this type, refer to Figure 4-14 and:

1. Remove end cover (2), gasket (10) plates (11) with packing (7). Be careful not to damage gasket and packings.
2. Slide pillow block housing (1), bearing (3) and one plate (11) onto shaft. Position bearing on shaft making sure that the cam end of inner ring (5) points out.
3. To position the floating bearing in its housing, measure to determine the length "A" of the pillow block housing. (See Figure 4-14). Place the bearing centerline at the location A/2 so that maximum movement can occur.
4. Bolt pillow blocks securely in position on their mounting surfaces after shimming and aligning. The outside diameter of shaft and housing bore should clear equally all around. Pillow blocks should be mounted so fan wheel and shaft do not strike any part of fan housing.
5. Slide locking collar (4) into position against bearing inner ring (5). Turn collar in direction of shaft rotation until it grips shaft and inner ring. Tighten collar with a drift pin. Tighten set screw in collar.
6. Replace gasket (10), and cover (2), packing (7) and plate (11) on end cover. Bolt on end cover.

7. Draw up screws holding plates just enough so packing rings are retained without undue deformation.
8. Fill with oil in top cup (6) until overflow cup (9) is full. Fill only when fan is not running.
9. Occasionally, bearings can be converted to grease lubrication per bearing manufacturer's instructions.
10. To disassemble, reverse this procedure. *Be sure* to remove burr on shaft (caused by the set screw) with a honing stone *before* removing the pillow block from the shaft.

Set Wheel and Shaft in Bearings

Proceed to install the wheel and shaft in the bearings. Extreme caution must be used as the thrust collars are placed over the held (fixed) liner. Mishandling can damage the bearing metal beyond repair. This is difficult for large bearings because the rotors are very bulky to move with only a few mils tolerance in their final position. The technique we suggest is to lower the rotor to just above the bearings and then lift the held (fixed) bearing liner up around the shaft. As the bearing liner is much lighter than the rotor, it is easier to guide into place between the thrust collars. Next, fasten the bearing liner to the shaft so that the liner can be lowered into the bearing housing. If clearance is restricted, some erectors assemble both bearings to the shaft completely before setting the bearings on the pedestals. If the bearings are completely assembled to the shaft, be sure that the free end bearing is secure so it does not slide off. Set rotor in bearings.

Arrangement 5 and 8 fans feature an overhung wheel that causes the outboard bearing to be top loaded. Use a chainfall to pull the outboard shaft extension so that the shaft seats tightly in the bearing. Do not tighten the plunger screw on sleeve bearings until the shaft is seated. The coupling alignment should follow.

On Arrangement 1 and 3 fans with a bearing on each side of the fan housing, a small deflection of the shaft will occur due to its own weight and that of the wheel. Level the drive-end bearing (held bearing), making sure the fan shaft passes through the center of the housing inlets. The fan shaft extension will then be level with the motor shaft and allow easier coupling or V-belt drive alignment. The outboard bearing is set at a slight angle due to the shaft deflection. The bearing surface must lie properly against the shaft especially if a sleeve bearing is used. On all direct-connected fans, the motor must be on the same horizontal centerline as the fan shaft, except on high temperature applications, where an adjustment is made for expansion. Refer to Figure 4-15.

For a dual drive unit, the driver that will operate longer is set perfectly level on the held bearing end of the shaft and the driver that will operate for shorter durations is placed on the free bearing end of the shaft.

Housing Completion

With the wheel and shaft in place in the bearings, erect the remaining housing sections. Place the gasketing material shipped with the unit at the split sections and assemble. If necessary, use a drift pin (moderately) to align bolt holes. No gasketing is to be used at the intersection of the inlet boxes and housing or at inlet cones. Gasket material is to be used at the split section of the housing only. If a special housing section is to be welded, the matching sections have bolt-through position clips that ensure a correct match and hold the section until seal welding is completed.

Both bolted and welded inlet boxes are used. For welded styles that cannot be shipped attached because of freight limitations, positioning clips on the fan housing orient the box until seal welding is completed.

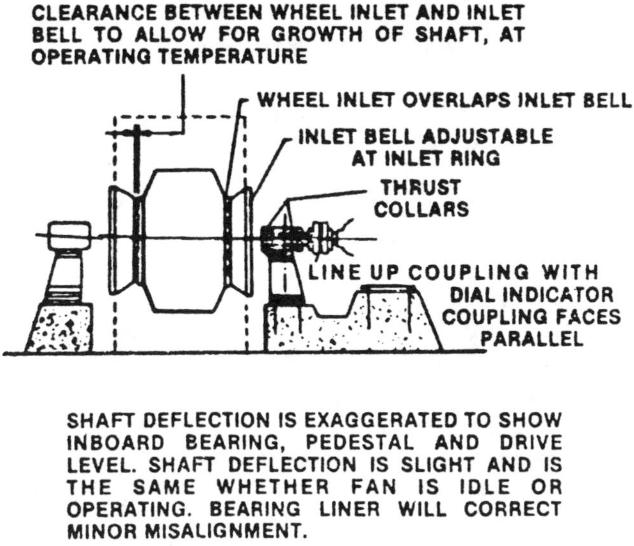

Figure 4-15. Expansion adjustment for high-temperature fan application. (Courtesy CB/CP Ltd.)

Inlet Bells or VIV's

Attach inlet bells to the housing (bolting from the outside of the hous ing). Do not tighten the bolts completely. Adjust the wheels and inlet bells for correct alignment with one another. For included draft service, the wheels should overlap in the inlet bell on the drive side to allow clearance on the opposite side for shaft growth at operating temperatures. For each 100°F increase in temperature, a steel shaft expands .008 in. per ft of shaft length. For example, a 5 ft, 6 in. fan shaft operating at 570°F would lengthen .22 in. (Room temperature at start is 70°, so increase is 500°. Expansion would be .008 in. × 5 × 5.5 in. or .22 in.) Thermal expansion clearances are shown on your assembly drawing. For forced draft service fans, check the assembly drawing for the correct gap on each side. For fans which use clearance fit, standard taper, or taper lock collar hubs, the wheel can be moved on the shaft slightly to align the inlet bells.

Proceed to pull down the foundation bolts on the housing and inlet box support legs. If the foundation is slightly high or low, these support legs can deflect the housing, making alignment of the wheel and inlet bell difficult. Shim if necessary. Double check inlet bell location and tighten inlet bell bolt completely.

Now, tighten the hub set screws. First tighten the set screw over the key, which should be in the six o'clock position before tightening. Next, tighten the set screw that leads rotation.

Stuffing Box Installation

Packing is usually factory-installed for fans that use stuffing boxes. However, stuffing boxes require the following break-in procedure:

1. With a feeler gauge, check that the clearance between the housing stuffing box and fan shaft is uniform. The shaft must be centered in the box.
2. Check for lubricant in the grease cup when a grease ring is supplied. Use a good grade of temperature resistant packing grease.
3. If the stuffing box is water-cooled, connect flexible water line with a valve in the drain return line to regulate flow. Discharge the cooling water into an open funnel so visual inspection will show water flowing. Adjust water flow to conditions.
4. Tighten the gland nuts finger tight. The packing should not be so tight that the shaft cannot be turned by hand.
5. After all erection procedures have been followed, start the fan and run for 15 minutes. If the stuffing box gets too warm or you see smoke, stop the fan and loosen the gland.

6. If the gland cannot be loosened further, take out one row of packing. Replace the gland, finger tightening the bolts. Run the fan for a few hours until you can take up on the gland. Coat the row of packing removed earlier with light oil and then replace.
7. Periodically inspect the stuffing box and replace packing as necessary.

Install Drive

Install the fan before the drive because installation is easier and alignment is simplified. Read the drive manufacturer's instructions before installing the drive.

Installing Couplings

If the fan incorporates gear couplings and if either coupling half has not been mounted on its shaft, the following procedure can be used:

1. Place coupling covers or sleeves over shaft ends.
2. Insert keys.
3. Install hubs on shaft with faces flush with shaft ends. If the hubs do not go on when tapped lightly with a soft lead hammer, expand them by heating not over 300°F in oil or oven.
4. Set the motor on its magnetic center. This is marked on many motors. The magnetic center must be known to properly adjust the clearance between the face of the hubs. If the driver is a sleeve-bearing motor, its magnetic center must be found before aligning the coupling to prevent the motor side of the coupling from moving against the fan coupling. To find the axial movement of the motor shaft:

   a. Run motor and mark a line on the shaft; this is the magnetic center.
   b. Push the shaft as far as it will go into the motor housing. Mark line on shaft at housing. Then pull the shaft out as far as possible and scribe another line. Half the distance between the two marks is the mechanical center.

Notes: (1) Flexible disc couplings may be available. Install per manufacturer's instructions.

(2) Finding the magnetic center is not necessary on ball bearing motors as their thrust bearings prevent movement. Axially soft couplings must be used on dual drive units where there is ther-

mal expansion of the shaft. If motor incorporates sleave bearings use a limited end float coupling to restrict movement.

5. With motor rotor on its magnetic center, locate motor on its base with coupling faces at the proper axial clearance between faces of hubs as shown on the fan assembly sketch in Figure 4-16.

Install Dampers and Connect Ductwork

Attach the outlet dampers or the inlet box dampers using suitable gasket material. Use drift pins for positioning only and not to force the damper into place. These parts will fit as they have been completely assembled at the factory unless they have been damaged in erection or shipping.

If an inlet box damper control shaft is used, it is shipped in a separate box with the dampers mounted and with the levers pinned in place on shafts. Mount the entire jack shaft and connect the individual dampers. See your fan assembly drawing for positioning.

It is recommended that ductwork be supported independently of the fan. Imposition of heavy duct loads can result in distortion of the fan casing and possible rubbing contact between housing and rotor.

Variable Inlet Vane Control Mechanisms

For single wheel units, the variable inlet vane controls mount on the side of the fan adjacent to the inlet unless ordered with a jack shaft and bearings for remote linkage.

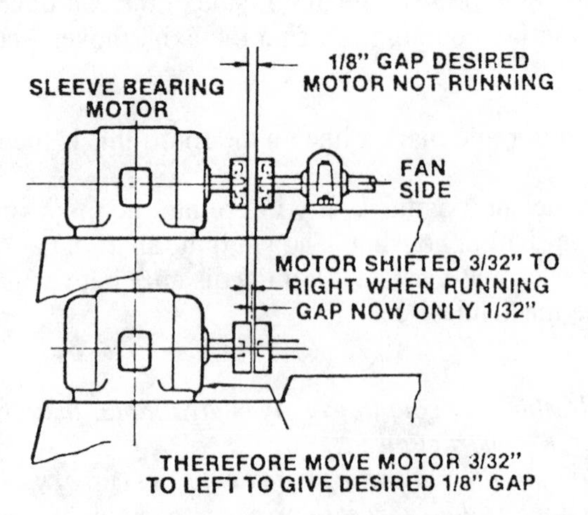

Figure 4-16. Determining coupling axial clearance. (Courtesy CB/CP Ltd.)

For double-width fans, the linkage from the vanes on each side is connected to the jack shaft that mounts on the fan housing. On double-width fans, if the jack shaft levers connecting each set of vanes do not line up with the connecting rods, the lever has been moved to protect it during shipment. Remove the pin from the jack shaft, slide levers back into position, carefully lining up the pin holes in levers, and replace the pin. Attach the connecting rods from individual mechanisms.

Expansion Ducts and Joints

If the fan being erected is to operate at high temperatures, expansion joints are absolutely necessary at the fan connections. They allow for expansion of the fan housing and connecting ductwork without distorting each other. Remove any shipping braces from the expansion joints before operating the fan and provide for fan expansion in setting the housing foundation. Do not use drift pins, "come-alongs," or any other means to force connections of ducts, fan housings, or inlet boxes.

Circulating Oil Lubrication Systems

If you specified a circulating oil lubrication system, detailed drawings, and erection, startup, operation and maintenance instructions should be included in the shipping data package.

During erection of the oil piping, be sure that the system is free from dirt, grit, weld spatter, or shavings. The piping system must be thoroughly cleaned and flushed before connecting to the bearings. Clean the filters before initial start-up.

After the selection of which pump is to be used (if a two-motor/pump system), the motor is turned on to activate the pump. Oil passes from the tank through the filter element, pump, pressure relief valve, oil cooler, sleeve bearings, and then by gravity back to the reservoir tank.

The circulating oil system must be operating before the fan can be started. Depending on the system specified, any malfunction in the system, low oil pressure, high temperature, etc., may either sound an alarm or shut down the fan. The oil system must operate 30 minutes after the fan is shut off or until the heat in the bearings has dissipated to an acceptable level.

Temperature Detectors

Dial thermometers screw into a tapped hole in the housing so that the oil sump temperature of bearings can be measured. See Figure 4-17.

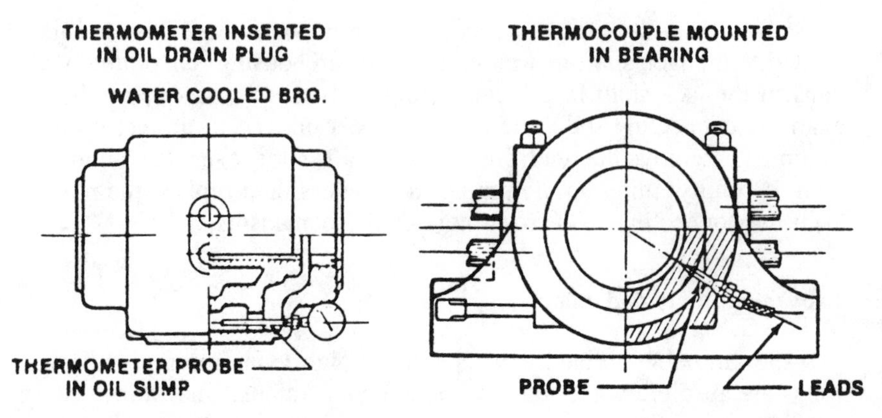

Figure 4-17. Dial thermometer installation.

Figure 4-18. Thermocouple installation.

Both thermocouples and electrical resistance detectors are mounted by inserting the end of a probe through a tapped hole in the pillow block that reaches into the liner.

Many major blower manufacturers furnish thermocouples from several vendors (see Figure 4-18). If the probe does not fit easily, do not force it. The bearing may have been assembled to the drilled liner on the wrong side. Probe leads are wired at the job site to the alarm or visual indicator. No electrical input is needed because the bimetallic strip in the probe generates current to trigger the metering system.

Electrical resistance temperature detectors do require an electrical input. Assemble and wire per the manufacturer's instructions.

Heat Slingers

Assemble heat slingers by placing both halves over the shaft between the fan-side pillow block and fan housing or shaft seal. Bolt together. For fans with oil-lubricated, anti-friction, or sleeve bearings, the inlet should face the fan housing. The inlet should face the bearing if for grease-lubricated. Your assembly drawing should show their proper location.

After assembling, check that the heat slinger is tight on the shaft to prevent it from rotating. If housing or inlet boxes are to be insulated and a heat slinger is to be used, see Figure 4-19.

Insulation Clips

When specified, insulation clips are furnished on fan housings and inlet boxes on 12 or 18 in. centers and the bracing drilled for insulation wires. Usually studs with holes drilled in the end are furnished for ease

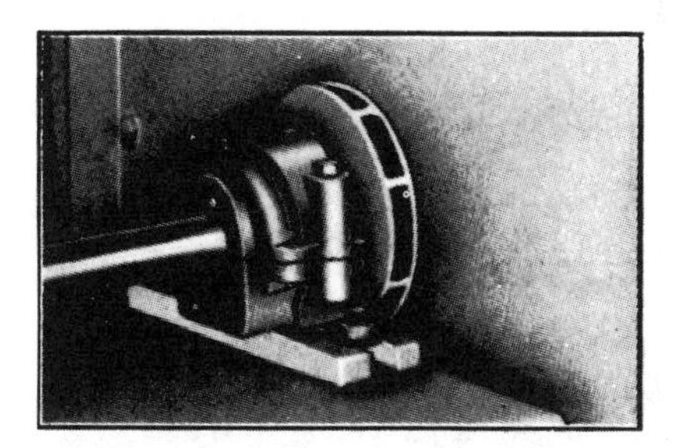

Figure 4-19. Heat slinger installation. (Courtesy CB/CP Ltd.)

of installation. Wire is placed through the insulation clip (the normal length of wire is about 24 in.) and bent together. Holes are punched in block-type insulation and the insulation placed over the extended wires. Ends of the wires are bent to hold the insulation against the fan.

Consider future disassembly when applying insulation. Do not run blocks across the housing splits. In addition to preventing heat loss, insulation protects operating personnel against burns.

When specified, insulated access doors are provided that have outside surfaces flush with the applied insulation.

Startup

Before startup, perform the checklist procedures outlined in the applicable operating and maintenance manual. Compare with other startup checklists given in Chapter 1 of this Volume and other applicable checklists in Volume 3 of this series.

Preventive Maintenance

To ensure trouble-free operation and long life expectancy, a schedule of preventive maintenance and lubrication must be set up. Frequency of inspection and lubrication depends upon operating condition and the amount of fan use. Daily inspections are recommended after the fan is first erected. Recommended periodic inspection procedures are listed in Table 4-1.

Fan Balancing

All heavy duty fans and blower wheels are balanced both statically and dynamically at the factory (Figure 4-20). A very elementary method of checking rotor balance is shown in Figure 4-20. If the wheels have not been damaged or repaired, no additional balancing should be required.

Table 4-1
Fan Preventive Maintenance

Fan Component	Check For
Air Flow	Obstructions, dirt, rags, etc. in inlet or outlet duct work. Bird and protective screens and louvres must be cleaned.
Housings, wheel and shaft	All bolts tight? Wheel clean? Dirt can unbalance a wheel. Cover bearings tightly with plastic film and clean with steam, water jet, compressed air or wire brush. Cracks in wheel? Fan must be put out of service until proper repairs are made. Badly worn wheel blades, wear strips or blade liners? In most cases, eroded areas can be repaired by welding. Contact fan manufacturer for the correct weld procedure for your wheel. Be sure to electrically ground the wheel before welding to avoid damaging the bearings. Be careful not to contaminate welds from wheel coatings or protective overlays. Repair all structural welds with rod that meets original specifications. Grind and repair all cracks. **Caution** After welding, balance should be checked. If it is necessary to disassemble the wheel hub from the shaft, see applicable section in operating manual.
Circulating oil lubricant systems	Filter clean? Reservoir level adequate?
Alignment of fan bearings, flexible couplings, wheel and inlet bells	Check alignment of fan bearings, flexible couplings, wheel and inlet bells regularly. Misalignment causes bearing or motor overheating, wear to bearing dust seals, bearing failure and unbalance.
V-belt drives	Check belt wear, alignment of sheaves and belt tension. Replace belts with a complete set of matched belts, as new belts will not work properly with used belts because of length differences. Belts must be free of grease.
Dampers and VIV's	All linkage connections tight? Check all automatic dampers for freedom of movement. Blades should close tightly in closed position. Make adjustments as required. Observe operating motors and controls through a cycle. Clean dampers and VIV's and inspect for corrosion and erosion.
Surface Coatings	Surface coatings or paints in good condition? Repainting of exterior and interior parts of fans and ducts extend the service life. Select paints to withstand operating temperatures. For normal

Table 4-1. Continued

Fan Component	Check For
	temperature, a good machinery paint may be used. If moisture is excessive, or if fans are exposed to the weather, bitumastic paints are available. Corrosive fumes require all internal parts to be wire brushed, scraped clean and repainted with an acid-resistant paint. Seek competent advice when corrosive fumes are present.
Scroll and Housing Liners	Worn? Replace because damaged liners can break free and severely damage the wheel. Liners are either bolted or welded to the housing.
Bearings	Excessive temperature or chatter? High-speed fan bearings are designed to run hot (100°F to 200°F). Do not replace a bearing simply because it feels hot. Check the pillow block temperature with a pyrometer or contact thermometer. Ball or roller bearing pillow blocks that are operating normally can have surface temperatures of 200°F. Ring-oiled sleeve bearings operate up to 170°F (oil film temperature) before the cause of high temperature need be investigated. For water-cooled bearings, check that exit cooling water temperature is about 100°F (unless your system is designed for higher water temperature). If the water temperature is too cool, condensation in the oil is a possibility. If roller or ball bearings are to be removed, follow the procedure in applicable section of operating manual.
Flexible Couplings	Lubricate all metal couplings (Falk Steelflex, Fast gear and similar types). Other types of flexible couplings such as the Thomas disc, and rubber insert styles such as T. B. Wood and Poole, do not need lubrication but must be inspected for pin and bushing wear.
Motors	Blow out open motor windings with low-pressure air to remove dust or dirt. Air pressures above 50 psi can cause insulation damage and blow dirt under loosened tape. Dust can cause excessive insulation temperatures. Check load motor against amperage rating on manufacturer's nameplate. Keep motors dry. Lubricate. When motors are idle for a long time, single-phase heating or small space heaters might be necessary to prevent water condensation in the windings. Excessive starting of large motors may burn out the motor. Consult the manufacturer for maximum allowable number of starts per hour.

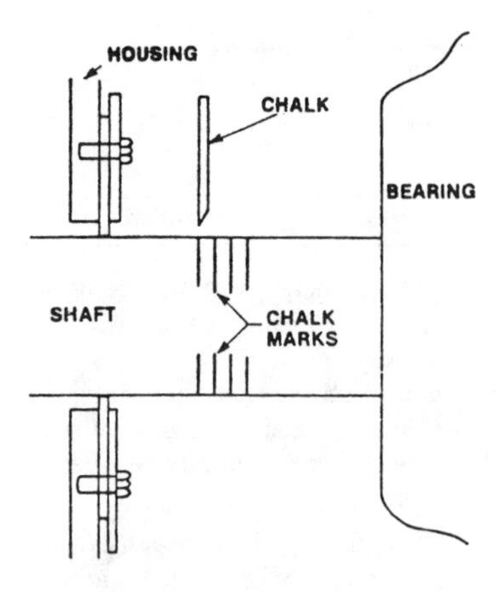

Figure 4-20. Chalking fan shaft to determine balance.

Before balancing a wheel for any reason, check the troubleshooting procedures in the operating and maintenance manual and also in Volume 2 of this series.

Portable instruments are available that indicate vibration displacement in mils (1 mil = .001 in.) or microns in metric system (1 micron = 1×10^6 mm). Use the manufacturer's manual or generalized data from Volume 2 to determine when a fan is operating with too much vibration. Table 4-2 illustrates vibration guidelines as a function of fan speed. Note that vibration velocities give constant parameter independent of shaft speed, whereas allowable displacements vary with speed.

Wheels can be balanced by using methods described in Volume 3 of this series.

Wheel Hub Disassembly

Depending upon the application for which the fan is intended and the operating speed and temperature, major blower manufacturers employ at least six different methods of hub-shaft attachment.

If the fan wheel must be disassembled from the shaft for service, obtain information on the type of hub-to-shaft fit and proceed as follows:

1. Standard Clearance Slip Fit—Used for most moderate applications. Wheel and shaft are fitted together in the factory, but shipped apart for field assembly unless otherwise specified. Set screws hold the hub to the shaft.
2. Close Tolerance Fit—While not an interference fit, this design requires assembly at the factory because the very tight clearance between bore and shaft requires hydraulic jacks for mounting. To ease assembly, the hub bore and shaft have stepped diameters with the

large diameter usually on the drive side. (See on assembly drawing.) If disassembly is required, observe special procedures available from the manufacturer to prevent damage to the mating surfaces.

3. Standard Taper—Fans with overhung wheels (Arrangements 5 through 8) often are equipped with a tapered bore which fits into a matching taper of the shaft. Disassembly and assembly in the field is practical, but before assembly a very thin coat of antiseize compound must be evenly applied to the bore to prevent galling.
4. Taper Lock Collar—Disassembly and assembly of these collars in the field is easy. Torquing and lubrication instructions for reassembly should be found on a separate drawing furnished by the manufacturer of taper lock collars.
5. Shrink Fit—Wheel/shaft assemblies are shipped as one unit. Separation is not usually possible, but could be accomplished with the help of the manufacturer.
6. Rapid Temperature Change Assembly—This design is always assembled at the factory. No field work should be done without contacting the original manufacturer.

Cleaning Fan Bearings

1. When roller or ball bearings are disassembled for service, the following procedure is recommended: Remove bearing races from

Table 4-2
Allowable Vibration at Fan Operating Frequency
(Courtesy CB/CP Ltd.)

RPM	INITIAL OPERATION DISPL. (MILS) P-P	INITIAL OPERATION VEL. (IPS) PK	ALARM DISPL. (MILS) P-P	ALARM VEL. (IPS) PK	SHUT DOWN OR STOP DISPL. (MILS) P-P	SHUT DOWN OR STOP VEL. (IPS) PK
3600	0.5	.1	1.6	.3	2.4	.45
1800	1.1	.1	3.2	.3	4.8	.45
1200	1.6	.1	4.8	.3	7.2	.45
900	2.1	.1	6.4	.3	9.4	.45
720	2.7	.1	8.0	.3	12.0	.45
600	3.2	.1	9.5	.3	14.3	.45

NOTES:

1) DISPLACEMENT IS A MEASURED VALUE WITH THE PROBE OR PICKUP (SEISMIC TYPE) POSITIONED FIRMLY AGAINST THE FAN BEARING HOUSING IN THE DESIRED PLANE OF MEASUREMENT. DISPLACEMENT IS A PEAK-TO-PEAK (FULL WAVE) VALUE.

2) INITIAL OPERATION VALUES ARE THE EXPECTED VALUES FOR CLEAN, WELL-MAINTAINED AND BALANCED FANS OPERATING AT STEADY STATE CONDITIONS AFTER THE TRANSIENT CONDITIONS OF START UP, E.G. ACCELERATION, TEMPERATURE CHANGES, ETC., HAVE PASSED. THE VALUES (FOR MEASUREMENT OF EQUIPMENT UNBALANCE) MUST BE TAKEN FOR THE EXACT FAN OPERATING FREQUENCY FILTERING OUT EXTRANEOUS VALUES THAT CAN BE MEASURED FOR DIFFERENT FREQUENCIES. THESE VALUES MAY NOT BE OBTAINED ON NEW, INITIAL INSTALLATIONS DUE TO FIELD CONDITIONS BEYOND THE CONTROL OF CB/CP LTD.

3) OPERATION OF ANY FAN ABOVE ALARM LEVELS FOR A PROLONGED PERIOD OF TIME OR OPERATION ABOVE SHUT DOWN FOR ANY PERIOD OF TIME, MAY CAUSE EQUIPMENT FAILURE AND EXTENSIVE DAMAGE, AS WELL AS ENDANGERMENT TO PERSONNEL. A CORRECTIVE MEASURE WOULD BE TO RETAIN AN AUTHORIZED CB/CP LTD. SERVICE REPRESENTATIVE (AT SERVICES RATES IN EFFECT AT TIME OF VISIT) TO INSPECT THE INSTALLATION, SUGGEST CORRECTIVE MEASURES AS NECESSARY AND BALANCE THE FAN IF REQUIRED. EFFORT SHOULD BE MADE TO MAINTAIN VIBRATION LEVELS AS CLOSE TO THE INITIAL OPERATION VIBRATION LEVELS AS POSSIBLE. THIS WILL HELP TO ASSURE OPTIMUM EQUIPMENT LIFE EXPECTANCY.

4) CB/CP LTD. CANNOT CONTROL INSTALLATION VARIABLES, SUCH AS (BUT NOT LIMITED TO) FOUNDATION AND MOUNTING PROVISIONS. HOWEVER, CB/CP LTD. GUARANTEES THE FAN ROTOR RESIDUAL UNBALANCE WILL NOT EXCEED QUALITY GRADE G6.3 OF ANSI S2.19.

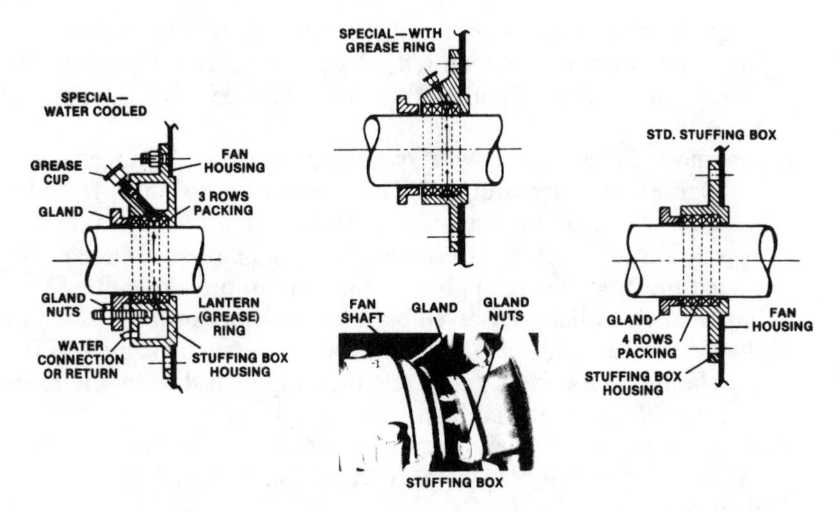

Figure 4-21. Repacking fan stuffing boxes. (Courtesy CB/CP Ltd.)

shafts, place in a suitable container with a clean petroleum solvent or kerosene and soak. Slowly and carefully rotate each bearing by hand to help dislodge any dirt particles.

2. Remove all old grease and oil from the housing and clean the housing with white kerosene or other suitable solvent. Carefully wipe all parts dry with a clean cloth to prevent dilution of the new lubricant by solvent.
3. When bearing grease is badly oxidized, soak in light oil (SAE 10 motor oil) at 200–240°F before cleaning as discussed in the prior steps. Flush the clean bearing in light oil to remove any solvent.
4. Reassemble and add lubricant to the proper level.

Repacking Stuffing Boxes

1. Clean out all old packing including that below the lantern ring. Flexible hooks are available for packing removal. See Figure 4-21 for typical stuffing box details.
2. Remove any nicks or score marks found on the shaft with emery cloth.
3. Clean the stuffing box housing, and the channels and holes of the lantern ring.
4. Cut new packing rings to length so that the ends meet but do not overlap.
5. Immerse the entire ring in oil before installing. Start by installing one end of the ring and bring it around the shaft until it is completely inserted in the stuffing box.

6. Use a split bushing to push the packing to the bottom of the box. Seat the ring firmly by replacing the gland and taking up on the bushing. Seat this bottom ring hard, because this first ring does most of the sealing.
7. Repeat with each packing ring, making sure to stagger the joints 90° apart.
8. If a lantern ring is used, position it properly under the grease or purge hole(s) in the box.
9. After the last ring is installed, position the gland and finger tighten the bolts.
10. Break the packing in (refer to Volume 3 of this series).

Checking Variable Inlet Vane Position

Variable inlet vanes (VIV's) are usually factory set to the maximum vane opening of approximately 75° (see Figure 4-22) except for heavy duty industrial airfoil design, which generally open a full 90°. Changing the factory-installed stops can result in under-design conditions or in the vanes ramming one another. To check the maximum open position:

1. Loosen the wing nut that locks the control lever to the quadrant and open variable inlet vane as far as possible.
2. Lay a straightedge or long bar across the inlet.
3. At right angles to this bar, extend another straightedge across the flat of one of the vanes.
4. Measure the angle between the straightedge and the fan shaft centerline.
5. Our sketch shows a counter-clockwise rotating fan. For a clockwise fan, the 15° will be to the left of the centerline instead of the right.
6. Maximum open angle of the vanes must be 75° or as specified by the manufacturer.

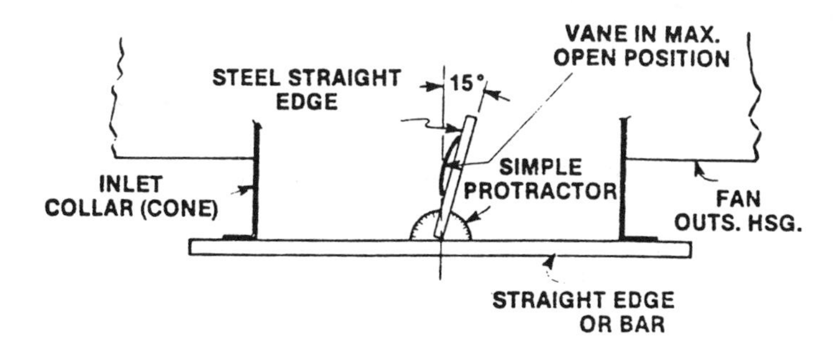

Figure 4-22. Variable inlet vane angle setting. (Courtesy CB/CP Ltd.)

Chapter 5

Mixer and Agitator Maintenance*

Mixing and agitation are universal in the petrochemical process industry. There are very few performance criteria for any given agitator and it seems that no one cares how a mixer works, as long as it works. Mean time between failures is usually long, but our experience has been that because of this, plant maintenance crews are unfamiliar with mixer and agitator details, and frequently another failure will occur shortly after a repair. Detailed repair instructions seem therefore in order. Consider the case of a top entering turbine-type mixer in the 150 horsepower range. Figure 5-1 illustrates this machine while Figure 5-2 depicts the very similar side-entering mixer.

Upper Shaft Removal

1. Remove cover (121), gasket (109) and all bearing member cover plates (Figure 5-3).
2. Remove impellers and disconnect any detachable lower shafts from the shaft extending through the bearings.
3. Insert an eyebolt into the end of the upper shaft (117) for support during removal procedure.
4. Loosen the lower bearing (See Figure 5-4):
   a. Bend the locknut washer tang free of slot so that locknut can be turned.
   b. Loosen and back off the locknut (126) approximately 1/4 in.

* Courtesy Mixing Equipment Co., Inc., Rochester, N.Y.

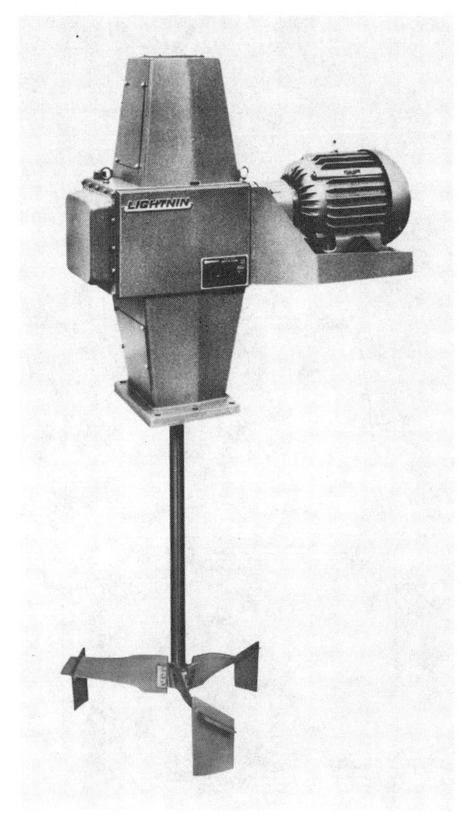

Figure 5-1. Top entering turbine-type mixer for large tanks. (Courtesy Mixing Equipment Co. Inc.)

Figure 5-2. Side entering propeller mixer. (Courtesy Mixing Equipment Co. Inc.)

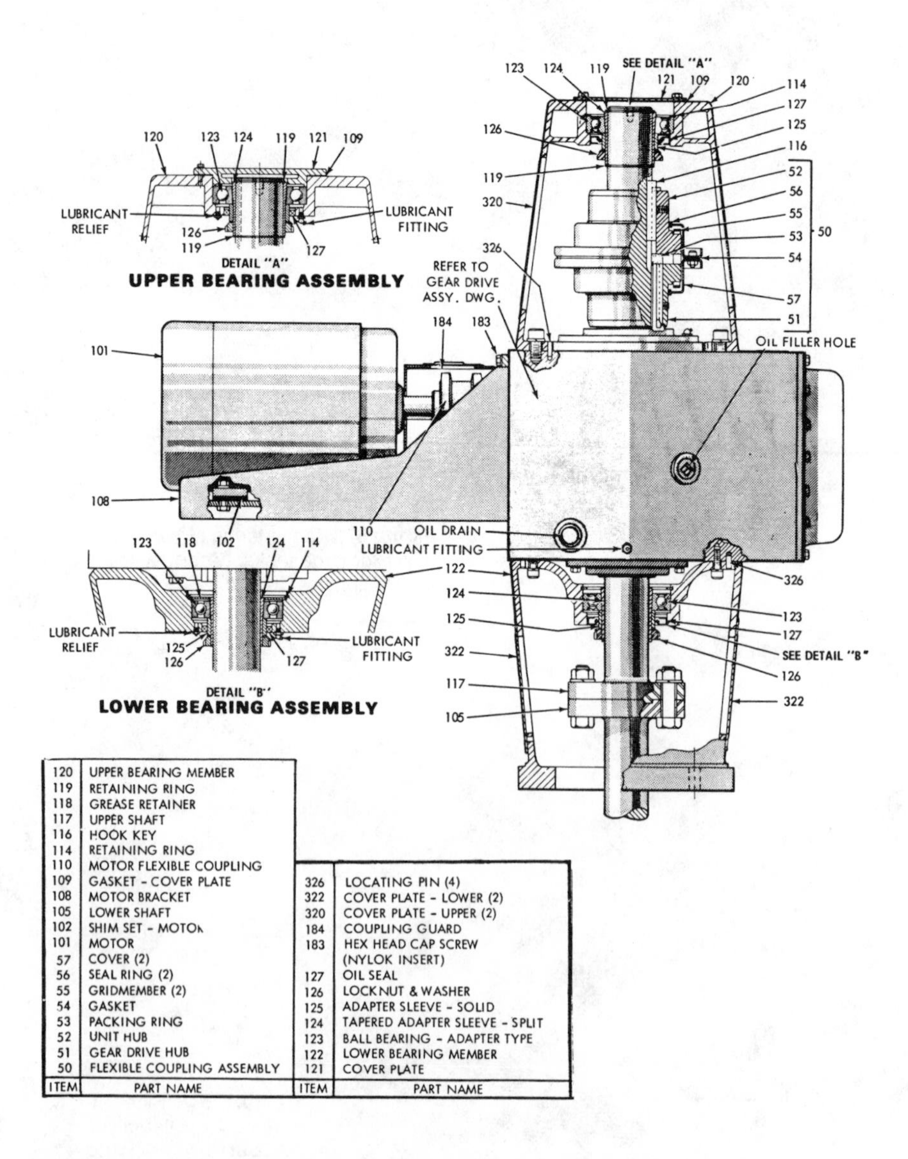

ITEM	PART NAME
120	UPPER BEARING MEMBER
119	RETAINING RING
118	GREASE RETAINER
117	UPPER SHAFT
116	HOOK KEY
114	RETAINING RING
110	MOTOR FLEXIBLE COUPLING
109	GASKET - COVER PLATE
108	MOTOR BRACKET
105	LOWER SHAFT
102	SHIM SET - MOTOR
101	MOTOR
57	COVER (2)
56	SEAL RING (2)
55	GRIDMEMBER (2)
54	GASKET
53	PACKING RING
52	UNIT HUB
51	GEAR DRIVE HUB
50	FLEXIBLE COUPLING ASSEMBLY

ITEM	PART NAME
326	LOCATING PIN (4)
322	COVER PLATE - LOWER (2)
320	COVER PLATE - UPPER (2)
184	COUPLING GUARD
183	HEX HEAD CAP SCREW (NYLOK INSERT)
127	OIL SEAL
126	LOCKNUT & WASHER
125	ADAPTER SLEEVE - SOLID
124	TAPERED ADAPTER SLEEVE - SPLIT
123	BALL BEARING - ADAPTER TYPE
122	LOWER BEARING MEMBER
121	COVER PLATE

Figure 5-3. Mixer assembly drawing and parts list. (Courtesy Mixing Equipment Co. Inc.)

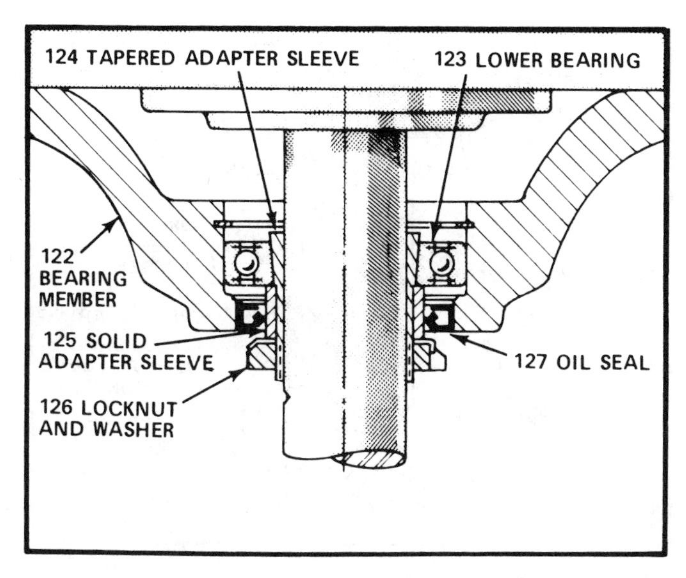

Figure 5-4. Typical lower (floating) bearing assembly. (Courtesy Mixing Equipment Co. Inc.)

c. Using a suitable device such as a brass bar, strike the loosened locknut firmly upward until the tapered adapter sleeve (124) breaks free from the bearing taper.

5. Loosen the fixed upper bearing (Figure 5-5):

a. Remove the upper retaining ring (119).
b. Bend the locknut washer tang free of slot so that locknut can be turned.
c. Loosen the back off the locknut (126) approximately 1/4 in.
d. Using a suitable device such as a brass bar, strike the loosened locknut firmly upward until the tapered adapter sleeve (124) breaks free from the bearing taper.

6. Loosen the set screws in the flexible coupling hub (52).
7. Remove the lower retaining ring (119).
8. Some units use a split ring in the bottom of the stuffing box as a packing retainer. This ring must be removed along with the stuffing box components before the upper end of the shaft will pass through the stuffing box.

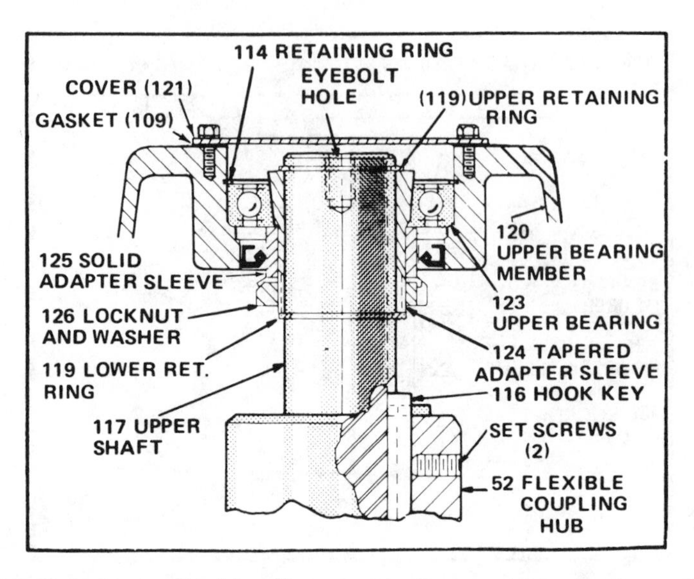

Figure 5-5. Typical upper (fixed) bearing assembly. (Courtesy Mixing Equipment Co. Inc.)

9. Lower the shaft out of the unit by working the one-piece lower retaining ring (119) over the end of the shaft while lowering.

Upper Shaft Installation

1. Before installing shaft (117):

   a. Inspect all oil seals for damage and replace if necessary. Coat oil seal lips with grease.
   b. If the bearings have been removed from the unit for inspection and/or replacement, thoroughly clean the bearings, components, and housing cavities.

2. Insert the shaft up through the unit assembly and while doing so:

   a. Hold the hook key (116) in place with the projecting pin resting on hub (52) while raising the shaft through the drive assembly.
   b. Place the lower retaining ring (119) on the shaft (117) before the shaft enters the upper bearing assembly. Position the ring just below the lower shaft groove.

3. Raise the shaft until the upper retaining ring groove is above the top of the bearing member (120) and install the upper retaining ring (119).
4. Lower the shaft until the upper retaining ring seats firmly against the bearing adapter sleeve (124).
5. Install the lower retaining ring (119) in the shaft groove.
6. Tighten the bearings and complete assembly of the unit.

Tightening Ball Bearings

1. Tighten the upper bearing assembly locknut (126) and set the lock-washer tang.
2. Tighten the set screws in the unit hub (52).
3. Tighten the lower bearing assembly: It is important that the lower bearing (123) is properly located midway in the housing to allow positive float axially in either direction.

   a. Force the entire bearing assembly to the bottom of the housing by applying an axial load against the top of the locknut (126) with a pry bar or other suitable device.
   b. Tighten the locknut (126) handtight.
   c. Mark a reference line on the shaft locating the bottom face of the tapered adapter sleeve (124).
   d. Tap the bearing assembly upward 3/16 in. above the reference line.
   e. Tighten the locknut (126) securely and set the lockwasher. Make sure the adapter sleeve does not move while the locknut is being tightened.
4. Replace the gasket (109) and coverplate (121).
5. Lubricate the flexible coupling (50). (See Figure 5-3).
6. On models with 4 1/4 in. diameter shafts, lubricate the upper and lower open type bearings per the bearing lubrication section. Models with smaller shafts have prepacked sealed bearings.
7. Replace all bearing member cover plates and safety covers.

Mixer Gear Drive Maintenance*

Usually these gear drives (Figures 5-6 and 5-6A) are precision manufactured and assembled to provide trouble-free service. If it becomes necessary to disassemble these units, careful and precise reassembly is necessary.

---

* Lightnin® 80 Series Model

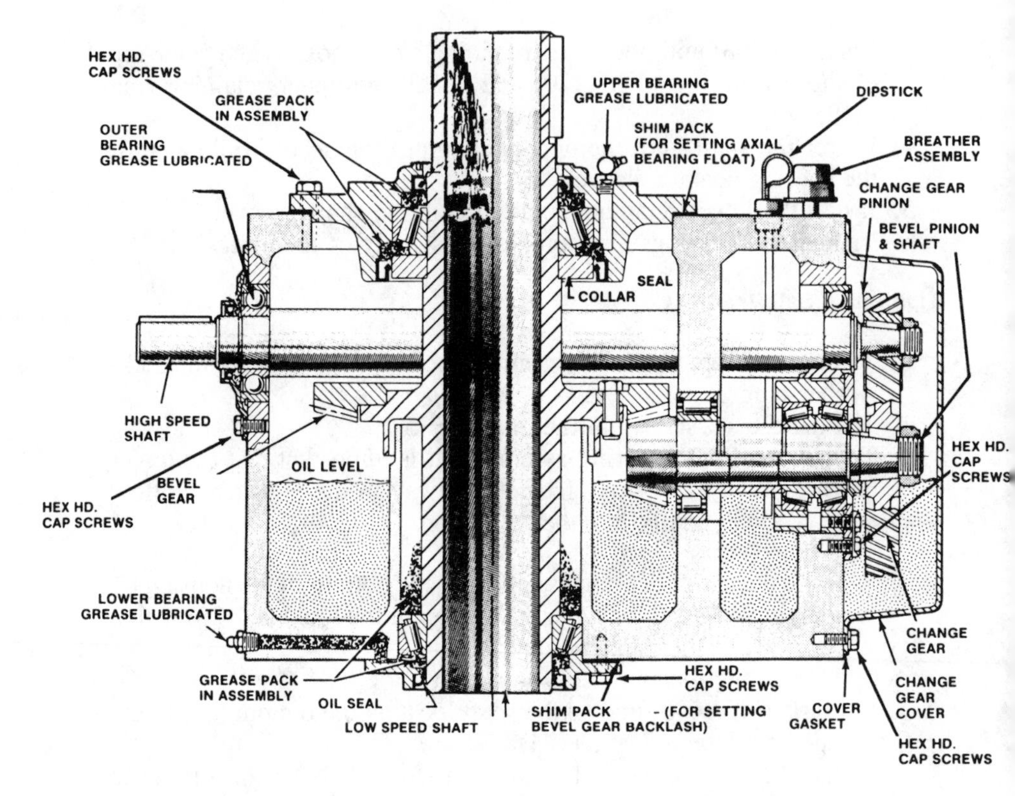

Figure 5-6. Typical double reduction gear drive for top entering mixer. (Courtesy Mixing Equipment Co. Inc.)

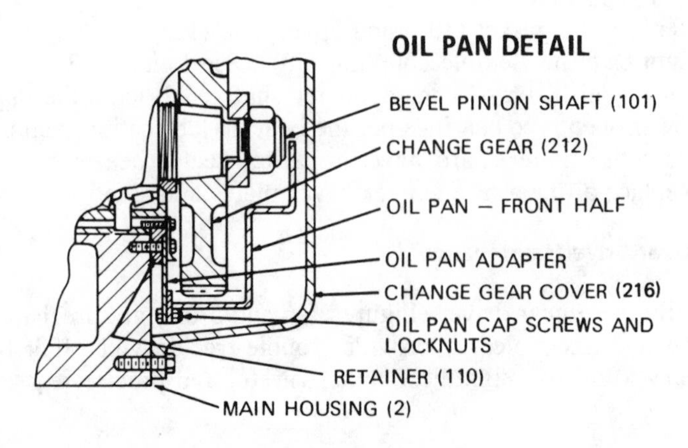

Figure 5-6A. Oil pan detail—double reduction mixer gear drive. (Courtesy Mixing Equipment Co. Inc.)

Equipment that may be required to service a unit, in addition to standard mechanics tools, are hoists, slings, arbor press, wheel pullers, torque wrench, feeler gauges, dial indicator and micrometers.

When disassembling a unit, clean external surfaces adjacent to covers to prevent dirt from entering the housing.

During disassembly keep old shim-pack sets with their respective cages and retainers for reference when reassembling.

It is recommended that oil seals, O-rings, and nonmetallic gaskets be replaced when units are reassembled.

Before dismantling the unit, drain the oil. To speed up drainage, remove the dipstick.

Seal Replacement

We recommend that oil seals and O-rings be replaced when the unit is dismantled (Figure 5-6). Carefully check for nicks, gouges and deformities if they are not being replaced. Drive out old oil seals and remove accumulations of sealing compound. Reinstall oil seals as outlined later in the section on assembly.

Bearing Replacement

Inspect bearings carefully and replace if necessary (Figure 5-6).

1. Remove worn bearings with a puller or at an arbor press.
2. Preheat bearings before pressing on shafts to make installation easier. Maximum oil bath or oven temperatures for heating bearings are:

   - Roller bearings (taper and cylindrical) .... 275°F
   - Ball bearings .... 200°F

When preheating bearings, do not apply direct flame or rest bearings on the bottom or near walls of ovens.

1. Thoroughly coat bearing surfaces and shaft seats with oil.
2. Make sure bearings are tightly seated against shaft shoulders with no clearance. Check with feeler gauges.

Change Gear Replacement

The majority of change gears are taper bored (Figure 5-7). The pinion (210) and change gear (212) can be removed with wheel pullers, wedges or a pry bar and brass hammer.

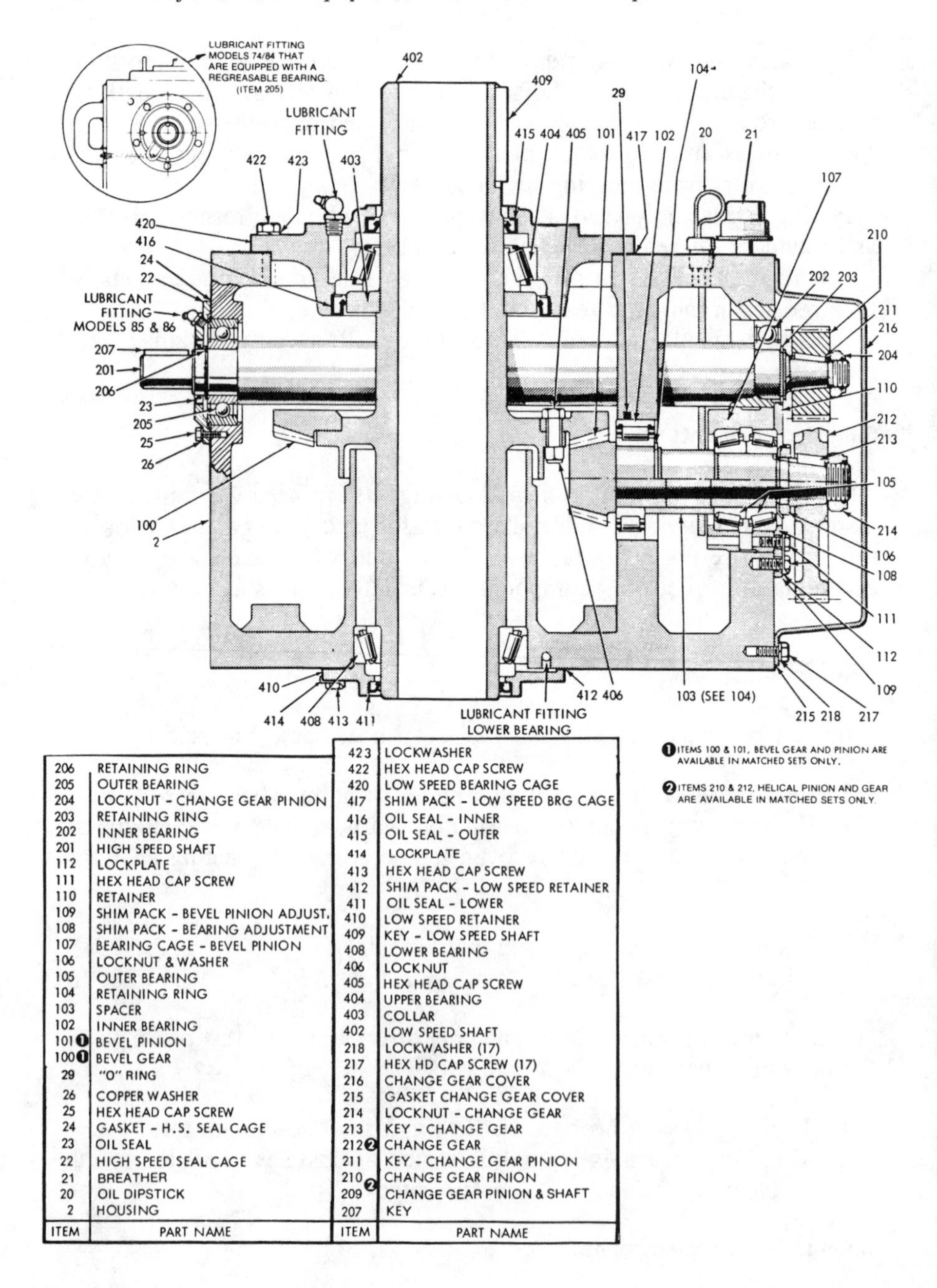

ITEM	PART NAME	ITEM	PART NAME
		423	LOCKWASHER
206	RETAINING RING	422	HEX HEAD CAP SCREW
205	OUTER BEARING	420	LOW SPEED BEARING CAGE
204	LOCKNUT - CHANGE GEAR PINION	417	SHIM PACK - LOW SPEED BRG CAGE
203	RETAINING RING	416	OIL SEAL - INNER
202	INNER BEARING	415	OIL SEAL - OUTER
201	HIGH SPEED SHAFT	414	LOCKPLATE
112	LOCKPLATE	413	HEX HEAD CAP SCREW
111	HEX HEAD CAP SCREW	412	SHIM PACK - LOW SPEED RETAINER
110	RETAINER	411	OIL SEAL - LOWER
109	SHIM PACK - BEVEL PINION ADJUST.	410	LOW SPEED RETAINER
108	SHIM PACK - BEARING ADJUSTMENT	409	KEY - LOW SPEED SHAFT
107	BEARING CAGE - BEVEL PINION	408	LOWER BEARING
106	LOCKNUT & WASHER	406	LOCKNUT
105	OUTER BEARING	405	HEX HEAD CAP SCREW
104	RETAINING RING	404	UPPER BEARING
103	SPACER	403	COLLAR
102	INNER BEARING	402	LOW SPEED SHAFT
101 ❶	BEVEL PINION	218	LOCKWASHER (17)
100 ❶	BEVEL GEAR	217	HEX HD CAP SCREW (17)
29	"O" RING	216	CHANGE GEAR COVER
26	COPPER WASHER	215	GASKET CHANGE GEAR COVER
25	HEX HEAD CAP SCREW	214	LOCKNUT - CHANGE GEAR
24	GASKET - H.S. SEAL CAGE	213	KEY - CHANGE GEAR
23	OIL SEAL	212 ❷	CHANGE GEAR
22	HIGH SPEED SEAL CAGE	211	KEY - CHANGE GEAR PINION
21	BREATHER	210 ❷	CHANGE GEAR PINION
20	OIL DIPSTICK	209 ❷	CHANGE GEAR PINION & SHAFT
2	HOUSING	207	KEY

Figure 5-7. Double reaction mixer drive and parts list. (Courtesy Mixing Equipment Co. Inc.)

1. Wedge a cloth or leather strap between the mesh of the pinion and gear and loosen the locknuts.
2. If a pry bar or wedge is used, apply pressure behind the gear in line with the keyway and sharply tap the gear 90° from the keyway between the outside diameter and the hub. A brass hammer or mallet should be used.
3. The change gears are provided with jack screw holes to facilitate removal. Two high strength bolts (Grade 5 or equivalent) may be threaded into these holes and tightened securely against the retainer. Then sharply tap the gear 90° from the bolts, between the outside diameter and hub. A brass hammer or mallet should be used.
4. If removal is stubborn, apply heat evenly around the circumference of the gear hub with a torch or other device, but do not allow the hub temperature to exceed 275°F. The heat should be applied quickly to the hub to prevent the shaft from heating.
5. Remove straight bore pinions from shafts at an arbor press.
6. To replace straight bore pinions, preheat to 225°F and press on shaft with chamfered side tight against shaft shoulder. Check with feelers for zero clearance between shaft shoulder and pinion before tightening locknut.

If the unit is further dismantled, make sure the oil pan as shown in Figure 5-6A is reinstalled.

Removal of High Speed Shaft (Figure 5-7)

The high speed shaft can be removed from the motor end only by dismantling the pinion (210) and pulling through. Be cautious when proceeding to slowly maneuver the shaft until the bearings clear the bevel gear (100) and low speed shaft (402).

Bevel Gear Removal

1. Remove high speed shaft (201) assembly from unit to provide clearance for bevel gear removal.
2. Remove the cap screws (422) from the low speed bearing cage (420), install 2 eyebolts and lift cage straight up. Use caution so as not to damage the oil seal (415 and 416) lips.
3. Wrap and interlock 2 web straps around the low speed shaft (402) directly below the seal collar (403) and carefully lift the shaft out.
4. Remove the low speed retainer (410).
5. Remove the bevel gear (100) from the low speed shaft (402). Remove the cap screws (405) by restraining the elastic nuts (406) and remove the gear.

Bevel Pinion Removal

1. Remove the outer set of cap screws (111) in the retainer (110) (Figure 5-7).
2. Insert jacking screws in the retainer (110) and remove the bevel pinion assembly. *Do not use a power wrench on the jacking screws.*
3. Remove inner set of cap screws (111) and the retainer (110).
4. Remove outer bearing cup (105) by pushing bearing cage (107) inward.
5. Remove the locknut and washer (106).
6. Press the bearings (102) and (105) off the shaft.

Bevel Gear Assembly

1. Place the bevel gear (100) on the shaft (402) flange and insert the cap screws (405) (Figure 5-7). Use new elastic stopnuts (406) and coat the capscrew threads (405) with oil.
2. Always use an open end wrench to restrain the elastic stopnuts (406) and tighten the capscrews (405) to proper torque.

Bevel Pinion Assembly

1. Insert the inner cup of the inside tapered roller bearing (105) in the bearing cage (107) (Figure 5-8).
2. Mount the inner bearing (102) on the shaft:
   Mount the 2 roller bearing cones with ground spacer between the 2 cones.
3. In order, mount the spacer (103) or the retaining ring (104), the bearing cage (107), the two tapered roller bearing cones (105) and the locknut-lockwasher (106) on the shaft (101).
4. Insert the outer tapered roller bearing cup and snugly attach the retaining plate (110) without the shims (108) or lockplates (112).
5. Measure the gap between the retaining plate (110) and the bearing cage (107) and add shims (108) equal to measured gap +.003 to .005. Follow these steps:

   a. Measure gap and add shims equal to gap + .003 in. to .005 in.
   b. Lightly bump bearing cage (107) downward.
   c. Raise cage upward evenly (do not tip) and read axial float on indicator.
   d. Obtain .003 to .005 axial float.

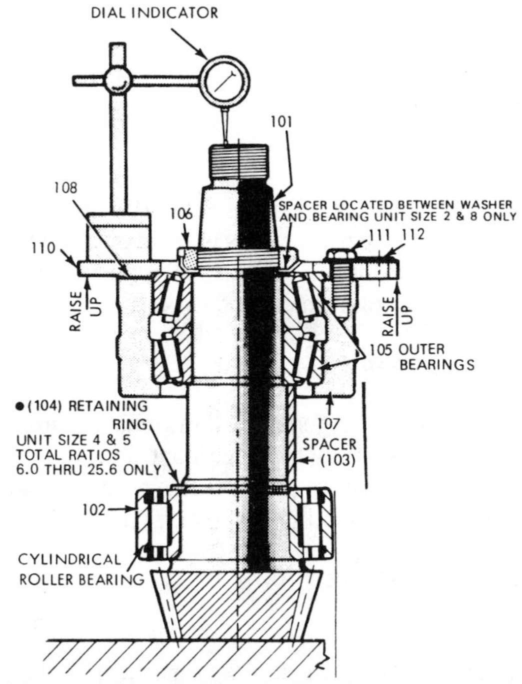

Figure 5-8. Bevel pinion assembly—double reduction mixer gear drive. (Courtesy Mixing Equipment Co. Inc.)

6. Add lockplates (112) and capscrews and tighten. Do not set lockplates (112) at this time.
7. Raise the bearing cage upward and recheck bearing float (.003 to .005). Reshim if necessary.

Preparation for Setting Bevel Gear Mounting and Backlash

Prior to setting the bevel gears as outlined under "Setting the bevel gears" below, follow these preliminary steps:

1. Record the mounting distance (MD) and the backlash (BL) values etched on the periphery of the bevel gear.
2. Measure the outside diameter of the machined portion of the oil dam, divide by two and record for reference.
3. Place the lower roller bearing (408) cup in the housing so that the face of the cup is recessed 1/16 in. into the housing.

4. Check alignment of the grease channel in the L.S. retainer (410) with the grease holes in the housing (2) and shims (412).
5. Add the following tentative shim pack (412).
   (1) .015 in. thick
   (1) .009 in. thick
   (1) .007 in. thick
6. Mount the shims (412) and retainer (410) to the housing, making sure the face of the bearing cup is in full contact with the raised face of the retainer. Do not install the oil seal (411) or the capscrew lockplates (414) at this time.

Setting Bevel Gears (Figure 5-9)

Bevel gears are precision generated and custom matched. They are precision lapped for optimum tooth contact and should always be replaced as *matched sets.* They require precise assembly with accurate setting of the bevel pinion mounting distance (MD) and backlash (BL).

Setting The Mounting Distance

1. Insert the bevel pinion subassembly into the housing without shims (109) and tap lightly until tight against face of the housing.
2. Apply pressure to the end of the bevel pinion shaft (101), and measure the gap between the oil dam and the toe of the bevel pinion (Dim. "C"). Record the measured gap.

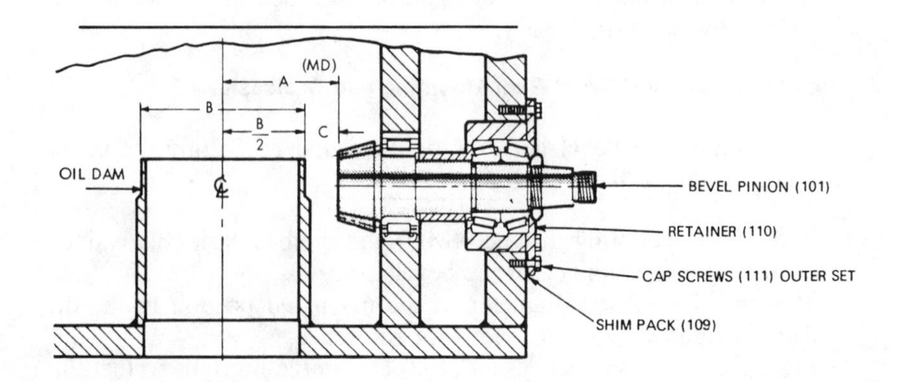

Figure 5-9. Setting mounting distance on bevel gears—double reduction gear drive for top entering mixer. (Courtesy Mixing Equipment Co. Inc.)

3. Determine the shims (109) required by referring to the mounting distance (MD) and oil dam diameter B ÷ 2 recorded previously according to the method shown in Figure 5-8.

   A (MD) minus (B/2 + C) = SHIM PACK REQUIRED ± .001 in.

4. Back off the bevel pinion sub-assembly approximately 1″ from the housing. (Use jacking screws in the holes provided.)
5. Split the calculated amount of shims and insert under the face of the retainer (110).
6. Push the assembly back into the housing and tighten all the outer capscrews (111).
7. Recheck the mounting distance after tightening. Bend up the lockplates if satisfactory.

Setting the Backlash

1. Carefully lower the low speed shaft (402) assembly into the unit.
2. Install the low-speed bearing cage (420) without the shim pack (417) and tighten with 4 capscrews (422).
3. Check the actual backlash against the value previously recorded from the bevel gear.
4. To measure the backlash, see Figure 5-10.

   a. Wedge a key into the key seat and set up a dial indicator with the indicator tip located at Dim. "D."
   b. Rotate the bevel pinion shaft (101) back and forth while holding the low speed shaft immobile and read the backlash on the indicator. The indicator reading should be two times the backlash etched on the bevel gear.
5. To obtain the required backlash, add or remove shims from the lower shim pack (412) equal to 1/2 of the indicator reading. Loosen and retighten the 4 upper bearing cage capscrews (422) and tap down the low speed shaft (402) each time the shim pack (412) is adjusted. When the proper backlash is obtained, measure the total shim pack. Allow approximately 3 to 10 percent of the total pack for final shim compression.
6. The final backlash (BL) can vary +.004 in. − .000 in. from the value etched on the bevel gear.
7. If backlash is satisfactory, remove the capscrews (413), install the lockplates (414) and retighten. Recheck the backlash and set the lockplates if satisfactory.

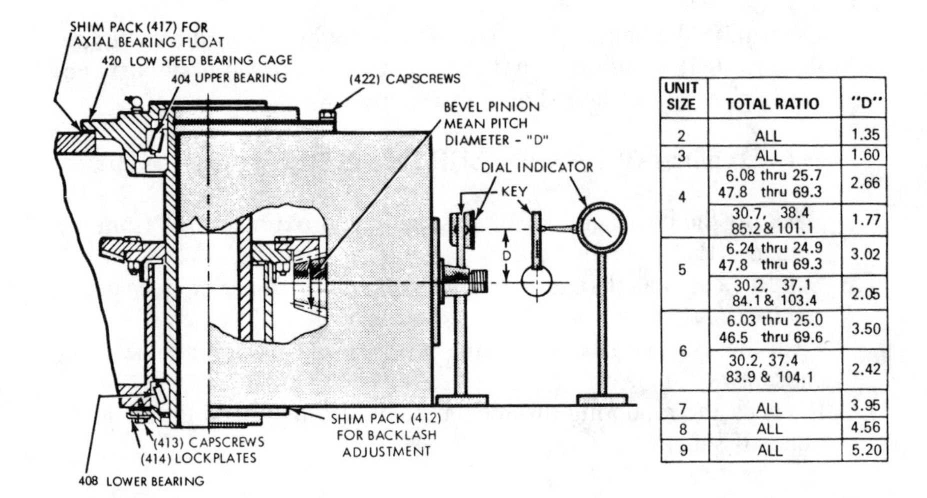

UNIT SIZE	TOTAL RATIO	"D"
2	ALL	1.35
3	ALL	1.60
4	6.08 thru 25.7 47.8 thru 69.3	2.66
	30.7, 38.4 85.2 & 101.1	1.77
5	6.24 thru 24.9 47.8 thru 69.3	3.02
	30.2, 37.1 84.1 & 103.4	2.05
6	6.03 thru 25.0 46.5 thru 69.6	3.50
	30.2, 37.4 83.9 & 104.1	2.42
7	ALL	3.95
8	ALL	4.56
9	ALL	5.20

Figure 5-10. Backlash adjustment—double reduction gear drive for top entering mixer. (Courtesy Mixing Equipment Co. Inc.)

Setting the Axial Float in the Low Speed Shaft Roller Bearings

1. Measure the gap between the upper low speed bearing cage (420) and the housing with feeler gauges.
2. The gap measurement plus the float value below is the required thickness of compressed shims required to obtain the shaft axial bearing float of .003 to .005 in.
3. To check the axial float:

    a. Install the calculated amount of shims (417), tighten down the bearing cage (420) and tap the low speed shaft (402) downward.
    b. Lower a sling through the shaft and insert a bar through the sling across the bottom end of the shaft.
    c. Place a dial indicator on the top face of the shaft and carefully raise the shaft with a hoist.
    d. Read the axial bearing float on the indicator and readjust if necessary.

Final Assembly

1. Remove the low speed bearing cage (420) and the low speed shaft (402) subassembly.

2. Clean and repack the upper (404) and lower (408) low speed bearings and cavities as shown in Figure 5-11 with high quality #2 ball and roller bearing grease.
3. Install the oil seals (415) and (416).

   a. Coat the OD of the outer oil seal (41) with "Permatex #3" or equivalent and install with the appropriate seal driver. Do not use Permatex® 3 on the inner oil seal (416).
   b. Be sure to install all oil seals with the seal lips and springs facing the direction indicated by the arrows in Figure 5-11.
   c. Coat the bearing cage cap screw (422) threads with "Permatex® #3" or equivalent before installing.
   d. Reinstall the low speed shaft (402) and bearing cage (420) assembly.
4. Install the high speed shaft (201) and change gear pinion (210).
5. Set the high speed shaft axial end float as outlined earlier.

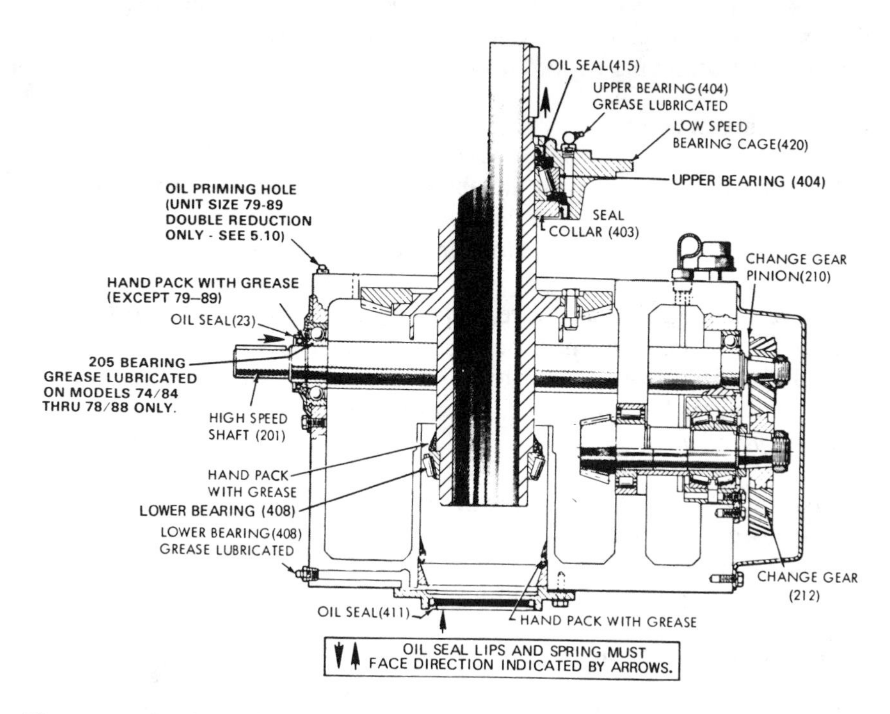

Figure 5-11. Final assembly—double reduction gear drive for top entering mixer. (Courtesy Mixing Equipment Co. Inc.)

6. Install high speed oil seal (23) and lower oil seal (411). Coat the seal OD with "Permatex® #3" or equivalent sealing compound.
7. Tighten all fasteners and set lockplates.
8. To ensure that grease channels are full and purged of air, add grease through the fittings provided at bearings (404), (408). Rotate the shafts while adding grease. Add until grease flows into the cavity above the upper oil seal (415).
9. When installing change gears, refer to change gear locknut and the appropriate torque table.
10. Prior to initial start, remove the pipe plug and prime the outer H.S. bearing (205) by adding 1 quart of oil.
11. Fill the unit with oil through the large opening with the socket head plug in the side of the housing to the dipstick full mark. The approximate oil capacity is shown on the nameplate.
12. Rotate the gears a few revolutions by hand before reassembling mixer and applying load.

Mixer Impeller Assembly

Mixer impellers occasionally have to be replaced. A good example is the axial flow mixer impeller shown in Figure 5-12. To assemble this impeller the following procedure is recommended.

Impeller Assembly

Mate the three blades with the hub ears and install hardware with the lockwashers (6), if furnished, located under the hex nuts (7). If fins (8) are furnished, mount them before installing lockwashers and hex nuts (Figure 5-12).

Before securing the hardware, apply pressure to the blade so that its edge is firmly seated against the raised shoulder on each hub ear. After tightening the hardware, check to make sure that the blade has not shifted away from the hub ear shoulder. *Proper blade positioning is important to impeller function.*

It is essential that the hardware securing the blades to the hub is tightened to specific torques. It is important that tight connections are maintained as impellers are usually subjected to a wide range of adverse loading conditions imposed by fluid force reactions.

Impeller Assembly To Shaft

1. Coat the shaft with a lubricant to facilitate movement and slide the impeller/s up the shaft slightly above the desired location.

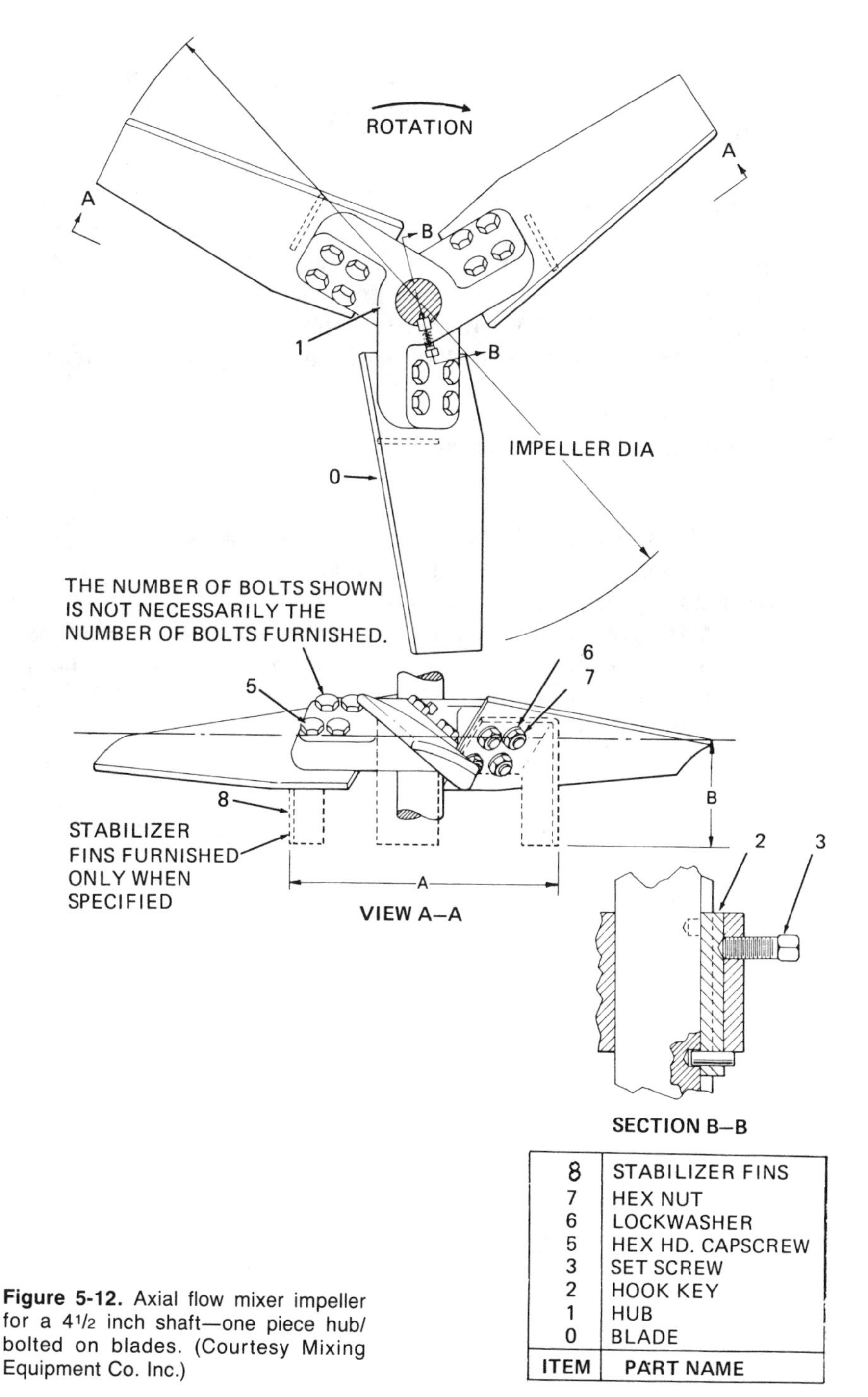

ITEM	PART NAME
8	STABILIZER FINS
7	HEX NUT
6	LOCKWASHER
5	HEX HD. CAPSCREW
3	SET SCREW
2	HOOK KEY
1	HUB
0	BLADE

Figure 5-12. Axial flow mixer impeller for a 4½ inch shaft—one piece hub/ bolted on blades. (Courtesy Mixing Equipment Co. Inc.)

All shafts are provided with pin holes in the shaft keyway for the impeller hook key and most shafts are multiple drilled for impeller field adjustment. Refer to the installation drawing for correct impeller positioning.

2. Insert the hook key (2) pin into the shaft keyway pin hole.
3. Lower the impeller *gently* over the hook key (2) until the hub is resting on the protruding pin. *Do not allow the impeller to drop on the pin.* The pin is a safety device designed to support the weight of the impeller, *but not* to withstand impacts.
4. Install the set screw (3). After the set screw (3) is properly seated in the countersunk hole in the hook key, tighten the set screw against key (2).

Repositioning the Mixer Impeller (Figure 5-12)

1. Remove the set screw (3).
2. Raise the impeller until the hook key (2) is exposed.
3. Relocate the hook key (2) in the desired position.
4. Retighten all hardware as noted previously.
5. It is good practice to *retighten* all bolted connections after the equipment has been in operation. It is recommended that all hardware be checked for tightness after one week, but before three months of operation.

Chapter 6

Reciprocating Gas Engines and Compressors*

Introduction

The degree to which engines and compressors are maintained today varies from a "wait until destruction" type of negative thinking to a "complete" but extravagant program. The former extreme has always existed but, of course, has never been justified; the latter extreme is costly but it has some strong points in its favor. The ideal preventive maintenance procedures, however, should be economical but give the equipment good coverage, and the purpose of this section is to describe such procedures.

Regardless of past thinking, modern economic practices are squeezing the appropriations for maintenance. The drive for better profit margins forces managers to cut costs everywhere, and in certain highly competitive industries, drastic cuts in maintenance budgets require a streamlined, but still effective, maintenance program.

The maintenance tips or pointers that have been used for years are basic and all could be used today, but in order to cope with the modern economic trend we will deal with the ones that best apply to modern equipment. It should be remembered that any maintenance program will not fit two or more installations exactly, placing the burden of forming a specific program on the operators. Therefore, we will discuss preventive maintenance (PM) procedures in general terms, explaining why each point is essential and how it may or may not apply to a large or small reciprocating machine. The various arrangements of compressors should be familiar to all, but the terminology may differ.

* By permission of Cooper Energy Services, Mt. Vernon, Ohio 43050

Figure 6-1. Integral gas engine compressor.

Figure 6-1 is a photograph of an internal gas engine compressor. Figure 6-2 illustrates a unit with compressor cylinders connected to a frame. An electric motor drives the crankshaft in the frame. This is known as a "motor-driven" unit. The driver could also be a turbine or an engine. Regardless of the arrangement, the maintenance of any unit can best be considered by dealing with the compressor and the power units separately. Since compressors generally require more maintenance than engines, we will discuss compressors first.

Compressor Cylinder Maintenance

The following malfunctions can occur in a compressor cylinder regardless of the gas pumped and whether or not it is double or single-acting, large or small in diameter or used as one of the stages of a multi-stage unit:

- Exceeding allowable rod load.
- Accelerated wear and scuffing.
  - Piston to liner.
  - Piston rings.
  - Piston rod and packing.

Figure 6-2. Reciprocating compressor connected to electric motor.

- Valve breakage.
- Knocks, noises and vibration.

Exceeding assigned rod load. It is essential that operators and mechanics understand rod loads before they attempt to start a compressor. Most major casualties, such as broken rods, damaged crossheads, pin bushings and frame failures, are caused by exceeding the maximum rod load. The frightening aspect of this is that the wreck can happen within just a few revolutions after the infraction. By way of explanation, let us consider a double-acting cylinder. As the piston pumps toward the head end (Figure 6-3) the discharge pressure force (P_2) on the pistons tends to compress or buckle the piston rod. At the same time, gas is entering the cylinder behind the piston, putting a suction pressure force (P_1) on the back side of

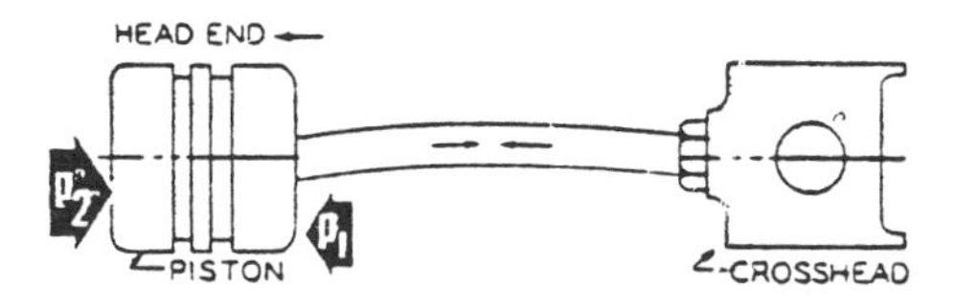

Figure 6-3. Reciprocating compressor piston-rod-crosshead system in compression.

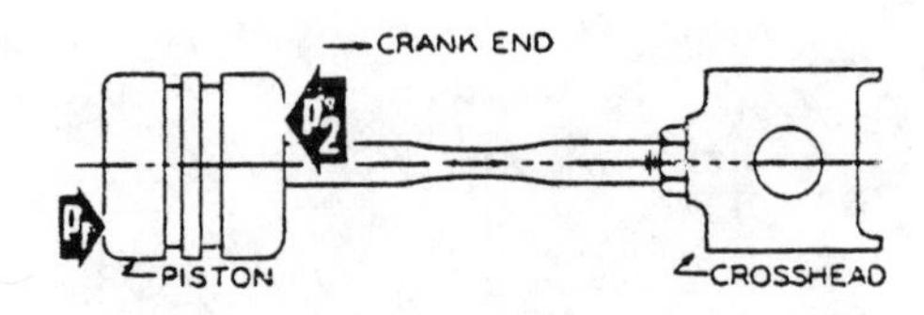

Figure 6-4. Reciprocating compressor piston-rod-crosshead (in tension).

the piston. The two forces are opposite in direction, but since the discharge pressure is larger, the net push tends to compress the rod. This is called "rod load compression." It is basic that as the suction pressure is decreased or the discharge increased, the net compression on the rod increases. Therefore, if the operator, at start-up, shut-down or during operation, lets the suction or discharge pressures deviate too far from design conditions, the maximum permissible compressive load may be exceeded. As the piston discharges toward the crank end on the return stroke (Figure 6-4), the net force of the suction and discharge pressures results in a tension load on the rod. This is known as "rod load tension," and the operator can damage the machine by decreasing the suction or increasing the discharge pressure too far above the design pressure.

Although the tension and compressive forces are absorbed by the rod, other parts such as head bolts, piston, connecting rod and bolts, crosshead and shoes, bushings, bearings, etc., are likewise stressed. In other words, the most highly stressed part determines the rod load assigned by the compressor builder. This value is different for each compressor model.

Rod loads can be calculated by simple arithmetic, but in operation suction and discharge, pressure can change so fast that the operator does not have time to calculate. There is, however, a safe and simple way to stay away from rod loads, by using a graph similar to Figure 6-5A.

In this example, the cylinder involved is 34 in. in diameter and has a design suction of 32 psig and a discharge pressure of 145 psig. The maximum rod load assigned by the compressor builder is 125,000 lbs in compression and 115,000 lbs in tension. If the operator reads the actual suction and discharge pressures on the cylinder as 32 and 145 psig, respectively, these readings, when projected on the graph, locate point "A." Since this point is below the line, the machine is safe.

If conditions change to a suction of 25 psig and discharge of 158 psig, those readings define point "B," which is above the line and indicates danger. Information for preparing a graph for any cylinder is given in Figure 6-5B. It is recommended that such a graph be made for every cylinder of a compressor. In many cases it will be found that the slope of the line will not be drastic, which means that the suction pressure can be reduced or discharge pressure increased to extremes before getting into trouble. In those cases, the graph can possibly be disposed of but the effort will still be worthwhile if for no other reason than to be confident that the cylinder will not be critically sensitive to rod loads and pressure changes.

Once the operator understands and respects rod loads, it is a simple matter to check critical cylinders by use of a graph. He should check his starting and shutdown procedures to make sure that pressures do not exceed the limits during these operations.

Loading and unloading cylinders of multi-stage units changes interstage pressures, and if the operator deviates from the established sequence of unloading, or if he is given the incorrect information, the rod load may be exceeded in changing steps. Therefore, each time the capacity is changed by loading or unloading, the operator should check the graph on those cylinders that have been sized close to the assigned rod load.

Attention to rod loads can be summarized by these rules:

1. Never exceed the value of design discharge pressure stamped on the cylinder nameplate unless checked.
2. Don't let suction pressure fall below the design value stamped on the cylinder nameplate.
3. The maximum pressure figure stamped on the cylinder nameplate is the maximum working pressure as customarily assigned to pressure vessels. Staying below it does not guarantee safe rod load.

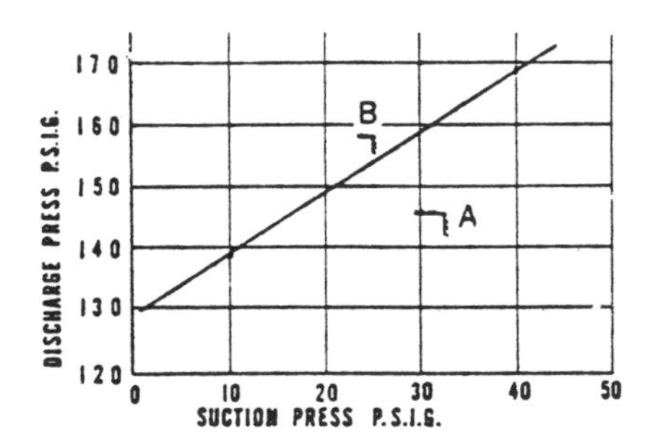

Figure 6-5A. Rod load diagram.

Method of preparing compressor cylinder rod-load check graph such as Figure 6-5A for any compressor cylinder

$$P_2 = \frac{1}{A}[CRL + P_1(A - a)]$$

Where:
A = cylinder area,
a = rod area,
CRL = compression rod load assigned to unit,
P_1 = suction pressure (psig),
P_2 = discharge pressure (psig).

Compression Rod Load. Known data for cylinder covered by Figure 6-5A:

Cyl. Dia. = 34 in.
Compressor Rod Dia. = 4 in.
CRL = 125,000 lbs
Choose any suction pressure (say 10 psig) and solve for P_2:

$$P_2 = \frac{1}{907.9}[125{,}000 + 10(907.9 - 12.56)] = 147.5 \text{ psi.}$$

The P_1 of 10 psi and P_2 of 147.5 psi will locate one point on a graph.

Choose another suction pressure higher than design, say, 40 psig, and substitute in above formula, P_2 then is 177 psi. The P_1 of 40 psi and P_2 of 177 psi established a second point for the compression-rod load line.

Tension Rod Load. The formula for tension rod load is

$$P_2 = \frac{1}{A - a}[TRL + P_1 \bullet A]$$

Known data are same as compression rod load except the assigned tension rod load, TRL, is 115,000. To determine load line for it, substitute P_1 of 10 psi. P_2 would be 138.5 psi, which would give one point.

Next choose a P_1 of 40 psig. For above formula, P_2 would be 169 psi. Insert this point on the graph and draw the tension rod load line. Since the compresson rod load line is above the tension rod line it should be disregarded. Use the graph as described in the text.

Figure 6-5B. Example of rod load calculation.

4. In multi-stage machines, check the load chart before operating valve lifters or unloading pockets.
5. Never bypass gas between stages unless the equipment is designed to operate that way.
6. When a relief valve blows, shut down and determine the cause.
7. Don't start a unit against discharge pressure.

Accelerated wear or scuffing. Normal wear of pistons, rings, liners, packing and rods is not a significant problem. Generally, the normal life of those parts is long and satisfactory. Furthermore, this type of wear is gradual and can be observed or detected by progressive fall-off of capacity. Accelerated wear, scuffing and sudden seizures, however, amount to a major portion of over-all compressor expense because they come during peak periods and when least expected. Yet these failures are for the most part avoidable because there is a precise way of spotting them, even while the machine is running.

The key to the check is the vent line (3) in Figure 6-6. Figure 6-6 is a section through a typical pressure packing compartment. The oil supply line (1) enters the compartment and connects to the packing flange (2). Oil is directed from the flange to the front end of the packing. The vent for the packing comes out of the bottom of the flange and is piped to the outside by the vent line (3), to the end of which a valve (4) is attached.

On low-pressure cylinders very little, if any, vapor will come out the vent when the packing is sealing, but a good portion of the oil that is being fed to the packing will drop out of the valve (4).

On high-pressure packing, there will be some vapor even if the packing is sealing. The vapor will have a higher velocity and will contain an oil mist.

It is very important that the amount and velocity of vapor emitted from the line be observed while the packing is in good condition. The color of

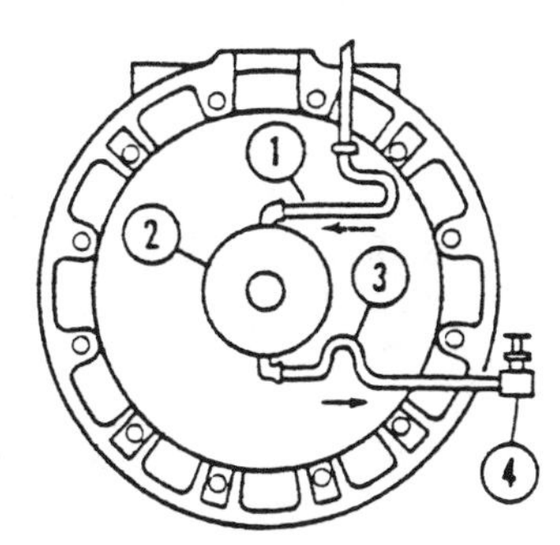

Figure 6-6. Section through reciprocating compressor pressure packing compartment.

the oil mist or drops of oil should be inspected. This can be done by collecting it on the tip of the finger. If the color is similar to that of new oil, the operator can be sure that the parts inside the cylinder are in satisfactory condition. A darkened oil, however, should be further scrutinized by holding the finger up to the light to check for metal particles, which will glisten like crystals. If any bronze, cast iron, or steel parts are starting to wear, the metal particles or dust will almost immediately darken the oil coming out of the vent. Consequently, this line should be checked immediately after start-up of a new installation, after overhaul, and twice daily during operation. If discoloration is noticed, it is an indication of distress, and the compressor should be shut down.

A vast amount of information for evaluating the condition of the cylinder can be obtained simply by removing a top compressor valve from the head and spotting the piston toward the crank end. The liner can then be seen. The surface of a healthy liner will be bright, and at first glance the observer may be convinced that the liner is too dry or has insufficient lube oil. Rubbing a finger across the surface will hardly wet the skin with oil. The only way the film can be positively checked is by wiping a facial tissue across the liner (the tissue will be stained with new oil). This is all the film needed to keep the two rubbing surfaces apart.

There are two vital spots in a compressor cylinder that require an oil film and one is between the piston and the liner. In cases where the piston is supported by a wear band, the film is maintained between the wear band and the liner. The first sign of distress between the piston and liner shows up in the bottom as a narrow score mark the length of the liner. A facial tissue wiped on either side of the score mark and held toward the light will substantiate the presence of darkened oil and small particles of metal dust or cuttings. The cuttings are a result of oil film breakdown and they discolor the oil. Contrary to widespread opinion, the discolored oil and cuttings from the liner and piston find their way through the packing and out the vent.

The significant point to remember about darkened oil appearing at the packing vent is that it happens in a matter of minutes after the scuffing starts. If the indication is not spotted or used as a PM check, the rubbing contact between piston and liner will make the narrow score mark progressively wider until the liner and piston are badly scored over a 180° arc. By that time, the piston may start to knock, dictating a shutdown, but then the piston, liner, and rings are such a blackened mess that the cause of failure cannot be determined. If the packing vent is used as a PM procedure, however, the scuffing will not only be observed before the failure, but it may be possible to find the cause of scuffing.

For example, excessive liquid in the gas will initiate scuffing because it dilutes the thin supporting oil film. Sometimes the liquids drop out within the cylinder and escape the attention of the operator. In addition, liquid being carried in the pipe to the cylinder is difficult to detect by checking blow-downs, but is likely to be spotted after the unit is shut down. The liner can now be observed through a valve port. Although the liquid may not be noticed at first, as the diluted oil drains down the sides of the liner and leaves a small stream of liquid and oil in the bottom of the liner. An inspection of this sort is very likely to disclose small amounts of harmful liquid. In cases of large quantities, detection is easy, because the cylinder may knock and, if allowed to continue, the heads or pistons may crack. The liquid also will come out the packing vent and may spit out between the rod and flange of the packing.

The other vital spot that requires a lubricating oil film is between the face of the piston rings and the liner. As in the case of the piston-to-liner interface, any scuffing of the rings will discolor the oil coming out the packing vent. This problem can be differentiated from that of the piston-to-liner interface because the valve-port inspection will show the liner to be marked or scuffed over the entire 360° contour, even in the early stages of failure.

The value of using the packing vent as a PM check is that it catches the scuffing in the initial stages. Generally speaking, once two mating surfaces start to scuff, they never heal, even if the oil film is reestablished, and the scuffing will continue to destruction. But if detected early, it may be stopped merely by wiping the discolored oil from the cylinder and spot-honing the scuffed area of the liner. The unit then can be put back on the line and the vent carefully watched. It will not take too long for the discolored oil left in the vent piping to be purged. If the cause has been corrected the oil will clear up; if not, it will be necessary to remove the piston from the cylinder and do a thorough job of cleaning and honing the liner.

Piston distress always occurs at the bottom of the piston. Therefore, the piston can be smoothed and rotated. Piston rings likewise can be saved, but their edges should be carefully checked—a small chamfer, or radius, was machined on these edges at the time of manufacture, but wear may have made the edges sharp. A metallic ring with a sharp face will start cutting and wearing and will progress. The sharp edges should be broken with a three-cornered scraper. Another critical area for scuffing is between the packing rings and piston rod. As mentioned before, the amount of vapor coming from the packing vent depends upon the ratio and pressure of the cylinder. Of course, excessive blowing is an indication the rings are not seating. Packing problems are common at the start-up of new installations and after overhauls. Dirt and liquid are the reason for problems at start-up, and, on rare occasions, there has been an improper choice of packing materials and design. Generally, packing maintenance falls into the routine category. For that reason, comment will be divided between the two different phases.

Packing Maintenance

At start-up or after overhaul, one inspection cover (if provided) should be removed from the compartment between the crosshead guide and the compressor cylinder. If this compartment is filled with vapors from the pressure packing, there should be concern, because the rod and packing are hot and some action will have to be taken. This is a point over which there is considerable controversy: one school of thought holds that the packing oil feed rate should be increased to its maximum, contending that the added oil both cools and seals; the opposing view is to momentarily cut off the oil supply, on the assumption that the packing will thus seat faster. We, however, agree with neither extreme—the added oil inside the packing is not enough to cool; and if dirt was the cause of trouble or if the material has started to cut, the extra oil will make a slurry of carbon and sludge in the cups. We have never eliminated the problem by increasing the amount of oil, but have had some success in reducing the feed rate to what it should have been in the first place. However, anything that can be done to cool the rod—by pouring oil, directing an air stream or even water on it as it comes out the packing—might hold the temperature level down long enough for the materials to seat. Where it is possible to reduce the speed of the unit or decrease pressure until seating starts, this will help too.

These are only temporary measures, and if the vapors increase the unit should be shut down and the entire assembly checked, cleaned, and perhaps lapped. This vigilance is vital, because the rod is bound to score, and not only are new rods expensive, but delivery on them sometimes is not very prompt.

During the period of heavy vapors and the state of confusion on go or no-go, any scuffing or roughness of the rod can be felt while the machine is running. Naturally, there will be marks or streaks on the rod, but if it is smooth, hold off shutting down. If scuffing starts, shut down immediately.

Once it has been established that there are no vapors in the packing compartment, the inspection cover can be replaced, but the packing vent line should be watched. Any increase in vapor will indicate packing scuffing. As mentioned before, the color of the oil from the vent will be an indication of the liner and piston condition. If the vent packing oil begins to darken and a check on the liner is satisfactory, then the scuffing is originating in the packing. In multi-stage air machines, there is water fallout after the second stage. Separators are installed before the suction of those stages, but it is difficult for the designer to be sure they are always performing as guaranteed. An inefficient separator will allow water to get into the cylinder and break down the oil film. However, before any damage is done, the water will emulsify the oil in the packing vent line and form a yellow emulsion. It is customary to run a test on separators in every new installation, but this is not necessary if the packing vent is observed as an indication. If emulsion is found at the packing vent of air machines, or if liquid is found there in the case of other gases, the effects of the inefficient separator can be temporarily nullified by slightly opening the blow-down line of the cylinder suction drum. The pipe nozzle extends up into the drum, and since the drum will act as a separator, the liquid can be drained off before it reaches the top of the nozzle.

If it appears that we have deviated from the immediate subject of packing maintenance at start-up or post-overhaul, it should be stressed that water and liquids will also seriously affect packing. Once packing seats and does not blowby, trouble-free service and long life can be expected.

Its expected life depends, of course, on the pressure ratio, the type of gas being pumped and the amount of dirt or liquids in the gas. The liner and piston will tolerate and pass small amounts of dirt, but each day a portion of the dirt will lodge in the packing cups. All of it does not get flushed out by the oil, and the remaining portion continues to build up. Generally, this build-up and fouling is the first thing that happens to packing in the long run; however, by using the packing vent as a PM measure, the exact time of trouble can be determined.

We have placed a great deal of emphasis here on checking the color of packing vent oil for PM. At some installations, however, crankcase oil from the power end is used to fill the compressor lubricators. This practice is followed for certain reasons and has merit, but this oil naturally is already darkened. However, a drop from the vent line can be placed on facial tissue, and if any metallic particles are present the oil will filter away and leave the particles.

Nonmetallic materials and modern design have made nonlubricated cylinders commercially sound. As a consequence, the packing vent will not be available as a PM indicator, and cylinders in this category do not have any good signs or indications that can be used to determine their condition or causes of trouble while they are running.

It is true that noises, increase in discharge temperature, fall-off of capacity and changes in interstage pressures will direct attention to trouble and will red-flag serious smash ups, but they will not save the materials involved. Excessive packing compartment vapor and discolored compressor rods will be a good check on packing, but for absolute protection of liners and pistons, a valve assembly will have to be removed for visual inspection at specified periods.

Brightness of the liners is a healthy sign. Any distress to the rings and wear bands will make dark streaks on the liner. There are two very important measurements that can be taken through the valve ports that will give exact information for detecting wear of liners and wear bands. The wear band is always larger in diameter than the piston, which means there will be a space between the liner and the piston. When the piston is installed, that space should be measured with feelers and then rechecked after a specified number of hours of operation.

Following start-up or overhaul, the first inspection should take place after no more than 48 hours of operation.

Subsequent periods of inspection will differ for each installation. Rotation of the piston will be necessary if the feeler clearance decreases. The liner diameter can be measured through the valve port. This measurement will check not only for wear but for out-of-roundness from distortion, which would indicate that the liner has been hot.

Valve Breakage

Valve failures have created much ill will toward reciprocating compressors. Excessive breakage, especially at start-up, has left a bad impression in the minds of many responsible people. Installation dirt, liquid, and off-design temperatures and pressures necessary to get started are largely responsible for failures that many times are mistakenly analyzed as incorrect materials, misapplication, corrosion, and incorrect design. On failures subsequent to start-up, the same reasons are given. This thinking sometimes creates unnecessary change of design, materials and sometimes a switch in suppliers. The point here is not to repudiate these statements, because in some cases a change is required. Furthermore, new ideas are healthy for the industry, provided they do not detour from the very issues that are giving trouble. However, if proper practices and basic principles are not recognized and observed, true progress on new ideas will be blocked.

The proper practices and basic principles can be described by following the sequence of events for start-up operation of a new installation.

An increase in the normal discharge temperature of the gas is a sure sign of poor valve condition. Although there are others, e.g., noise and reduction of capacity, discharge temperature is a very adequate indicator. Generally, the start-up crews watch it very carefully and shut down for repairs when it becomes too high. On the other hand, there are some who put off replacing hot valves. They justify this by stating that the process cannot be disturbed or they do not have time and it can wait. In addition to direct damage to the valve assembly, a faulty valve can score compressor rods and cause packing failure. This is especially true if the defective valve is in the crank or frame end of the cylinder. The leaky valve seems to have a torch effect on the rod and packing. A faulty valve in either end of the cylinder can scuff the piston and liner and actually cause a serious piston seizure and possible subsequent crankcase explosion. The length of time a hot valve can be operated without damage cannot be established for all installations. Furthermore, the period of time cannot be set by assigning a limit to the discharge temperature. This all depends on the size, class, and design of the cylinder, the gas being pumped, and the pressure ratios involved. In addition to installation dirt, another common occurrence is trouble with the cap gasket (Item 11, Figure 6-7). Where corrosive gas is involved, a poor choice of gasket material has caused trouble. Also, regardless of material, this gasket is sometimes incorrectly installed. It is also common for some people to use the old gasket after overhaul. Regardless of what made the gasket defective, one of the very serious consequences of leakage at that point is that the operator often further tightens the cap nuts (Item 10, Figure 6-7). When the leak continues, the tendency is to put an extension on the wrench and further tighten the nuts. The result is a distorted valve seat (illustrated in the exploded view to the left in Figure 6-7). This distortion is also characteristic of a valve installed in a cocked position. A valve plate cannot conform to a distorted seat without fatigue and breakage.

These examples are a few known reasons for valve failures but, continuing the description of proper practices, we can now address the topic of removing valves for repair. Most plates are steel, and the seats and guards (Items 2 and 4, Figure 6-7) are either steel or cast iron and, in almost every case, the seat or the guard or both are marked with a noticeable dent caused by breakage of the metal plate. At this point, there are three different approaches to restoration of the valve assembly. One is to put in a new plate (Item 3, Figure 6-7) and leave the guard and seat as is. In another, when the operator understands that a plate cannot seat on a damaged seat without fatigue and breakage, he machines or laps it. What most people fail to realize is that the impact of the plate on the guard upon opening is as severe as the action of the plate returning to the seat.

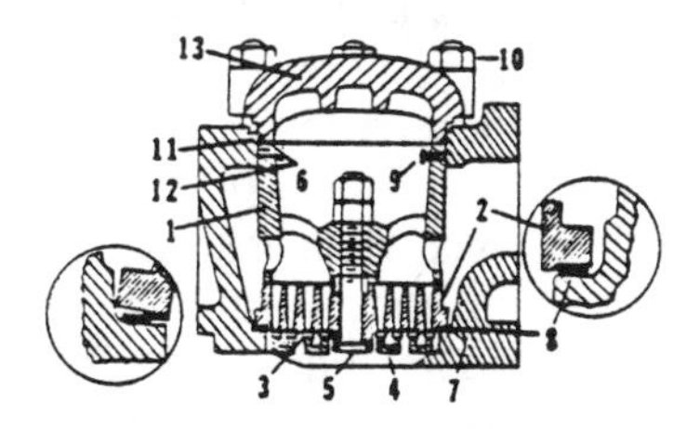

Figure 6-7. Reciprocating compressor valve.

Therefore, it is just as important to repair and square the guard. The third, and correct, approach is to repair both seat and guard. The first and second approaches are admittedly popular; however, there is no escaping the fact that regardless of design, materials, etc., valve failures will multiply after the first normal expected failure if the assembly is not repaired correctly.

It is worth mentioning here that Bakelite, Nylon, and other nonmetallic plates cannot withstand high discharge temperatures and pressure ratios, but since they do not damage seats and guards when they break, they should be used when possible.

The practice of lapping valve plates and seats to square and remove dents is highly recommended, but many people fail to realize the importance of removing the sharp edges formed in the lapping operation. Contrary to general belief, a plate does not lift straight off the seat as it opens; instead, one side lifts first. Thus, if sharp edges are left on either the valve or the seat, they will dig into each other. For that reason, all sharp edges should be broken after lapping.

There are many other examples of bad practices but they apply to specific types of valve designs and cannot be covered here. The point to remember is that valve assemblies are subjected to adverse conditions at start-up and after overhaul.

After the initial start-up pains, the valve troubles usually level off, but since anything that moves is prone to fail ultimately, it is only a matter of time until problems will develop and will need attention. Perhaps the best known tool for dealing with valve failures is a complete history on every valve assembly. The expense of keeping a history on an installation that is giving good service is not justified, but once anyone issues a valve complaint on the unit, it will be profitable to go back in with new valves and start keeping accurate records. Information such as the life of each plate, its exact location (i.e., crank end, head end, suction or discharge), and whether there have been any changes in capacity, pressure and temperature conditions is valuable. The number of times that such a history has revealed the trouble to us is almost unbelievable.

An interesting point often brought out in such a history is that there are more valve assemblies in the plant than realized and that the actual life is as much as three years. Sometimes the failures will be repeated in a certain stage, cylinder, or end, or perhaps in the same valve pocket of the cylinder. A close inspection may reveal that the seat in the cylinder that receives the valve assembly was damaged. In one specific case of repeated failures in one location, the entire cylinder was found to be distorted due to piping strain. The notation on the positions of the unloaders is important in the case of some variable piston unloaders. The valve plate may tend to flutter at very high clearance volumes.

Valve flutter, due to improper design can lead to valve failure. Experience shows that true valve flutter is not a frequent occurrence. When it does occur, valves can be destroyed in a matter of hours. It takes advanced instrumentation to verify and solve valve flutter, however.

Pressure pulsations in suction and discharge piping of compressor cylinders have been responsible for valve failures. Pulsations can be suspected as the cause if there also is rapid wear, and possibly breakage of the piston rings. This can be verified if there are bright marks on the entire circumference of the liner the exact width of the ring and located at the end of the ring travel. The trouble results because the pulsating pressure gets behind the ring when the piston stops at each end of travel and literally beats the ring against the liner. An exact study of piping size and drum location is required to eliminate this problem.

Dirt and Liquid

Dirt and liquid are scourges to reciprocating compressors; unfortunately, not much can be done about them. It is true that a better job is being done in reducing installation dirt, thanks to better cleaning methods and a respect for the problem. But process dirt or fouling is something else again. One thing that concerns us is that screens are becoming popular. They are being used both for start-up of new units and as permanent installations. Screens are in some ways beneficial, but some people are beginning to think they are the cure. They contend that with better materials and design, the screens can be made fine enough to protect cylinders. True, coarse particles that can be stopped by a screen are not good for the equipment, but a cylinder will digest a certain amount of it. However, it is the fine dirt that no screen will stop that has the worst effect on the liner because it mixes with the oil and makes a perfect lap for wearing parts. Furthermore, screens may load up and, if they break, cause more damage than would have occurred without any screens at all.

In regard to polymerizing of certain gases, some chemical companies claim success in injecting compounds into the suction nozzle of the cylinders. The amount injected is about ½ gallon per hour per cylinder and it increases the time between shutdowns for cleaning valves and cylinders by four times. The same beneficial results may be obtained from using diester-based synthetic lubricants for cylinder lubrication.

Knocks, Noises, and Vibration

Knocks, noises, and vibration are good indications of trouble. On the other hand, normal noises are sometimes misinterpreted by even the ex-

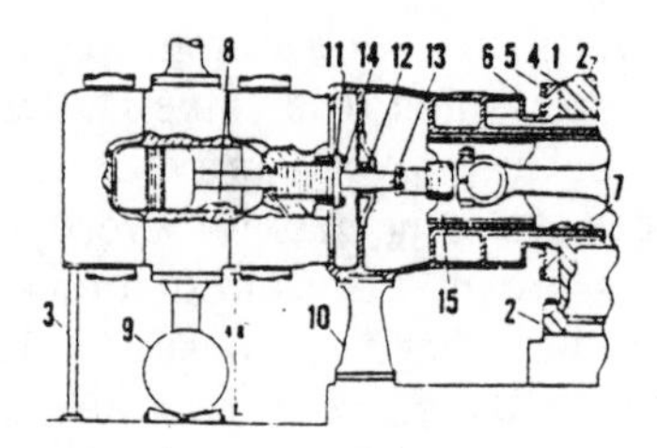

Figure 6-8. Cross-sectional view of reciprocating compressor cylinder and crosshead guide.

perienced operator. In order to keep from initiating panic, and to create confidence in the machinery, the operator should become familiar with the natural beat of the machine by spending a great deal of time around the unit after it is installed.

A common type of knock is caused when the piston of a double-acting cylinder hits either the head or crank end of the cylinder. This can happen if the piston is incorrectly spaced at assembly. The piston may clear during the cranking or barring-over check but, due to take-up of all clearances (bushings, rod, and main bearings) by inertia and expansion of rod and piston due to heat, the piston may strike the head during operation. The knock is very easy to analyze because it is a distinct metallic thud and can be felt by placing the hand on the cylinder head. Rapid wear of the liner and piston also can cause the piston to strike the head. From Figure 6-8 it can be seen that there is a large radius on each end of the piston that also matches the contour of the heads. If the piston is lowered by wear, it will strike the radius of the head. Some pistons are bullet-nosed and are more likely to strike other parts when wear sets in.

Note (in Figure 6-8) that the piston rod (14) is threaded into the crosshead (15) and secured by a nut (13). This nut has been known to come loose, permitting the piston to turn and hit the head.

The knock that has caused more confusion than any other, however, is the one that results when the piston nut is loose. This is the nut that secures the piston to the rod. If it becomes loose by as little as .003 in., it will knock very loudly; however the noise will appear to be in the crankcase of the unit. In many cases the knock has been eliminated merely by tightening the nut one or two flats. If this nut continues to come loose after tightening several times, there is good reason to believe that the cylinder is not in alignment. If the cylinder is high on the head end, the piston will rise as it comes to outer center, thereby setting up a severe vibration that affects the tightness of the nut. This condition can be further substantiated by loose grout under the crosshead guide support (Item 10, Figure 6-8) and perhaps under the frame of the unit. In some cases, the grout has loosened to such a degree that the guide-support hold-down bolts break. Vibration at the frame (2) with increasing intensity at the support (10) and end of the cylinder also will be noted. Misalignment has many possible causes. In some installations there is only one support, i.e., the crosshead guide support (10), which leaves the cylinder overhung. However, there are usually some wedges under the drum (9) and it is the incorrect usage of the wedges that sometimes causes the trouble. When the cylinder is aligned and the support (10) grouted, strain is sometimes mistakenly taken on the drum wedges. If the unit is of the bottom discharge type, the drum will push the cylinder up as it is heated by discharge temperature. Assuming a 48-in. distance and a discharge temperature of 300°F, it is possible for the expansion of the drum to raise the cylinder .066 in. The real purpose of the drum wedges, however, is to keep the drum from vibrating; therefore, the proper method of adjusting the wedges is to leave them loose until the machine gets up to operating temperature and only then snug up on them.

In those installations that have the added support (3) at the head end, it is a common error to adjust it too high and also snug up the drum wedges before starting.

Some millwrights think they are setting the unit correctly by adjusting the support (3) until the level (8) reads level. That would be correct if the level (7) placed in the crosshead guide also indicated level. Sometimes the level of the crosshead guide is slightly down, and that cannot be changed because it is determined by the squareness of the guide and frame at (4). Therefore, the support (3) should be adjusted in height until the bubbles in both levels (8) and (7) read the same.

Excessive clearances in the crosshead-to-guide, crosshead bushings, connecting rods and main bearings will initiate load knocks but they will be dealt with during discussion of maintenance on the frame or power end. Other causes of knocks are liquids, loose valve assemblies and packing glands. Loose assemblies are not too difficult to locate. In regard to liquids, the knock is spasmodic as the slugs pass through the cylinder.

Now that the importance and method of using each sign or indication has been explained, it takes very little time to pass through the list of checks, which are as follows:

1. Checking for excessive rod load is required only when changes in capacity and pressures are made—the operator need not bother with non-critical cylinders.
2. Check the packing vents for liquids and discolored oil twice daily.
3. If a packing vent does not clear, conduct a cylinder inspection through a valve port.
4. Check suction and discharge temperatures and pressures twice a day.
5. Listen for knocks and noises, and check vibration twice a day.

This small number of checks does not appear to be much of a preventive maintenance program for reciprocating compressor cylinders but if it is followed, every critical and moving part can be scrutinized and protected without shutting the unit down (except for cylinder inspection through the valve port). Furthermore, no instruments are required other than the standard gauges and thermometers on every unit.

The program just described is the bare minimum that will fit the smallest pocketbook. We would not argue with anyone who would want to spend more, but it is interesting that some would consider it too extensive. However, for this or any program to be successful, a well-trained, conscientious and cost-minded person must conduct the tests and evaluate the operational warning signs and indications, and he must be given the authority to shut the unit down when the warning signs indicate it should be done.

Procedures

Although the expendable parts of engines, such as pistons, rings and valves, have been the subject of many maintenance writeups and probably are understood by almost everyone, the lower part of the engine, such as foundation, grout, frame and crankshaft, has not received enough coverage in the past. Perhaps it is because this area has been relatively trouble free; however, in the last decade, troubles there have increased. For this reason, let us look closely at the lower portion of the engine installation.

Crankshaft deflection, determined by web gauge or inside micrometer, is the most important indication or test of the condition of the foundation, grout, frame, crankshaft and main bearings. But in order to be used to full advantage, the method of conducting this test has to be understood.

Taking Crankshaft Deflections

Figure 6-9 shows the exact position for locating the web gauge (A). Note, in the right-hand view, that it is installed at the midpoint of the web. If the gauge were installed at the edge of the webs, it would not follow true web deflection as the crank is rotated. Note in the left-hand view that the gauge is located a definite distance from the centerline of the connecting rod journal. The reason for this is that the engine builder has assigned a maximum deflection to each engine. If the operator located the gauge at "B," which is out on the counterweights, the deflection recorded there would be twice the actual deflection measured at "A." In some installations the rod cap interferes with the gauge as the crank is rotated, and the instrument has to be located out on the counter-

Case 1

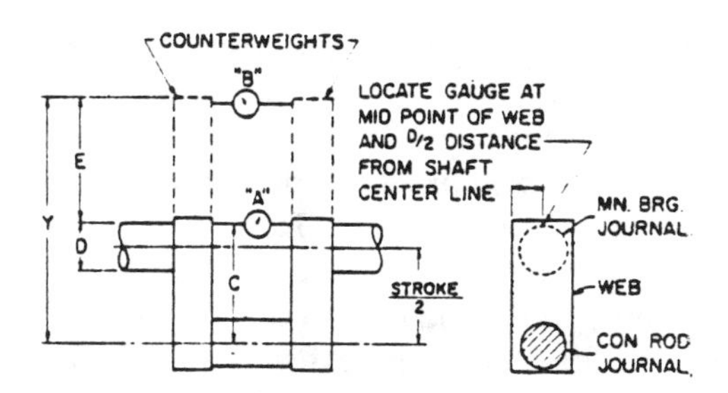

ZERO (STARTING) POSITION OF CRANK

Formula for converting "B" to "A"—A = B reading x C÷Y where "C" = ½ stroke + D/2
Y = C + E

CRANKSHAFT DEFLECTIONS AT "A"						
CRANK POSITION	THROW					
	1	2	3	4	5	6
0°	000	000	000	000	000	000
90°	000	.00025	-001	000	000	-0005
180°	000	-001	-005	-002	-00025	-001
270°	000	000	-001	-00025	000	-00025

DATE______ CRANKCASE TEMP ______ OBSERVER___

Figure 6-9. Taking crankshaft web deflections. Exact position for locating the web gauge. Note that the crank pin is down. This is designated the "zero" or starting position.

weights at "B." In these cases, the reading at "B" will have to be proportioned back to position "A" by the formula shown in Figure 6-9.

In engines where there is only one or no counterweight on the webs, the gauge must be installed out from the "A" position. The bottom centers of the webs have tapped holes (if the webs are not tapped they should be) for receiving threaded rods which will support the gauge.

Observe that in Figure 6-9 the crank pin is down. This is designated the "zero" or starting position. The reason for starting with the pin down can be seen in the following example: Let us assume that the main bearing journal to the right is low due to a bad main bearing. The webs would then be spread apart. In locating the gauge, the dial is set at zero. When the crank is rotated to the up, or 180°, position, the webs moved inward, which registers a minus (−) reading on the dial. However, if the operator used the up, or 180°, position as a starting point, the webs would be moved inward, and at that starting position the dial is set to zero; but when the crank is rotated to the down position, the webs would spread and the dial would register a plus (+) movement. The magnitude of deflection in both cases, for the same cause (bad bearing), is identical, but the signs (+) or (−) would be reversed. Starting with the throw down is standard, and the signs tell the person analyzing the readings whether the shaft is bowed or sagging.

Note in Figure 6-9 that there is a space for recording crankcase temperature. It is impossible to check or repeat readings from period to period unless they are taken at the same engine temperature. Deflections will change from hot to cold, and since hot readings are the conditions under which the shaft operates, the readings should be taken hot.

Figure 6-9 can be used as a form for recording deflections each period—which should be every six months and certainly no more than every year. In cases of foundation problems, the deflections should be taken every three months.

In regard to a "V" machine with a horizontal compressor on the throw, it is impossible to rotate the crank more than 180° without the connecting rod striking the gauge. In these cases, inside micrometers may be used to measure the distance between the webs at the 0°, 90°, 180°, and 270° positions. The webs of most large crankshafts are not too smooth, so the ends of the micrometers will have to be located in exactly the same spot for each reading. Bench marks on the webs will be helpful in assuring exact location.

Analyzing Crankshaft Deflection Readings

In order to analyze the readings, some experience and sound thinking are required, but the effort will be worth it. Because problems with the lower end of the machine are never the same, the best way to deal with instructions on analyzing the data is to go through several hypothetical cases. The reader can better follow the examples and subsequent problems by using a model crankshaft made from wire or a paper clip.

The readings listed in the table of Figure 6-9 will be used as Case 1 and were obtained from a machine with a 22 in. stroke. The engine builder assigned a maximum deflection figure of .004 in. The −.005 at the 180° position for No. 3 throw is above the specified limit, indicating that something is wrong. The 90° and 270° positions are normally used to determine whether the main bearings are out of alignment in a horizontal plane. However, when the 180° position has excessive deflection (caused by one journal being low), it carries up to the 90° and 270° positions, which in this case results in the −.001 reading. Furthermore, if the bearing saddles were out of alignment in a horizontal plane, the signs at the 90 and 270° positions would be reversed. Therefore there is nothing wrong with the horizontal alignment. (Actually, the 180° position readings are the most significant, because rarely will main bearing saddles be found in sidewise misalignment.)

CRANKSHAFT DEFLECTIONS—CASE NO. 2

CRANK POSITION	THROW					
	1	2	3	4	5	6
0°	000	000	000	000	000	000
90°	—.00025	—.0005	—.0015	—.001	—.0005	000
180°	—.001	—.003	—.006	—.004	—.0015	000
270°	000	—.00025	—.0015	—.001	—.00025	000

CASE NO. 3—SAME AS ABOVE EXCEPT ALL SIGNS (+)

CASE NO. 4

CRANK POSITION	THROW					
	1	2	3	4	5	6
0°	000	000	000	000	000	000
90°	—.00025	—.0015	000	+.00025	+.0015	+.00025
180°	—.001	—.004	—.0005	+.001	+.005	+.001
270°	—.00025	—.001	000	000	+.001	000

Figure 6-10. Crankshaft deflections illustrating different types of deflection problems that might be encountered.

Returning to the example, since the No. 3 throw 180° reading is the only one that is excessive, it is apparent that the bearing to the right of No. 3 throw is wiped. Note that this low bearing has caused distortion in the shaft past No. 3 as indicated by the −.002 reading of No. 4 throw. In this case the correction is simple, because it is only a matter of replacing the bearing.

Figure 6-10 shows a set of crankshaft deflections that will be used to explain Case 2. Here, the 180° deflections get worse from No. 1 to No. 3 throws and better from No. 3 to No. 6, and all signs are minus (−). A condition such as this means that the shaft is in a continuous bow. This can be verified by bending your wire model crankshaft into a bow and by rotating it as is done in taking the readings. It will be seen that all signs would be (−), and the highest separation of the webs would be in the middle throw. This situation is not characteristic of one or more bearings being wiped, because it is improbable that both end bearings would be wiped, leaving the center high. A typical cause for this condition is for the bond between the frame and grout at each end of the engine to have broken loose. The horizontal couple forces cause the frame to move relative to the grout, which, over a period of a year, can actually wear it down.

If this is the problem in Case 2 it can easily be checked by inserting long feelers (about 8 in.) between the frame and grout. If the feeler thickness is too great (up to .025 in.), the situation is actually worse than the deflections indicate because the frame is not supported. There are many installations in which feelers can be inserted all the way at the end of the frame, but the gravity of the circumstance is determined by how far the feelers can be moved from the end toward the middle once they are inserted. Regardless, the deflections are excessive in Case 2, and if there is a loosening of the grout, with frame movement, the unit may have to be regrouted. A common error is to tighten the foundation bolts to restrict movement. Such tightening is useless because once the bond is broken the foundation bolts cannot hold the engine down. The amount by which the maximum deflection can be exceeded will be discussed in subsequent paragraphs.

If the inspection just described indicates that the bond between the frame and grout is satisfactory and the grout has not broken up, then the bowed condition of the shaft could be caused by a change in the shape of the foundation. There is a possibility that it may be cracked. This can be verified by a thorough examination of the foundation. Almost all concrete structures have hairline cracks, which should be ignored; but open cracks, regardless of the width, are a good indication of trouble. A sketch showing the exact location of the open cracks is sometimes useful in correlating their location to the crankshaft deflections.

In regard to Case 3, if the deflections were exactly the same as Case 2 but the signs were all plus (+), then the grout or foundation is in a bad sag. Comments for this condition are the same as Case 2.

In Case 4, the changes in signs of the deflections show the shaft to be in a reverse bend. This could be caused by bad bearings, grout, foundation or frame. In this case, as well as in the preceding three, the analysis should not be confirmed or acted on until all main bearings have been inspected.

Maximum Deflection Specifications

The number of variables involved and the complexity of the problem make it impossible for an engine builder to predict the deflection at which shaft failure will occur. Therefore, a very tight maximum figure has to be assigned to any shaft so that all situations will be covered. It is for this reason that failures have happened to shafts with deflections slightly above specifications while other engines have run for years with deflections much higher than engine builders' limits. Furthermore, there are some locations that make it very difficult to keep the engine level enough to stay within the limits. The problem is to decide how far one can go beyond recommendations. The following discussion might help in making that decision.

In regard to Case 1, the change in deflection from throws No. 2 to No. 3 is very abrupt. In that case the web stress is very high, and it is recommended that the specified maximum deflection not be exceeded. This can be demonstrated by holding adjacent main bearings of the wire-model shaft and creating a bending motion. This would break the shaft quicker than by holding it at the end main bearings.

Case 4 is also a very undesirable situation in that there is a reverse bend, or "S," indicated by a change from plus to minus signs. The stress concentration in the throw between the change of signs can become very pronounced if the deflection is much above the engine builder's standards.

Case 2, which is a bow (all plus), should allow more deviation from standards than the other examples, because the stress concentration, as in the case of the sag, is not as dangerous. Also, a bow is better than a sag because in the former the deflection is minus. Where a minus reading is involved, the webs are inward from the neutral position when the throw is up. The up position is when the peak firing pressure exerts maximum force on the journal and tends to spread the webs apart. Since the webs are already inward, the peak firing pressure does not contribute as much to web stress as it does in the situation of a plus reading, where the webs are spread apart before the firing force is exerted.

It can be seen that it is difficult to assign a maximum deflection to any engine, but if the value specified by the engine builder is not exceeded under any conditions, experience has shown that the shaft should not break. It is always wise to consult the manufacturer when deflection limits are reached.

Crankcase Inspection

The preceding paragraphs have covered the foundation, grout, frame, crankshaft, and main bearings. It should be noted that those important items can be checked without disassembly of any parts, except for removal of the crankcase doors. Once the doors have been removed, the operator should take advantage of one of the most revealing inspections available to him, i.e., the crankcase inspection.

As each door is removed, its back should be inspected for foreign material thrown there by centrifugal force of the connecting rod. Bronze cuttings from a faulty wrist-pin bushing will adhere to the door. The same is true for babbitt from bearings and cast iron from liners, pistons, etc. The walls of the crankcase as well as the bottom should also be scrutinized for particles of those metals. The condition of the oil can be checked by looking for lacquer formations on machined surfaces or deposits of sludge that could come from trouble with valves, rings, or pistons. All nuts and bolts should be tapped with a hammer for the familiar ring common to tightness. Each piston should be moved to top center and the liner checked for scuffing.

Generally, two mating parts that have had a tendency to seize while in operation will generate enough local heat to discolor the casting supporting them. This is particularly true of main and connecting rod bearing caps or the wrist-pin end of the connecting rod. Consequently, the entire crankcase should be observed for any blue discoloring, and if any is found it should be thoroughly investigated. Any parts that have been hot enough to become discolored will normally be warped, cracked, or both. Therefore, they should be Magnafluxed or dye-checked for cracks. Connecting rods and main bearing saddles should be measured for warpage.

The inspection outlined so far has not consumed any more time than it takes to look at every square inch of the crankcase. The operator should make these observations every time a door is removed and certainly at intervals of not more than every three months. At all intervening crankcase inspections, the main and connecting rod bearings should be checked for clearances. As the following explanation will show, these bearings do not have to be dismantled for this check, but it will take at least two hours to complete, depending on the size of the engine.

Determining Bearing Clearances

The crankshaft web deflection test and crankcase inspection are good indicators of main bearing condition, but they must be supplemented by a clearance check.

Excessive clearance in all main bearings, which could be caused by abrasives in the oil, may not show up in web deflection tests. Excessive clearance should not be ignored in any engine, but it is less dangerous in a two-cycle engine than a four-cycle engine.

The method of determining the clearance is very controversial. Although many people use the lead wire method, this method is not recommended due to two main factors: lead wire expands after removal from the cap, or the wire can become embedded in the babbitt, especially in soft, high-lead-base babbitt bearings.

Plasti-Gauge® wire, which can be purchased in any auto parts store, is a very good medium and it is generally considered to give reliable readings, Its use suffers three disadvantages, however:

1. It is time consuming.
2. The bearing has to be dismantled at least twice, which betters the chances for human error.
3. It cannot be used on vertically split bearings common to engines where the crankshaft is removed through the side.

The dial indicator method can be used on main bearings, i.e., a dial indicator is clamped to the main bearing cap with the button of the indicator set on the shaft. The shaft can be raised by a very small hydraulic jack until it contacts the top of the bearing and the clearance is recorded on the dial. The feel of the shaft hitting the top of the bearing and the return to the bottom is quite pronounced. Several cycles of this movement will repeat the indicator reading.

Some people prefer to use a jack on each side of the journal being checked. If this method is used, caution is necessary in placing the jack to assure a solid base support for the jack. Users have had good success with this method and highly recommend it. Difficulty in raising the shaft will be encountered when the alignment is such that the shaft is in a reverse bend. There is no need for controversy in using the dial indicator to check connecting rod clearances. It has all the advantages, it is fast and precise, and it does not require disassembly. Any readings recorded by this method that are above or below specifications or a previous reading are reasons for disassembly and inspection.

Maintenance of Upper Engine

A review of what has been covered will show that the lower half of the engine has received complete attention. This was accomplished by simple inspections that required a minimum of downtime or disassembly and without expensive instruments. Everything from the connecting rod to the top of the engine will be dealt with in a similar fashion.

Piston Rings—Piston ring trouble is one of the most dreaded problems of an internal combustion engine. Engines operating with faulty rings are vulnerable to cracked pistons, cracked heads, worn liners or piston seizures with subsequent crankcase explosions. Faulty rings will also reduce the life of lubricating oil. Consequently, it is very important to be continually alert for any indicators that will point up ring condition. In large two-cycle engines, ring trouble caused by problems in the port area will result in a distinct clicking noise at the base of the cylinder. Therefore, the piston sounds should be checked daily for any unusual sounds or excessive piston slap. Excessive vapors from the crankcase breathers will undoubtedly be the first sign of ring "blow-by." When "blow-by" increases to a dangerous level, the vapor will escape past the crankshaft seals at the drive end of the frame. Other indicators of ring or liner trouble are increased lube oil consumption, breakdown of lube oil, decreased life of the lube oil filter elements, increased crankcase pressure, high exhaust temperatures and the inability of the engine to carry load without detonation.

Inlet, Exhaust, and Gas Valves—The tappet clearance of any valve should remain constant (after temperatures have leveled out) once it is set. If it becomes necessary to readjust clearances after short periods of operation, this is an indication of dangerous wear. The wear could be anywhere in the valve operating gear, such as the cam, roller, roller bushing, push rod ends, or tappet. When tappet clearance keeps changing, these parts should be inspected immediately, because their failure can result in a complete engine wreck. If, on a four-cycle engine, inspection shows these parts to be in good condition, the trouble will be found at the valve seat or valve insert. Where hydraulic lifters are used, a noisy tappet is a definite "red flag." Burned or leaky valves can result in a rough-running engine, increased exhaust temperatures, decrease in compression pressure and detonation. On four-cycle, turbocharged engines, the first sign of valve trouble is unstable air manifold pressure accompanied by intermittent misfiring and muffled detonation. This is caused by burned combustion gases leaking into the air intake manifold, thereby raising the pressure. The manifold regulator senses the increased pressure and reduces the amount of air at a time when actually more air would be needed to counteract the burned gases.

Nearly everyone is conscious of what valve trouble can do to a four-cycle engine but some do not realize what faulty gas injection valves can do to a two-cycle engine. One indication is a very hot pipe jumper between the gas header and the gas injection valve. This will be followed by rough running, missing, and detonation. Some mechanics do not appreciate the fact that a good seat is required. The safe way to ensure against gas valve leakage is to pressure-test them by air on the bench. Soap and water can be used to check the lapped seat and valve. If it is difficult to get the seat to hold, the usual cause is excessive valve bushing wear. Insufficient valve tappet clearance will burn valve seats; too much clearance is detrimental to cams, rollers, and tappets. There have been several cases of operators experimenting with gas injection valve tappet clearance which have led to cracked heads and liners and piston seizures. It should be noted that gas injection valves are not exhaust valves; they are designed to operate only in the cool part of the stroke. If allowed to be held open or leak, they will burn during the combustion and expansion periods, and compression pressure will charge the inlet jumper and header with air. When the gas valve does open, the gas charge is diluted with air or burned combustion gases.

Cylinder Heads—Cracking of cylinder heads is a problem that does not often occur, but when it does it is costly. In certain installations, a given cylinder head design may be subject to cracking problems, while in other types of installations the problem may be totally nonexistent even with the same head design. (This comment, naturally, does not apply to those installations where engines are subjected to long periods of overload and heavy, continuous detonation.)

In many instances, design and operation of cooling-water systems have been found to contribute to cylinder head cracking. It is known that, in controlling temperatures, if the spread between inlet and outlet of the engine exceeds 15°F, cracked cylinders, heads and/or exhaust manifolds are quite likely to show up. It is also reasoned that if adequate provisions for removing entrained air are not made, cracked heads are likely to appear. Some designers use traps in the high locations, while others contend that a vented standpipe which will slow the water to 0.5 ft per second is required to release the air. The relationship between entrained air and cracked heads is difficult to determine but air appears to have been a contributing factor in many failures.

Ignition—Space is too limited here to cover this subject in detail, but it is so important that a few words of advice will have to be included. Everyone knows that faulty ignition can contribute to most of the failures mentioned earlier. The problem is that not everyone has the necessary equipment to quickly analyze and locate the trouble. An operator can, by the process of elimination, find the trouble in a one- or two-unit installation. However, in multi-unit installations, it is difficult to keep ahead of troubles in magneto or interruptor systems without costly instruments. About all that a person can do without these instruments is make sure that the plugs, points, coils, condensers, gap settings, battery voltage, grounds, connections and wire insulation are satisfactory. This, of course, can be accomplished visually or by simple tests.

The ignition analyzer is recommended as a very valuable instrument for checking out ignition system components. These instruments, of which there are a number on the market, are highly versatile and can be used with pressure pick-ups to locate such varied troubles as bad valves, rings, pistons, etc., in both the power and compressor areas. Although some companies feel they cannot justify the cost of an analyzer and the man-hours required to operate the unit, analyze readings, and keep records, experience shows that this cost is well paid for in the reduction of engine and/or compressor outage. The reader may refer to Volume 2 of this series for details on instrumentation and analysis.

Ignition has for years been the critical part of an engine, but with the pulse generator and transistorized equipment, problems will be fewer in the future. As a matter of fact, if progress could be made on the life and reliability of spark plugs, the trouble-free day for ignition systems would not be too far off.

Turbochargers and Blowers—Any part of the engine involved in furnishing air for combustion is very important. Most modern engines have rotating equipment for that purpose and, although they operate at high speeds, their life and service are very acceptable if they are properly maintained. When trouble does occur with turbochargers or blowers, vibration is one of the first signs. However, this can be detected long before damage sets in by checking each month with any one of the many instruments available today. If an instrument is not allocated to the installation, the ends of the fingernails are sensitive enough to feel vibration before damage is done. This check does not require any time; therefore, it can be done every day.

A common problem with turbochargers is carbon formation on the turbine end that finally takes up end-thrust clearance, resulting in complete damage to the very expensive rotor. To guard against this, the rotor end play should be checked at least every three months. If an inspection window is provided, the check can be made without any disassembly. Oil leaks at the front end and air-impeller fouling sometimes cause problems, but this can be determined through the inspection window. The limiting factor for any turbocharger is exhaust inlet temperature. Therefore, the engine builder's maximum should not be exceeded. The turbocharger output remains fairly constant (depending on atmospheric conditions) with engine load and speed. Any decrease or increase in turbocharger speed is reason for concern. The instrument used for checking vibration should be one that records speed as well as amplitude of vibration.

The spin-down test requires very little time and is so convenient and informative that it is a "must" for maintaining this type of equipment. This test amounts to recording the time it takes the rotor to come to rest after the engine throttle is shut off. This reading can be taken any time during a scheduled shutdown. It is better to do it at no load and rated engine speed. As the throttle is moved, the stop watch can be started. The operator will have plenty of time to get around to the inspection window of the turbocharger because it generally takes four to five minutes for the rotor to stop.

Engine Balance—The term "engine balance" means that each power cylinder should produce its equal share of power. Everyone in the trade is very conscious of its importance, but there seems to be confusion among operators and mechanics as to how to accomplish it.

Some people periodically check the exhaust temperatures and peak firing pressures of each cylinder and, if they are not normal, adjust the air-to-fuel ratio controls or change the amount of gas supplied to certain cylinders. This practice of making adjustments without first determining the reason for unbalance is not sound. There are others who will change the same adjustments to relieve a detonating engine with the assertion that the cause will be investigated later. This approach can be dangerous and in the meantime causes unnecessary work. An engine installation should be viewed as an engine with a certain number of adjusting knobs on it. The simple installations may have one or two adjustments. In order to get the "last squeal out of the pig," modern engines have several adjusting knobs. The point is that there is only one position for each knob and, once they are set in that position, they should never need changing because adjusting screws of gas valves or air-to-fuel ratio controls do not wear. A change in air-to-fuel ratio or balance is not caused by adjustments changing but is due to some malfunction of the engine.

For example, in the case of a two-cycle engine, an increase in air manifold pressure indicates carbon in either the intake and/or the exhaust ports. Carbon does not form in equal amounts in all cylinders, which upsets engine balance. The balance may be restored by a change in adjustments but it will last for only a day or two. In other words, it is impossible to keep an engine in balance with carbon in the ports. The proper procedure is to watch for signs that carbon is forming so that port cleaning can be scheduled. In addition to carbon in the ports, other malfunctions such as faulty ignition, valves, turbochargers, blowers, and piston rings will upset engine balance. Consequently, the correct approach to engine balance is to repair the assembly that is causing the unbalance.

Engine Safety Devices—The trend in new installations is toward complete or partial automation—in these cases, plant designers include a safety shutdown device for every engine function. The instruments are called on to do a very difficult job because they do not operate for long periods. In the meantime they collect rust, dirt, moisture, and in some instances oxidized oil—and when called on to protect the engine they may not function. Also, the problem of false shutdowns is sometimes temporarily put off by blocking out the instrument. This has proved costly and embarrassing in many instances. It should be a hard and fast rule in all installations that the engines be shut down at least every six months by actual operation of every shutdown device.

This section may impart the feeling that if engines are to be maintained as suggested, there would not be any time left for running. However, remember that throughout, the theme is to watch for signs and make inspections that require a minimum of downtime and disassembly. Most can be done while the unit is on the line and require few and inexpensive instruments. Some are only observations. They are elementary and well known and have been used by the old-timers for many years. If used with judgment, these simple indicators can be used to maintain any modern installation.

The modern trend, however, is to forget these basic items and to deal only with exotic instrumentation for maintaining engines, and it is not only a trend that can damage the equipment but can also be quite expensive in terms of operational downtime. Instruments such as the ignition analyzer, balance pressure indicator, etc., are here to stay and are highly recommended as a part of the basic maintenance equipment for any installation.

There is also a tendency for some engine-builder field representatives, as well as operators, to make field design changes before making sure of correct installation and basic adjustments. The redesign approach is healthy, as it keeps the engine builder on his toes and does give him valuable information. However, when overdone it deters from necessary basic maintenance procedures, and the equipment will not function as intended.

Reciprocating Compressor Component Overhaul and Repair

Usually we will try to inspect the reciprocating compressor whenever the opportunity arises. Each plant has its favorite "hours elapsed" which would indicate that it is time to do an inspection. The box at right shows a typical checklist used during a reciprocating compressor inspection. This checklist would be used in conjunction with an appropriate maintenance data sheet similar to the one shown later in Figure 6-17.

While there could be a substantial or near endless array of possible repair procedures we would like to concentrate on three important activities around our reciprocating compressors:

1. Valve repairs
2. Packing replacement
3. Cylinder honing

Valve repairs are likely to occur most around reciprocating compressors. The following is a typical example of a procedure used to remove, repair, test, and reassemble compressor valves.

RECIPROCATING COMPRESSOR OVERHAUL CHECK LIST

1. Check clearance of main and connecting rod bearings. Inspect crank pin, crosshead pin bushings, crosshead bearings.
2. Inspect main bearing and shaft alignment, and crankshaft alignment.
3. Check alignment of crosshead and piston and readjust shoes, if necessary, using level on crosshead guide and cylinder base.
4. Check alignment of cylinder to crosshead and frame
   (a) Mount dial indicator on cylinder housing and take reading on rod in horizontal and vertical direction through stroke.
   - if greater than 0.003″ in vertical correct by shimming crosshead shoes.
   - if greater than 0.003″ in horizontal, check various fits and joints and correct as necessary.

   (b) Alternate method: Use alignment wire through centers of cylinder bore and crosshead guide bores.
5. Inspect all stud nuts for tightness—crankcase cover, distance piece and foundation bolts.
6. Inspect, measure and record piston and cylinder diameters.
7. Inspect, measure record piston rings, rider rings, piston ring grooves.
8. Inspect valves.
9. Inspect oil coolers, water jackets, intercoolers.
10. Inspect lubrication devices, oil filters, sumps and piping.

Disassembly of API style, single deck plate valve* as shown in Figure 6-11:

1. Remove valve cage from valve assembly.
2. Remove Drake nut (8) from valve stud (7), and remove seat (11), if suction valve, or guard (10) if discharge valve, for access to valve springs (13) and valve plates (4 and 5). Remove stud bolt (7) if necessary.

We would now proceed to perform a bench inspection of the valves. Here are the steps to be followed:

1. *Guard and Seats.* Use an approved cleaning solvent. Clean and blow dry, using compressed air if available. Check for cracks,

* API = American Petroleum Institute Standard 618

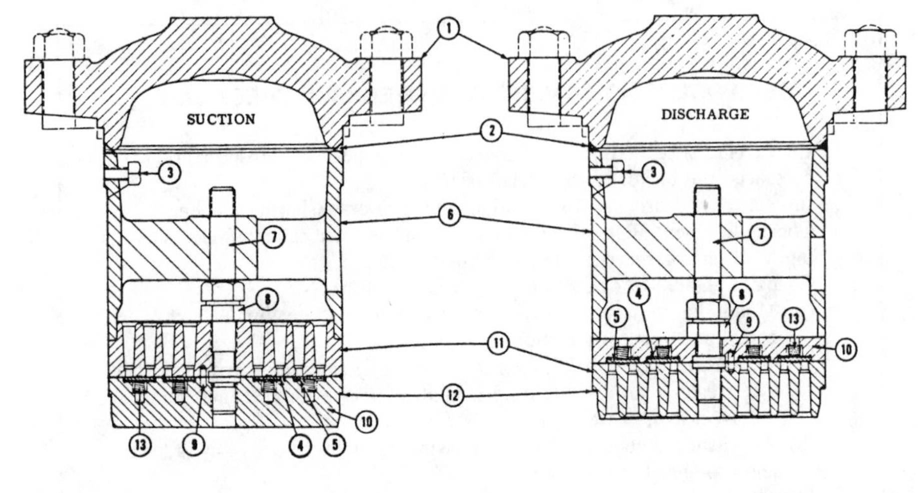

1. Valve Cap
2. Valve Cap Gasket
3. Set Screw
4. Valve Plate (Inner)
5. Valve Plate (Outer)
6. Valve Cage
7. Valve Stud
8. Drake Locknut
9. Seat to Guard Dowel
10. Valve Guard
11. Valve Seat
12. Valve Seat Gasket
13. Valve Plate Spring

Figure 6-11. Single deck plate valve assembly for reciprocating compressor—API style.

warpage, and wear. Seat and guard can be checked by placing a metal straight-edge on the mating surfaces of seat and/or guard. If warpage or indentations of the seat are minor, the valve seat can be reground. If guard is worn or warped, it must be replaced. Check valve gasket surfaces for burrs or roughness that would prevent seating of the gasket. One caution—if valve seat surface has been hardened to reduce wear, excessive grinding will reduce or destroy the seat hardness.

Carefully check damped plate and channel valve guards for wear. Proper damped plate valve operation depends upon a close clearance between the plate and groove in the guard. Excessive clearance will require replacement of the guard.

Valve plates should be considered for replacement with new ones when doing a valve inspection. There are some good reasons for this: Valve plate material for each compressor cylinder is selected to give the best operation possible. Valve plates are available in chrome vanadium steel, stainless, nylon or thermosetting plastics.

Valve plates can be installed with either side facing the valve seat. Valve plate failure always results in breakage and new plates must be installed.

Damped valve plates must be checked for freedom of movement and clearance in the guard grooves and channel valve plates must be checked on the guard ribs. Rotate each plate one full revolution. If plates do not turn freely they must not be used. After assembly of the damped plate into the guard, move the plate sideways as far as it will go. The clearance at any point between plate and guard should not exceed .011 in.

After assembly of the channel plate over the guard rib, move the channel sideways as far as it will go. The clearance at any point between guard rib and channel should not exceed .007 in.

One reputable compressor valve repair shop offers the following inspection service:

1. Each valve is identified.
2. Each valve is dismantled and inspected for wear or breakage. An inspection report is issued.
3. Seats, guards, bolts, etc. are cleaned with a vapor blaster.
4. Seats and guards are inspected for cracks.
5. Seats and guards are remachined to OEM* specifications and, if necessary, metalized before machining. Seats are lapped or concentrically ground.
6. Each valve is assembled. Only OEM parts are used.
7. Each valve is function tested, plus two leakage tests: (a) liquid-fill, observe for 1.5 minutes and slots of seat must still be half full (b) connect to metered air supply system.
8. If the valve contains steel parts subject to corrosion, it is dipped into rust-preventing oil and wrapped.
9. Each valve is identified by a sticker on the wrapping.

Reassembly of the valve takes place in the following order:

1. Replace valve stud bolt—if removed—by turning it firmly into guard (suction valve), or seat (discharge valve).
2. Reassemble the valve by first inserting springs in recess in guard.
3. Position valve plates (4 and 5) on springs (13) and depress plates. Hold plates and springs in place by installing special valve plate clips. See Figure 6-12—or, if special retaining T-bolts are used, see Figure 6-13.

---

* Original Equipment Manufacturer

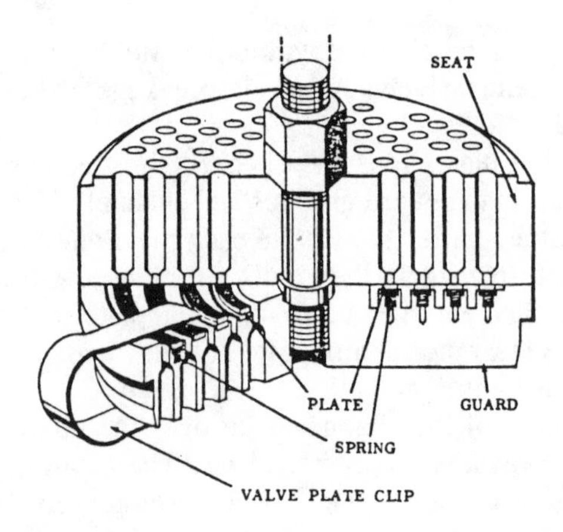

Figure 6-12. Application of valve plate clip during plate valve assembly.

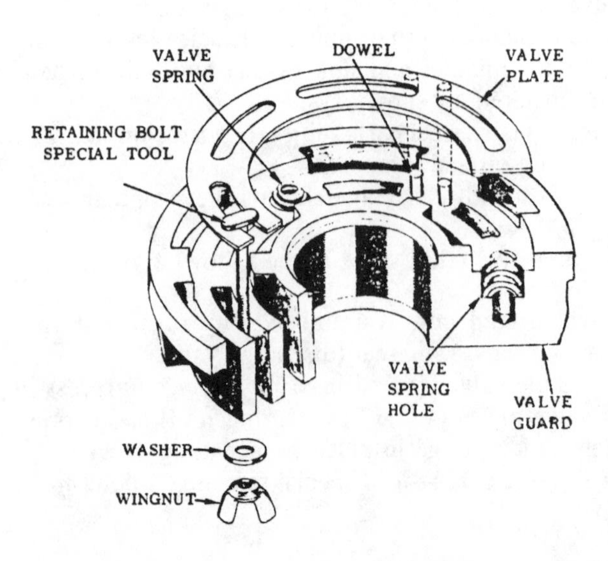

Figure 6-13. Application of "T-bolts" during plate valve assembly.

One should watch that ported plates are located on guard so that locating dowels are on each side of "land" between milled ports on plates. Dowels prevent plate from rotating (Figure 6-13). When an indexing dowel is used, the dowel must be aligned with index hole in valve plate to prevent plate rotation.

4. For suction valves, hold plates in position on the guard (10) with special valve clips or T-bolts and install the valve seat (11) over stud (7) onto the guard. For discharge valves, install valve guard (10) over stud (7) onto valve seat (11). Alignment dowel (9) in guard (10) must align with hole in seat (11).
5. Install Drake locknut (8) on valve stud finger-tight and remove special clips or T-bolts. Tighten Drake locknut as recommended in parts list drawing if given, or per manufacturer's torque recommendations.
6. Check valve plate operation by inserting a screwdriver or similar tool through the valve seat ports, forcing the valve plates off the seat. Repeat this at several points on each plate to assure that plates are free to open and close. Failure of the plate to unseat indicates that plate is not aligned with dowels in guard, and the valve must be disassembled.
7. Install assembled valve in cylinder.

Installation of the valves in the cylinder must be executed with the greatest care. Do not install a suction valve in a discharge pocket or vice versa. Such an installation could result in excessive internal cylinder pressures, creating an extremely hazardous condition.

One should proceed as follows:

1. Install a gasket on the valve seat or guard.
2. Carefully enter the valve into the cylinder pocket. Special tools are available for lifting and installing large valves.
3. Single deck valves are usually installed with the cage. Double deck valves and gas-operated plug-type unloader valves require installation of the cage after the valve is located in the pocket unless valve is threaded into cage or cage is part of valve assembly.
4. With the valve and cage installed, tighten two set screws near the top of the cage to hold the assembly in position. Do not tighten the set screws excessively; this will damage the cylinder pocket. Safety wire the set screws to prevent loosening during operation.
5. Place a gasket on the cage and install the valve cap, clearance bottle or unloader bonnet. Tighten nuts to parts list drawing specifications, if given, or refer to torque recommendations.
6. Some valve caps may have the 45° chamfer while some may have a "J" groove seal. Neither design should be over-tightened, but with

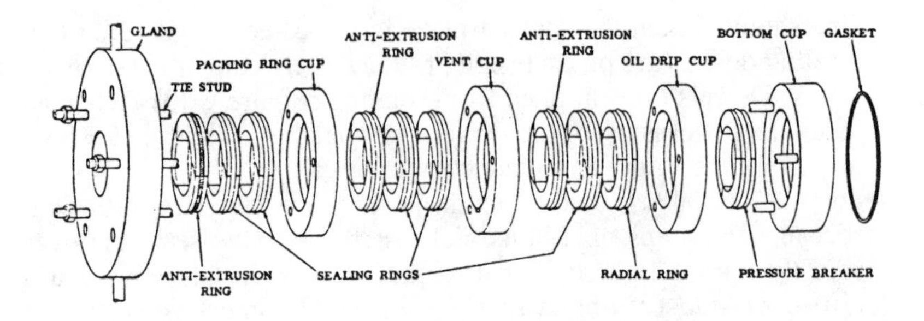

Figure 6-14. Typical TFE piston rod packing—exploded view.

particular reference to the "J" groove seal, additional torque applied to valve cap stud nuts will not result in further sealing since the cap is already in metal-to-metal contact with the cage.

Packing Replacement. Rod packings are furnished in a variety of designs and materials to cover a wide range of operating conditions. Accordingly, the following instructions are general, but should prove helpful in the application and maintenance of a plant's particular packing.

Minimal and nonlubricated packings are shown in Figure 6-14 as an exploded view of a typical TFE* cup type packing with the various parts named to identify them with the functions and the terms used. Rod packings consist of a series of TFE filled packing rings held in place around the piston rod and enclosed by a metal case. In low pressure cylinder applications, a combination of radial and tangential rings is used. High pressure applications use an anti-extrusion ring in addition to the conventional radial and tangential rings. The purpose of the metallic anti-extrusion rings is to prevent, as a result of pressure, the extrusion of the sealing rings between the annular space formed by the packing case and the piston rod.

In order to reduce the heat generated by the sealing forces of the rings on the piston rod, water is sometimes circulated through the packing case as illustrated in Figure 6-15.

A note of caution! When stopping the unit, coolant—where applicable—must be shut off and drained from the packing before the cylinder is

* Tetrafluorocarbon (Teflon®)

depressurized. Also, before installing the packing, record all identifying markings from the case and packing for reference when replacing individual components.

Operational break-in of a packing is very important. Improper break-in can result in faulty sealing and damage to the piston rod. The following points will in most cases assure proper break-in and operation. Break-in time will vary depending upon conditions.

1. *Cleanliness*—The packing and rod must be clean! If the rod has any nicks or corrosion in the area that will enter the packing, carefully remove nicks and clean the rod.
2. *Pressure*—During the break-in period, it is advisable to gradually increase the cylinder pressure and vary the operating speed if possible. This allows the packing to "wear-in" and to conform to the piston rod before being subjected to full load pressure. Wear-in time will be shortened if it is possible to operate at a low load, rather than at no load. If during break-in, leakage or overheating appears to be increasing, or out of control, reduce pressure and/or speed to allow packing to stabilize. If this does not reduce leakage, the unit should be stopped for packing inspection to determine the cause.

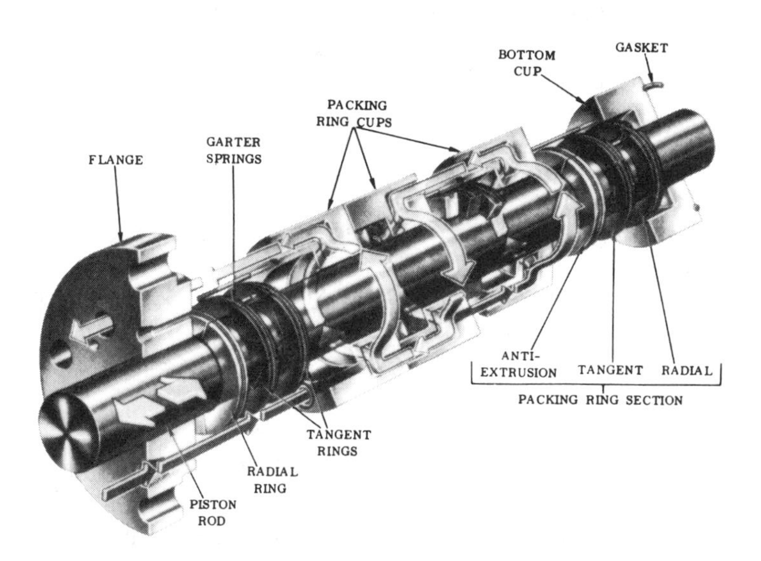

Figure 6-15. Typical nonlubricated piston rod packing showing coolant flow.

A typical detailed break-in procedure for tungsten carbide coated rods is described in the following:

- Machine made ready to operate by process personnel.
- Set load at 25 percent and run machine for *five minutes.*
- Shut machine down and allow piston rod to cool down to ambient temperature.
- Set load at 25 percent and run machine for *20 minutes.*
- Shut machine down and allow piston rod to cool down to ambient temperature.
- Set load at 50 percent or less and run machine for *one hour.*
- Shut machine down and allow piston rod to cool down to ambient temperature.
- Run-in is now completed.
- Put machine on line and set for desired operation.

The purpose of this run-in procedure is to allow the TFE packing to warm up gradually and "flow" to the contour of the piston rod. The theory is that putting the machine on line, as one might have done in the past, just "melts" and burns the packing so it can never seal.

Removal of piston rod packing consists of these steps:

1. Remove crosshead guide side covers.
2. Position compressor piston at extreme crank end of its stroke. If the packing case does not require removal, it is not necessary to disconnect the piston rod from crosshead.
3. Remove crosshead diaphragm packing
4. Remove crosshead diaphragm—Figure 6-16—and move diaphragm toward crosshead as far as possible.
5. Disconnect piping from the packing case flange.
6. Remove nuts from flange studs and slide case from pocket in cylinder head.
7. Remove nuts that hold flange and packing cups together and slide flange toward crosshead.
8. Separate packing cups, then remove garter springs and packing ring segments.

Packing Cleaning and Inspection. Clean all metal parts in a suitable solvent and dry thoroughly with a lint-free cloth. Inspect all parts for cracks, breaks, scoring and wear. Inspect for weakened garter springs. Do not break the corners where any two surfaces of a packing ring set match. Do not file TFE packing.

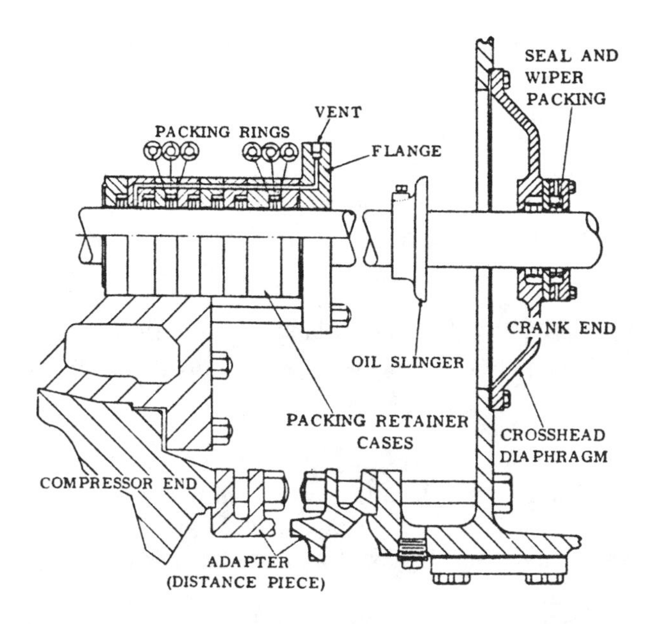

Figure 6-16. Typical nonlubricated crosshead diaphragm and compressor packing.

Inspect the piston rod for evidence of damage. If new packing rings are to be installed, check the rod for wear and misalignment. All minor defects in the surface of the rod should be removed by lapping. If the rod is deeply scored, or if shoulders are present, the rod should be refinished or replaced.

Inspect the surface of the cylinder head against which the packing rests. It should be cleaned and free of defects.

Special size packings may be required for undersized rods.

Pressure Breaker Ring. The pressure breaker ring is sometimes used to stabilize the pressure in the packing case during the suction and discharge cycle. This prevents damage to packing rings and garter springs due to the shock effect of the differential pressures involved.

Anti-Extrusion Rings. Since filled TFE components do not have the physical strength of their metallic counterparts, anti-extrusion rings are used with TFE packing. Their primary purpose is to prevent, as a result of pressure, the extrusion of the packing ring between the packing case and piston rod. The metallic anti-extrusion ring also helps to dissipate heat from the packing ring to the piston rod and packing cup.

The general practice is to allow .010 to .015 in. diametrical running clearance between the anti-extrusion ring and the piston rod.

A word of caution: In each cup having 3-ring construction with a radial cut ring as part of a sealing element and a radial cut anti-extrusion ring, the radial cut sealing element when assembled on the rod will have clearance at its joints and is to be placed nearest the pressure source when positioned in packing cup. The anti-extrusion ring when placed on the rod will not have clearance at its joints, will be free to turn on the rod and is placed farthest from the pressure source when positioned in the packing cup.

TFE Packing Installation. When servicing and installing piston rod packing, the following points should be observed to help prevent packing failure:

1. The cups, rings, and piston rod must be clean. When assembling packing on the rod, cleanliness is imperative. Dirt on the ring seating surface will cause the rings to stick, and dirt on faces of cups will throw the packing case out of line and permit leakage between cups.
2. The faces of the cup must be parallel.
3. Position the packing gasket or O-ring in the bottom cup or cylinder as applicable. Always use a new packing case gasket or O-ring at this location.
4. The face of the pocket on which the packing case gasket seats must be clean and free from burrs or scars to provide a flat seat for the packing.
5. Position the pressure breaker ring in the bottom cup, if applicable.
6. If water cooling is used, see that O-ring seals are properly installed.
7. Position the next cup on piston rod and butt it against the bottom cup. Install the radial ring, tangential ring, and anti-extrusion ring, if applicable. See that alignment dowels are properly aligned, if applicable.
8. Make sure packing rings are assembled so that match-marks are properly aligned and that the proper side of the ring is facing pressure. Refer to Figures 6-14 and 6-15.
9. Continue installing packing rings, one at a time, until the complete packing is in place.
10. Tighten the packing case stud nuts evenly to assure proper sealing. Refer to parts list drawing for tightening recommendations, if given, or torque recommendations.
11. While tightening the packing check with feelers between the bore of the packing case gland and the piston rod to be sure packing is being centered over rod. After the case has been installed, the

clearance between the rod and the bore of packing case should be checked with the compressor piston at both ends and in the center of its stroke.

12. After installing the packing case, connect the oil, vent and coolant lines where applicable.

Cylinder Honing. After an accident a cast iron cylinder can often be salvaged by honing it. Table 6-1 reflects the operations involved. This procedure can be practiced in the shop as well as in the field. It is applicable to both compression and power cylinders.

Reciprocating Unit Preventive and Predictive Maintenance

In Volume 2 of this series and in the preceding pages we described how reciprocating machinery of the type shown in Figures 6-1 and 6-2 as well as large diesel engines and other large internal combustion engines lend themselves uniquely to preventive or predictive maintenance activities. We would like to conclude this section by listing some of the important steps in preventive and predictive measures around reciprocating units. Here, in summary of most of the foregoing, is what should be regularly done for maximum unit availability. For safety's sake, remember: "Don't touch, just look."

Lower Portion

Foundation

- Correct a loose grout problem as soon as possible. Use an oil-resistant grout, normally a resin epoxy. Refer to Volume 3 of this series for details.
- Make sure torque on foundation bolts is adequate.
- Look for any movement between engine base and grout. If any is found, consider fixing with rails and shims.
- Check compressor cylinder supports and bottle wedges. Confirm that all support and bottle wedges are tight and have not been broken. If any are found to be loose or broken, confirm that distance piece studs have not broken.

Base, Frame and Crankshaft

- *Take main-bearing bridge readings and crankshaft-deflection readings periodically to detect misalignment problems early.* Record web deflections.

Table 6-1
Routing of Field/Shop Honing of Reciprocating Compressor Cylinder

No.	Operation	Description
1	Wash out cylinder.	Thoroughly wash cylinder with kerosene or varsol; remove all deposits of carbon and oil.
2	Dry cylinder.	Dry cylinder with a clean, lint-free rag.
3	Determine condition.	Check cylinder to determine exact honing requirements.
4	Remove build-up and dress down cylinder.	Cylinder Has Been Scored to Depths of Over 0.005″: Use a coarse stone (#136) to remove the extra build-up and dress down the cylinder walls; leave a depth of 0.002″ to be finished with a #236 stone. Scoring Is Less Than 0.002″ Deglaze cylinder
5	Deglaze cylinder.	To deglaze use a #236 stone or a #200 series stone—follow procedure outlined in Step 9. Using Stones: All stones must be used dry on cast iron cylinders. Keep stones and guide blocks together, and use as sets; never use the same pair of guide blocks with different sets of stones. Hone Driving Motor: For best results, use an air-driven slow speed drill (250–400) rpm). The drill must have right-hand rotation, and have a ¾″ or larger chuck size.
6	Insert and expand stones and guides against cylinder walls.	Insert and expand stones and guides firmly against cylinder walls by turning clockwise the winged collar on the hone. Note: During this adjustment stones should not extend more than ½″ out of the cylinder.
7	Hone cylinder.	a. Push stone to bottom of cylinder; allow stones to go through lower end of the bore ½″ to 1″. b. Start stroking at bottom of cylinder using short strokes to concentrate honing in the smallest and important section of cylinder. c. Gradually lengthen stroke as metal is removed and stones contact higher on cylinder walls. Stroke all the way to top of cylinder; maintain a constant steady stroke of about 30 cycles per minute. Cylinder Condition—An excellent indication of cylinder condition is speed of the drill. A reduction in drill speed during honing indicates a smaller diameter; localize stroking at such sections until drill speed is constant over at least 75% of cylinder length.
8	Check for cylinder appearance.	After free stroking (including no binding) for approximately one minute, remove stone and check for cylinder appearance. Finished Cylinder—For ideal seating, finished cylinder should indicate a diamond-shaped hatch pattern as follows: For Bronze Piston Rings: 24–30 micro-inches. For Teflon Rings: 16–20 micro-inches. DO NOT OVER-HONE. If finish is too fine, oil consumption will be excessive. Use roughness comparators to check finish.
9	Clean Cylinder	Use a soap and water solution to remove small particles off the cylinder and hone stone which would cause rapid wear.

- *Check Condition and Clearance of Main Bearings.* Inspect bottom of crankcase for signs of bearing materials. Visually inspect each main and crank bearing for signs of bearing failure without removing. Use this information as a guide to determine additional work, if any. Measure and record radial clearance on all main and crank bearings. Use feeler gauge and/or dial indicator gauge with jack as needed. Determine that all lateral lines from main oil header to bearings are in good shape and that each fitting connection is tight.
- *Check Flywheel for Tightness on Crankshaft.* Torque all flywheel bolts. Visually inspect contact area for fretting. If fretting is occurring, remove flywheel, clean and inspect contact surfaces. Reassemble and torque all bolts.
- *Check Condition and Clearance of Crosshead Slippers.* Visually inspect each crosshead slipper for signs of bearing deterioration. Use feeler thickness gauge to check clearance between upper slipper and crosshead guide without removing. Record this clearance.
- *Check Condition and Clearance of Crosshead Pin and Bushing.* Without removing, visually inspect crosshead bushing for signs of deterioration. Using dial indicator and bar, shift crosshead and record movement or slack in bushing. If excess clearance is found, pull pin and bushing as needed.

Upper Portion/Cylinders

- *Check Condition and Clearance of Power Piston Articulated Pin Bushings.* Visually inspect bushings for signs of deterioration. Use dial indicator and bar to measure clearance. If excessive clearance is indicated, change out bushing and/or pin.
- *Check Camshaft and/or Layshaft Drive Chain.* Adjust idler to tighten chain when needed.
- *Check Water Pump Drive Chain Tightness.*
- *Check Lobes, Camshaft, or Crankshaft.* It is not intended that any disassembly of the camshaft box be made to complete this preventive maintenance step. Instead, only measure and record the normal lift of the valve stem at the top of each power cylinder head. Measurement is visual, using a steel scale. Observation of such measurements should be recorded to the nearest 1/16 in. Note: "Normal" or "standard" lift should be the same for all valves for a specific model and type engine. Deviations from this value are an indication of worn lobes on the camshaft, and appropriate repair steps should be initiated.

- *Inspect Power Cylinders with Boroscope.** Remove spark plug or gas/air injection valve. Insert boroscope into cylinder, and examine cylinder wall, valves and/or ports for abnormal conditions.
- *Check Condition and Clearance—Master Rod Bearings.* Visually inspect bearing for signs of deterioration. Check bearing clearance with dial indicator and jack.

Valves

- *Check Fuel Gas Injection Valves.* One could plan that a spare set of valves would be installed once each year under this preventive maintenance step. Valves removed would be repaired as needed and then warehoused for another unit, or for the next changeout.
- *Check Air Starting Valves.* It could be planned that a spare set of valves would be installed once each year under this preventive maintenance step. Valves removed would be repaired as needed and then warehoused for the next unit, or for the next changeout.
- *Check Compression on All Power Cylinders.* Remove spark plugs, insert compression tester, rotate engine, and record compression pressure on each power cylinder. This test confirms condition of rings and/or valves. Repair as needed.
- *Run Bridge Gauge on Power Valves.* Use special bridge gauge available from engine manufacturer, and with thickness gauge, determine amount of wear of existing valve.
- *Turbocharger—Clean Blades and Inspect Bearings.* Remove existing turbocharger and install spare. Disassemble existing turbocharger and repair as needed.

Compressor

- *Inspect Compressor Piston Rod.* After removing distance piece covers, visually inspect or feel with hands the compressor piston rod for signs of wear or scuffing.
- *Inspect Compressor Valves—Suction and Discharge.* All suction and discharge valves should be removed from the cylinder, tested with a solvent, repaired as needed, and returned to the cylinder.
- *Inspect Compressor Piston Rings.* Remove compressor piston from cylinder and confirm that compressor piston rings are in good operating condition. Measure amount of wear on each ring; if wear is excessive, replace rings.

---

* Endoscope

- *Inspect Compressor Piston Ring Lands.* While piston is out of cylinder inspect and measure width and depth of compressor piston ring grooves and record measurements. See also Figure 6-17.
- *Inspect and Measure Compressor Cylinder.* During the previous steps, use inside micrometers to measure wear on compressor cylinder bore. Measurements at three points along the axial length of piston travel, and both vertically and horizontally at each point, should be made. Record all six inside diameter measurements. See also Table 6-2.

Oil System

- *Clean Crankcase Breather System and Engine.* Clean all foreign materials from vent line.
- *Change Crankcase Oil—Engine.*
- *Change Governor Oil.* Drain and replace oil in reservoir with recommended oil.
- *Change Oil Filter—Turbocharger.*
- *Check and Record Turbocharger Oil Pressure.*

Routine Checks and Adjustments

1. *Check Compressor Piston Rod Packing.* Visually inspect packing box for excessive gas leak. Be certain that distance piece covers are replaced, and that vents and drains are working properly.
2. *Run Vibration Check—Turbocharger.* Complete with special equipment.
3. *Inspect Rocker Arms and Push Rods.* Visually inspect to be certain they are properly lubricated and each shows no excessive wear.
4. *Adjust Lifters.*
   - *Solid Lifters.* Adjust tappet to allow for proper clearance. See manufacturer's recommendations in engine operating manual.
   - *Hydraulic Lifters.* Physically examine each lifter to be certain that each is operating on a hydraulic cushion. Be certain that lifter is operating near mid point of full range travel.
5. *Check Timing.* Set ignition timing per manufacturers recommendations to achieve maximum fuel gas economy. All ignition magnetos should be equipped with adjustable cradle to permit timing change without engine shutdown.
6. *Check, Clean, and/or Change Spark Plugs.* Check plugs with spark indicator to be certain that proper spark is generated and plug is not grounded. Clean existing plugs with abrasive blast and reset gap.

7. *Balance Power Cylinders.* Prior to completing this step, confirm that engine timing is satisfactory; fuel gas pressure is normal; scavenging air pressure is normal; and lifters are operating normally. Use Pi meter and/or BMEP* indicator to balance load on each power cylinder by adjusting fuel gas valve. Be certain that at least one power cylinder fuel gas valve is fully open when work is completed.
8. *Check Safety Shutdowns.* Refer to safety shutdown test manuals for procedures to be followed in testing each shutdown. Be certain all appropriate test information is included on the test sheets.

---

* BMEP = Break Mean Effective Pressure

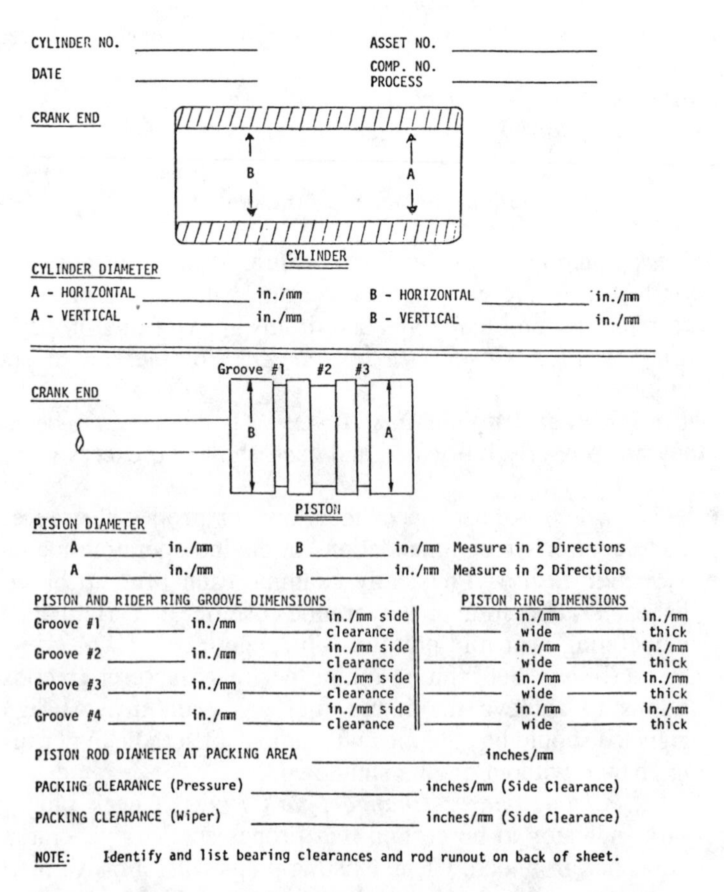
CYLINDER NO. ________ ASSET NO. ________

DATE ________ COMP. NO. PROCESS ________

CYLINDER DIAMETER

A - HORIZONTAL ________ in./mm B - HORIZONTAL ________ in./mm

A - VERTICAL ________ in./mm B - VERTICAL ________ in./mm

PISTON DIAMETER

A ________ in./mm B ________ in./mm Measure in 2 Directions

A ________ in./mm B ________ in./mm Measure in 2 Directions

PISTON AND RIDER RING GROOVE DIMENSIONS / PISTON RING DIMENSIONS

Groove #1 ________ in./mm ________ in./mm side clearance ‖ ________ in./mm wide ________ in./mm thick

Groove #2 ________ in./mm ________ in./mm side clearance ‖ ________ in./mm wide ________ in./mm thick

Groove #3 ________ in./mm ________ in./mm side clearance ‖ ________ in./mm wide ________ in./mm thick

Groove #4 ________ in./mm ________ in./mm side clearance ‖ ________ in./mm wide ________ in./mm thick

PISTON ROD DIAMETER AT PACKING AREA ________ inches/mm

PACKING CLEARANCE (Pressure) ________ inches/mm (Side Clearance)

PACKING CLEARANCE (Wiper) ________ inches/mm (Side Clearance)

NOTE: Identify and list bearing clearances and rod runout on back of sheet.

Figure 6-17. Reciprocating Compressor Maintenance Data Sheet.

9. *Turbocharger—Check Cooling Water ΔT.* Monitor temperature rise of cooling water across the turbocharger and record.
10. *Service Auxiliary Belts and Bearings.* This means that belts are tightened properly, all bearings greased or lubricated, and all sheaves are properly aligned.
11. *Change Oil Filter—Engine.*
12. *Check Force Feed Lube System.* Disconnect tubing at each lubrication point and confirm that lube oil is reaching that point.
13. *Change and Clean Inlet Air Filter*
    - *Oil Bath Type.* Change oil and clean reservoir. If differential pressure remains excessive, steam clean the mesh pads.
    - *Dry Type.* Change out filter elements.
14. *Run Jacket Water Analysis.* If a laboratory is available, complete chromate analysis and determine pH of water. If a laboratory is not available, take quart sample of water to the nearest commercial laboratory for analysis.

Part II

Maintenance for Power Generation and Transmission

Chapter 7

Power Transmission Gears*

Gear drives have always been a necessary part of industry. Consequently, since industrial personnel need to have a good working knowledge of gearing, some of the basic principles of operation, installation, lubrication, maintenance, troubleshooting, and repair of power transmission gears are outlined in this chapter.

Four basic types of gears comprise most of the heavy duty industrial gearing in use today. Therefore, due to time and space limitations, the main gear types covered here are the most common ones used to transmit high torques: single and double helical, spiral bevel, and spur gearing. However, most of the information presented here can be applied to other gear types with slight modification.

Introduction

Gears are among man's oldest and best recognized mechanical devices. They create the impression of positive action—coordinated, interlocked, precise effort to secure a desired result. Gears were once primarily used for navigation, timekeeping, grinding, etc. Now, the automobile transmission is probably the most common use of gearing that the everyday citizen sees.

Gears are machine elements that transmit motion by means of successfully engaging teeth. Of two gears that run together, the one with the larger number of teeth is called the "gear." The "pinion" is the gear with

* By James R. Partridge, Chief Engineer, Gear Division, Lufkin Industries, Inc., Lufkin, Texas. Reprinted by permission.

the smaller number of teeth. A rack is a gear with teeth spaced along a straight line and is suitable for straight-line motion. Many kinds of gears are in general use. For each application, the selection will vary depending on the factors involved. One basic rule of gearing is that to transmit the same power, more torque is required as speed is reduced. The torque is directly proportional to speed, and therefore, the input and output torques for power transmission are directly proportional to the ratio if efficiency is neglected.

Gears are usually used to change the speed of the driven equipment from that of the driver, to alter the direction of power flow, or to change rotation direction. In very few cases are the most efficient design speeds of a driver and driven machine identical. Probably the most common example we see of this is the modern automobile where we use gearing to change both the speed and the direction of power flow from the engine to the wheels. In this case, the reciprocating engine would be extremely large should gearing not be available to change the speed.

Today's best reason for using gearing is to conserve energy due to its scarcity and high cost. In most cases, more efficient drivers and driven machines can be used when a gear is available for a better speed match. For instance, steam turbines operating at speeds available for reciprocating compressors would be very inefficient. In addition, the use of gears enables a reduction in the size of driving and/or driven machines and comparable conservation of materials since higher-speed machines tend to be smaller than lower-speed ones for the same amount of work produced.

Other reasons for using a gear unit are to change the direction of power flow and to change the direction of rotation between the driving and driven machines. Were gears not available to perform all of these important functions, designing and constructing compact, efficient machinery systems would be virtually impossible. Gearing gives the engineer the flexibility to make the machine system fit the job, not the other way around. The power loss of 1.25 to 4.0 percent is a small price to pay for the advantages obtained.

Gear Types

Some of the common gear types are listed below:

1. *Spur*—Cylindrical in form and operate on parallel axes (See Figure 7-1). The teeth are straight and parallel to the axis.
2. *Helical*—Cylindrical in form and have helical teeth—teeth set at an angle to the axis (See Figure 7-2).

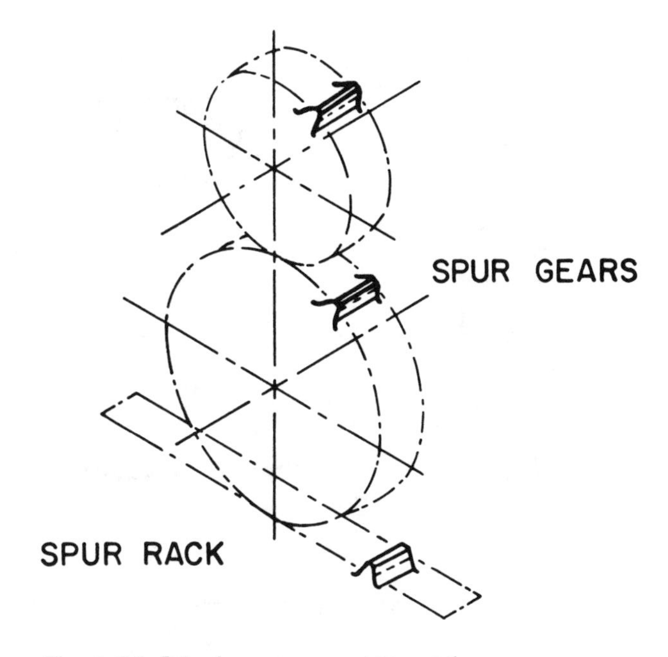

Figure 7-1. Pair of spur gears and spur rack.

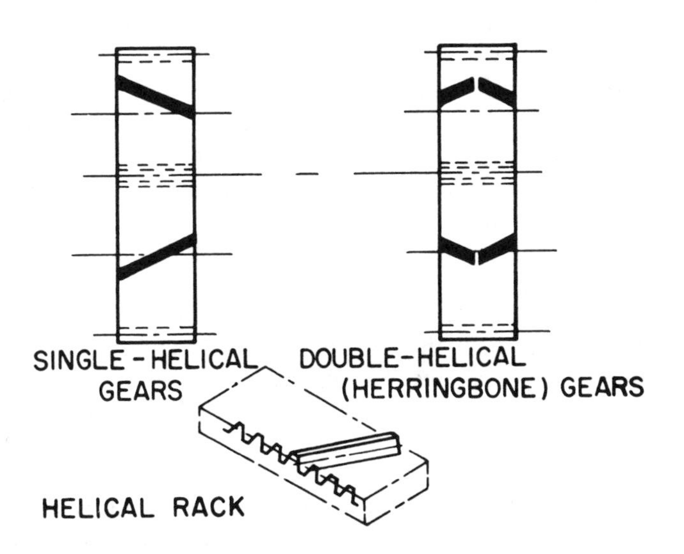

Figure 7-2. Pairs of single helical and double helical gears and helical rack.

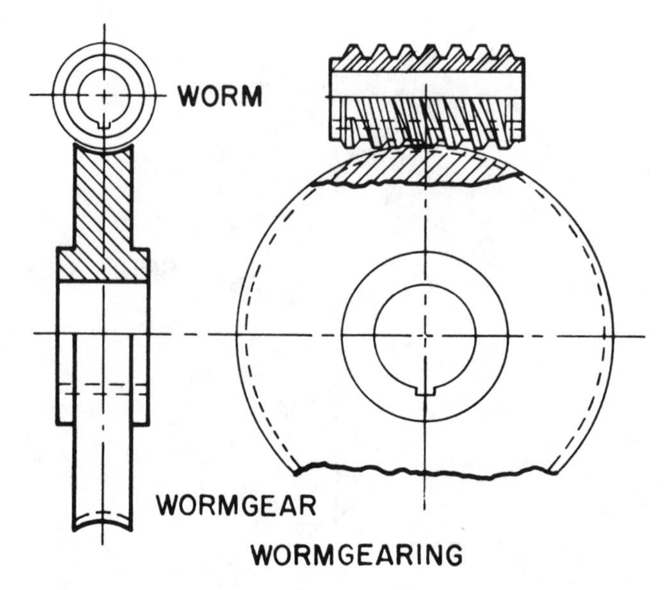

Figure 7-3. Single-enveloping and nonenveloping wormgearing.

3. *Single-helical*—Have teeth of only one hand or direction on each gear (See Figure 7-2).
4. *Double-helical*—Have both right-hand and left-hand helical teeth on each gear and operate on parallel axes (See Figure 7-2). These are also known as herringbone gears.
5. *Wormgear*—Mate to a worm (See Figure 7-3). A wormgear that is completely conjugate to its worm has line contact and usually is cut by a counterpart of the worm. Some forms of hourglass worms and gears are called double-enveloping. A spur gear or helical gear used with a cylindrical worm has only point contact.
6. *Wormgearing*—Includes worms and their mating gears (See Figure 7-3). The axes are usually at right angles.
7. *Straight Bevel*—Have straight tooth elements which, if extended, would pass through the point of intersection of their axes, which are usually at right angles (See Figure 7-4).
8. *Spiral Bevel*—similar to straight bevel but have teeth that are curved and oblique (See Figure 7-5).

There are many types of gearing: hypoid, Zerol, face, angular bevel, elliptical, planetary, and crossed helical. Each of these types occupies a unique place in the world of gearing. Of these types, the most used is probably the hypoid since it is the main drive gear in an automotive differential.

Worm gears and straight bevel gears have limited application due to size limits on bevels and sliding velocity limits on worm gears. Spur gearing is also limited since, for a given situation, it must be much larger than helical gearing to transmit the required horsepower, and in addition, it is not well suited for higher speed applications. Spur gears are very common for high torque low speed drives such as kilns, ball mills, and sugar mills.

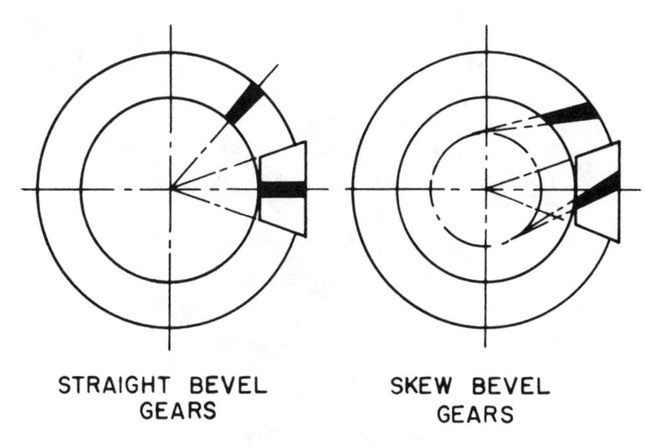

Figure 7-4. Drawing illustrating straight and skew bevel gear sets.

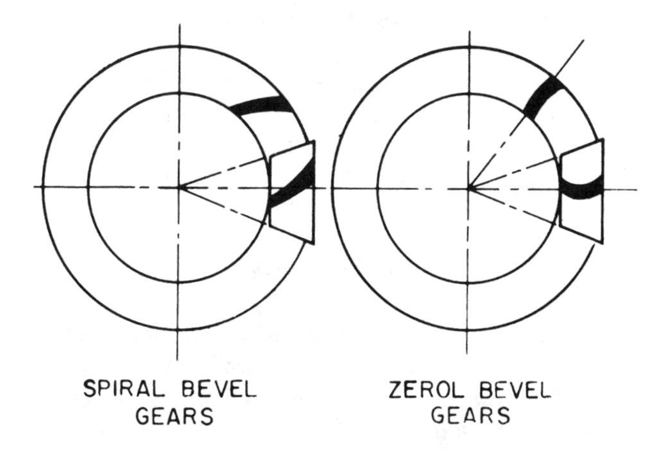

Figure 7-5. Spiral and Zerol bevel gears and pinions.

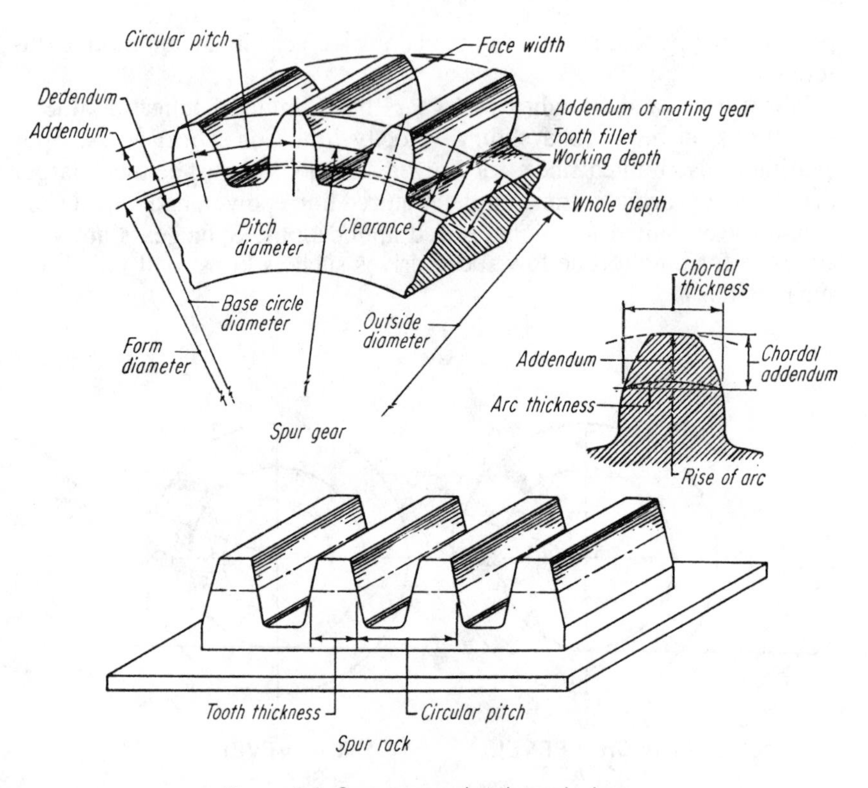

Figure 7-6. Spur gear and rack terminology.

Gear Terminology

Figures 7-6, 7-7, 7-8, and 7-9 illustrate the standard terminology often referred to in the world of gearing. Every gear user should be familiar with at least some of the terms represented in these figures. The following is a brief listing of the definitions of some of the more commonly used gearing terms:

1. **Addendum**—*The radial distance between the pitch circle and the addendum circle.*
2. **Addendum Circle**—*The circle which bounds the outer ends of the teeth. Usually referred to as outside diameter.*
3. **Angle of Action**—*The angle through which the gear turns from the time a particular pair of teeth come into contact until they go out of contact.*

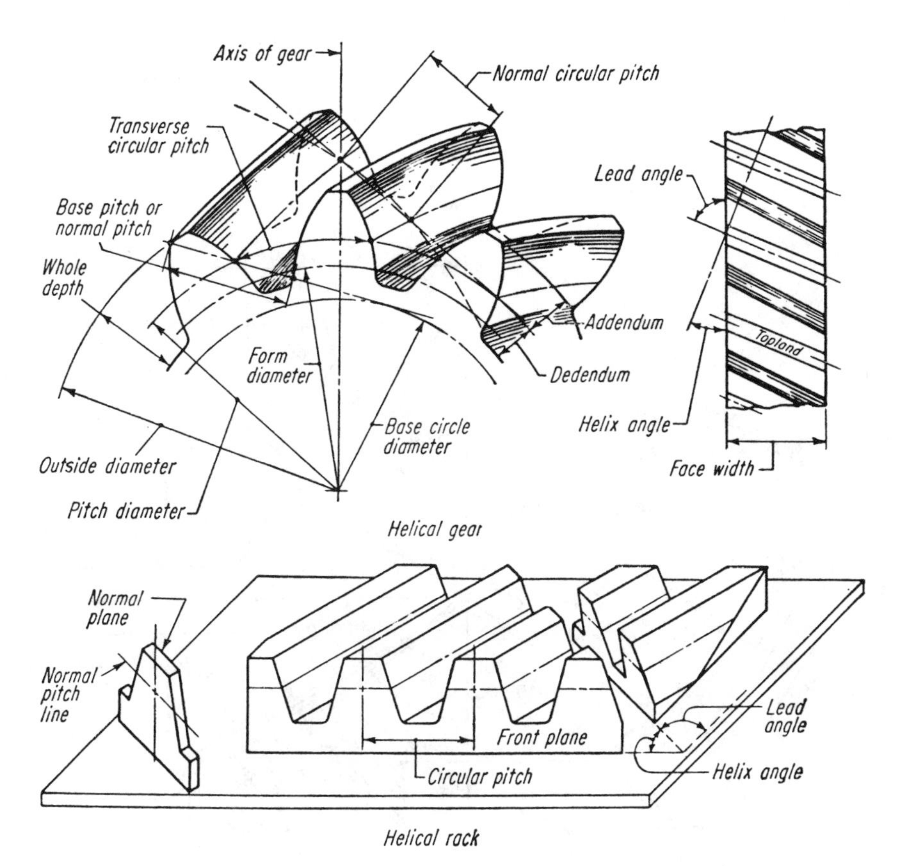

Figure 7-7. Helical gear and rack terminology.

4. **Angle of Approach**—*The angle through which the gear turns from the time a particular pair of teeth come into contact until they are in contact at the pitch point.*
5. **Angle of Recess**—*The angle through which the gear turns from the time a given pair of teeth are in contact at the pitch point until they pass out of mesh.*
6. **Backlash**—*The difference between tooth thickness and the space width in which the tooth meshes. The different backlash terms are normal backlash, transverse backlash, radial backlash, and axial backlash.*
7. **Base Circle**—*The circle from which the involute tooth form is generated.*
8. **Base Pitch**—*The distance measured along the base circle from a point on one tooth to the corresponding point on the adjacent tooth.*

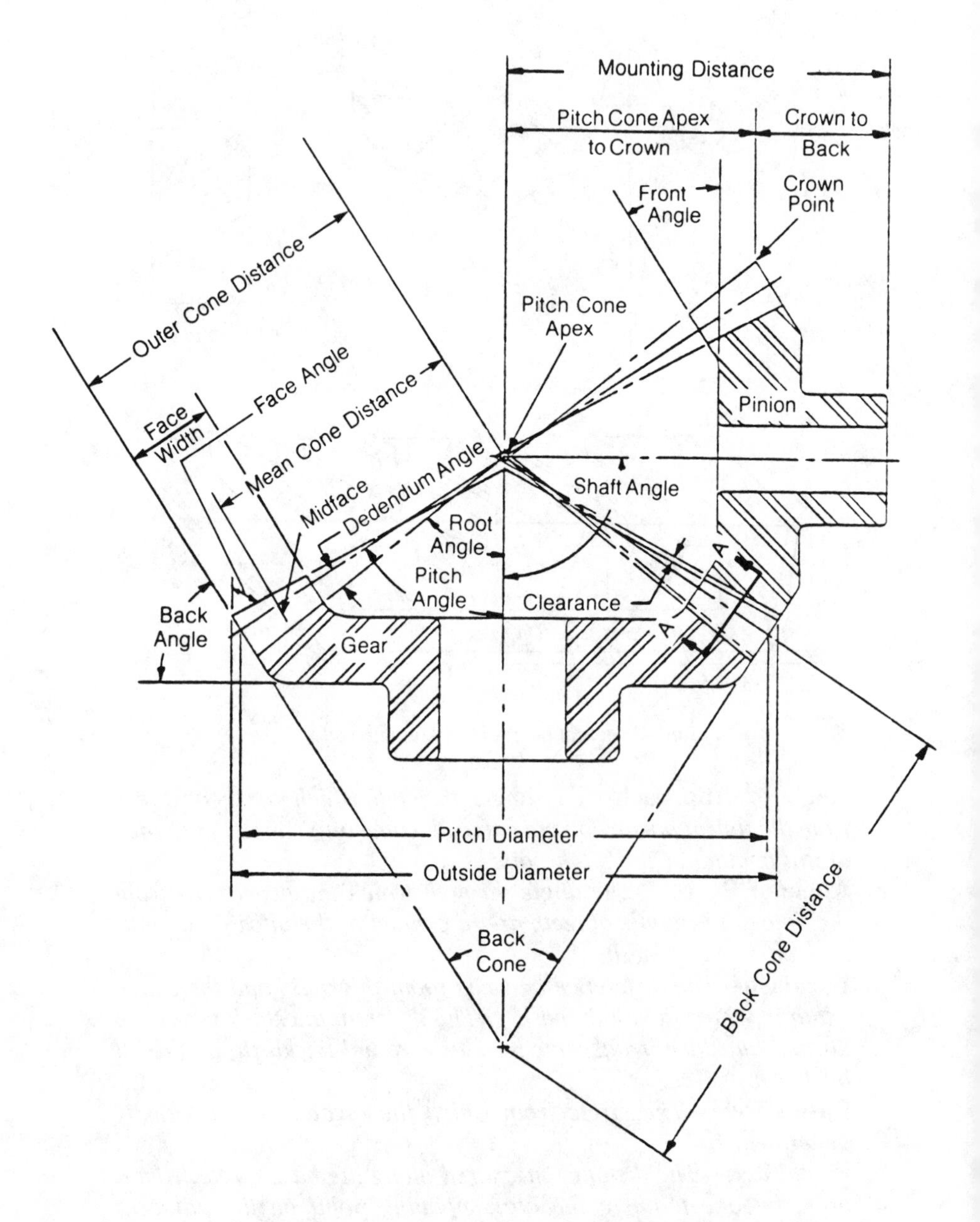

Figure 7-8. Bevel gear nomenclature (axial plane).

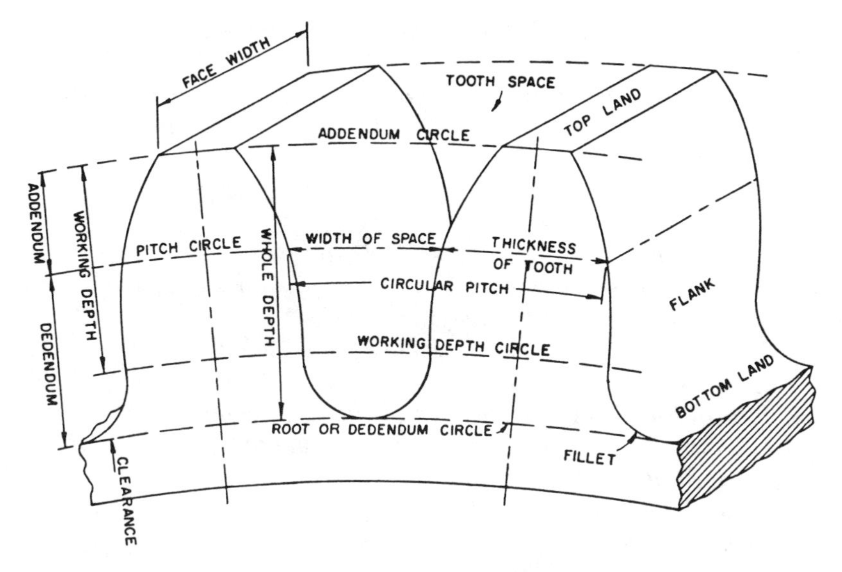

Figure 7-9. Gear tooth nomenclature (transverse plane).

9. **Bottom Land**—*Surface of the bottom of the tooth space.*
10. **Chordal Addendum**—*The distance from the outside diameter to the point where chordal thickness is measured on the pitch circle.*
11. **Chordal Thickness**—*The straight line thickness of the gear tooth measured on any circle. Usually measured on the pitch circle unless otherwise stated.*
12. **Circular Pitch**—*The distance measured along the pitch circle from a point on one tooth to the corresponding point on an adjacent tooth.*
13. **Clearance**—*The radial distance between the working depth circle and the root circle. Also, the distance by which the tip of the tooth of a particular gear will clear the bottom land of the mating gear.*
14. **Cone Distance**—*(bevel gears) The distance along the pitch cone from the apex to any given point on the tooth.*
15. **Contact Ratio**—*The ratio of the length of the line of action to the base pitch.*
16. **Dedendum**—*The radial distance from the pitch circle to the root circle or to the root diameter.*
17. **Dedendum or Root Circle**—*The circle that bounds the bottom of the teeth. Usually referred to as root diameter.*

18. **Diametral Pitch**—*The ratio of the number of teeth per inch of gear pitch diameter.*
19. **Face of the Gear or Face Width**—*Width of a gear measured along an element of a tooth of a spur gear or the width of the pitch surface measured axially.*
20. **Flank**—*The surface of the tooth between the addendum circle and root fillet.*
21. **Gear**—*The larger of two meshing gear elements.*
22. **Helix Angle**—*The angle between the tooth on a gear and the axis of rotation measured at the pitch circle.*
23. **Lead**—*The amount a tooth would advance in one complete revolution along the tooth helix.*
24. **Lead Modification**—*The manufacture of a gear with a modified helix angle to account for the bending and torsional deflection of the teeth under load.*
25. **Line of Action**—*A line drawn tangent to the base circles of a pair of mating gears. All points of contact between mating teeth lie somewhere on this line.*
26. **Mounting Distance**—*(bevel or hypoid gears) The distance from the intersection of the two axes of the gears to a locating surface on one of the gears.*
27. **Normal**—*The term applied to dimensions of the tooth made perpendicular to the tooth flank.*
28. **Pinion**—*The smaller of two meshing gear elements.*
29. **Pitch**—*A measure of the spacing and usually the size of a gear tooth.*
30. **Pitch Circle**—*The circle formed in the transverse plane by the point on each tooth at which the meshing action is purely rolling with no sliding. The pitch circles of two mating gears are tangent to each other. It is the base measurement of a gear. A gear size is the diameter of its pitch circle.*
31. **Pitch Cone**—*(bevel gears) The cone with the intersection of the axes of two mating gears as its apex and the pitch circle of a gear as its base.*
32. **Pressure Angle**—*The angle the line of action makes with a line drawn perpendicular to the line of gear centers.*
33. **Spiral Angle**—*(spiral bevel gears) the angle formed between a tooth on a gear and the axis of the gear.*
34. **Thickness of Tooth**—*The thickness of a gear tooth measured along the pitch circle.*
35. **Top Land**—*The surface of the top of the tooth.*
36. **Transverse**—*Measurements made in the plane of rotation of the tooth.*

37. **Width of Space**—*The tooth space width measured along the pitch circle.*
38. **Working Depth**—*The radial distance from the addendum circle to the working depth circle. The depth of engagement of two mating gears.*
39. **Working Depth Circle**—*The imaginary circle on a gear formed by the deepest points of the tips of the mating gear teeth as the gears pass through mesh.*

How Gears Work

Figure 7-10 is a cross section of a pair of gears in mesh. This cross section is in the transverse plane (plane of rotation) and applies to spur, helical, or spiral bevel gears. All contact is along the line of action, which is the heavy broken line. Note that the pressure angle, which is often referred to on gears, is measured between a perpendicular to the line of centers and the line of action. All contact will be on the line of action normal (perpendicular) to the tooth profile; it commences at point "A" and is completed at point "B." The relationship between the length of the line of action and the base pitch is referred to as the transverse

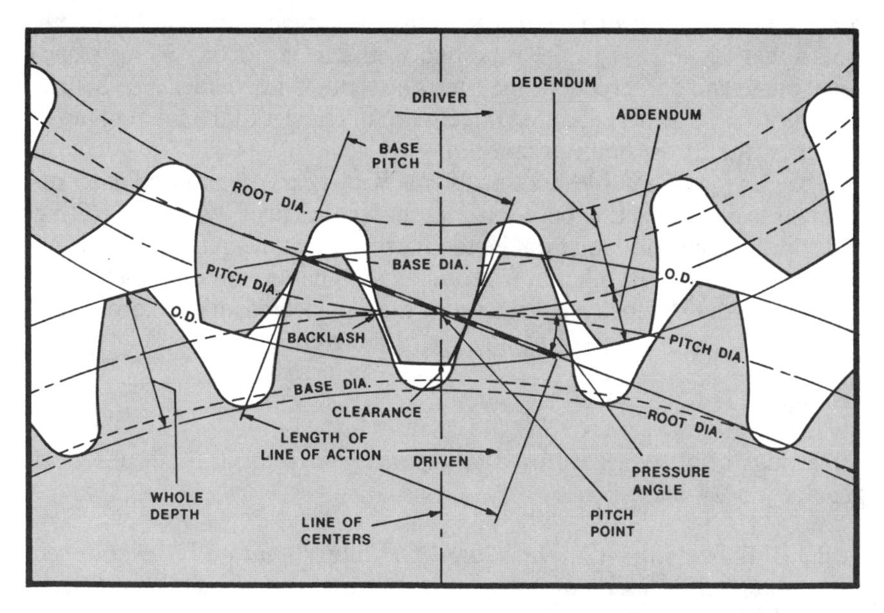

Figure 7-10. Transverse section through a gear and pinion in mesh.

Figure 7-11. Ball bearing (assembled).

contact ratio. For instance, if a gear has a contact ratio of 1.45, this means that 45 percent of the time two teeth are in contact in the transverse plane and 55 percent of the time one tooth is in contact. For helical and spiral bevel gears, additional teeth are in contact in the axial plane at all times due to the helix or spiral angle.

When trying to troubleshoot problems with gears, a good point to remember is that the action between gear and pinion profiles is both rolling and sliding. The only point on the tooth profile that has pure rolling is on the pitch line, and all other areas have a combination of rolling and sliding. The sliding velocity is greatest at the tip, decreasing to zero at the pitch point, and increasing again toward the root.

Bearings

Bearings of all types are used to support gear rotors. The types most generally used are:

1. **Ball Bearing**—*Generally used for lighter load and higher speed applications. Can also handle moderate thrust loads. (See Figure 7-11.)*

Figure 7-12. Straight roller bearing with inner race taken out of bearing assembly. Note roller shape.

2. **Straight Roller Bearing**—*Highest radial load capacity for the space used. Limited allowable shaft deflection through bearing—seldom used for thrust. (See Figure 7-12.)*
3. **Spherical Roller Bearing**—*Very common in large drives where large misalignment capability or medium thrust capacity or a combination of both is required. Often applied as a combination radial and thrust bearing. (See Figure 7-13.)*
4. **Tapered Roller Bearing**—*Very common in all gear drives. High radial and thrust capacity. Often supplied as a pure thrust bearing. (See Figures 7-14 and 7-15.)*
5. **Plain Journal Bearing**—*Sliding type of bearing. Works on either thin film or hydrodynamic film lubrication. Usually made from bronze, zinc, or babbitt-lined steel or bronze and may be either solid cylindrical or split shell. The rolled insert type is also common. (See Figure 7-16.)*
6. **Modified Plain Journal Bearing**—*Can be modified for high speed stability by cutting in a pressure dam, making an elliptical bore, or cutting in longitudinal grooves. (See Figure 7-17.)*

Figure 7-13. Double row spherical roller bearing with inner race and roller/cage assembly turned to show roller form.

Figure 7-14. Tapered roller bearing with outer race removed from inner race and roller/cage assembly to demonstrate roller shape.

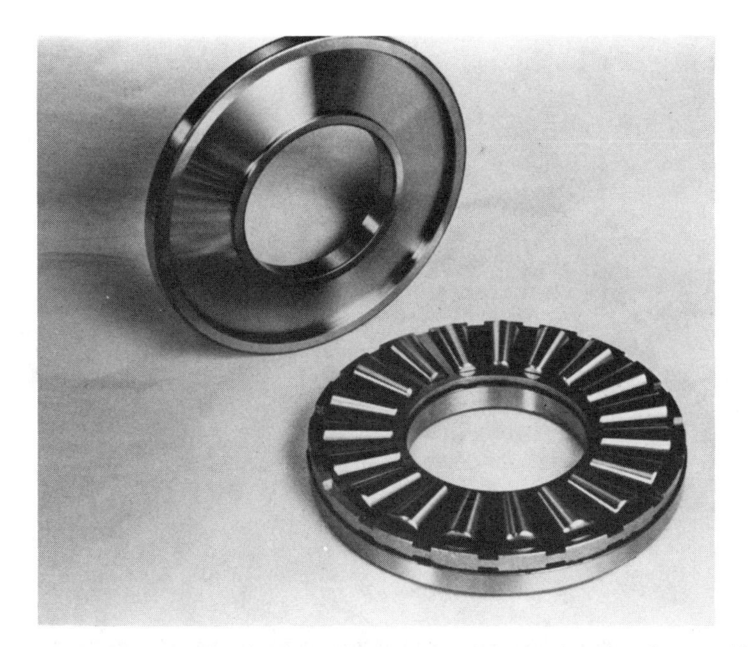

Figure 7-15. Tapered roller thrust bearing—carries thrust loads only—disassembled.

Figure 7-16. Plain journal bearing halves, disassembled.

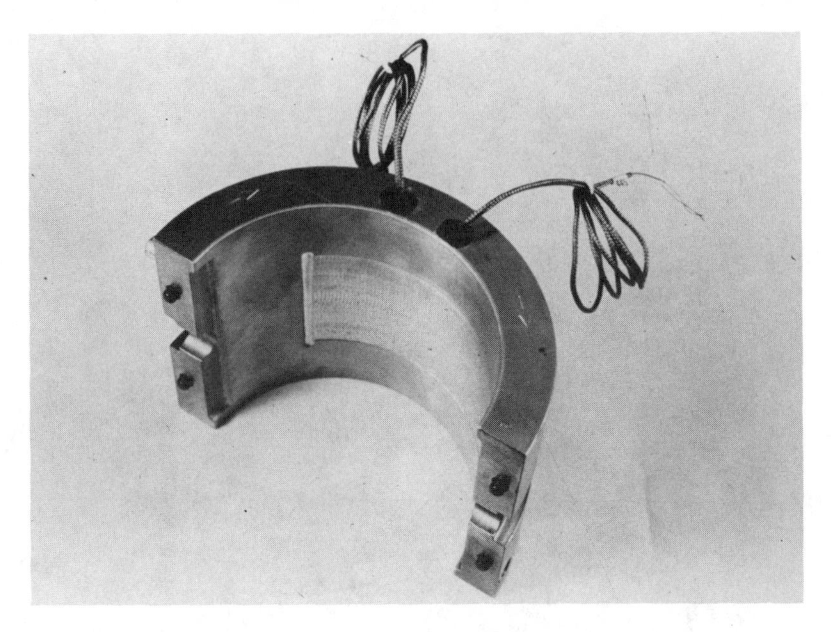

Figure 7-17. Modified journal bearing half with pressure dam cut into bearing bore.

7. **Tilting Pad Journal Bearing**—*Consists of several babbitted bearing segments (five is probably the most common number) that are free to pivot in the circumferential direction. Is generally used for high speed shaft stability. (See Figure 7-18.)*
8. **Hydrostatic Journal Bearing**—*Uses oil pressure to lift the shaft so that there is almost no wear. Loaded half has pressurized pockets to generate oil film.*
9. **Flat Face Thrust Bearing**—*Similar in construction to the plain journal bearing except has flat babbitted face to absorb thrust. Will not develop a hydrodynamic film on face. Can be used in all speed ranges. (See Figure 7-19.)*
10. **Tilting Pad Thrust Bearing**—*Sometimes referred to by the trade name Kingsbury. Has a number of babbitted tilting shoes to absorb thrust. Can be designed with a self-equalizing (load sharing) feature. (See Figures 7-20 and 7-21.)*

Roller bearings are almost always applied to lower speed applications. However, they can be designed for extremely high speeds where relatively light loads are involved.

Journal-type bearings are used for almost all applications, from railroad axles to extremely high speed turbines. They have infinite life if

Figure 7-18. Tilting pad journal bearing halves and shoe (disassembled).

Figure 7-19. Flat face thrust bearing halves with radial oil grooves cut in face.

Figure 7-20. Tilting pad thrust bearing, shoe and collar (disassembled).

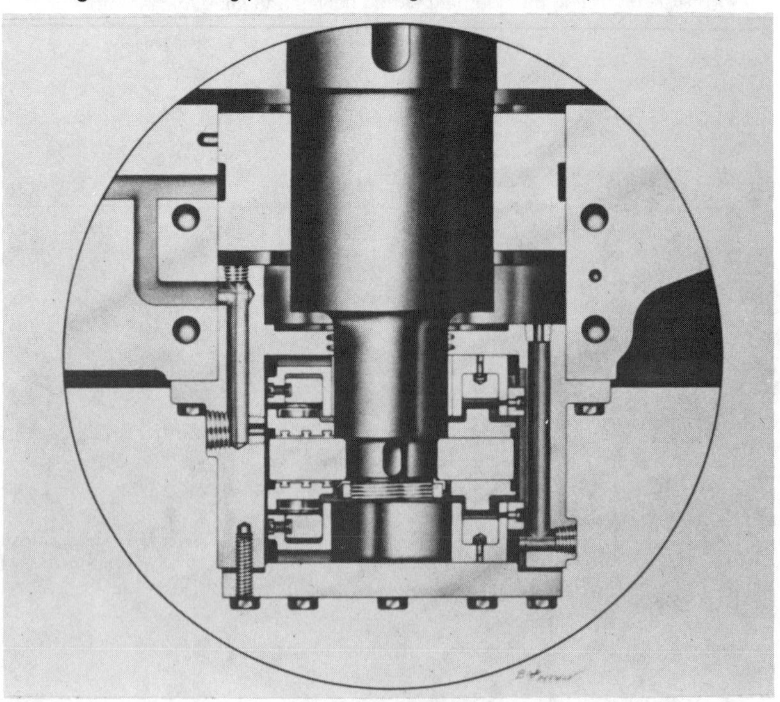

Figure 7-21. Cross-sectional drawing of tilting pad thrust bearing installed on low speed shaft of gear unit.

there is no starting and stopping, adequate clean oil is supplied, and the speed is high enough to generate a hydrodynamic oil film. The journal bearing can easily be manufactured in a split configuration for effortless removal. Also, it can be repaired with a roll of solder and an acetylene torch when absolutely necessary, and machined in almost any shop where an engine lathe of sufficient size is available.

Unlike some machine bearings, gear bearings have imposed operating loads in addition to loads due to rotor weights. These operating loads are directly proportional to the transmitted torque, and since gears are basically constant torque machines, the bearing loads are basically constant. Figure 7-22 is included to show the approximate bearing load directions for speed reducers and increasers with single stage gearing. Bearing load magnitude and direction can be determined from many handbooks and are different for each type of gearing. Thrust loads are produced by both single helical and spiral bevel gearing; the direction of thrust changes with the direction of rotation.

Installation

Handling

A gear unit should always be moved by rolling on bars or skates or by lifting it with slings through the lifting lugs or eye bolts found on all gear units. Never lift or sharply pound on the shaft extensions or lubrication piping as serious damage may result.

Most manufacturers' gears are test run with a break-in oil that contains a rust preventive which will protect the internal parts for at least six months under normal storage conditions after they leave the factory. Do not store gear units outdoors unless covered. If the inoperative period is greater than six months, special treatment is required (see "Inoperative Periods" under "Lubrication").

Foundation

The foundation under a gear unit has great bearing on the unit's operation and life. First of all, proper alignment is absolutely essential for long, trouble-free operation, and obviously, to maintain the alignment required for satisfactory operation, the gear unit must be securely mounted to a suitable rigid foundation. Two of the more commonly used foundations are the concrete foundation and soleplate combination, and the common bedplate (baseplate).

The concrete foundation and soleplate combination is semi-permanent and thereby allows for the removal of the gear housing at a later date

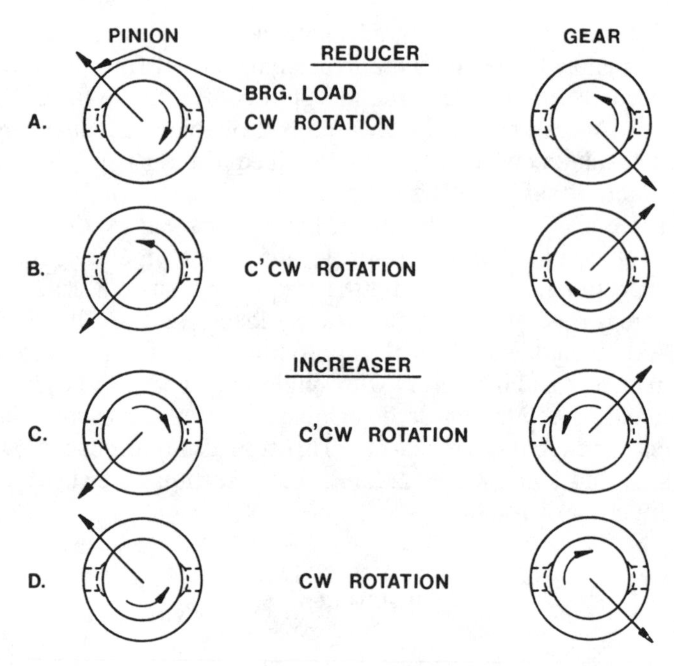

Figure 7-22. Bearing loading directions for gear and pinion bearings, speed increasers and reducers.

without disturbing the permanent mounting pad. The elevation of the concrete foundation should allow for final grouting of the machinery once accurate alignment of the shafts is complete. Most manufacturers recommend that a minimum of 1/16 in. of shims be used between the gear and soleplate to allow for final alignment.

The common baseplate or bedplate is a rigid structural steel foundation common to both the gear unit and either the prime mover or the driven equipment. This type of foundation is quite permanent. The gear manufacturer occasionally supplies the baseplate with the gear roughly aligned to the prime mover or driven equipment. In this case, final alignment is necessary after the baseplate has been grouted in.

Mounting of Couplings

Before attempting to mount the coupling, inspect the coupling bore and shaft diameter with a micrometer to determine that the coupling bore and shaft are correct. Also, inspect the key and keyseat for proper fit, making certain that the key sits at the bottom of the shaft keyway and that there is clearance at the top of the key. If the keyway extends past the coupling hub, the key should be split on the protruding end to fill only the keyway to maintain proper shaft balance.

If an interference fit is used, heat the coupling hub uniformly until there is sufficient clearance to place it on the shaft. It is important that the coupling be heated uniformly and that care is taken so that localized spots are not overheated (maximum 500°F). An ideal way to do this is by using a heated oil bath. In placing the coupling hub on the shaft, do not pound directly on the coupling with a steel hammer, but use lead or rawhide mallets so as not to damage the coupling hub or shaft. A temporary spacer block placed between the coupling hub and the gear housing is helpful in preventing the coupling hub from sliding too far onto the shaft.

Many gear drives are being furnished with keyless, hydraulic dilation coupling-hub-to-shaft fit. These coupling hubs must be properly mounted and fitted to develop enough torque-carrying capacity between the shaft and hub. The amount of advance on the shaft is dependent on the amount of taper and transmitted load. Improper mounting can cause these hubs to turn on the shaft and destroy the fit. In some cases, repair is not possible and the gear or pinion is destroyed. Obtain instructions from the vendor for the particular keyless coupling fit before mounting; also, refer to the section on coupling installation in Chapter 9, Volume 3 of this series.

Basic Installation Procedures

Generally, after a suitable foundation has been laid and coupling hubs mounted, an installation procedure should start with securing the driving or driven machine (whichever is more permanently settled) and rough-aligning the gear unit to it. Jacking screw holes are provided on the base flange for bringing the gear unit to the same horizontal plane as the connecting shaft. Once there, the gear unit should be supported on broad, flat shims located adjacent to and on each side of the foundation bolt holes. Next, move the unit on its shims until the gear shaft is in the same vertical plane as the connecting shaft with the correct spacing between coupling hubs.

At this point, the operating positions of driven and driving shafts, which will be different from the cold static positions, should be anticipated, and desired cold positions for each established based on that information. Final alignment is established by moving the shaft of the machine that is not tied down from its present cold position to its desired cold position by adjusting the machine very slightly at the base.

After final cold alignment is achieved and before tightening the foundation bolts, be sure that the base sets evenly on all shims so that there will be no distortion when the foundation bolts are fastened. After tightening the bolts, check for distortion by placing a dial indicator on the gear housing foot near the bolt to be checked. If the housing foot moves when the bolt is loosened, then distortion is present, and the housing needs more shims around the bolt. The gear housing should be checked for distortion using this method at each foundation bolt.

After the gear unit is properly aligned to the first component and bolted down free of distortion, a soft blue tooth contact check should be performed. If the contact pattern is satisfactory, then proceed with the installation. However, in the event that the soft blue check indicates poor tooth contact, shaft and coupling alignment and housing distortion should be rechecked before proceeding with installation.

Before operating the gear unit, dowel it to the base or soleplate as recommended by the manufacturer, leaving room to redowel if necessary. Doweling a gear unit before operating is very important since, as a gear unit is heated and cooled, it will crawl on the mounting surface and an alignment change may result. If instructions are not available from the manufacturer, dowel the unit under the shaft for which alignment is the most critical.

After doweling the gear unit, apply layout blue to the teeth as directed for a hard blue tooth contact check, and then start the unit up and operate it for a short period of time. *NOTE:* Care should be taken in joining the two coupling halves to observe any coupling match marks. Lubricate the coupling and check for free axial movement of pinion and gear. After shutting the unit down, check the tooth contact pattern as evidenced by the wear-off of the bluing. If a satisfactory contact pattern is obtained, proceed with installation. However, should the contact pattern be poor, corrective measures should be taken as recommended by the manufacturer before going on with installation.

At this time, a hot alignment check should be made by running the gear package until temperatures stabilize, shutting it down, and taking indicator readings while the gear package is hot. Any corrections necessary should be made while the unit is still hot. After complete correct hot alignment is obtained and before the package cools off, the gear unit should be redoweled to the foundation or base. When final coupling alignment has been established and the gear has been redoweled, place the coupling guards in position. *Warning:* Failure to use coupling guards may result in serious injury to personnel.

Succeeding the installation of the gearbox, the installation of the third machine element should follow along the same basic lines. Again, after alignment has been completely finalized, coupling guards should be installed for protection.

When the entire machinery package has been completely and satisfactorily installed, it should be started up and operated. Special attention should be paid to several checks that need to be made before and after start-up to help ensure trouble-free operation. During the first day or two of operation, special emphasis should be focused on bearing and oil temperatures and on housing and shaft vibration to catch any potential problems.

Alignment

Securing proper shaft alignment is one of the most important phases of setting up a gear unit. Even though flexible couplings are used on the shaft extensions, any appreciable amount of misalignment can cause a multitude of gear problems ranging from nonuniform bearing and gear tooth wear to vibration and coupling problems. Therefore, it is essential that good alignment be established and maintained and that thermal growth and bearing clearances be anticipated in shaft alignment.

Gear units experience both thermal and mechanical movements when they are operating under load. The thermal movements are due to temperature changes, both environmental and operational. Mechanical movements are due to internal gear loads causing the shafts to move within the bearing clearances; in severe cases, operating torques and thrust loads may also distort the housing causing apparent shaft movement. All of these factors should be accounted for when aligning a gear to other equipment.

As is evident, machinery alignment is one of the most important factors contributing to satisfactory gear operation. However, it is too complex a subject to be covered in full detail in this chapter. Although the following discussion of alignment procedures is very brief, it does cover the high points and give a good general procedure that may be followed. The reader should refer to Volume 3, Chapter 5, for a more complete description of current alignment practices.

Checking Indicator Arm Sag

Prior to taking indicator readings for coupling alignment check, the spanner arm that clamps to the shaft and supports the indicator arm and indicators must be checked for possible gravity sag. This spanner device must be rigid enough to minimize deflection due to the weight of the arm and indicators. This check can be made by using the following method, as illustrated in Figure 7-23.

1. Rotate clamp while attached to shaft; observe change in outside diameter indicator reading at 90° positions.

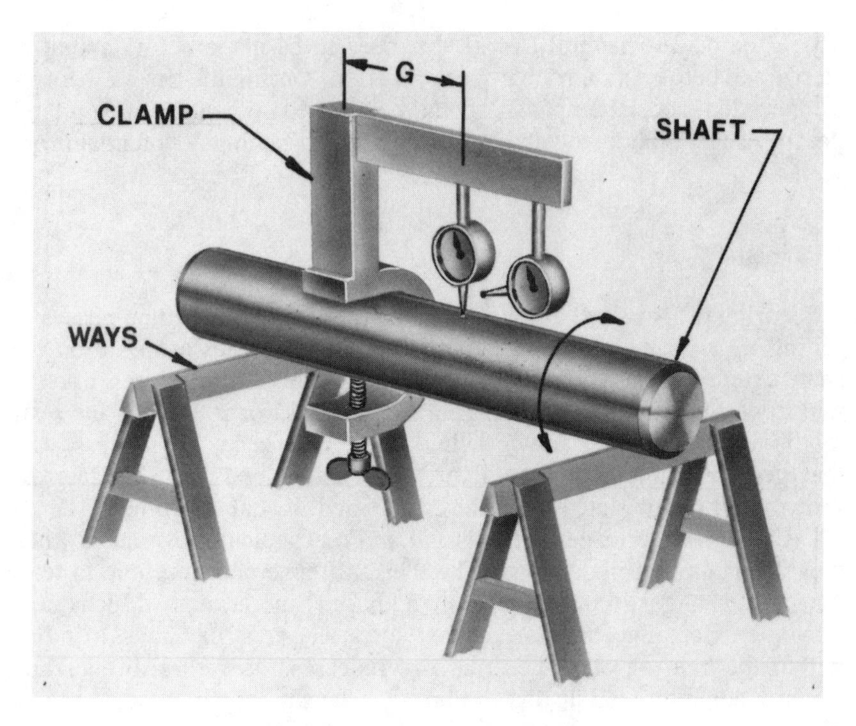

Figure 7-23. Checking indicator arm sag.

2. When changes in indicator readings dictate, reduce sag by keeping "G" dimension as short as possible, using light-weight indicators, and using one indicator at a time.

Establishing and Checking Final Alignment

An acceptable method of establishing and checking alignment in the ambient condition is by clamping a spanner arm to one of the shafts (shaft "B" in Figure 7-24) and spanning the indicator arm across to the mating shaft or coupling hub. For flanged couplings, use two dial indicators for reading outside diameter and face alignment as illustrated in Figure 7-24. (Other types of couplings may require set-up variations.) Proceed in the following manner:

1. Perform check for indicator arm sag (see Figure 7-23 and related section).
2. Record approximate "R" dimension.
3. Record ambient temperature.

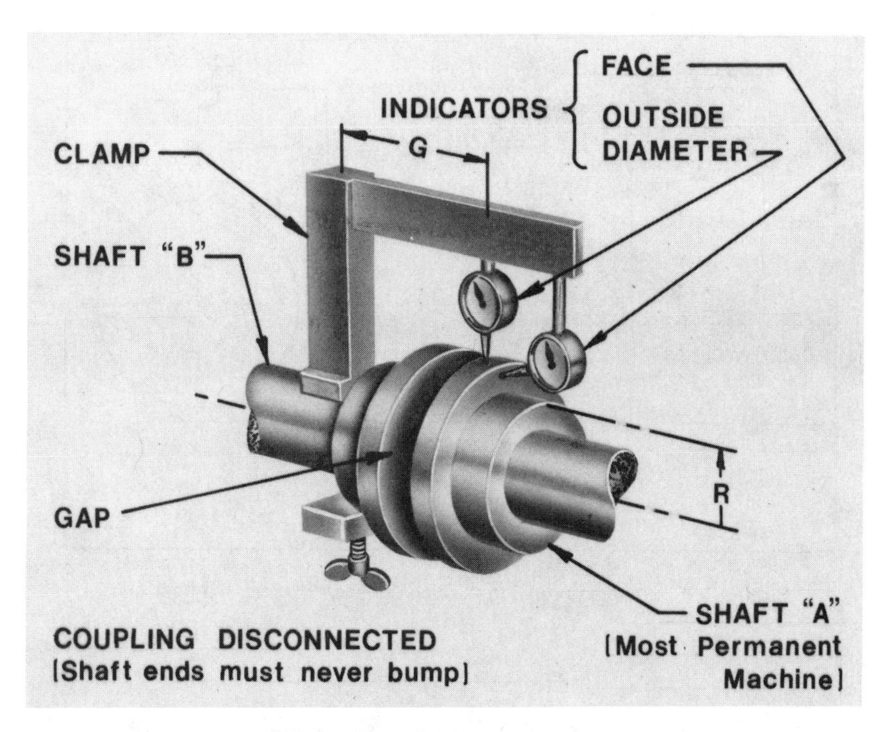

Figure 7-24. Coupling alignment.

4. Hand rotate shaft "B" around shaft "A." Always rotate the gear shafts in the direction they will turn under load. Do not reverse rotation during alignment check.
5. Record the outside diameter and face indicator readings at 12 o'clock, three o'clock, six o'clock, and nine o'clock positions. Shift shaft "B" into the same relative axial position prior to each face reading; that is, with prize bar, move shaft axially in one direction until all axial clearance is removed.
6. Correct alignment as needed. Consider the anticipated thermal and mechanical movements as discussed on page 257.

Shaft Operating Positions

At operating load and temperature, the final positions of the shafts of a gear unit will differ from their positions under no load and ambient tem-

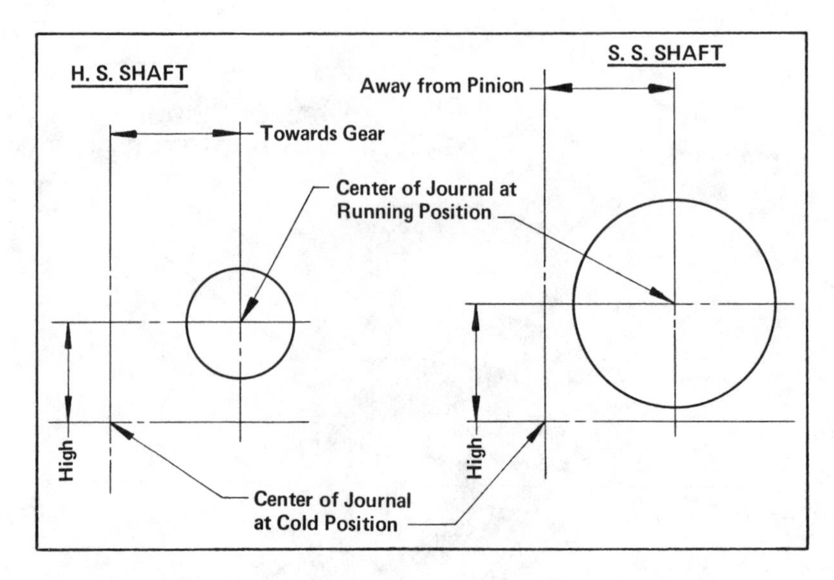

Figure 7-25. Static and running positions of shaft centers in a gear unit.

perature. This phenomenon, illustrated in Figure 7-25, is due to thermal expansion of the gear housing, and the magnitude and direction of the mechanical loading imposed by the gear action. The driven and driving machines also have thermal movements which must be either added to or subtracted from the gear movements, depending upon the directions of those movements.

Axial Shaft Positioning

Normally, gears are located axially in the gear housing by thrust bearings. These thrust bearings are located on either side of the slow speed gear. Sufficient clearance is usually provided to allow for normal thermal expansion of both the high and slow speed shafts. However, if axially rigid couplings are used or excessive thermal expansion is anticipated, additional clearance can be provided by the manufacturer.

Usually, during alignment procedures, the slow speed shaft should be axially positioned as far toward its mating shaft as possible before the coupling clearance is obtained. However, if the prime mover is an electric motor that has a magnetic center, the magnetic center should be located and the gear positioned so there is equal clearance on either side of the magnetic center. Similarly, limited end float couplings are sometimes used to axially position the driving shaft. Here the running position should be determined and the clearance split on either side of that running position. On single helical gears, the gear and the pinion both are equipped with thrust bearings; consequently, this procedure applies to each.

For double helical gearing with a thrust bearing on the low speed shaft, care must be taken when axially locating the high speed pinion. This is accomplished by moving the pinion as far as possible in both directions axially and measuring this movement. The pinion is then centered, and this is the resulting alignment position. For units running at elevated temperatures, the axial growth of the pinion must also be taken into account.

Thermal and Mechanical Movement

Thermal movements of gear shafts are caused by both environmental and operational temperature changes, whereas mechanical movements are caused only by internal gear loads which force the shafts to move within the bearing clearances. Figure 7-25 shows the positions of the journal centers for a single stage gear after the gear unit reaches operating temperature and load. To properly determine the operating positions, it is necessary to calculate the operating loads and the gear weights, compute the movements due to these mechanical forces, and then add these movements to those due to the expansion of the gear housing caused by the elevated operating temperature. When the alignment procedure is started, these mechanical and thermal movement values should be calculated and recorded on a worksheet and kept for future reference.

Thermal movement can be calculated by multiplying the coefficient of thermal expansion (.0000065 in. per in. per °F for ferrous materials) by the distance involved (either vertical height or horizontal offset) and by the anticipated temperature rise above ambient. Normally, the temperature rise will range from 30 to 70°F when an ambient temperature of 60 to 70°F is present. Before making an assumption on the direction of thermal expansion, the location of the dowel pins which hold the gear mounting feet on the base or soleplate must be established. All movement is calculated assuming a properly fitted dowel pin and is away from the pin.

Mechanical movement must be estimated from bearing clearances and directions of rotation. A rough rule of thumb for estimating mechanical movement is to assume a minimum running clearance of .003 in. or .001 in. per in. of shaft diameter, whichever is greater, at the bearing. This should only be used for journal, straight roller, spherical roller, and double row tapered roller bearings. For single row tapered roller bearings, a clearance of .003 in. can be assumed for all shaft sizes. If possible, the mechanical movements should be obtained from the manufacturer of the gear unit.

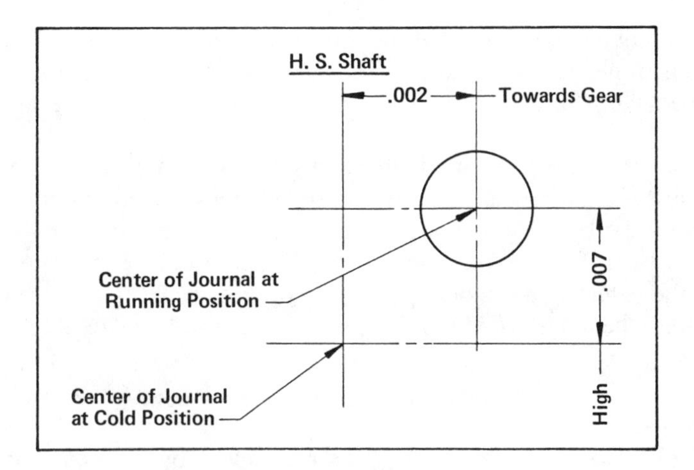

Figure 7-26. Example of thermal movement.

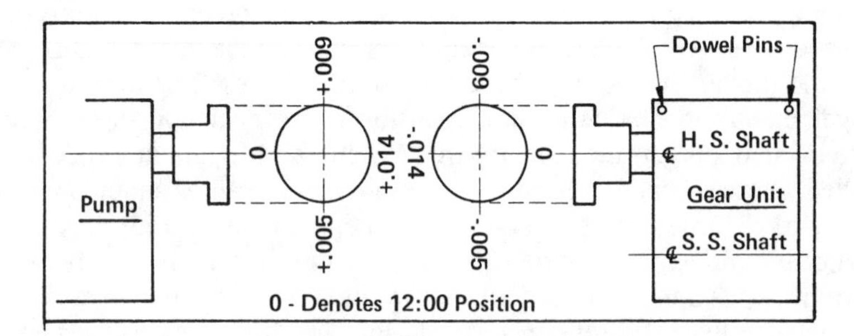

Figure 7-27. Ideal cold alignment example.

In the example illustrated in Figure 7-26, the thermal and mechanical movements have been determined as .007 in. vertical and .002 in. horizontal toward the gear. Using these values, the cold indicator reading should be as shown in Figure 7-27. In this example the assumption is made that the pump remains in the ambient condition. If this supposition is not true, the indicator readings must be corrected for the pump movement.

Tooth Contact Check

In gearing, a gear tooth must have an even load across the entire face width if the stress on that tooth is to be minimized. The type of contact between gear teeth is instantaneous line contact; therefore, the alignment between the rotating elements (pinion and gear) is critical. The alignment is controlled by the accuracy of the rotating elements, the housing, and the bearings. Twisting the housing either during shipment or because of poor foundation conditions will cause poor tooth contact; incorrectly installed rotating elements or bearings will cause poor contact, and of course poorly manufactured parts will also cause poor contact. Therefore, tooth contact should be checked on all new installations, after any disassembly of the gear unit, and after any housing-to-foundation change.

How to Check Tooth Contact

Gear tooth contact can be checked two ways. Soft machinist's blue or transfer blue can be applied to the teeth of one gear and that gear rolled by hand through mesh with its mating gear. The transfer of the blue from one gear to the other is read as the contact. Another method is to paint the gear teeth with hard blue or layout blue and run the gear unit for a short while. Then stop the unit, and observe the pattern of "wear-off" of the bluing. The term "bluing" is used for convenience. Some of the layout dye or layout blue used is red in color. Some people claim that using this color makes it much easier to see the contact pattern.

The soft blue method of checking gear tooth contact is usually done first. Since the unit is not running, this check does not give true contact. It does, however, give a good indication of what contact will be. If it indicates poor contact you may choose not to start the unit until the contact is corrected. If the unit has been disassembled, then a soft blue check before the housing cover is installed may save you a tear-down later to correct contact. This is especially important if a new set of rotating elements is installed.

Soft blue is usually applied to three or four teeth on the pinion in two places 180° apart. The contact should be checked at three or four places around the gear; however, you must reapply and resmooth the blue on the pinion after each meshing. If time is very critical, two checks at 90° apart on the gear will suffice.

First of all, clean the teeth thoroughly with solvent, and spread the blue on thinly and evenly. A one in. wide good quality paint brush with the bristles cut off to a length of about one in. makes a good blue application brush. The blue still will not be even enough, so with a shop rag smooth it to a very thin and even layer. If the gear set is double helical, center the pinion and gear meshes. If the gear set is single helical, position the pinion and gear against their proper thrust faces. Now hold a drag on the gear and roll the pinion blue area through mesh with the gear. Rotation direction is not important in itself but should be adhered to, since it is important to check the loaded tooth flank and not the unloaded tooth flank. Now observe the blue that transferred from the pinion to the gear. This transfer is the contact pattern. An acceptable contact pattern for helical gears without lead modification is blue transfer for approximately 70 to 80 percent of the length of the tooth on each helix.

A piece of cellophane tape can be used to remove this blue pattern from the gear and save it for maintenance records. After the check, place a piece of tape on the gear tooth flank and press it firmly on the tooth. Remove the tape, and then place the tape on a clean sheet of white paper. Be sure to record where the check was made. The exact position should be marked on the gear using a light punch mark on a part of the tooth checked that will not be contacting another tooth surface.

Hard blue or layout blue, the second method of checking tooth contact, is sprayed or brushed on both the gear and the pinion. First clean the areas to be blued thoroughly. Just cleaning with a solvent such as naphtha is not sufficient, since this procedure will not completely remove the lubricant. Additional cleaning with a volatile solvent such as electrical contact cleaner, lacquer thinner, or Freon® is necessary. If the teeth are not absolutely free of oil, the blue will not adhere properly, and large flakes will chip off making the contact check difficult or impossible.

The layout blue is applied to a three or four-tooth-wide area at three or four places on the gear and at two on the pinion. The unit is then started and run at full speed. Running conditions may vary from no load to full load. The best procedure is to run the unit at a very light load (approximately 20 percent of full load if possible) for twenty minutes or so and then shut down and check the contact. With higher loads you should run the unit a shorter time before checking contact. The trick is to run the unit just long enough to wear the blue off the areas of lower contact stress. High loads can mask poor contact and give false readings by deflecting the gear teeth enough to indicate better contact than that actually present.

If the soft and hard blue checks show satisfactory contact, operate the unit at full load until temperatures have stabilized in the system. Shut the unit down, and re-check the blue wear-off to be sure the contact is still acceptable.

If poor contact is indicated by the soft or hard blue checks or both, housing distortion and coupling alignment should be checked. In addition, any corrective measures recommended by the manufacturer should be taken before proceeding with installation.

Helix Angle (Lead) Modification

Many wide face width single and double helical gears have helix angle modifications to correct for torsional twist and bending deflections of the teeth due to operating loads. When these modifications are made, the helix angles on two mating gears are purposely cut differently so that they will be the same when the gear teeth deflect as the design load is applied. In some cases, only the ends of the teeth are relieved (end ease-off), or a gentle curve is cut into the tooth flank (crowning).

In cases where the helix angles have been altered, the face contact will not approach the 70 to 80 percent usually recommended on nonmodified leads. Therefore, when modified leads are furnished, it is necessary to determine from the manufacturer what blue transfer or wipe off is expected under different load conditions. Figure 7-28 illustrates possible contact patterns for double helical gears with and without modified leads under no load conditions. Several possible contact patterns for single helical gears under no load conditions are depicted in Figure 7-29.

Spiral Bevel Tooth Contact

The tooth contact checking procedure for spiral bevel gears is the same as that described for helical gears except the expected contact pattern is different. Spiral bevel gears are generally manufactured in matched sets, and the contact will vary with the mounting position. These gears are usually mounted on tapered roller bearings with shims to make it possible to position the gear and pinion axially and thereby to obtain proper tooth contact. Figure 7-30 illustrates one contact pattern to be expected on spiral bevel gears.

Doweling

After completing the alignment procedures and before starting the drive, install dowel pins as specified by the manufacturer. To start a unit and allow the temperature to change without the dowel pins in place will cause misalignment; alignment cannot be maintained without pinning.

Each drive train should be evaluated, and the best dowel pin locations should be determined. Most manufacturers furnish starter holes for the dowel pins, and they have selected what they feel is the best location for the zero movement point on the gear unit. A good rule to follow is that on high speed units, the dowel pins should be located as near as possible to the high speed pinion since this shaft has the most critical coupling alignment. Many large, heavy duty, low speed high torque units will require the dowel pins be placed under the low speed output shaft since this shaft has the most critical alignment point. On a unit which has rigid couplings

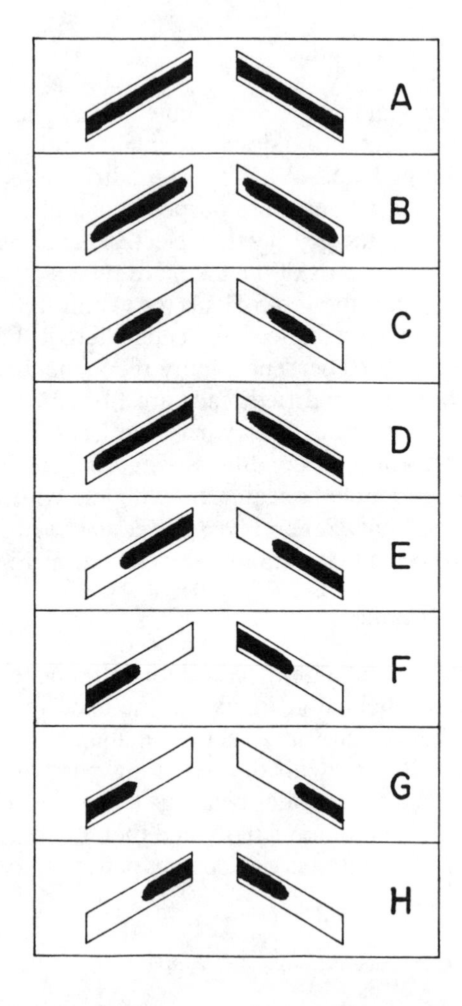

Figure 7-28. Typical tooth contact patterns for double helical gearing under no load.

A. Perfect tooth contact, no modification.
B. Gear with end ease-off modification (slight crowning)—acceptable.
C. Gear with crowning modification—acceptable.
D. Gear shafts slightly out of parallel—acceptable.
E. Gear shafts out of parallel—not acceptable—must be corrected before operating gear unit.
F. Gear shafts out of parallel—not acceptable—must be corrected before operating gear unit.
G. If no lead modification, gear or pinion or both miscut—not generally acceptable—one or both parts should be corrected before operating gear unit. Will result in premature wear and failure if not corrected.
H. If no lead modification, gear or pinion or both miscut—not generally acceptable—one or both parts should be corrected before operating gear unit. Will result in premature wear and failure if not corrected.

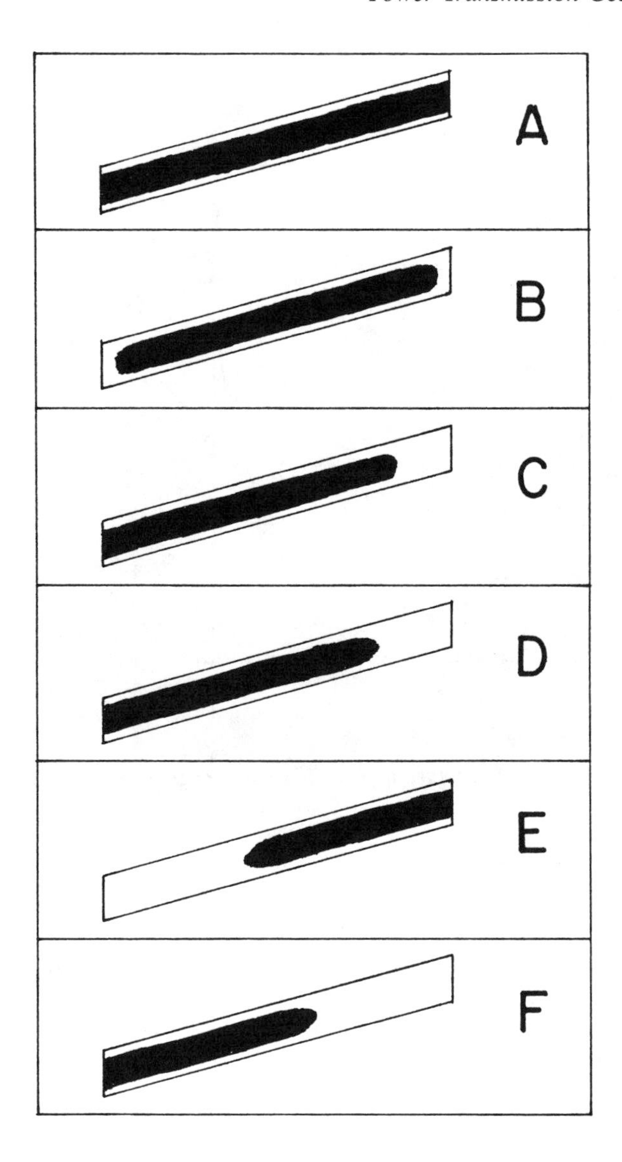

Figure 7-29. Typical tooth contact patterns for single helical gearing under no load.

A. Perfect tooth contact, no modifications.
B. Gear with end ease-off modification (slight crowning)—acceptable.
C. Gear with helix angle (lead) modification cut into part—acceptable.
D. No lead modification, gear shafts slightly out of parallel—acceptable.
E. No lead modification, shafts out of parallel—not acceptable—must be corrected before operating gear unit.
F. No lead modification, shafts out of parallel—not acceptable—must be corrected before operating gear unit.

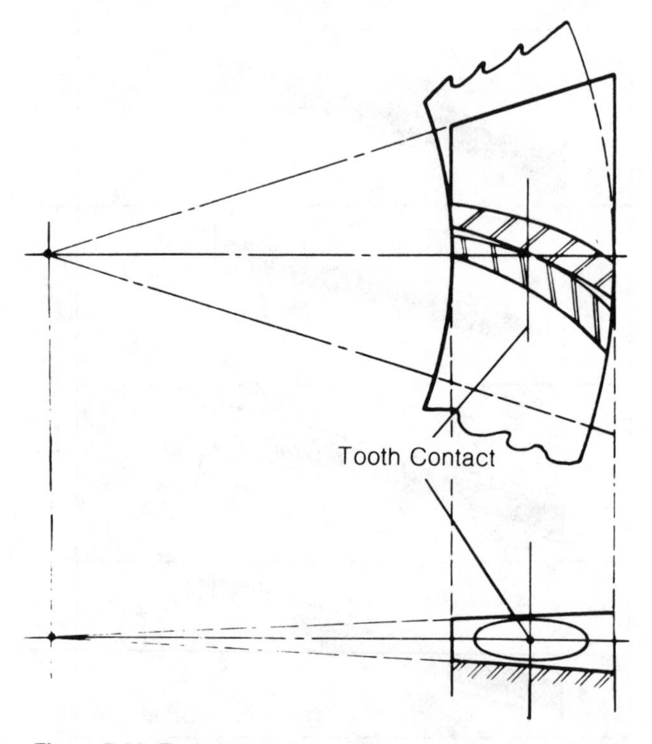

Figure 7-30. Typical tooth contact pattern for spiral bevel gearing.

on one shaft, this rigid shaft alignment will be the most critical, and dowels should be installed under the rigid shaft.

A gear unit should be redoweled after any housing-to-foundation change, no matter how slight. Proper location and installation of the dowel pins are essential to maintaining the good alignment and satisfactory operation of a gear unit.

Hot Alignment Check

After all components have run long enough to stabilize operating temperatures, a hot alignment check should be made as quickly as possible. The coupling halves should be separated, the alignment checked, indicator readings recorded, and any necessary corrections made before the shafts and machines cool down. After this initial hot alignment check is made, the unit should be redoweled to the foundation if any corrections were necessary. Then an additional hot alignment check should be made to ensure as correct an alignment as is possible.

Indicator clamps for bridging across the coupling sleeves and/or hubs should be installed as shown in Figure 7-31. This method allows the final hot alignment check to be made without disconnecting the coupling halves. Since no two couplings can be aligned absolutely perfectly, suggested maximum allowable runout values for use in alignment procedures are listed in Table 7-1. Proceed with the hot alignment check in the following manner:

1. Record ambient temperature and oil sump temperature on worksheet.
2. Rotate shafts in the direction they were designed to operate under load conditions.

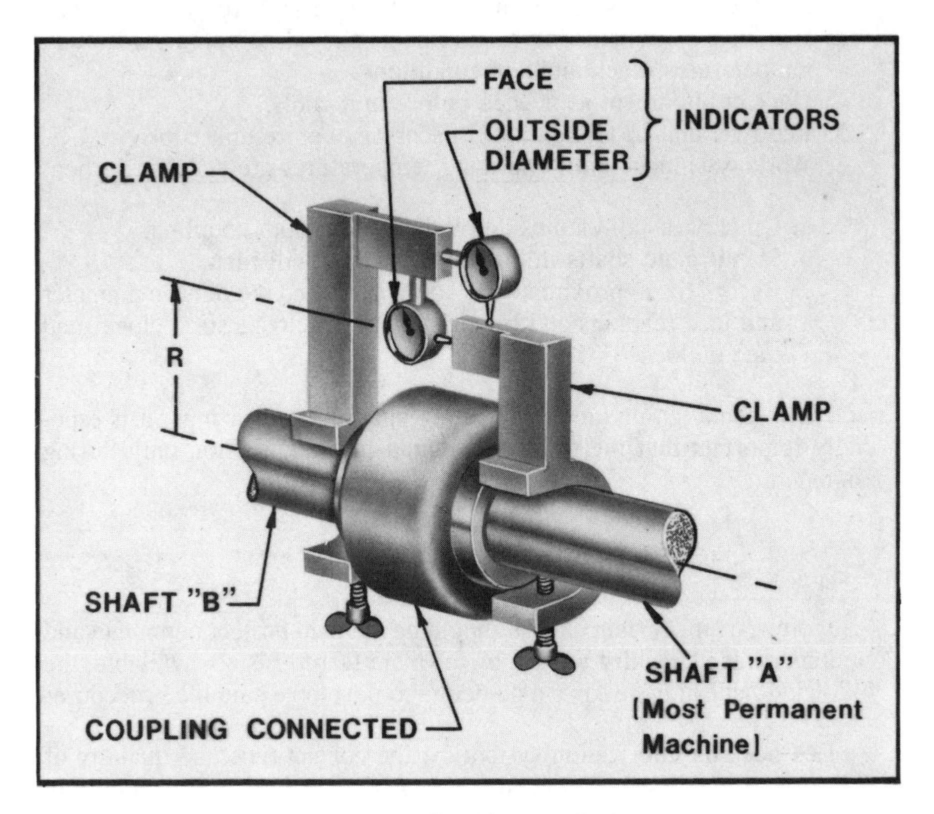

Figure 7-31. Hot alignment check.

Table 7-1
Recommended Maximum Allowable Runout During Coupling Alignment, In. TIR

Shaft Surface Velocity, feet per minute	Maximum Allowable Run-Out, Inches, TIR: Outside Diameter of Coupling	Coupling Face
5000 and up	.002	.0004 per in. of radius
3000 to 5000	.004	.0005 per in. of radius
1500 to 3000	.006	.0006 per in. of radius
500 to 1500	.008	.0008 per in. of radius
500 and down	.010	.0010 per in. of radius

3. Record the approximate "R" dimension and the outside diameter and face readings at 90° intervals starting at 12 o'clock. NOTE: Readings and correction requirements should be noted before unit temperatures reach ambient conditions.
4. Correct alignment as needed before unit cools.
5. Redowel unit to foundation if necessary before unit cools.
6. Work machinery until operating temperatures are stabilized, then:

    a. Quickly install clamps and indicators at both couplings.
    b. Hand rotate shafts in the direction they will turn.
    c. Record the approximate "R" dimension and the outside diameter and face readings at 12 o'clock, three o'clock, six o'clock, and nine o'clock.

NOTE: When aligning units with tilting pad journal bearings, it is especially important that the shafts be rotated in one direction only during alignment.

Checklist Before Startup

Before startup, certain checks should be made to protect personnel and equipment. If a checklist written by the manufacturer is not available, the following list can be used as a guideline to develop a suitable procedure:

1. Check the lubrication system for the correct type and quantity of oil. If a splash system is used, be sure the gears or slingers dip in oil.

2. If a pump is used, be sure that it is primed to minimize the time necessary to build up a positive oil pressure. Pressure should build up in 10 to 15 seconds. If pressure does not develop, stop unit and determine the problem. When an electric-driven oil pump or some other remote pump provides lubrication for the gear, it is a good idea to run that pump a few minutes prior to startup to provide initial lubrication during startup.
3. Check rotation. Be sure that the gear unit will rotate in the direction for which it is intended. The shaft-driven oil pump is uni-directional and must rotate in the direction indicated by the arrow. If the other direction of rotation is desired, it will be necessary to replace the pump with one of the opposite rotation from manufacturer.
4. Make sure the gears have backlash.
5. Check for free turning of the shafts.
6. Check for correct coupling alignment.
7. Check to see that all necessary piping and accessory wiring is complete.
8. Check for correct water flow and temperature through the oil cooler.
9. Check for foundation bolt tightness.
10. Check tooth contact.
11. Make sure there is running clearance around all moving parts.
12. *Warning:* Coupling guards and inspection covers should be secured *before* startup.

Checklist After Startup

Each gear installation requires that different operating checks be made depending on the instrumentation furnished, the size of the unit, and how the unit is equipped. The following checklist can be used as a rough guide:

1. Run gear unit at light load and reduced speed if possible while checking for proper lubrication. After the unit has been running approximately 15 seconds, the oil should be circulating. If there is a noticeable drop in oil pressure on pressurized systems after several hours of operation, clean the oil filter. Occasionally, lint will clog the filter after initial startup.
2. Watch the bearings for a sudden high temperature rise which could indicate a bearing problem. In general, the bearing temperature should be no more than 50°F above the inlet oil temperature or a maximum of 195°F. These temperatures are very conservative and

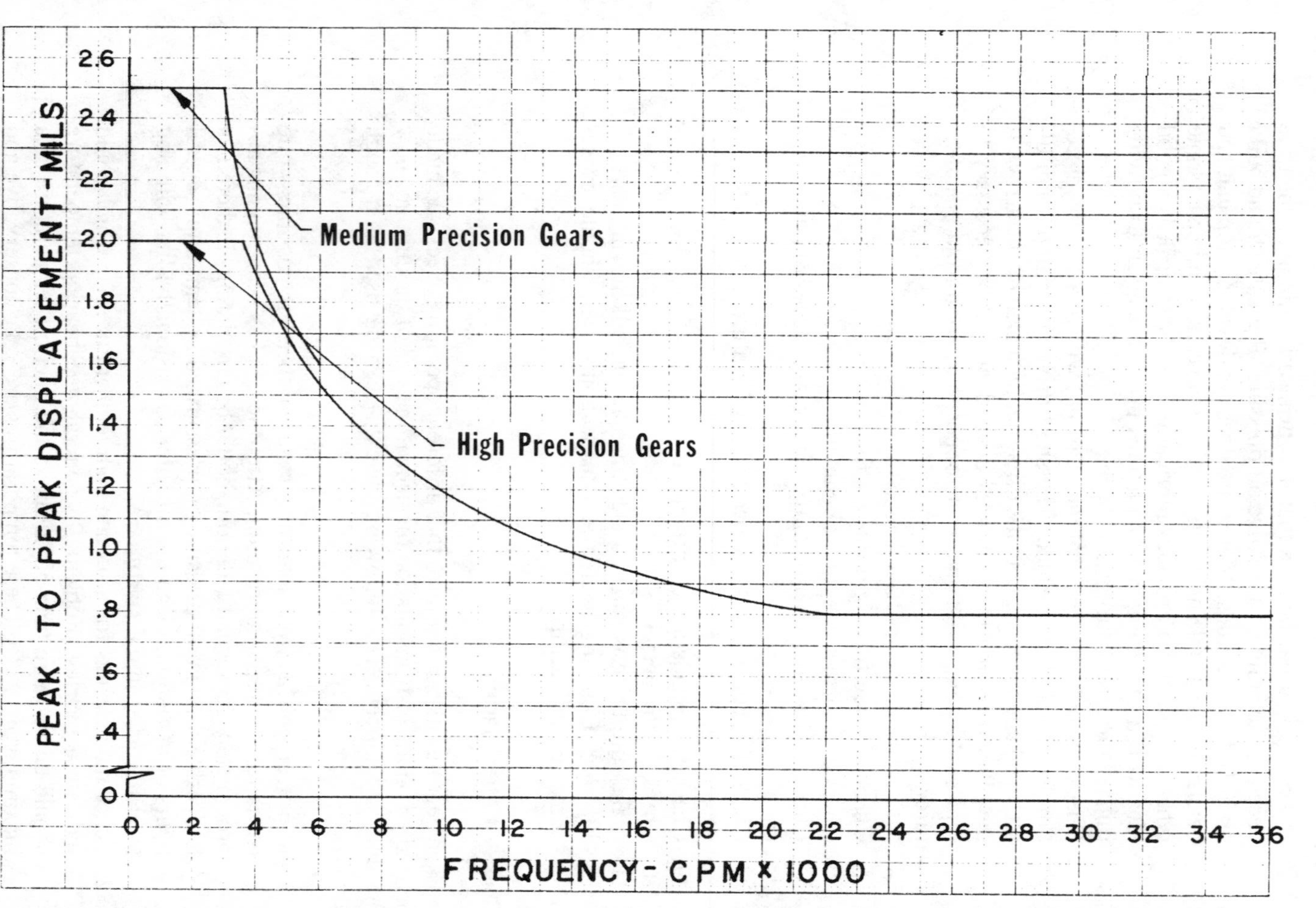

Figure 7-32. Expected maximum shaft vibration levels (peak-to-peak displacement) with good alignment and balance for medium and high precision gearing.

may be exceeded on many high performance designs. If in doubt, refer to the instruction manual furnished with the gear. Also, measured bearing temperature rise will depend on where and how the measurement is made.

3. Run gear under full load and speed and check for unusual noise, vibration, oil temperature, and bearing temperature. After temperature stabilization, the oil temperature downstream from the oil cooler (if used) should generally be no greater than 140°F. However, some special units are designed to operate with oil inlet temperatures up to 180°F or more.
4. After the unit has run several hours (six or eight) under load, shut it down, check coupling alignment, tighten any bolts that may be loose, and recheck tooth contact.

Vibration Levels

A well manufactured and properly installed gear unit should run with very little vibration; actual vibration levels may vary depending on the type of foundation used. Expected vibration in terms of peak-to-peak displacement levels for gears on permanent foundations and in good condition are shown in Figures 7-32 and 7-33 for the shafts and housings respectively. Housing vibration levels should be measured on the bearing caps, and areas that may be in resonance should be avoided. In addition, vibration velocity levels can be measured on the housing; acceptable levels are shown in Table 7-2. Very low speed units with high shock loading can have vibration levels several times those shown in Figures 7-32 and 7-33 and still be acceptable. Refer also to chapters on vibration monitoring in Volume 3, and later in this volume.

Acceleration Levels

To monitor gear tooth condition using accelerometers, a reasonably stable area of the housing should be selected, such as a bearing cap. Record base-line acceleration levels when the gear is in good condition if possible. Levels can be monitored continuously on a regular schedule. Slight changes will indicate that something in the system is different. Large increases at tooth mesh frequency will usually indicate gear wear or deterioration.

Very often large increases in acceleration will indicate that changes have taken place in the alignment, bearings, or couplings, causing uneven loading on the gear teeth. If external forces are causing the change, early detection and correction can prevent gear failure.

There are no good rules as to what are acceptable acceleration levels for gear drives. These levels will vary with design, installation, coupling

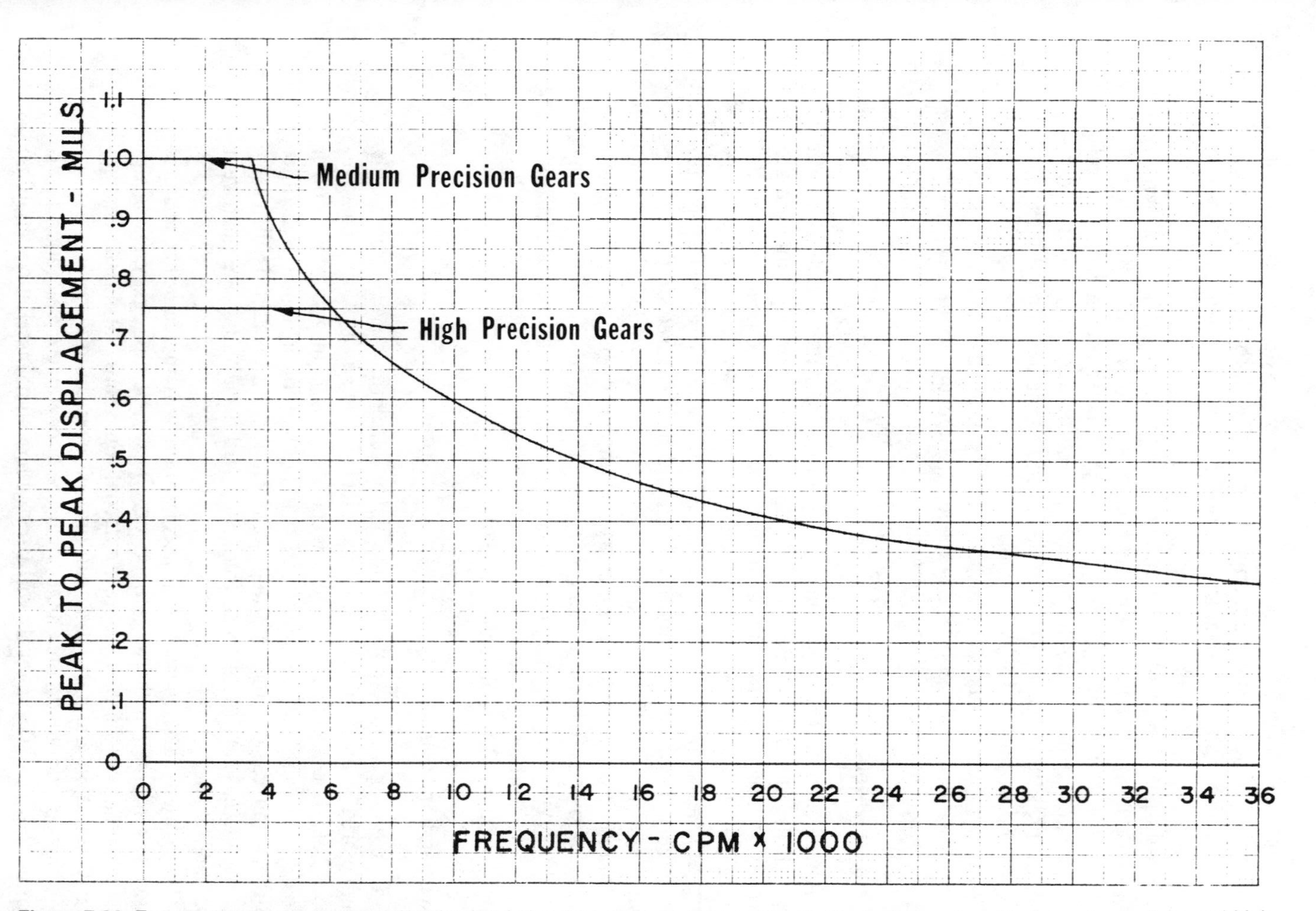

Figure 7-33. Expected maximum housing vibration levels (peak-to-peak displacement) with good alignment and balance for medium and high precision gearing.

Table 7-2
General Guidelines for Housing Velocity Vibration Levels—Gear Units*

Peak Velocity Level, Inch per Second (ips)**		
High Speed, High Performance Gear Units (e.g., per API 613 Specification)	**Gear Units for Low Speed Drives, Extruders, Bandbury Mixers, Reciprocating Machinery, Etc.**	**Vibration Classification and Recommended Action*****
Less than 0.1	Less than 0.2	Smooth; no correction necessary
0.1 to 0.2	0.2 to 0.3	Acceptable; correction not necessary (wastes money)
0.2 to 0.3	0.3 to 0.4	Marginal, action taken or not depending on circumstances
0.3 to 0.5	0.4 to 0.6	Rough; planned shutdown for repairs
Greater than 0.5	Greater than 0.6	Extremely rough; immediate shutdown necessary

* *The data in this table apply to gear units only; they are not applicable to other types of machinery.*

** *The peak velocity levels listed represent housing velocity vibration levels as measured on the bearing caps of the gear unit.*

*** *Vibration classifications and recommended courses of action listed in relation to peak housing velocity levels are intended as general guidelines for evaluation only. There are no absolutes in vibration severity analysis. All environmental factors, such as the peculiarities of adjacent equipment and the gear unit foundation, in addition to the basic characteristics of the gear unit itself, must be taken into account whenever attempting to evaluate vibration severity.*

types, gear accuracy, condition, and many other variables too numerous to mention. Acceleration levels will usually also vary with load, and to monitor trends, this variable must be accounted for. Furthermore, when measuring tooth mesh frequencies in the acoustic range, the mounting of the accelerometer is very important in that the accelerometer base and mounting can greatly influence the accuracy of the readings obtained.

Gear Lubrication

Gear lubrication is something to which gear suppliers have a lot of exposure, but most of their knowledge of this subject is secondhand. Very seldom do manufacturers have the opportunity to witness long-term lubricant performance. A maintenance crew may know more about lubrication than a gear designer because gear manufacturers have an opportunity to be an integral part of lubrication selection and witness its performance only when a gear problem exists. Even when there is a problem, the manufacturer cannot assume the fault lies with the lubrication. Very often a gear problem may be due to unknown factors in the system, installation errors, or poor maintenance, and yet it appears as a lubrication problem. Also, poor manufacturing workmanship and engineering design can initially resemble lubrication problems.

Lubricant Function

Lubricants in gear units have basically two functions: to separate the tooth and bearing surfaces, and to cool these surfaces. On low speed gear units, the primary function is lubrication; on high speed units, the primary function is cooling. This statement does not imply that both functions are not important but rather refers to the relative quantity of oil required to perform each function.

On low speed gear units, the quantity of oil necessary is determined by the amount required to keep the gear tooth and bearing surfaces wetted. On high speed units, oil quantity required is generally determined by the amount of heat loss (or inefficiency) in the bearings and mesh. As a general rule, one gallon per minute must be circulated for each 100 horsepower transmitted; this quantity would result in a temperature rise of approximately 25°F. Higher horsepower units use a 40 to 50°F temperature rise and require 0.5 to 0.6 gallons per minute per 100 horsepower transmitted. These figures are based on the assumption of 98 percent gear unit efficiency.

Modes of Gear Tooth Lubrication

Three different lubrication conditions that can be present between the teeth of two meshing gear elements are boundary lubrication, hydrodynamic lubrication, and elastohydrodynamic lubrication. Depending on load, speed, temperature, tooth design, and tooth finish, any or all of these lubrication modes could exist in the same gear drive. The goal is to have either hydrodynamic or elastohydrodynamic lubrication present between meshing gear teeth. Unfortunately, all too often boundary lubrication is present and damage to the gear teeth results. In other words, gear life is determined by wear and consequently, by the mode of lubrication present.

Boundary Lubrication

Boundary lubrication most often is found at slow to moderate speeds, on heavily loaded gears, or on gears subject to high shock loads. This mode of lubrication exists when the oil film is not thick enough to prevent some metal-to-metal contact. This condition usually shows up as early wear and pitting on the teeth due to irregularities in the tooth surfaces. When boundary lubrication is encountered, extreme pressure oils should be used to minimize wear and possible scuffing.

Hydrodynamic Lubrication

Hydrodynamic lubrication occurs when two sliding surfaces develop an oil film thick enough to prevent metal-to-metal contact. This type of lubrication usually only exists on higher speed gearing with very little shock loading.

Elastohydrodynamic (EHL) Lubrication

Elastohydrodynamic theory of lubrication is now accepted as very common in gear teeth. The formation of EHL films depends on the hydrodynamic properties of the fluid and deformation of the contact zone. This flattening of the contact area under load forms a pocket that traps oil so that the oil does not have time to escape and results in an increase in oil viscosity. This increase makes possible the use of light oils in high speed drives and usually only occurs above 12,000 feet per minute pitch-line velocity.

Lubricant Selection

Six factors affecting lubricant selection for gear units are listed in Table 7-3 along with the lubricant properties that should be considered in relation to each. Viscosity is probably the single most important element in lubricant selection and is determined by load, speed, and temperature variations. All of these factors should be reviewed and evaluated to determine the exact lubricant properties necessary for satisfactory gear performance. Final selection of the lubrication oil for the gear unit should be based on the best combination of all of the required lubricant properties.

Lubricant Types

A good rule to follow when evaluating the type of lubricant to use is to consider the least expensive one available that will perform well in that situation. If a specially blended type of oil is to be tried, determine its stability by selective use before making major changes. Lubricant failures are expensive!

Table 7-3
Important Gear Factors and the Lubricant Properties Related to Each[*]

Factor	Related Lubricant Property
Load	Viscosity EP Additives
Speed	Viscosity EP Additives
Temperature (Operating and Ambient)	Viscosity Viscosity Index Fluidity Oxidation Stability EP Additives
Contamination	Demulsibility Corrosion Protection Oxidation Stability
Life	Oxidation Stability Additive Depletion
Compatibility	Synthetic (Paint and Seals) EP Additives

[*] *Equivalent viscosities of different classification systems are included for reference only in Table 7-4.*

There are many brand name lubricants available on the market today, but all fall into five basic types. The following discussion is a brief summary of the characteristics, advantages, and disadvantages of each of the different categories.

Mineral Oils

Mineral oils are still the most commonly used type of gear lubricant. Containing rust and oxidation inhibitors, these oils are less expensive than the other types, readily available, and have very long life. When gear units operate at high enough speeds or low enough load intensities, a type of mineral oil is probably the best selection.

Extreme Pressure Additives

Extreme pressure (EP) additives of the lead-naphthenate or sulphur-phosphorus type are recommended for gear drives when a higher load capacity lubricant is required. As a general rule, this type of oil should be used in low speed, highly loaded drives with medium operating temperatures. EP oils have the disadvantage of being more expensive and they must be replaced more often than straight mineral oils. Some of these EP oils have a very short life above a temperature of 160°F.

A good gear EP oil should have a Timken OK load above 60 lbs. and pass a minimum of 11 stages of the FZG test. The Timken OK test is considered acceptable for determining whether a lubricant has extreme pressure properties, but is considered questionable in evaluating levels of extreme pressure capacity. The FZG test is widely used in Europe for evaluating gear oils and is being used with increasing frequency in the United States. This test uses spur gears in mesh under load, and the amount of wear is determined by weight loss. The FZG test procedure is very sensitive to scoring and is considered capable of evaluating the extreme pressure properties of industrial lubricants.

Boron compounds as EP additives are being tested, and these products show promise as extremely high load capacity lubricants. The compounds being tested exhibit Timken OK loads greater than 100 lbs. and pass 14 stages of the FZG test. This type of additive is nontoxic and highly stable but sensitive to water.

Synthetic Lubricants

Synthetic lubricants are not usually recommended by gear manufacturers for general gear applications due to high cost, limited availability, and lack of knowledge of their properties. Nevertheless, they are used with good success in applications with extremely high or low temperatures, where fire protection is required, or where very high speeds or high wear rates are encountered. The user must be careful when selecting these lubricants since some of them remove paint and attack rubber seals. The new synthesized hydrocarbons (SHC) have many desirable features such as compatibility with mineral oils and excellent high and low temperature properties. They are excellent selections when EP lubricants along with high temperature operation are required.

Compounded Oils

Compounded oils are available with many different additives. The most commonly available is a molybdenum disulfide compound that has been successfully used in some gear applications. It is very difficult for a gear manufacturer to recommend these oils at this time since some of these additives have a tendency to separate from the base stock. In many instances, however, compounded lubricants are the only solutions to gear lubrication problems. These oils can be blended for extremely high load-carrying capacity and high temperature operation. Most of these "super" properties can be obtained, but sacrifices must be made in other lubricant properties such as life or corrosion protection.

Viscosity Improvers

Viscosity improvers in gear drives should be used with great care. These polymer additives make great textbook improvements in the viscosity index and extend the operating temperature range of an oil. However, these polymers are non-Newtonian fluids, and the viscosity of these fluids reduces with shearing. A gear drive is a very heavy shear application, and as a result, the viscosity is reduced rapidly if too much polymer is used. These lubricants are seldom recommended in long life gear drives.

Methods of Supplying Lubricant

Several different techniques of supplying lubricating oil to the gears and bearings in a gear unit are available to the gear manufacturer. The three primary methods in use today are splash lubrication, forced-feed lubrication, and intermittent lubrication. Each of these methods has identifying characteristics which are described in the following sections.

Splash Lubrication

Splash lubrication is the most common and foolproof method of gear lubrication. In this type of system, the gear dips in oil and in turn distributes that oil to the pinion and the bearings. Distribution to the bearings is usually obtained by throw off to an oil gallery or is taken off the sides of gear by oil wipers (or scrapers) which deliver the oil to oil troughs.

When using the throw-off system, care must be taken that the operating speed is high enough to lift and throw off the oil. In this system, the minimum speed required may be determined using the following formula:

$$n_p = (70{,}440/d)^{.5}$$

n_p = Minimum speed, RPM
d = Pitch diameter, inches

Oil wiper systems can operate at much lower speeds, which are usually determined by test or through experience.

The splash system can be used in gear units with up to 4,000 ft per minute pitch line velocity. Higher speed gear units can be splash lubricated with special care.

Forced-Feed Lubrication

Forced-feed lubrication is pressurized lubrication and is used on almost all high speed gear drives, on spiral bevel drives, and on low speed drives when splash lubrication cannot be used due to gear arrangement. A simple forced-feed system consists of a pump with a suction line and supply lines to deliver the oil; the gear housing serves as the reservoir. In contrast to this simple arrangement, more complicated lubrication supply systems for high speed drives may include many of the following components:

1. Large reservoir
2. Filters (duplex or single)
3. Shaft-driven pump
4. Auxiliary pumps (motor- and steam-driven)
5. Heat exchangers (single or duplex)
6. Accumulators
7. Pressure control devices
8. Safety alarms and shutdowns (temperature and pressure)
9. Temperature regulators
10. Isolation valves
11. Heaters (steam or electric)
12. Purifiers (to remove water and oxidation products)
13. Flow indicators

Many of these lubrication systems are well designed and constructed not only to lubricate the gears and bearings of the gear unit but also to enhance performance of the driving machine, gear unit, and driven machine. Figure 7-34 illustrates one such system.

Intermittent Lubrication

Intermittent lubrication is exactly that: a system in which the lubricant is not available continuously but is supplied periodically to the gears or bearings, or both. This type of lubrication system is the least common and is primarily suited for low speed applications. Of the three methods of applying the lubricant—brushing or pouring, hand spray, and mechanical spray—mechanical spraying is by far the most commonly used.

1. Brushing or pouring: In this method an extremely heavy lubricant is brushed or poured on the gears by hand. It is used when a pan or any form of flooded lubrication is impractical for gears operating at very low tip speeds. This lubricant can be applied either cold, if the viscosity allows, or hot when preheating is required for application.
2. Hand spray: Pressure lubrication is obtained when the lubricant is placed in a container and sprayed similar to applying paint. This method is better than the brushing or pouring application as it provides more uniform distribution of the lubricant.

Figure 7-34. Three views of high speed gear unit and turbine mounted on integral baseplate with complicated pressurized lubrication system which supplies oil to gear unit and turbine. Lube console consists of shaft-driven main oil pump, motor-driven auxiliary oil pump, dual oil filters, dual heat exchangers, relief valve, pressure and temperature gauges and switches, oil level switch, interconnecting piping, and all turbine, gear unit, and lubrication system instrumentation mounted on control panels. The oil reservoir is a drop-in type located in the baseplate. *(Continued on next page.)*

Figure 7-34. Continued.

3. Mechanical spray: This is intermittent lubrication usually performed by an automatic timer where the oil is supplied to the gears or bearings in limited amounts at certain intervals.

Lubrication of High Speed Units

The oil furnished to high speed gears has a dual purpose: lubrication of the teeth and bearings, and cooling. Usually, only 10 to 30 percent of the oil is used for lubrication and 70 to 90 percent is used for cooling.

For high speed gear units, a turbine-type oil with rust and oxidation inhibitors is preferred. This oil must be kept clean (filtered to 40 microns maximum, preferably to 25 microns), must be cooled, and must have the correct viscosity. Synthetic oils should not be used without the manufacturer's approval.

For some reason, the high speed gear unit makes all the compromises when the oil viscosity for the machine system is determined. Usually a viscosity preferable for compressor seals or bearings is selected, and gear life is probably reduced. The bearings in a gear unit can use the lightest oils available, but gear teeth need a much heavier, more viscous oil to increase the film thickness between the teeth.

When lighter viscosity oils (such as light turbine oil which has a viscosity of 150 SSU at 100°F) are necessary, inlet oil temperatures should be limited to 110 to 120°F to maintain an acceptable viscosity. In addition, the oil should be supplied in the temperature and pressure range specified by the manufacturer. See Table 7-4.

In high speed gears with a pitch line speed of up to approximately 15,000 ft per minute, the oil should be sprayed into the out-mesh. This procedure allows maximum cooling time for the gear blanks and applies the oil at the highest temperature area of the gears. Furthermore, a negative pressure is formed when the teeth come out of mesh, and this pressure pulls the oil into the tooth spaces.

Above approximately 15,000 ft per minute, 90 percent of the oil should be sprayed into the out-mesh and 10 percent into the in-mesh. This additional spraying of the in-mesh is a safety precaution to assure the amount of oil required for lubrication is available at the mesh. In addition, for gears in the speed ranges from 25,000 to 40,000 ft per minute pitch line velocity, oil should be sprayed on the sides and in the gap area (on double helical types) of the gears to minimize thermal distortion.

The primary detrimental effect of inadequate lubrication is scoring or scuffing (adhesive wear) which is caused when the oil film does not pre-

Table 7-4
Equivalent Viscosities of Different Lubricant Classification Systems (for reference only)

AGMA Lubricant No.	Equivalent ASTM-ASLE Grade No. (Average SSU @100°F)	Equivalent ISO Grade No.	Metric Equivalent Viscosity Ranges cSt at 37.8°C (100°F)
Light turbine	S150	32	28.8 to 35.2
1	S215	46	41.4 to 50.6
2, 2 EP	S315	68	61.2 to 74.8
3, 3 EP	S465	100	90 to 110
4, 4 EP	S700	150	135 to 156
5, 5 EP	S1000	220	198 to 242
6, 6 EP	S1500	320	288 to 352
7 Comp., 7 EP	S2150	460	414 to 506
8 Comp., 8 EP	S3150	680	612 to 748
8A Comp.	S4650	1000	900 to 1100

vent contact between mating surfaces. Small areas touch each other due to load and surface irregularities. This results in welding of the two surfaces. As sliding continues, these surfaces break apart and particles adhere to the surfaces, causing rapid adhesive wear to occur.

Gear lubrication at the present time is not a highly developed technology in general industrial applications, and as a result, the ultimate capacity of gearing is partially determined by lubricant load limits. Hopefully, a much higher capacity lubricant will be developed in the future that will solve our gear lubrication problems by generating thick films without the pitfalls now associated with the extremely heavy oils. However, today, due to the higher prices and disposal problems of lubricating oils, we must make greater efforts to obtain maximum benefits from the oils now available.

Inoperative Periods

In new gear units shipped from the factory, the rust inhibitor adhering to internal exposed surfaces should prevent corrosion of interior parts for at least six months. Exterior preservatives should last at least six months, but this protection will depend on handling and exposure to the elements. A new gear unit should be stored inside if possible, but if not, covered outside storage can be used. It is always a good idea to use a dry nitrogen purge during storage to prevent or minimize condensation inside the gear housing.

When the recommended lubricant is used and the reducer has been operating for a period of time, the lubricating oil should protect interior parts for inoperative periods up to 30 days since most of these oils have rust and oxidation inhibitors added.

If additional downtime or storage time is required, one of the following methods can be used to protect the internal parts of the gear unit:

1. The unit can be operated for a short period of time every 30 days to redistribute the oil to the nonsubmerged parts and gain another 30 days protection.
2. If extended downtime is expected and it is impractical to spin the unit, a rust preventative oil should be brushed or sprayed on the gear teeth and bearings through the inspection opening. Any opening such as breathers or labyrinth seals should be sealed with masking tape. A quality rust preventative oil should give 12 months protection against corrosion when applied in this manner. This oil should be compatible with the operating oil, and it should not be necessary to remove the rust preventative when the unit is re-

started. When this method of protection is used, a dry nitrogen purge is recommended.
3. For adverse conditions or very long term storage, seal all openings and fill the unit completely to the top with the lubricating oil. When the equipment is to be operated, the seals must be removed, and the oil level dropped to the proper operating level.
4. The most permanent method for long term storage of equipment is to disassemble the gear unit and coat each part with a preservative such as Cosmoline® or an equivalent. Before the unit can be placed in service, special cleaning with solvents will be necessary to remove all preservatives from the internals and the lube components.

When a unit is inoperative, most gear manufacturers recommend that it be inspected every thirty days to six months depending on the method of protection. Any areas of the preservative not performing properly should be removed with solvent and recoated.

Maintenance

Gear unit maintenance, whether preventive or after-the-fact, is one of the most important aspects affecting satisfactory unit performance. Preventive maintenance can help to keep problems from occurring. If they do occur, timely maintenance can prevent problems from getting worse and can even correct them in some cases if performed properly.

Preventive Maintenance

Good preventive maintenance habits will prolong the life of the gear unit and possibly help in detecting trouble spots before they cause serious damage and long downtime. *Warning:* When working near rotating elements, be certain the prime mover is turned off and locked. The following schedule is recommended for most operating conditions:

1. Daily Maintenance
   a. Check the oil level.
   b. Check the oil temperature and pressure against previously established norms.
   c. Check for unusual vibration and noise.
   d. Check for oil leaks.
2. Weekly Maintenance
   Inspect oil filter for possible flow obstructions.

3. Monthly Maintenance
   a. Check operation of auxiliary equipment and/or alarms.
   b. Clean air breather.
   c. Check tightness of foundation bolts.
   d. Clean oil filter.
4. Semi-Annual Maintenance
   a. Check gear tooth wear.
   b. Check coupling alignment.
   c. Check zinc pencils in heat exchanger.
5. Annual Maintenance
   a. Check heat exchanger for erosion, corrosion, or foreign material.
   b. Check bearing clearance and end play.
   c. Check tooth contact pattern.
   d. Check condition of oil and change if necessary.

Journal Bearing Maintenance

The most commonly used journal bearings in high speed gear units are split, steel or bronze backed, babbitt-lined journal bearings; however, when operating speeds or loads make oil-whirl possible, the manufacturer can use pressure dam journal bearings or some other type of stabilized bearing such as tilting pad journal, elliptical bore, or longitudinal groove bearing. The pressure dam bearing is designed for a particular direction of rotation; therefore, care should be taken at assembly to assure correct placement. The pressure dam grooves are normally found on the pinion bearings only. However, the gear bearings on some of the smaller units may also require pressure dam grooves in some cases. The grooves are positioned on the unloaded side of the bearing journal as shown in Figure 7-35.

To axially locate the gear train and to take any nominal thrust created by external loads, the manufacturer normally uses flat face thrust bearings with radial grooves, tapered land, Kingsbury, ball, or tapered roller thrust bearings. In units with double helical or herringbone gearing, only one thrust bearing is required to locate the gearset as this type of gearing generates no external thrust. The thrust bearing is usually placed on the low speed shaft allowing the pinion to center itself with respect to the low speed gear. Where single helical, spiral bevel, or other types of gearing that generate external thrust loads are used, thrust bearings must be positioned on each shaft.

Carefully inspect bearings and journals for uneven wear or damage. If required, polish journals using belt-type crocus cloth and remove high

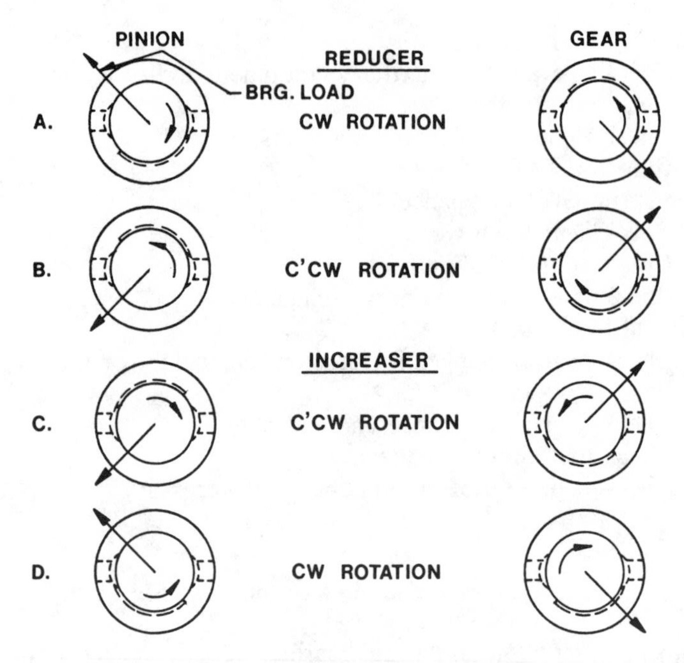

Figure 7-35. Pressure dam bearing groove location.

spots. Sleeve and thrust bearings should be thoroughly inspected for correct clearance, high spots, flaking of babbitt, scoring, and wiping using the following procedures.

Bearing Clearance

The journal bearings used in high-speed gears must have clearance between the shaft journal and the bearing. The amount of clearance necessary depends on the oil viscosity, the journal speed, and the bearing loading. Manufacturers' engineers consider each of these parameters in calculating a bearing clearance that will produce hydrodynamic lubrication as well as a flow of oil sufficient to cool the bearing. The required clearance for your particular unit may be found in the technical data furnished with the unit or may be available from the manufacturer. A rough rule of thumb (when better data are not available) is .001 in. per in. of journal diameter plus .0015 in. up to 9000 ft per minute journal velocity. Above this speed, clearance should be .002 in. per in. of journal diameter for shafts larger than 2.5 in.

Measurement of bearing clearances may be done by lifting the shaft and measuring the distance traveled by the shaft with a dial indicator. Also, feeler gauges or plastic gauge material can be used. Some wear should be expected, especially on a gear unit that is stopped and started or slow-rolled frequently. The bearing may be considered operational as long as the measured clearance does not exceed the design clearance by more than .004 to .005 in. for bearings in the 3 to 8 in. diameter range. This value for clearance increase is acceptable for most applications and can be used when the manufacturer's recommendation is not available. *Note:* If high shaft vibration develops, this clearance increase may not be acceptable.

The thrust bearing clearance provides room for the formation of an oil film between the bearing face and the gear hub and for thermal expansion of the shafting. Normally, wear of the thrust bearing is not very critical unless it is enough to cause loss of oil pressure in the lubrication system. Measurement of thrust bearing clearance wear may be done by pushing the slow speed shaft axially to one side of the unit, setting a dial indicator, and then pulling the shaft axially to the other side of the unit against the dial indicator.

On gear units with double helical gearing where there is one thrust bearing located on the low speed shaft, the end play of the high speed shaft may be determined by holding the slow speed gear stationary and measuring the axial movement of the pinion against the gear. This value should be added to the thrust bearing clearance to obtain the total end play of the high speed shaft. When both the high speed and low speed shafts have thrust bearings, the high speed thrust bearing clearance may be measured in the same manner as the low speed.

Bearing High Spots

Evidence and location of high spots in the bearing are indicated by bright spots or areas. These spots are caused by wear resulting from a ruptured oil film around the high point. Bright spots should be lightly scraped and polished with fine steel wool or crocus cloth until they blend in with the rest of the bearing. *Caution:* Do not use sand paper.

To check the bearing contact, install the lower half of the bearing in the housing with the journal and thrust face clean and dry. Check outside diameter of bearing with a .0015 in. feeler gauge to be sure the lower half is seated in the housing. Apply a light coat of soft blue to the journal and to each thrust face. The journal should show blue transfer for a minimum of 80 percent of bearing length. Thrust faces should show a minimum contact of 60 percent of load area. Contact may be spread out to the desired amount by removing high spots using a bearing scraper and fine steel wool or crocus cloth for polishing. Repeat the checking process until the contact area is satisfactory.

Flaking of Babbitt

Flaking of babbitt in the loaded area of the bearing is caused by vibration or shock loading of the bearing material, which causes the babbitt to fatigue and break loose from the steel shell. Not only do the flakes cause scoring as they pass through the bearing, but they also contaminate the oil. In the advanced stages of flaking, the load-carrying area of the bearing is destroyed, and the bearing must be replaced. However, if flaking is caught in the early stages, the bearing may be repaired by scraping and polishing. Whatever the case, the cause of vibration or hammering should be corrected before the unit is put back into service.

Scoring

Scoring is the scratching, or marring, of the bearing babbitt or the journal riding in the bearing, or both. It is caused by dirt or metal particles present in the oil that passes through the bearing. A little scoring is not serious, and the bearing may be polished with fine steel wool to remove any rough edges caused by scoring. Any foreign particles imbedded in the babbitt which could score the journal should be carefully picked out, and that area should then be polished smooth. Scoring becomes serious when it significantly reduces the bearing area. In this case, the bearing should be replaced, and the gear unit drained and flushed out with a solvent.

Wiping

Wiping is the melting and wiping away of a spot or area of the babbitt due to the bearing temperature rising above the pour point of the babbitt. Abnormally high bearing temperatures can be caused by one or more of the following conditions: insufficient bearing clearances, insufficient oil pressure, excessively high oil temperature in the bearing, a high spot in the bearing, extreme bearing loading caused by poor bearing contact, or a gear mesh failure. If wiping is localized in a small spot, the bearing may be repaired by scraping and polishing the spot until it blends in with the remainder of the bearing; otherwise, the bearing must be replaced.

Before replacing a wiped bearing, determine the cause of the wipe and take corrective measures. If new bearings are necessary, the following precautions should be taken:

1. Remove all nicks and burrs from the housing and bearing shell.
2. Be sure that journal and thrust faces are free of nicks and high spots. These spots can be removed using a fine hone and polishing with crocus cloth.
3. Obtain the proper bearing contact as described previously.
4. After the bearings are fitted and the lower halves are installed in the housing, check the radial clearance using plastic gauge material available in most supply stores. Check thrust clearance by moving shaft axially in both directions while an indicator pointer is positioned against the shaft. In many cases it may be necessary to use a small hydraulic jack to move the large gears axially to check the clearance.

Rolling Element Bearing Maintenance

Probably a majority of the gear units in factory operations use rolling element bearings to support the gear rotors and absorb the external and internal loads associated with gear drives. These bearings are extremely reliable and give long trouble-free service if they are not abused and are properly installed and maintained. The two primary disadvantages of rolling element bearings are that they are very seldom furnished split for ease of installation and they cannot easily be repaired or manufactured in local repair shops.

Volume 3 of this series contains excellent data on the causes of bearing failure and describes proper failure analysis. Many factors can contribute to bearing failures in gears; however, most rolling element bearing failures can be attributed to one or more of the following causes: defective bearing seats; misalignment; faulty mounting practice; incorrect fits; inadequate lubrication; ineffective sealing; vibration, and electric current.

Rolling element bearings should be inspected if practical for damage that could indicate incipient failure. However, as bearing inspection is not usually practical since the gear unit must be disassembled, special attention should be focused during routine maintenance and inspection on factors that can contribute to premature bearing failure such as the presence of dirt in oil, oil condition and quantity, vibration, and electrical current.

Defective Bearing Seats on Shafts and/or in Housings

For bearings to have long trouble-free life, the thin inner and outer rings must be mounted on shafts or in housings that are as geometrically true as modern machine shop techniques can produce. In other words, the shaft and bore must be round and free of taper and must completely support the inner and outer rings. Also, the sizes of each must be correct so as not to reduce the internal clearances in the bearing nor allow the bearing to fret on the shaft or in the housing bore. When bearing fits are damaged, they should be remachined and returned to their original sizes by metalizing or plating.

Misalignment

Misalignment is a source of premature spalling of the bearing. This condition generally occurs when the inner ring of a bearing is seated against a shaft shoulder that is not square with the journal or when the outer ring is seated against a housing shoulder that is out of square with the housing bore. Also, misalignment can occur when the bores from side to side of a gear housing are not parallel and square to the centerline.

Faulty Mounting Practice

Faulty mounting practice contributes greatly to premature bearing failure and usually results from abuse and neglect during the mounting procedure. The most common faulty mounting practices are foreign matter in the bearing (not properly cleaned), impact damage during handling or mounting resulting in brinelling, and overheating when expanding the bearing to slip on the shaft. When heating a bearing for mounting, an oil bath should be used if available; if a torch is used, care should be taken not to overheat in one spot since localized overheating will actually soften the bearing material.

Incorrect Shaft and Housings Fits

A bearing may need to be fitted either with an interference fit or a slip fit on the shaft and in the housing depending on the conditions present. The degree of tightness or looseness in the bearing is governed by the magnitude of the load, the speed of the journal, and the arrangement of the bearing. In gear unit bearings, the inner ring usually rotates relative to the load, and therefore, it will have an interference fit on the shaft and a light slip fit in the housing. When it is incorrectly fitted, a bearing will creep on the shaft or in the housing and cause wear to the journal or the bearing seat.

Inadequate Lubrication

Any load-carrying contact between the rollers and the inner and outer races in a bearing requires the presence of lubricants for reliable operation. All bearing rollers undergo varying amounts of sliding motion in addition to the primary rolling motion present as they transmit the load between the inner and outer races. In addition, the rollers must carry the bearing cage as the bearing rotates, so they also slide on the bearing cage. This sliding motion can be very detrimental to a bearing unless the lubricant film is thick enough to prevent contact between the sliding parts.

The viscosity of a lubricant is the most important characteristic of the oil either as oil itself or as the oil in grease lubrication. An oil with too low a viscosity allows metal-to-metal contact between the rollers and the inner and outer races, which results in bearing failure. Also, an insufficient quantity of lubricant at medium to high speeds generates a temperature rise which in turn can cause lubricant failure. Lubricant failure generally causes surface damage in the bearing ranging from frosting to spalling, discoloration, glazing, or smearing.

Ineffective Sealing

The effects of dirt and other abrasives in bearings can result in changes in bearing internal geometry. Freedom from abrasive matter is so important that some bearings for very high precision equipment are even assembled in air conditioned white rooms. In addition to abrasive matter, corrosive agents must be excluded from bearings. Water, acid, and other agents that deteriorate lubricants result in corrosion and premature bearing failure. Acids formed in the lubricant with water present etch the bearing surfaces and reduce the load-carrying capacity.

Vibration

Rolling element bearings exposed to vibration while the shafts are not rotating are subject to a damage referred to as false brinelling. This is usually indicated by either bright polished depressions at each roller or a corrosive type stain or fretting. The vibrating load causes minute sliding in the area of contact between the rolling element and raceways and sets free small particles of material that are oxidized and cause accelerated wear. Many bearing failures are probably caused by false brinelling which is never discovered since the unit is usually operated until the bearing is destroyed before it is inspected.

Passage of Electric Current Through the Bearing

In certain applications where electrical machinery or electrical equipment is in use, there is the possibility that electric current will pass through a bearing. Current that seeks ground through the bearing can be generated from magnetic fields in the machinery or can be caused by welding on some part of the machine with the ground attached so that the circuit is required to be completed through the bearing. There are many other causes of this phenomenon varying from electrostatic discharge and belts to manufacturing processes involving leather, paper, or rubber.

When the current is broken at the contact surfaces between rolling elements and raceways, marking results; this marking produces localized high temperature, and consequently, the surfaces are damaged. This damage is usually manifested as very small pits in the raceways and the rollers. Very moderate amounts of electrical pitting do not necessarily result in failure that may cause bearing life to be reduced. Severe electrical pitting will cause almost immediate bearing failure when the machine is started.

Gear Unit Disassembly and Assembly

Due to the large variety of gear housing or enclosure designs, it is not practical to describe disassembly and assembly procedures for all different types. Alternatively, the basic steps involved and some important precautions to observe will be discussed.

Disassembly

The first step in disassembly is to remove the housing cover. Be careful not to damage internal and external piping and instrumentation that may be routed from the housing to the cover. The upper halves of journal-type bearings also tend to stick in the cover half bores and may fall out as the cover is being lifted, thereby damaging the bearing halves.

After the cover is removed, disconnect the internal instrumentation and remove the upper bearing halves if present. On antifriction bearings with bearing carriers or capsules, it may be necessary to either remove or support the carriers before removing the gear elements.

Extreme care must be used when removing, lifting, and handling gear elements. Use soft slings and protect all gear surfaces when lifting and handling. Do not set gear or pinion on a hard surface if it is to be reused.

Preparation for Assembly

Before assembling a gear unit, several steps should be taken to prepare it. These steps will help to ensure trouble-free, satisfactory operation of the gear unit after it is put back into service.

First of all, remove all of the old split line sealant from the machined surfaces of the housing and cover. Also remove the excess sealant that ran into the bores and oil passages. Many bearings have been failed by the presence of excessive sealant or old sealant. As a final step in preparing the split line, flat-file the machined surfaces to remove all sealant residue and nicks.

While a gear unit is disassembled, try to keep any dirt or trash out of the housing and off the parts. In addition, before assembling the gear unit, ensure that the parts and housing are as free as possible of dirt and trash. If feasible, the gear housing should be washed down with solvent during assembly, then the system should be flushed with oil after assembly is completed. During the flushing procedure, the oil temperature should be raised to 110 to 140°F and the shafts rotated in both directions by hand to dislodge any trash.

Cleanliness cannot be over-emphasized. With higher load and higher precision gearing, cleanliness is more important since very high precision gears operate with a lubricant film thickness at the gear mesh of a few microns. Any minute piece of foreign material present can pass through the mesh and damage the gear teeth. Furthermore, journal type bearings operate with film thickness of less than .001 in., and any trash present will become embedded in the bearing, causing damage to the bearing and the shaft. Rolling element bearings are a little more forgiving of foreign material than journal bearings are since any foreign particles tend to pass through the bearing rather than embed in it.

Before installing the bearings in the gear unit, it is important to be sure that they fit down in the bores and that the bores do not crush the bearings at the split line. For journal type bearings, it is often necessary to fit the bearing outside diameter to the housing bore and also to fit the bearing inside diameter to the shaft journal. While fitting the bearing to the journal and bore, gear tooth contact should be checked using the soft blue procedure. In the event of poor contact, the bearings should be refitted until the contact pattern is acceptable.

One of the aspects most influential on the satisfactory operation of a gear unit is the handling of the rotating elements. Many gear sets have been destroyed or have required extensive rework due to improper handling. Before assembling gear elements in a housing, they should be inspected visually and by feel for nicks and bruises. The importance of the sense of touch during this inspection should not be underestimated since many small areas of damage can be practically invisible yet still very harmful. Remove all nicks and burrs present by stoning or filing lightly.

Assembly

The first step in assembling a gear unit is to place the bearings and rotating elements into the housing. Proper precautions should be taken during this procedure. When the bearings and rotating parts are placed in the housing, be sure all fits are correct and the bearings are properly seated in the bores. Recheck tooth contact with the soft blue procedure. Also, using the hard type of layout blue (spray or brush), coat three or four teeth at three locations around the gear.

Next, apply a sealant to the machined split line of the housing. If a silicone-type sealant is used, care must be taken to prevent the excess from entering the oil system and clogging the orifices. If anaerobic sealers are used, make sure that the jacking screws are installed, since it may be very hard to disassemble the unit afterward. Do not use anaerobic materials for pipe threads or for stud locking on the split line since the strength of these sealants may be so high that future disassembly may not be possible without damage.

After applying the sealant, set the cover over the gearing, and insert taper pins or locating devices. Snug down bolts and studs by hand, and drive taper pins "home." Check the alignment of the bores at the split line to be sure any offset is minimal. If offset is present, raise the cover and reassemble. An allowable amount of offset cannot be given since this value varies with the size and design of the gear unit. If the offset present seems too great and the unit cannot easily be rebored, correction can be made by scraping or sanding the bore from the split line up (approximately 20 to 30°) to minimize the "pinch" on the bearings. Finally, tighten all bolts completely, and add end caps and auxiliary equipment.

Before operating the reassembled gear unit, be sure to hand turn the unit if possible and check the coupling alignment. After operating the unit for a short time, check the tooth contact by observing the wear-off pattern of the hard blue.

Use hands, ears, and available instrumentation to check for abnormal temperatures, noise, or vibration, especially during the initial period of operation. Observe the checklists for before and after start-up listed elsewhere.

Troubleshooting

The most common gear problems are noise, overheating, vibration, tooth wear, and tooth breakage. The following is a discussion of the most common causes of and remedies for each type of problem.

Gear Noise

Unfortunately, gear units are a noise source. Manufacturers are working on solutions, but they have not found any economical answers. Many factors, some of which are listed below, can contribute to gear noise. However, the causes are not limited to the items on this list.

1. *Tooth Spacing Errors*—Spacing errors are usually caused by manufacturing problems, damage to the teeth during assembly/disassembly, or occasionally during operation when foreign objects pass through the mesh. The spacing error may cover only one tooth space or a number of tooth spaces. This type of error usually will cause a bumping noise or a cyclic noise with a frequency equal to the rotating speed of the gear or pinion with the spacing error. In most cases, the gear will just run roughly and have reduced life.
2. *Involute Error*—Involute is the curve form used for the mating tooth surfaces. It is not necessary that the involute be "textbook" but the tooth flank curves must be conjugate, that is, matching. This error can occur when these surfaces are either manufactured incorrectly or destroyed by wear or scoring.
3. *Surface Finish*—Surface finish on gear teeth very seldom causes noise except in extreme cases of scoring or abrasive wear.
4. *Lead Error (Helix angle error)*—Lead error is only important if the leads (helix angles) of the gear and pinion are not matched. When the leads do not match, the gear may be quiet when new, and as wear occurs, the gear will become noisy.
5. *Wear on Tooth Flanks*—Wear only causes noise when it is severe and when the gear teeth do not wear evenly and maintain conjugate action.
6. *Pitting*—Tooth flank pitting is not a great influence on noise unless it is very severe.
7. *Resonance*—Exciting the natural frequency of gears, shafting, housing, or supports can produce high noise levels. The most troublesome resonances tend to be in the housings and support bases and can sometimes be corrected with additional stiffening.
8. *Tooth Deflections*—Under load, teeth deflect and tend to lose their conjugate forms. Gear manufacturers modify the involute form so that under load, interference does not occur, and conjugate action is maintained. Excessive tooth deflections can cause noise.
9. *Improper Tip or Root Relief*—This is one method of making the tooth deflection corrections as required in No. 8. The higher the load on a gear tooth, the more tip or root relief is required to prevent interference or to maintain lubricant film.

10. *Pitch Line Runout*—Pitch line runout is a form of tooth spacing error where each tooth is not an equal distance from the axis of rotation. This usually shows up as a cyclic noise with a frequency equal to the shaft rotation speed.
11. *Excessive Backlash*—Excessive backlash can only cause noise when the gear set has torque reversals. Large amounts of backlash should never be used as criteria for determining whether a gear is acceptable for use. On nonreversing drives, backlash is only important when it becomes so excessive that the tooth strength is reduced.
12. *Too Little Backlash*—A gear set without sufficient backlash to account for manufacturing errors and thermal growth is a disaster. These gears will be noisy and will fail in a very short time.
13. *Noise Transmitted From Driving or Driven Machine*—A gear housing is a large drum-like container and can amplify the noises emitted through the structure from the motors, compressors, pumps, generators, etc.
14. *Load Intensity*—As a rough rule of thumb, the higher the load intensity, the higher the noise level. As just pointed out, gear teeth require tooth form modifications to account for deflections under load. As a result, the higher the load intensity, the harder it is to make these corrections properly.
15. *Rolling Element Bearings*—This type of bearing tends to be more noisy than others due to the loose pieces in the bearing such as cages and rollers. Also, the roller passing frequency is quite high and will produce noise.
16. *Clutches and Couplings*—Couplings are some of the primary noise producers of rotating machinery. Windage noise is produced by the bolts and other openings in the coupling flanges. Clutches have all of the problems of couplings but in addition have a tendency to rattle.
17. *Face Overlap Ratio*—This is the number of teeth in contact at any time across the face of a gear. With more teeth in contact, gear errors tend to average out producing quieter gear operation. The best way to increase face overlap ratio is to use a higher helix angle during manufacture.
18. *Transverse Contact Ratio*—This is the number of teeth in contact in a transverse plane. Low pressure angle gears have higher contact ratios and are quieter but do not have adequate tooth strength. Generally, a compromised pressure angle is selected to give an optimum balance between contact ratio and bending strength.
19. *Lube Oil Pump and Piping*—Lube oil pumps can be extremely noisy if the piping is not properly designed. The most common

causes of pump noise are cavitation in the pump suction area and piping resonance.

Gear noise can be controlled to some extent by three measures: very careful design and super quality manufacturing, extra heavy cast iron or double wall steel housings, and acoustical enclosures. Any of these measures can be used singly or in combination with one or both of the other measures to effectively reduce gear noise.

Very careful design and super quality manufacturing is the most expensive way to control gear noise in addition to being the most difficult to apply. Contrary to some opinions, the perfect gear is useless for power transmission because of tooth deflections under load. The trick is to obtain a gear which has a perfect involute (or conjugate) form under load. The harder the gear, the more deflection there is due to higher allowable loading; as a result, good mesh conditions are more difficult to obtain since the involute produced is distorted more.

Secondly, extra heavy cast iron or double wall housings used with reasonably accurate gearing can be very effective in controlling gear noise. This method is less expensive than the first and is easier to apply consistently in manufacturing. Also, detuning techniques can be used on housings and gear blanks to reduce noise based on calculated and experimental data.

Using an acoustical enclosure is the least expensive way to reduce gear noise. Almost any noise level can be attained if space is not a problem. However, sound enclosures have a very definite disadvantage when maintenance is required. In addition, the inability of operators or maintenance personnel to actually place their hands on the equipment or hear the noises emitted if problems begin to appear may allow total destruction of the gear unit to occur instead of just minor damage. No matter how sophisticated the monitoring equipment, the senses of touch and hearing are still the best indications of a machine's condition.

Overheating

Before it can be determined whether or not a gear unit is overheating, the expected operating temperature must be determined. Very low speed gears will run near the ambient temperature, and some high speed drives may operate above 250°F.

Overheating in gear units may be caused by:

1. *Low Oil Level*—This condition will lead to both overheating and gear failure should the level fall below a point where the gear teeth or the oil pump can pick up the oil.

2. *High Oil Level*—High oil level will cause overheating due to the heat generated when the gear teeth run under too much oil. Many high speed units have an oil level just below the gear teeth, and to allow these gears to dip in oil causes overheating.
3. *Internal Failure or Poor Assembly*—Overheating can be caused by a multitude of things from gear tooth breakage to internal rubs or even gear teeth improperly meshed.
4. *Blocked or Reduced Oil Passages*—Many gear designs include either cast or machined oil passages and drilled orifices. These passageways can be blocked by oil deposits, sludge, or excessive sealants, thereby reducing oil flow and producing overheating.
5. *High Ambient Temperature*—Overheating can be due to the high climatic temperature in areas around furnaces, paper machines, and other machines that radiate heat. This situation can be corrected by external cooling or in some cases by the use of heat shields.
6. *Low Bearing Clearance*—Reduced bearing clearance can be caused by pinching the bearing at the bore split line, by not having sufficient clearance bored into journal-type bearings, or by having the wrong clearance in roller bearings. This condition will usually show up as localized overheating and can be detected with bearing instrumentation.
7. *Housing Coated with Foreign Material*—A foreign material on the exterior of the housing reduces the heat transfer to the air, and many gear designs depend on radiation and convection to the air for cooling.
8. *Internal Rubs*—This condition usually results from poor assembly and occurs when part of a gear rotating element actually rubs internal piping or housing walls.
9. *Insufficient Water Flow to the Cooler*—This condition can usually be checked by measuring the water inlet and outlet temperatures. If the temperature rise across the cooler is higher than the designed rise, insufficient water flow is usually indicated.
10. *Malfunction of Oil Heaters*—Oil heaters are used to raise the oil temperature for start-up or to maintain an acceptable oil temperature during extremely cold conditions. Should the thermocouples or thermostats on these oil heaters malfunction, they can overheat the oil.
11. *Contaminated Oil Filter*—This condition can cause overheating by reducing oil flow to the gear unit. If the contamination is excessive, the element can collapse and be carried into the oil passages, reducing oil flow and causing overheating.

Vibration

All gear units operate at certain vibration levels. Generally speaking, Figures 7-32, 7-33, and Table 7-2 depict vibration values expected of a gear unit properly installed and in good condition. Vibration levels above these may be perfectly acceptable but must be evaluated on a case-by-case basis.

Excessive vibration may be caused by:

1. *Unbalance*—This phenomenon is the most common cause of gear unit vibration and can be produced by broken teeth, couplings, key fitting practice, improper balancing during manufacture, poor assembly of gear to shaft, and even oil inside the gear blanks. Almost any vibration specialist can isolate the cause of unbalance vibration and either balance the parts or determine what must be corrected.
2. *Loose Foundation Bolts*—This condition is usually detected by inspection. When retightening loose foundation bolts, be careful that shims are not missing. Be sure that the housing foot is not "soft."
3. *Coupling Misalignment*—Misalignment is a serious problem with gear units, and many papers have been written on alignment control. A machinery train that is properly aligned today will change over the years due to settling of the foundation. Misalignment severe enough to cause high vibration levels will damage the gear set and shorten the life.
4. *Inadequate Foundation*—This cause of vibration is self explanatory and is most generally due to improperly designed and manufactured steel bases under gear units.
5. *Wear in Bearings and Gears*—Wear in gear teeth most generally shows up as an increase in the vibration or acceleration levels at tooth mesh frequency. Bearing wear can be detected by excessive clearance in journal-type bearings and pitting or spalling of rolling element bearings.
6. *Lateral and Torsional Critical Speed Response*—On high speed drives, lateral critical speeds of the shafts become very important, and users should be very careful when changing couplings to be sure that the weights and centers of gravity are the same as used during design stages. Torsional critical response is very important but is most common on reciprocating machines. In addition to causing vibration, both of these responses reduce gear life and in some extreme cases can cause immediate failure.
7. *Coupling Lockup*—Lockup is a form of coupling misalignment that occurs when toothed couplings are unable to shift axially to

account for thermal growth. This condition can be caused by full load starts, poor lubrication, wear, and centrifuging of the coupling lubricant.

8. *Coupling Wear*—On toothed couplings, wear can cause both coupling lockup and shifting of the loose pieces on the coupling. As an example, the teeth on the outer element of a gear tooth coupling can wear and allow the sleeves and spacer to shift off center. This shifting then produces an unbalance equal to the weight of the shifted coupling parts times the distance shifted.
9. *Lack of Coupling Lubricant*—Inadequate coupling lubricant will prevent the coupling from performing as required by the design and is equivalent to having a rigid shaft connection.
10. *Coupling Not as Designed*—The use of couplings with weights and stiffnesses different than the original design can cause encroachment on lateral and torsional natural frequencies of the system. On very low speed drives, coupling weight has much less importance than on pinions operating at high rates of speed.
11. *Improper Installation*—This subject covers a broad range of problems from foundation to lube oil and cooling water piping connections. When planning a gear unit installation, all environmental conditions must be carefully considered since operating conditions will vary with the cold wind, hot sun, and all other external influences.

Tooth Failure and Inspection

The most up-to-date work on gear tooth distress is ANSI/AGMA standard 110.04, "Nomenclature of Gear Tooth Wear and Failure." The term "gear failure" is in itself subjective and a source of considerable disagreement. One observer's "failure" can be another observer's "wearing-in." For a summary of this AGMA standard, refer to pages 131–155 of Volume 2 of this series, *Machinery Failure Analysis and Troubleshooting*.

Suffice it to say that during the initial period of operation of a set of gears, minor imperfections will be smoothed out, and the working surface will polish up, provided that proper conditions of design, application, material manufacture, installation, and lubrication have been met. Under continued normal conditions of operation, the rate of wear will be negligible.

Failure in a gear train can in many instances be prevented. When it does occur, the proper remedial action or redesign will ensure a trouble-free unit. Regardless of when the trouble is rectified, the most important faculty of those concerned with the problem is the ability to recognize the exact type of incipient failure, how far it has progressed, and the cause and cure of the ailment. Before a gear is shut down and replaced because of questionable or seemingly severe damage, periodic examination with photographs or impressions is recommended to determine if the observed condition is progressive.

Repair

The most common repair performed on a gear unit is the replacement of bearings. This procedure is normally straightforward, and the only basic difference from replacing pump, turbine, or compressor bearings is that the alignment between the gear teeth must be maintained. After installing the new bearings and before replacing the cover, the tooth contact should be checked. The importance of maintaining good tooth contact cannot be overemphasized.

When gear tooth failure occurs, different methods can be used to repair the gear set depending on the original design, hardness, and manufacturing method used.

When wear or pitting is the only problem, the gear and pinion can sometimes be recut or reground and returned to like-new condition. If wear or pitting is severe, the gear can be reduced on the outside diameter and recut and a new oversized pinion manufactured. This repair method can be used on almost all through-hardened gears and does not change the ratio.

Case-hardened gears cannot be recut due to the high hardnesses, and the outside diameters cannot be reduced since the hardened case is too thin. Regrinding is possible but risky for the same reasons. When the case is broken through by pitting, regrinding will probably only delay ultimate failure. In some cases, these gear blanks can be normalized, recut, reheat-treated, and reground.

When only a pinion tooth is broken, a new matching pinion can be made or the gear can be recut for an oversized pinion as just described. In many cases, if a gear tooth is broken, a new alloy steel band can be installed on the gear hub by shrinking or welding and a new pinion manufactured. This procedure is difficult to do on case-hardened gearing due to heat treating requirements. Also, this method of repair cannot be used safely at extremely high pitch line speeds.

In the event of a breakdown, these repair procedures can save time, materials, and money. The most important saving is usually in repair time when spares are not available for rapid replacement.

Bibliography

1. Dudley, Darle W., *Gear Handbook.* McGraw-Hill Book Company, Inc., New York, 1962, 1st ed.

2. Merritt, H. E., *Gear Engineering.* John Wiley & Sons, New York, 1971.
3. Thoma, Frederick A. (DeLaval Turbine, Inc.). "An Up-to-Date Look at Marine Gear Tooth Loading." The Society of Naval Architects and Marine Engineers, Philadelphia.
4. Shipley, Eugene E., (Mechanical Technology, Inc.). "Testing Can Reduce Gear Failures." *Hydrocarbon Processing,* Dec. 1973, pp. 61–65.
5. *AGMA Standard for Surface Durability (Pitting) of Helical and Herringbone Gear Teeth,* AGMA 211.02. American Gear Manufacturers Association, Arlington, Va, 1969.
6. *AGMA Standard for Rating the Strength of Helical and Herringbone Gear Teeth,* AGMA 221.02. American Gear Manufacturers Association, Arlington, Va, 1965.
7. *AGMA Information Sheet—Gear Scoring Design Guide for Aerospace Spur and Helical Power Gears,* AGMA 217.01. American Gear Manufacturers Association, Arlington, Va, 1965.
8. *AGMA Gear Handbook—Volume 1—Gear Classification, Materials and Measuring Methods For Unassembled Gears,* AGMA 390.03. American Gear Manufacturers Association, Arlington, Va, 1973.
9. *AGMA Standard—Nomenclature of Gear Tooth Failure Modes,* ANSI/AGMA 110.04. American Gear Manufacturers Association, Arlington, Va, 1980.
10. *AGMA Standard Specification—Lubrication of Industrial Enclosed Gear Drives,* AGMA 250.04. American Gear Manufacturers Association, Arlington, Va, 1981.
11. *Practice for Enclosed Speed Reducers or Increasers Using Spur, Helical, Herringbone and Spiral Bevel Gears,* AGMA 420.04. American Gear Manufacturers Association, Arlington, Va, 1975.
12. *Special Purpose Gear Units for Refinery Services,* API 613 Second Edition. American Petroleum Institute, Washington, D.C., 1977.
13. *AGMA Standard—Practice for High Speed Helical and Herringbone Gear Units,* AGMA 421.06. American Gear Manufacturers Association, Arlington, Va, 1969.
14. Calistrat, Michael M. "Hydraulically Fitted Hubs, Theory and Practice." *Proceedings of the Ninth Turbomachinery Symposium.* Texas A&M University, College Station, Tx, 1980, pp. 1–10.
15. Campbell, Al J.; Dodd, V. Ray; Essinger, Jack; Finn, Albert E.; Jackson, Charles; Murray, Malcolm G., Jr.; and Hollis, Don B., "Tutorium on Alignment Techniques and Practices." *Proceedings*

of the Ninth Turbomachinery Symposium. Texas A&M University, College Station, Tx, 1980, pp. 119–147.

16. Jackson, Charles, *The Practical Vibration Primer.* Gulf Publishing Co., Book Div., Houston, Tx, 1981.
17. *Acknowledgement*—My thanks to Tina Randolph and Midge Cooney for their editorial and word processing skills in putting this section into form.

Appendix 7-A

Helical Gear Formulae, Standard Gearing

1. $\cos \psi = \dfrac{N + n}{2 P_n C}$

2. $d = \dfrac{n}{P_n \cos \psi} = \dfrac{2C}{m_G + 1}$

3. $D = \dfrac{N}{P_n \cos \psi} = \dfrac{2C\, m_G}{m_G + 1}$

4. $d_o = d + \dfrac{2\, a_c}{P_n}$

5. $D_o = D + \dfrac{2\, a_c}{P_n}$

6. $d_R = d - \dfrac{2\, b_c}{P_n}$

7. $D_R = D - \dfrac{2\, b_c}{P_n}$

8. $C = \dfrac{d + D}{2}$

9. $h_t = \dfrac{a_c + b_c}{P_n}$

10. $V_t = \dfrac{\pi\, d\, n_p}{12}$

11. $W_t = \dfrac{126000\, P_{sc}}{n_p\, d}$

12. $W_r = \dfrac{W_t \tan \phi_n}{\cos \psi}$

13. $W_x = W_t \tan \Psi$

14. $m_G = \dfrac{N}{n}$

15. $\tan \phi_t = \dfrac{\tan \phi_n}{\cos \Psi}$

16. $P_t = P_n \cos \Psi$

17. $$p_N = \frac{\pi \cos \phi_n}{P_n}$$

18. $\sin \Psi_b = \sin \Psi \cos \phi_n$

19. $$d_b = \frac{n \cos \phi_n}{P_n \cos \Psi_b}$$

20. $$D_b = \frac{N \cos \phi_n}{P_n \cos \phi_b}$$

21. $$Z = \frac{\sqrt{d_o^2 - d_b^2} + \sqrt{D_o^2 - D_b^2}}{2} - C \sin \phi_t$$

Ψ = Helix angle
C = Center distance
P_n = Normal diametral pitch
P_t = Transverse diametral pitch
n = Number teeth, pinion
N = Number teeth, gear
a_c = Addendum constant of cutting tool
b_c = Dedendum constant of cutting tool
h_t = Whole depth of tooth
d = Pitch diameter, pinion
D = Pitch diameter, gear
d_o = Outside diameter, pinion
D_o = Outside diameter, gear
d_R = Root diameter, pinion
D_R = Root diameter, gear
n_p = Revolutions per minute, pinion
m_G = Gear ratio
V_t = Pitch line velocity (ft./min.)
W_t = Tangential load on tooth
W_r = Radial load on tooth (separating)
W_x = Axial load on tooth (thrust)
P_{sc} = Service or transmitted horsepower
ϕ_n = Normal pressure angle
ϕ_t = Transverse pressure angle
P_N = Normal base pitch
Ψ_b = Base helix angle
d_b = Base circle diameter, pinion
D_b = Base circle diameter, gear
Z = Length of line of action

Appendix 7-B

Typical Gear Unit Arrangements

Figure B-1. Single reduction low speed gear unit (cutaway view) featuring herringbone gearing and splash lubrication system which utilizes oil wipers on the sides of the gear to channel the oil into troughs that carry it to the bearings. Rolling element bearings support both shafts in this unit and absorb thrust on the low speed shaft.

Figure B-2. Single reduction high speed gear unit (cutaway view) with double helical gearing and a simple pressurized lubrication system. The gear housing acts as the lubricant reservoir, and a shaft-driven oil pump pressurizes the oil. The high and low speed shafts are supported by babbitted journal bearings, and any thrust loads are handled by babbitt-faced thrust bearings on the low speed shaft.

Figure B-3. Extremely high horsepower service rating (16,000 HP) single reduction gear unit (cover removed) with double helical gearing. This speed increaser features tilting pad journal bearings on the high speed shaft, babbitted journal bearings with pressure dams for added stability on the low speed shaft, a tilting pad thrust bearing on the low speed shaft, and specialized gear and pinion design to facilitate heat removal and thereby prevent excessive heat buildup.

Figure B-4. Two extremely high horsepower single reduction gear units (identical to the one in Figure B-3) undergoing full load, full speed back-to-back locked-torque testing. This test is fully instrumented for vibration and temperature monitoring. It provides the most reliable indications of gear accuracy and operating temperatures as compared to other test procedures and in general gives the best overall prediction of gear performance in the field. This test can be performed using a driver with a horsepower rating equivalent to the combined horsepower losses of the gear units.

Figure B-5. Double reducton low speed gear unit (cut-away view) featuring herringbone gearing and double-extended input and output shafts. This unit has the capacity to handle high overhung loads on the low speed (output) shaft.

Figure B-6. Double reduction two speed gear unit (cover removed) utilizing herringbone gearing and employing a shifter bar mechanism for changing speeds between the two ratios of gearing contained in the housing. This reducer powers a conveyor drive handling coal and salt.

Figure B-7. Double reduction high speed gear unit (cut-away view) with double helical gearing and a simple pressurized lubrication system powered by a shaft-driven oil pump. This unit exhibits gearing arranged in a "nested" design, where the high speed gear set is split and the low speed gear set is nested between the two halves. The main advantage of this particular arrangement is that it equalizes the loading on all bearings. It also utilizes the available space more efficiently than some other double reduction designs.

Figure B-8. High speed, high horsepower double reduction gear unit (cover removed) containing double helical gearing and utilizing a torque shaft and two flex-rigid couplings to transmit power from the high speed gear set to the low speed gear set. The use of this arrangement enables much more horsepower to be transmitted at much higher ratios and speeds than is possible with a simpler arrangement. It also makes possible torsional fine-tuning of the gear unit and the entire machinery train.

Figure B-9. Extremely high speed (input pinion speed of over 22,000 RPM) double reduction gear unit (cover removed) utilizing single helical gearing in a "foldback" design to conserve space. Intensive engineering design analysis and the incorporation of several specialized features enable this speed reducer to perform efficiently as well as satisfactorily at high speeds.

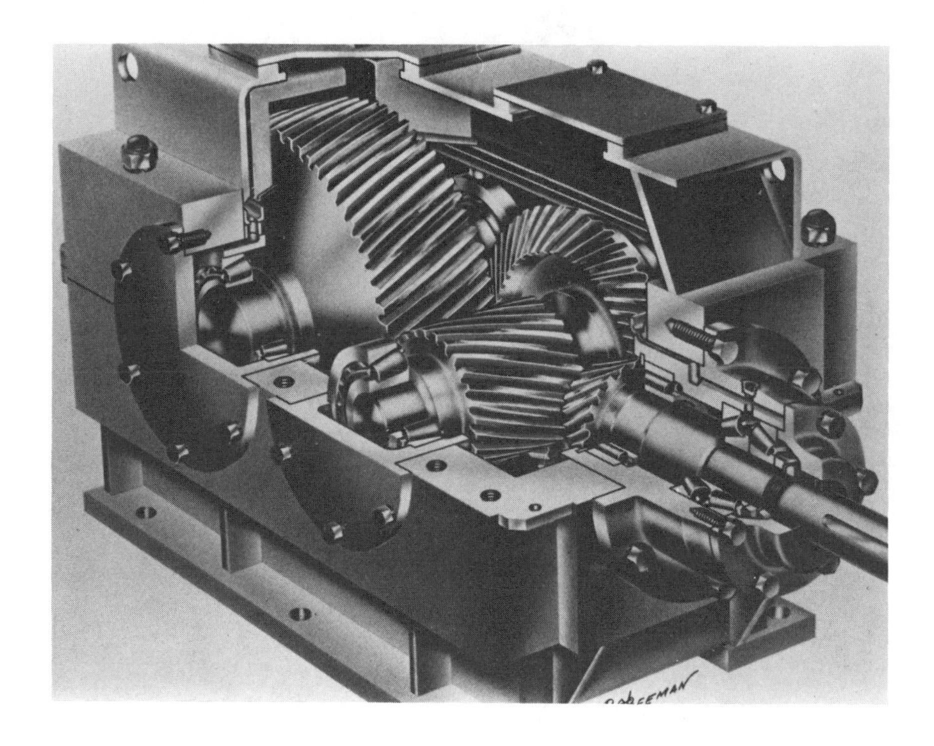

Figure B-10. Double reduction horizontal right-angle gear unit (cutaway view) utilizing spiral bevel gearing for the high speed reduction and single helical gearing for the low speed reduction. This particular unit has rolling element bearings on all shafts.

Figure B-11. Triple reduction articulated gear unit (cover removed) containing herringbone gearing and utilizing a pressurized lubrication system (not shown). Input torque is split between drive trains inside the housing for more efficient use of space and better design of gearing. This type of speed reducer is usually used for high reduction, low speed, very high torque applications such as sugar mill drives.

Figure B-12. Reverse reduction marine drive gear unit (cut-away view) equipped with double helical gearing, integral clutches, and anti-friction (rolling element) bearings. This gear unit sports two forward speeds in addition to one reverse speed.

Chapter 8

Installation and Maintenance of V-Belt Drives *

Well designed and properly installed V-belt drives are without question among the most reliable, trouble-free means of power transmission available. In general, except for an occasional retensioning, many will literally run for years without maintenance.

However, some V-belt drives do require periodic inspection and maintenance, both while the drive is running and while it is stationary.

Inspection While Running

A noisy V-belt drive is like a person with a fever. Both need attention.

V-drive noise can be caused by the slapping of belts against the drive guard or other obstruction. Check for an improperly installed guard, loose belts, or excessive vibration. Squealing of belts as a drive is started or while it is running is usually caused by a poorly tensioned drive and/or by a build-up of foreign material in the sheave grooves. But it can also be caused by oil or grease between the belt and the sheave groove.

If necessary, remove the belt guard and watch the drive while it is running under load. *(Caution: Observe only; stand clear of the running drive!)* Much can be learned by watching the action of the slack side of the drive. Each variation in the driven load causes a corresponding change in the tension of the slack side of the belt. During across-the-line starts or suddenly applied loads while running, the sag on the slack side of the drive will increase. If the sag under these conditions is excessive, tension should be increased.

---

* Source: T. B. Woods Company, Chambersburg, Pa. Their permission to use this material is gratefully acknowledged.

Any vibration in a system will cause the slack side of the belts to dance up and down. Excessive vibration will also induce a vibration in the tight side of the drive. The cause of the vibration should be determined and corrected.

If a set of belts is perfectly matched, all belts will have the same amount of sag. However, perfection is a rare thing and there will usually exist some difference in sag from belt to belt. It is more important to look at the tight side of a drive to be sure that all of the belts are running tight. If one or more belts are running loose, the drive needs to be retensioned, or the belts replaced with a matched set.

These conditions could also be caused by uneven wear of the grooves in the sheave. These should be checked with sheave groove gauges.

Inspect Sheaves Often

Keep all sheave grooves smooth and uniform. Burrs and rough spots along the sheave rim can damage belts. Dust, oil, and other foreign matter can lead to pitting and rust and should be avoided as much as possible. If sheave sidewalls are permitted to "dish out," as shown in Figure 8-1, the bottom "shoulder" ruins belts quickly by chewing off their bottom corners. Also, the belt's wedging action is reduced and it loses its gripping power.

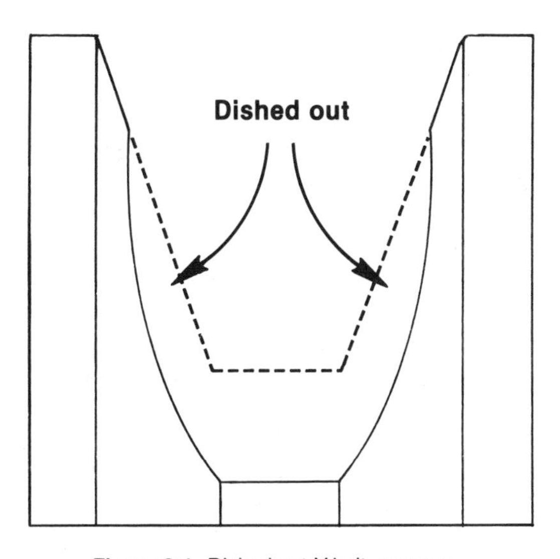

Figure 8-1. Dished-out V-belt grooves.

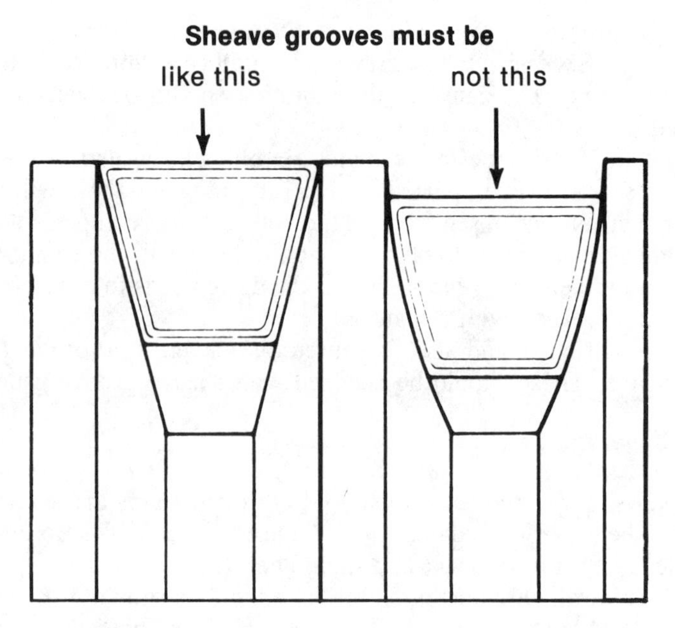

Figure 8-2. Good vs. bad grooves in a V-belt sheave.

A shiny groove bottom indicates that either the sheave, the belt or both are badly worn and the belt is bottoming in the groove.

Badly worn grooves cause one or more belts to ride lower than the rest of the belts, and the effect is the same as with mismatched belts. This is called "differential driving." The belts riding high in the grooves travel faster than the belts riding low. In a drive under proper tension, a sure sign of differential driving is when one or several belts on the tight side are slack. Note Figure 8-2 for groove details.

Check alignment of drive. Sheaves that are not aligned properly cause excessive belt and sheave wear. When the shafts are not parallel, belts on one side are drawn tighter and pull more than their share of the load. These overloaded belts wear out faster, reducing the service life of the entire set. If the misalignment is between the sheaves themselves, belts will enter and leave the grooves at an angle, causing excessive cover and sheave wear and premature failure.

Belt and Sheave Gauges

Belt and sheave groove gauge sets are available from your distributor and should be in your tool set.

You can use them to determine the proper belt section by trying the old belt in the various gauges until a proper fit is obtained. The cross section of conventional or narrow-wedge belts can be read from the gauge.

To check sheave grooves for wear, simply select the proper gauge and template for the sheave diameter; then insert the gauge in the groove until the rim of the gauge butts against the outside diameter of the sheave flange. Worn grooves will show up as illustrated in Figure 8-3. If more than 1/32 in. of wear can be seen, poor V-belt life may be expected.

Check Belt Fit

Conventional V-belts should ride in standard sheave grooves so that the top surface of the belt is just above the highest point of the sheave. In A-B

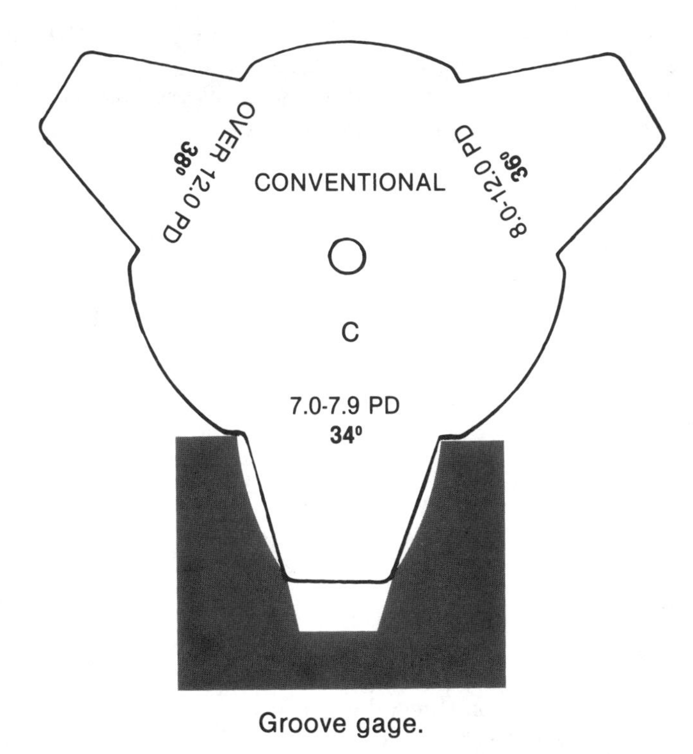

Figure 8-3. Groove gauge inserted in worn groove.

combination grooves, an A section belt will ride slightly low in the groove, while a B belt will be in the normal position. In special deep groove sheaves, belts will ride below the top of the sheave.

However, some V-belts are purposely designed so that the top of the belt will ride above the OD of the sheave. The tensile cords are located in the belt so that they ride almost at the OD of the sheave. This simplifies sheave identification and drive calculations.

No matter which V-belt section the sheave is grooved for, the belts should never be allowed to bottom in the groove. This will cause the belt to lose its wedging action, to slip and/or burn. Sheaves worn to the point where they allow a belt to bottom should be replaced immediately.

Keep Belts Clean

Dirt and grease reduce belt life. Belts should be wiped with a dry cloth occasionally to remove any build-up of foreign material. If the belts have been splattered with grease and/or oil, clean them with methyl chloroform or soap and water. Inflammable cleaners such as gasoline are to be avoided as a matter of safety.

Although all premium grade V-belts are of oil resistant construction, an occasional cleaning will help to prolong their life.

Under no circumstance is the use of belt dressing recommended on a V-belt. The remedial effect is only temporary. It is much better to keep the belts and grooves of the drive clean.

Use Belt Guards

Belt guards protect personnel and the drive itself. They should be definitely used in abrasive atmospheres to protect the drive from sand, metal chips, and other foreign matter. But they should be ventilated to avoid excessive heat.

Check them periodically for damage and for loose or missing mounting bolts. These could cause the belts to come in contact with the guard and cause failure.

Guards alone will generally protect belts from abrasion. But where abrasive materials are common—in rock processing machinery, grinders, foundries, etc.—drives should be inspected frequently for excessive belt and groove wear.

Check for Hot Bearings

When the drive has been stopped for inspection, check the bearings to make sure they are not running hot. If they are, it could be due to improper lubrication or improper drive tension. Hot bearings can be caused by belts that are either too tight or too loose. Check the tension carefully using the instructions furnished.

If the belts are slipping on your drive, retension the drive. Never use belt dressing to correct slipping belts.

Maintain Proper Belt Tension

Maintaining correct tension is the most important rule of V-belt care. It will give the belts 50 percent to 100 percent longer life.

Belts that are too loose will slip, causing excessive belt and sheave wear. V-belts that sag too much are snapped tight suddenly when the motor starts or when peak loads occur. That snapping action can actually break the belts, because the added stress is more than the belt was designed to take. This can be clearly demonstrated with a piece of string, as illustrated in Figure 8-4.

Figure 8-4. Belt tension analogy: Loosely-held string snaps easily, taut string can stand strong pull.

Selecting the Correct Belts

All the work and experience that goes into designing a V-belt drive is wasted if the specified belts are not used or the number of belts is changed. Over-belting is wasteful. Under-belting is even more expensive in the long run, because overloaded belts wear out faster.

V-belts are identified for size according to industry standards. A combination of letters and numbers as shown in Figure 8-5 indicates the width across the top of the belt (often referred to as "cross section") and the belt length. Conventional belts come in five widths: A, B, C, D, and E; while narrow V-belts are made in three widths: 3V, 5V, and 8V. In addition, there are the light-duty 2L, 3L, 4L, and 5L belts. If you are not sure which to use, measure the top width of the old belts carefully, or use the gauges described previously.

Be careful in measuring V-belts. The top widths of the B and 5V belts are very close; however, the 5V is considerably thicker, and the groove angles of the sheaves are different. Do not attempt to use these belts interchangeably. The 4L and 5L Light-duty belts are also very close in size to the A and B belts. But again, groove angles may be different. Light-duty belts should not be used on heavy-duty drives.

Explosive Atmospheres

Belts on drives in hazardous atmospheres should be kept reasonably free of encrusted accumulations of nonconducting materials. In addition, all elements of the drive must be interconnected and grounded as illustrated in Figure 8-6.

Store Belts Properly

V-belts should be stored in a cool, dry place out of direct sunlight. They should be kept away from ozone-producing equipment such as arc welders and high voltage apparatus. Temperature should be below 85°F, relative humidity below 70 percent. If belts are stored in piles, the piles should be kept small to avoid excessive weight which could distort the bottom belts. When belts are stored in boxes, the box size should be limited. Ideally, belts should be hung on saddle type pegs. With proper storage, belt quality will not change significantly within six years.

Assuming good storage practices, a decrease in service life of approximately 10 percent per year of storage beyond six years can be expected. From a norm of six years storage life at 85°F, it is estimated that the storage limit should be reduced by half for each 15°F increase in temperature. A significant increase in humidity may cause a fungus to form on belts, but any effect on the performance of the belt would be very slight.

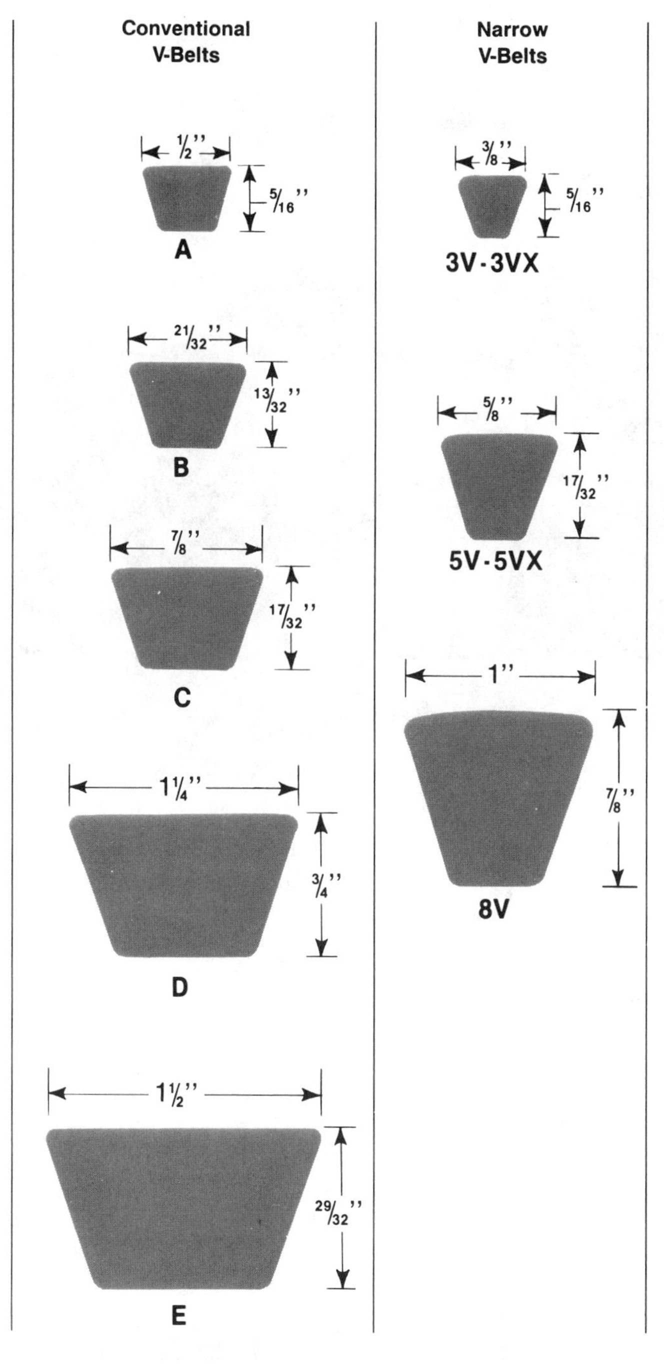

Figure 8-5. V-belt cross section dimensions.

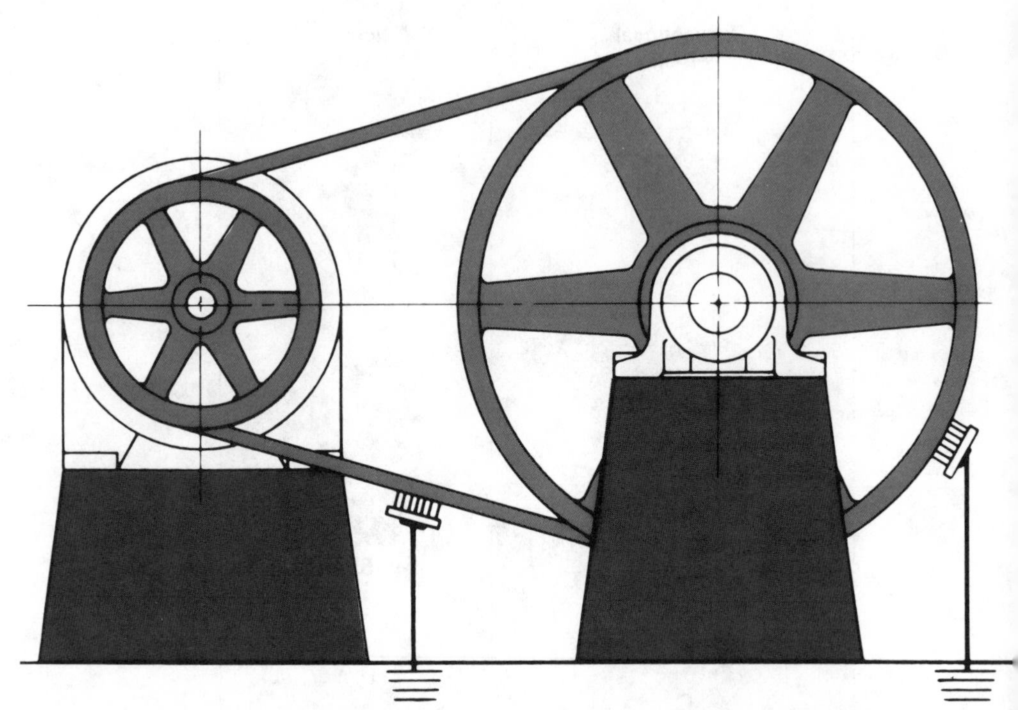

Figure 8-6. Proper V-drive installation in explosive atmospheres.

No matter where rotating machines are located or by what means they are driven, there is always a chance of personal injury unless they are installed and operated under safe conditions. This section is written with this thought in mind.

Guard All Drives Properly

All regulating agencies such as OSHA, State Departments of Labor and Industry, insurance firms and other safety authorities either recommend or insist on drive guards. We, also, strongly recommend that every V-belt drive be completely guarded. Do not be lulled into a sense of security by a temporary or makeshift guard.

Of course, provision can and should be made for proper ventilation and inspection by the use of grills, inspection doors and removable panels. But the guard should have no gap where workers can reach inside and become caught in the drive. Besides being a safety asset, a good guard helps make maintenance easier by protecting the drive from weather and foreign objects.

Check Safe Speed Limits

Safe speed limits for sheaves manufactured by reputable companies have been established by a rigorous burst testing program. The limit for cast iron sheaves has been established at 6,500 fpm; the maximum speed in rpm corresponding to 6,500 fpm is either cast or stamped on each sheave, as shown in Figure 8-7.

Before installing the drive, this safe speed limit should be checked against the speed of the shaft on which it is being installed. Operating sheaves above recommended speeds could result in serious damage to equipment and/or serious personal injury.

Typical Sheave and Bushing Installation Instruction

Tapered bushings are widely used, and have exceptional holding power that eliminates wobble. Standard and reverse mounting features provide greater adaptability. Quality bushings can be used interchangeably in many of a given vendor's products as well as those of other manufacturers.

Before installation, you should thoroughly inspect the bore of the mating part and the tapered surface of the bushing. Any paint, dirt, oil, or grease must be removed.

Assemble bushing into mating part in either the Standard or Reverse positions, as illustrated in Figure 8-8. (Since either the standard or the reverse mounting assembly can be rotated so that the bushing flange is toward or away from the motor, four ways of mounting are obtainable.)

Figure 8-7. Safe speed is cast into the arm of quality sheaves.

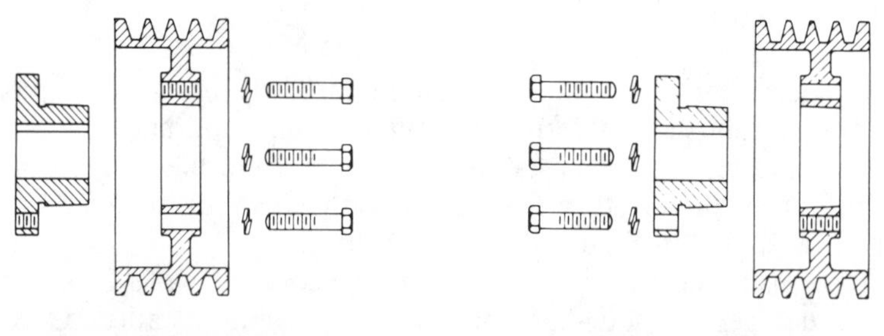

Figure 8-8. Standard and reverse mounting of typical sheave.

Loosely insert the cap screws into assembly, but *do not lubricate* cap screw threads. (Note: Install M through S bushings in the hub so that the two extra holes in the hub are located as far as possible from the bushing's saw cut.)

With key in keyseat of shaft, slide assembly to its desired position with cap screw heads to the outside, as shown below. (A few small sheaves may have to be installed with the cap screws on the inside.) If it is difficult to slide the bushing onto the shaft, wedge a screwdriver blade into the saw cut to overcome the tightness.

Position the assembly on the shaft so the belts will be in alignment when installed. Tighten each cap screw evenly and progressively until obtaining the torque value given by the manufacturer. There must be a gap between the bushing flange and mating hub when the installation is complete.

Typical Sheave and Bushing Removal Instruction

Loosen and remove cap screws.

As shown in Figure 8-9, insert cap screws in tapped removal holes and progressively tighten each one until mating part is loose on bushing. (Exception: If mating part is installed with cap screw heads next to motor, with insufficient room to insert screws in tapped holes, loosen cap screws and use wedge between bushing flange and mating part.)

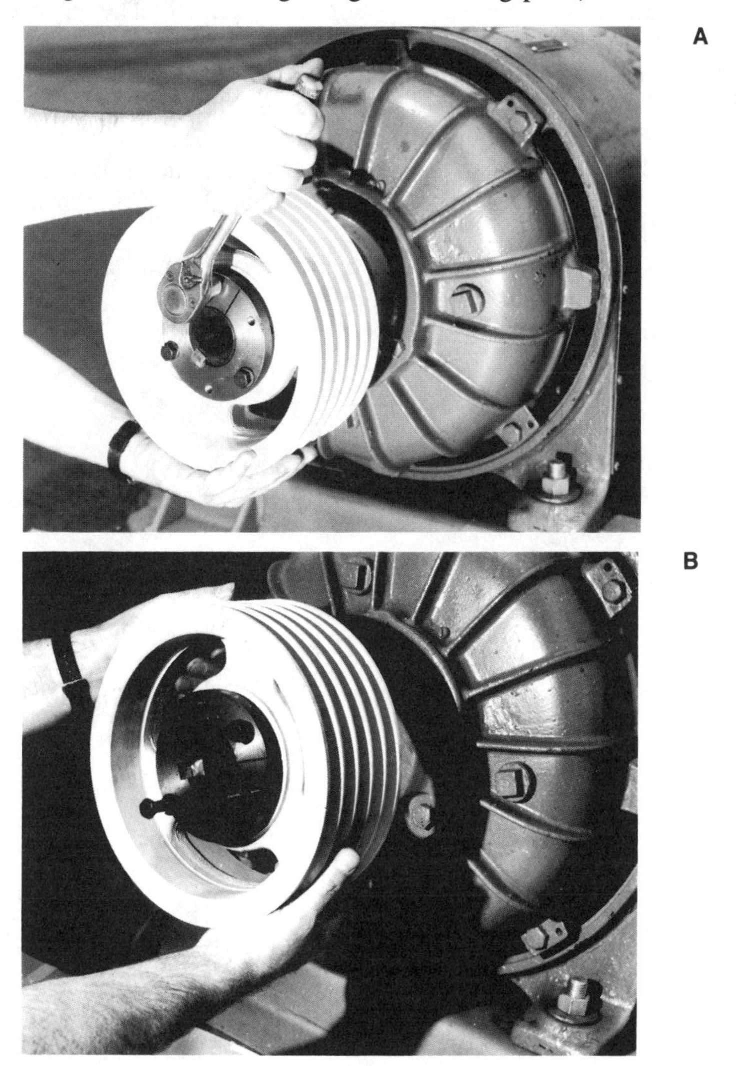

Figure 8-9. Sheave removal sequence: (a) tightening cap screws in removal and (b) actual removal of sheave.

Remove mating part from bushing and, if necessary, bushing from shaft. If bushing won't slip off shaft, wedge screwdriver blade in saw cut to overcome tightness.

Alignment Checking

Although alignment is not as critical in V-belt drives as in others proper alignment is essential to long belt and sheave life.

First, make sure that drive shafts are parallel. The most common causes of misalignment are nonparallel shafts and improperly located sheaves, as shown in Figure 8-10. Where shafts are not parallel, belts on one side are drawn tighter and pull more than their share of the load. As a result, these belts wear out faster, requiring the entire set to be replaced before it has given maximum service. If misalignment is in the sheave, belts will enter and leave the grooves at an angle, causing excessive belt cover and sheave wear.

Shaft alignment can be checked by measuring the distance between the shafts at three or more locations. If the distances are equal, then the shafts will be parallel.

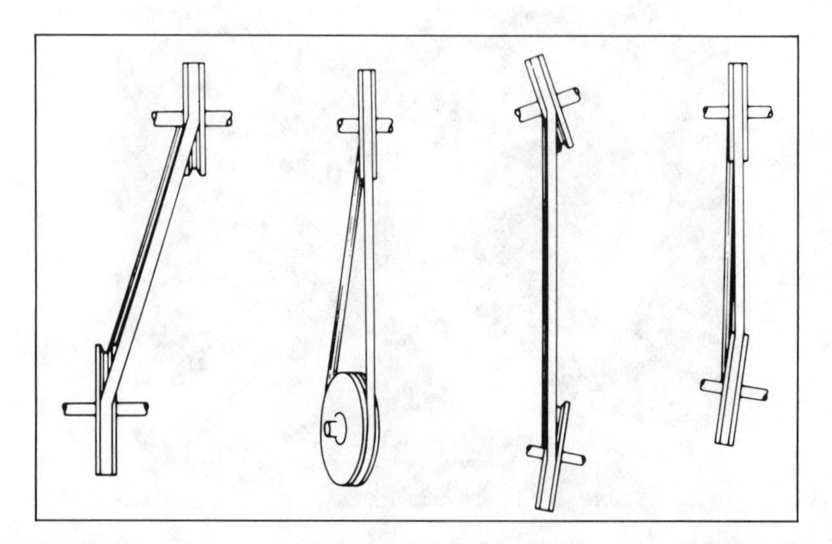

Figure 8-10. Effects of nonparallel shafts and improperly located sheaves on belt condition.

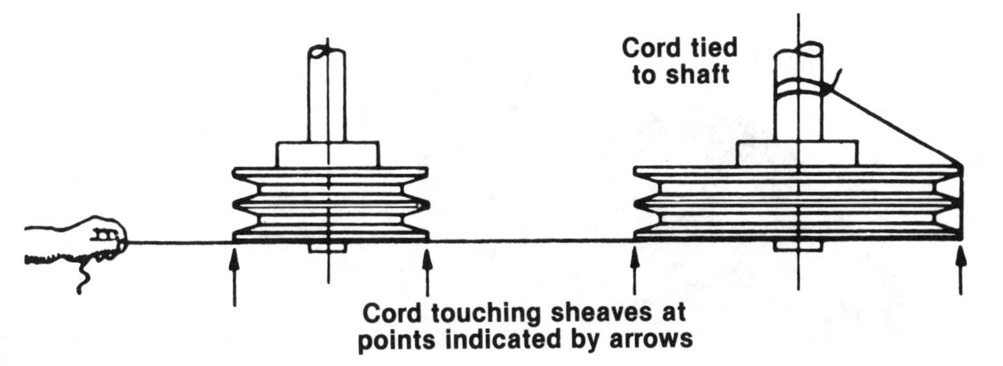

Figure 8-11. Checking sheave alignment.

To check the location of the sheaves on the shafts, a straightedge or a piece of string can be used as shown in Figure 8-11. If the sheaves are properly lined up, the string will touch them at the points indicated by the arrows in the accompanying sketch. Rotating each sheave a half revolution will determine whether the sheave is wobbly or the drive shaft is bent. Correct any misalignment.

With sheaves aligned, tighten cap screws evenly and progressively. Apply the recommended torque to cap screws as recommended by the manufacturer. Note: There should be 1/8 in. to 1/4 in. gap between the mating part hub and the bushing flange. If the gap is closed, the shaft is probably seriously undersized.

Installation of Belts

Shorten the center distance between the driven and driver sheave so the belts can be put on without the use of force.

While the belts are still loose on the drive, rotate the drive until all the slack is on one side. Then increase the center distance until the belts are snug. The drive is now ready for tensioning.

Note: Never "roll" or "pry" the belts into the sheave grooves. This can damage the belt cords and lead to belt turnover, short life, or actual breakage. Moreover, it is both difficult and unsafe to install belts this way. Note Figure 8-12!

Keep take-up rails, motor base, or other means of center distance adjustment free of dirt, rust, and grit. Lubricate adjusting screws and slide rails from time to time.

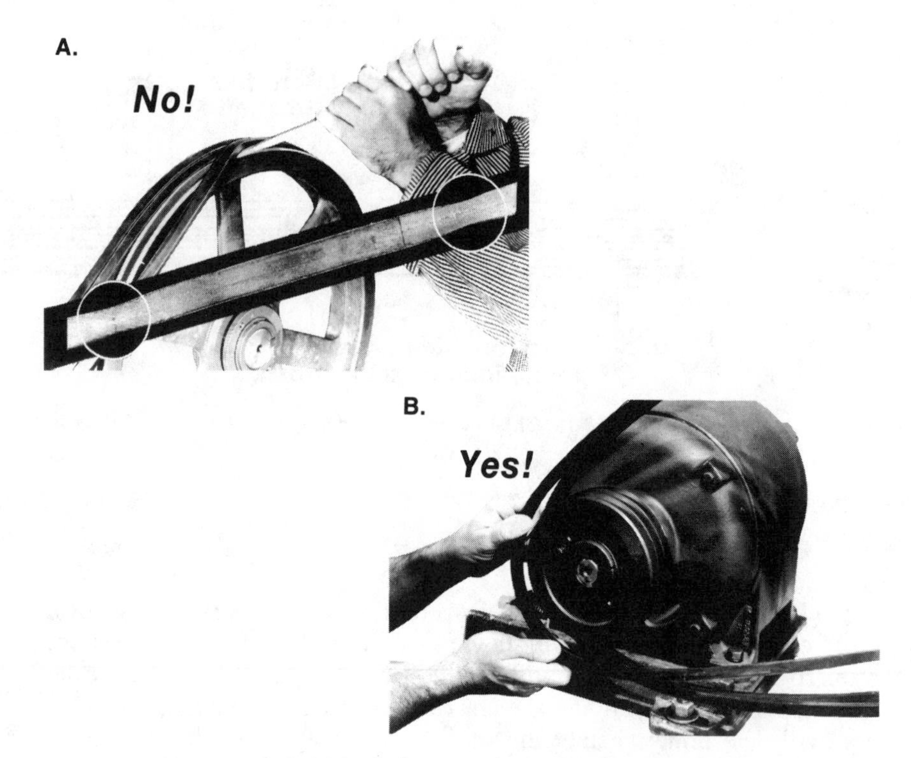

Figure 8-12. Belt installation procedures—(a) bad, vs. (b) good.

Tensioning V-Belt Drives

Without exception, the most important factor in the successful operation of a V-belt drive is proper belt-tensioning. To achieve the long, trouble-free service associated with V-belt drives, belt tension must be sufficient to overcome slipping under maximum peak load. This could be either at the start or during the work cycle. The amount of peak load will vary depending upon the character of the drive machine or drive system. To increase total tension, merely increase the center distance. Before attempting to tension any drive it is imperative that the sheaves be properly installed and aligned. If a V-belt slips it is too loose. Add to the tension by increasing the center distance. Never apply belt dressing as this will damage the belt and cause early failure.

Tensioning by General Method

The general method for tensioning V-belts should satisfy most drive requirements:

Step 1 Reduce the center distance so that the belts may be placed over the sheaves and in the grooves without forcing them over the sides of the grooves. Arrange the belts so that both the top and bottom spans have about the same sag. Apply tension to the belts by increasing the center distance until the belts are snug. See Figure 8-13.

Step 2 Operate the drive a few minutes to seat the belts in the sheave grooves. Observe the operation of the drive under its highest load condition (usually starting). A slight bowing of the slack side of the drive indicates proper tension. If the slack side remains taut during the peak load, the drive is too tight. Excessive bowing or slippage indicates insufficient tension. If the belts squeal as the motor comes on or at some subsequent peak load, they are not tight enough to deliver the torque demanded by the drive machine. The drive should be stopped and the belts tightened.

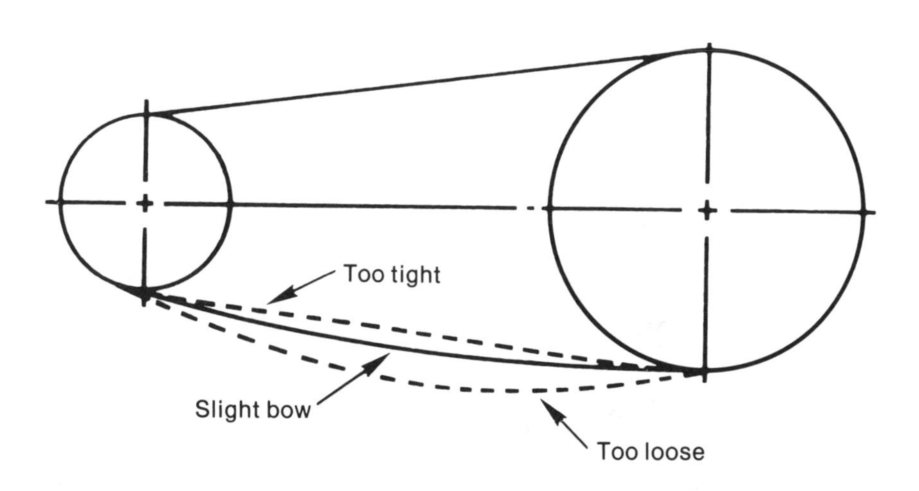

Figure 8-13. Belt tension diagram.

Step 3 Check the tension on a new drive frequently during the first day by observing the slack side span. After a few days' operation the belts will seat themselves in the sheave grooves and it may become necessary to readjust so that the drive again shows a slight bow in the slack side.

Force Deflection Engineering Formulas

For a more precise method, use the following engineering formulas to determine force deflection values.

Step 1 Determine span length (t) and deflection height (h). Refer to Figure 8-14.

Step 2 Calculate the static strand tension (Ts).

$$Ts = \frac{K \times DHP}{N \times S} + \frac{MS^2}{2}$$

Step 3 Calculate the recommended deflection forces (P) for drives using multiple belts or more than one V-band

$$P_{minimum} = \frac{Ts + Y}{16}$$

$$P_{maximum} = \frac{1.5(Ts) + Y}{16}$$

$$P_{initial} = 1.33 \text{ times } P_{maximum}$$

Explanation of Symbols

Ac = Arc of contact—smaller sheave, degrees
C = Center distance, inches
D = Larger sheave pitch diameter, inches
d = Smaller sheave pitch diameter, inches
DHP = Design horsepower based upon the recommended application service factor
h = Deflection height, inches (refer to Figure 8-14)
K = Value from Table 8-1 depending on $\frac{D - d}{C}$

$$\text{or } K = 16.5 \frac{2.5 - Ac}{Ac}$$

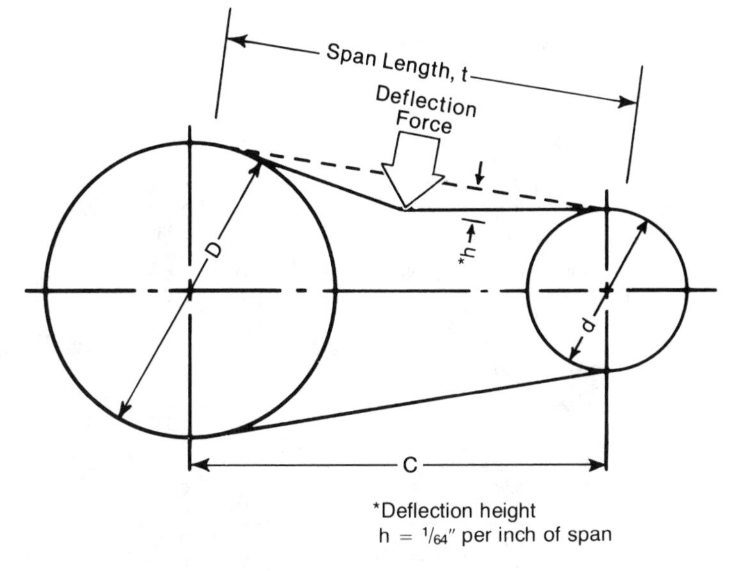

*Deflection height
h = 1/64″ per inch of span

$$t = \sqrt{C^2 - \left(\frac{D-d}{2}\right)^2}$$

$$h = \frac{t}{64}$$

where t = Span length, inches
C = Center distance, inches
D = Larger sheave diameter, inches
d = Smaller sheave diameter, inches

Figure 8-14. Determination of span length and deflection of belt drives.

L = Belt length, inches
M = Centrifugal constant, Table 8-1
N = Number of belts or V-band ribs
P = Deflection force, pounds
S = Belt speed, FPM/1000
t = Span length, inches (refer to Figure 8-14)
Y = Belt constant, Table 8-2

Table 8-1
K Factors and Arc of Contact

$\frac{D-d}{C}$	Arc Contact Degree	Factor Ac	Factor K	$\frac{D-d}{C}$	Arc Contact Degree	Factor Ac	Factor K
0.000	180	1.000	24.750	0.750	136	0.879	30.411
0.025	179	0.997	24.883	0.775	134	0.874	30.688
0.050	177	0.994	25.019	0.800	133	0.869	30.975
0.075	176	0.990	25.158	0.825	131	0.864	31.270
0.100	174	0.987	25.300	0.850	130	0.858	31.576
0.125	173	0.983	25.444	0.875	128	0.852	31.892
0.150	171	0.980	25.591	0.900	127	0.847	32.219
0.175	170	0.977	25.742	0.925	125	0.841	32.558
0.200	169	0.973	25.896	0.950	123	0.835	32.909
0.225	167	0.969	26.053	0.975	122	0.829	33.273
0.250	166	0.966	26.213	1.000	120	0.823	33.652
0.275	164	0.962	26.377	1.025	118	0.816	34.045
0.300	163	0.958	26.545	1.050	117	0.810	34.454
0.325	161	0.954	26.717	1.075	115	0.803	34.879
0.350	160	0.951	26.892	1.100	113	0.796	35.323
0.375	158	0.947	27.072	1.125	112	0.789	35.786
0.400	157	0.943	27.257	1.150	110	0.782	36.270
0.425	155	0.939	27.445	1.175	108	0.774	36.777
0.450	154	0.935	27.639	1.200	106	0.767	37.307
0.475	153	0.930	27.837	1.225	104	0.759	37.864
0.500	151	0.926	28.040	1.250	103	0.751	38.448
0.525	150	0.922	28.249	1.275	101	0.742	39.064
0.550	148	0.917	28.463	1.300	99	0.734	39.713
0.575	147	0.913	28.684	1.325	97	0.725	40.398
0.600	145	0.908	28.910	1.350	95	0.716	41.123
0.625	144	0.904	29.142	1.375	93	0.706	41.892
0.650	142	0.899	29.381	1.400	91	0.697	42.709
0.675	141	0.894	29.627	1.425	89	0.687	43.580
0.700	139	0.889	29.881				
0.725	137	0.884	30.142				

Table 8-2
Belt Constants M and Y*

Factors	Ultra-V			Ultra-V Cog		Sure-Grip Premium					Torque-Flex			
	3V	5V	8V	3VX	5VX	AP	BP	CP	DP	EP	AX	BX	CX	DX
M Single Belts	.46	1.23	3.28	.39	1.08	.66	1.08	1.98	3.74	5.85	.61	1.00	1.78	3.97
M V-Band	.51	1.32	3.80	—	—	—	1.40	2.33	4.29	6.26	—	1.28	2.10	4.56
Y	4.0	12.00	22.00	7.0	20.0	6.0	9.0	16.0	30.00	45.00	7.0	10.00	28.00	82.00

* *Belt constants and trade names refer to products made by T. B. Woods Company.*

Chapter 9

Steam Turbines and Turboexpanders *

Special Purpose Steam Turbines

While much attention has been devoted to new high-power, high-speed centrifugal compressors, steam turbine drives have more often been the cause of plant downtime.

Initially, problems peculiar to process drives were not given enough thought. Unplanned shutdowns resulted. The combined requirements of high speed, high power, and variable speed, associated with process drives, have led to some rethinking by steam turbine manufacturers.

Rigorous operating conditions demand careful optimization between process requirements and mechanical considerations when selecting, maintaining, and operating a special purpose steam turbine such as is illustrated in Figure 9-1.

Steam turbine types and applications are shown in Tables 9-1 and 9-2. The second table shows that all types of steam turbines have a place in process plants.

In some instances where the reason for the type in use is not clear-cut, the back-pressure turbine may be favored because it is:

- Lowest in capital cost.
- Most suitable for high speeds.
- Simplest in construction . . . and hence more reliable.

Condensing Turbine Disadvantages. The condensing turbine has several disadvantages compared to the back-pressure turbine.[1]

* Material related to special purpose steam turbines is edited from information furnished by Westinghouse Canada, Inc., Hamilton, Ontario, Canada. By permission.

Table 9-1
Steam Turbine Frame Capabilities

Model	Maximum Inlet Pressure (Psig/Kpag)	Maximum Inlet Temperature (°F/°C)	Maximum Exhaust Pressure (Psig/Kpag)	Approximate Power (Hp/Kw)	Maximum Speed	Number of Stages	Governor Valve Arrangement	Wheel Diameter (in/mm)	Maximum Inlet Size (in/mm)	Maximum Exhaust Size (in/mm)
1	700/4830	900/480	125/860	2000/1500	15000	1	Single	16/405	4/100	8/205
2	700/4830	900/480	125/860	3500/2600	C-6500 R-12500	1	Single	20/510	6/150	10/255
3	1500/10345	950/510	100/670	4000/3000	6000	1	Single	25/635	8/205	10/255
4	1500/10345	950/510	75/515	3500/2600	C-10500 R-12500	1	Single	20/510	4/100	10/255
5	1500/10345	950/510	75/515	4800/3600	C-8000 R-10000	1	Single	25/635	8/205	14/355
6	1500/10345	950/510	300/2070	3000/2200	C-10500 R-12500	1	Single	20/510	4/100	6/150
7	1500/10345	950/510	300/2070	4500/3400	C-8000 R-10000	1	Single	25/635	8/205	8/200
8	1500/10345	950/510	200/1380	5500/4100	12500	2 or 3	Single	20/510	6/150	20/510
9	1500/10345	950/510	200/1380	6500/4800	10000	2 or 3	Single	25/635	8/205	20/510
10	1500/10345	950/510	300/2070	7000/5200	12500	1–6	Single	20/510	8/205	As Required
11	1500/10345	950/510	300/2070	8000/6000	10000	1–9	Single	25/635	8/205	As Required
12	1500/10345	950/510	75/515	8000/6000	6000	1–9	Single	32/815	8/205	As Required
13	1500/10345	950/510	300/2070	12000/9000	12000	As Required	Multiple	20/510	8/205	As Required
14	1500/10345	950/510	300/2070	20000/15000	10000	As Required	Multiple	25/635	10/255	As Required
15	1500/10345	950/510	300/2070	30000/22400	6000	As Required	Multiple	32/815	12/305	As Required
16	As Required	As Required	75/515	50000/37300	8000	As Required	Multiple	Variable	As Required	As Required
17	1500/10345	950/510	600/4140	80000/60000	3600	As Required	Multiple	Variable	12/305	As Required

1. C-Curtis Stage R-Rateau Stage
2. Inlet area can be doubled on multivalve turbines for greater inlet flow and power.

Figure 9-1. Multivalve special purpose steam turbine.

- Larger blades due to high steam volumes. Therefore low blade natural frequencies and increased risk of blade excitation exist.
- Lower over-all reliability because of the need to provide a condenser, ejectors, extraction pumps, etc.
- High first cost caused by two factors:

  a. A larger turbine due to high specific volumes
  b. The extra cost of condenser, etc.

- Poor operating cost because about two-thirds of the steam heat content is used in heating condenser cooling water.
- Difficulty in measuring performance on site because of problems in finding the steam energy at exhaust.
- More costly boiler feedwater treatment to remove chlorides, salts and silicates which would otherwise produce deposits or corrosion of the blades.
- Blade failures are more likely. The last few rows of blades are subject to erosion by the water droplets which are present in condensing steam.

Table 9-2
Classification of Steam Turbines with Reference to Application and Operating Condition

Basic type	Operating condition	Steam condition	Application
Condensing	High-pressure turbine (with or without extraction for feedwater heating)	100–2,400 psig, saturated, 1050°F, 1–5 in. Hg abs	Drivers for electric generators, blowers, compressors, pumps, marine propulsion, etc.
	Low-pressure turbine	0–100 psig, saturated, 750°F, 1–5 in. Hg abs	Drivers for electric generators, blowers, compressors, pumps, etc.
	Reheat turbine	1,450–3,500 psig, 900–1050°F, 1–5 in. Hg abs	Electric-utility plants
	Automatic extraction turbine	100–2,400 psig, saturated, 1050°F, 1–5 in. Hg abs	Drivers for electric generators, blowers, compressors, pumps, etc.
	Mixed-pressure (induction) turbine	100–2,400 psig, saturated, 1050°F, 1–5 in. Hg abs	Drivers for electric generators, blowers, compressors, pumps, etc.
	Cross-compound turbine (with or without extraction for feedwater heating, with or without reheat)	400–1,450 psig, 750–1,050°F, 1–5 in. Hg abs	Marine propulsion
Noncondensing	Straight-through turbine	600–3,500 psig, 600–1050°F, atmosphere, 1,000 psig	Drivers for electric generators, blowers, compressors, pumps, etc.
	Automatic extraction turbine	600–3,500 psig, 600–1050°F, atmosphere, 600 psig	

Condensing Turbine Advantages. The condensing turbine has some advantages over the back-pressure turbine.

- It requires less change in live steam for different loads and so control is easier.
- It requires less live steam as the drop in heat energy is larger.
- It only affects one steam level for a change in power.

Extraction and Induction Turbines. Either of the basic types of turbine is in use with one or more intermediate nozzles for extraction or induction of steam.

The normal criterion used in operating an extraction turbine is whether 15–20 percent of the power required can be generated by the extracted steam. Various plant operating conditions are taken into account in this assessment and not just the design conditions.

Extraction/induction turbines afford the following advantages:

- Process steam requirements at two or more levels can be satisfied without having to provide boilers at different pressures, or alternatively having to throttle steam, without obtaining useful work, i.e., the power from the extraction steam is obtained cheaply.
- A process steam level can be controlled and maintained by the steam turbine.
- An over-all steam balance can be achieved more readily.
- Optimization of process steam and power demand.
- Allows flexibility under various plant conditions.

Extraction and induction turbines are slightly less reliable because:

- Disturbances are caused in the steam at the intermediate nozzle; therefore, blade vibrations can be excited.
- The possibility of starving part of the turbine of steam exists, resulting in overheating due to windage.
- Extra valves, etc., are required, if the intermediate pressure is controlled.
- A longer turbine shaft is required to allow for the extra nozzle, etc., thereby increasing the bearing span resulting in a more flexible shaft. This could produce difficulties with critical speeds.

The efficiencies of these turbines are about five points lower than those of basic turbines.

Review of Turbine Hardware

The steam turbine is a comparatively simple type of prime mover. It has only one major moving part, the rotor which carries the buckets or blades. These, with the stationary nozzles or blades, form the steam path through the turbine. The rotor is mounted on a shaft supported on journal bearings, and axially positioned by a thrust bearing. A housing with steam inlet and outlet connections surrounds the rotating parts and serves as a frame for the unit.

However, a great number of factors enter into the design of a modern turbine, and its present perfection is the result of many years of research and development. The following is a listing of special purpose turbine "hardware" as shown in Figure 9-2.

1. High pressure parts.

   a. Trip valve—trip and throttle.
   b. Governor valve(s).
   c. Steam chest, nozzle bowl, chamber or box.
   d. Nozzles—rings and diaphragms.

2. Low pressure parts.

   a. Casing or cylinder.
   b. Blade rings and covers.
   c. Exhaust hood.
   d. Glands or seals.

3. Rotor assembly.

   a. Blades or buckets.
   b. Disk(s).
   c. Shaft or drum.

      - Bearings.
      - Thrust control mechanism.

4. Auxiliaries.

   a. Overspeed trip device.
   b. Governor.
   c. Turning gear.

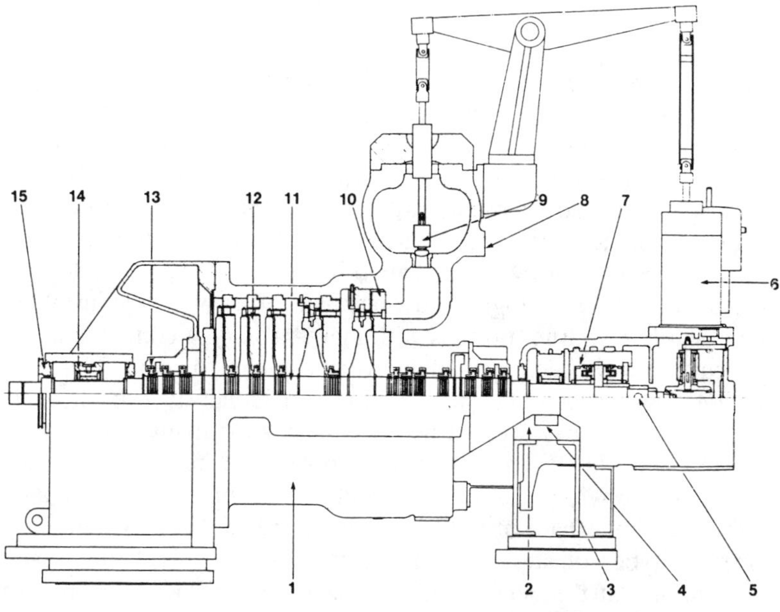

1 Steel inlet and exhaust casing
2 True centerline support
3 Flexible governor end support
4 Lubricated sliding keyed ways
5 Separate overspeed trip device
6 Woodward NEMA D governor
7 Double acting, tilting pad thrust bearing
8 Separate trip-throttle valve
9 Multiple valve lift bar arrangement
10 Stainless steel nozzles with overlapping exits
11 Precision balanced rotating assembly
12 Interstage nozzle diaphragms
13 Labyrinth steam seals
14 Tilting pad radial bearings
15 Labyrinth oil seals

Figure 9-2. Cross section of a typical multivalve multistage special purpose steam turbine (General Electric).

5. Lube oil system.
    a. Pumps.
    b. Filters.
    c. Reservoir.
    d. Lube oil conditioning system.
    e. Instrumentation.

Special Purpose Turbine Inspection and Repair

Inspection and overhaul have been traditional activities around special purpose steam turbines. These activities have gained importance as the possibilities of nondestructive testing and the complexity of large steam turbine installations have increased. Attempts to lengthen intervals between inspections and overhauls have been instigated by a number of isolated reports of steam turbines being operated without overhauls. Today, periods between major overhauls range from two to five years depending on the degree of technological advancement of a particular installation. In the case of new large turbines with many prototype components, individual casings have been opened up every one or two years. Thus far, no disadvantageous accumulations of failure incidents have been encountered after *partial overhauls,* where the inspection of individual turbine components during the available shutdown time has been practiced.[2] However, it is obvious that the suitability for complete and independent inspection and overhaul of individual components differs widely among the various types of steam turbines. Investigating this "suitability" or maintainability at the very beginning of a planned large steam turbine acquisition is therefore of the utmost importance.

Internal inspections must be scheduled to suit plant load demand. However, it is obvious for economic reasons that to reduce forced outage for corrective maintenance, general knowledge of the internal condition of the turbine at all times is desirable. A systematic check during operation to detect significant change in this condition is a valuable guide. Inspections may then be regarded as preventive rather than necessarily corrective.

A complete and detailed "case history" starting at the time of installation should be compiled for each turbine. This should include a description and analysis of any unusual circumstance during its operation as well as any noteworthy condition found during inspection: also a statement of the corrective measures taken or planned. The first complete inspection of a new turbine forms the most valuable datum point in its history and we recommend that a very thorough inspection be made at or near the end of the first year of operation.

Before taking the turbine out of service for inspection a number of parameters should be checked and the past "case history" reviewed to determine items requiring special attention and investigation. Here is a comprehensive listing:

On-Line Monitoring Results

Measurement of steam consumption. Comparison with the result of a previous measurement makes it possible to find out if there was a change in efficiency.

Determination of internal efficiency. If a measurement of steam consumption is too costly, measure inlet, intermediate and final steam pressures and temperatures.

Measurement of stage pressures. Measure pressures as a function of steam flow and compare with those obtained with a "clean" machine. This will give an indication of blade deposits or deformation.

Vibration monitoring. Review vibration history. Review bearing and casing vibration frequency spectra. Perform vibration signature analysis, diagnosis and prognosis.[3]

Review shaft position history.

Review bearing pressure and temperature readings. A reduction in inlet pressures may indicate increased bearing clearances. A rise in bearing metal temperature may indicate a change in bearing geometry.

Obtain answers to these questions:

1. Is there oil leakage?
   - In the piping?
   - At bearing oil seals?
   - At hydraulic lines?

2. Do the emergency and auxiliary oil pumps start when the oil pressure fails?

3. Is there steam leakage at

   - Joints?
   - Valve stems?

4. Do throttle and governor valves close promptly when tripped?
5. Does the throttle valve stop the unit when tripped?

6. Are extraction check valves in working condition?
7. Are turbine rotor glands sealing properly?
8. Has there been any change in control system readouts?
9. Is the control system stable?
10. Will governor hold speed at no load, full steam pressure and normal exhaust pressure?
11. Are other control devices operating satisfactorily?
12. Does automatic overspeed trip function at correct speed?

Off-Line Inspection

1. Without major disassembly

   - Determine "bump-to-bump" thrust clearances.
   - Alignment check—See chapter 5, Volume 3.
   - Boroscope (endoscope) inspection.
   - Inspection of casing keys and base plates.
   - Piping anchors—See chapter 3.
   - Determine radial bearing clearance.

2. Major dissassembly. Table 9-3 shows all necessary inspection operations following major disassembly.

Documentation

Written documentation and photographs of all inspection results are of the utmost importance.

Subsequent inspection schedules should be based on what is found at the last inspection, the "case history," and the operating log. Periodic checks of the lubricating system, control system, throttle valve, and automatic features are important. As we saw, test data at some fixed reference, load on steam flow, and intermediate stage steam pressures checked back to the early operation of the turbine may detect the presence and extent of blade deposits or mechanical damage. Similarly, variations in the stage enthalpy measured by steam pressure and temperatures in the superheat zone may detect any important change in internal stage efficiency. Table 9-4 suggests how often we should do all this, but our readers are encouraged to establish these frequencies for their own installation.

Overhaul Procedures

Each plant has established specific overhaul procedures. Typically they would feature points like:

1. Safety requirements
2. Worklist
3. Bar chart
4. Material list and equipment catalog
5. Tool list
6. Special service and equipment list
7. Procedures:
   - alignment
   - flexible diaphragm coupling
   - bearing assembly
   - casing
   - rotor assembly
   - internals
   - interstage seals
   - design clearances
8. Field notes
9. Clearance inspection forms
10. Critical path method chart
11. Photographs

Based on the steam turbine manufacturer's specific operating and maintenance instructions these documents are an invaluable part of the technical inventory of petrochemical plant maintenance departments.[4]

Special Purpose Steam Turbine Operation and Maintenance

As petrochemical process machinery increases in complexity, proper coordination of the operating and maintenance functions becomes an important aspect of machinery management. Someone once observed in an exaggerated way that if one could see the "gray line" between machinery operation and maintenance functions one would be in trouble. Large steam turbines are no exception. A good example would be the running in and startup of a special purpose turbine after an extensive overhaul. A good machinery maintenance person will not walk away after the overhaul—his job seems never done. For instance careful carbon ring break-in is often ignored or bypassed based on the justification of getting the turbine on line a few hours sooner.

Table 9-3
Inspection and Overhaul After Disassembly Special Purpose Steam Turbines

COMPONENTS	INSPECT FOR:																						INSPECTION					
	Deformation	Cracking	Fracture	Loosening	Embrittlement	Exfoliation	Corrosion/Erosion	Wear	Cutting	Gouging	Pitting	Scoring	Scratching	Wiping	Binding	Misalignment	Excessive Clearance	Lost Motion	Contamination	Plugging	Fouling	Deposits	Visual	Mirror/Boroscope Insp.	Dye Penetrant Insp.	Magnaflux Insp.	Radiographic Insp.	Ultrasonic Testing
1.1 VALVES & FITTINGS		●	●																				●	●	●			
Steam Strainers		●	●																	●	●		●					
Valve Bodies		●	●								●												●	●	●			
Valve Stems																						●	●					
Disks & Seats									●		●								●		●	●	●	●	●			
Guides/Bushings												●							●			●	●					
1.2 CONTROL VALVE GEAR																							●					
Linkage				●				●										●					●					
Bearings/Bushings								●							●	●							●					
1.3 STEAM CHEST		●	●								●												●	●	●	●	●	
1.4 NOZZLE BOWL																												
Nozzle Ring				●																		●	●	●				
Stationary Blades		●				●	●															●	●	●	●	●	●	●
Diaphragm	●																					●	●	●				
2.1 CASING/CYLINDER		●																					●	●	●	●	●	
Drain Connections																				●			●	●				
2.2 BLADE RINGS/COVERS		●																					●					
2.3 EXHAUST/EXH. HOOD																							●					
2.4 GLANDS/SEALS																												
Carbon Segments		●	●					●													●	●	●					
Labyrinths							●														●							
Leak-Off																				●	●		●					
3.1 ROTOR BLADING		●																			●	●	●					
Blade Edges		●					●															●	●		●	●		
Buckets		●					●															●	●		●	●		
Shrouds		●					●																●		●	●		
Shroud Rivets			●																				●		●	●		●
3.2 DISKS		●																					●		●			
3.3 SHAFT/DRUM		●								●	●												●		●			●
Journals								●				●	●	●									●					
Seal Areas								●	●			●	●	●									●					
Disk fits																							●					
Coupling Hub								●															●		●			
3.4 BEARINGS								●				●		●			●						●					●
Pedestals				●																			●					
Casing Keys															●								●					
3.5 ROTOR/CASING																●	●						●					
4.1 OVERSPEED TRIP																					●	●	●					
Trip Lever	●														●								●					
4.2 GOVERNOR																												
Drive Gears								●								●							●					
Internal Parts							●	●											●			●						
4.3 TURNING GEAR							●								●								●					
5.1 LUBE OIL PUMPS								●																				
5.2 L.O. FILTERS																												
5.3 L.O. RESERVOIR/PPG.																			●									
5.4 L.O. CONDITIONER																					●							
5.5 L.O. INSTR./CONTROL																												
6.1 PIPING & CONDENSER																												
Piping Supports															●	●												
Condenser		●					●																●	●				
Ejectors																				●	●		●					
Cond. Pumps								●															●					

METHODS														ACTION												
Eddy-Current Testing	Deposit/Coating Analysis	Bluing Check	Measurement-Clearances	Measur'mt.-Surface Finish	Measur'mt.-Geometry	Measur'mt.-Eccentricity	Measur'mt.-Mapping	Hardness Check	Oil Analysis	Oil Filter Insp.	Written Docum'tn.	Photographic Records	Replica Docum'tn.	Replace	Clean-Solvent	Clean-Abrasive Blasting	Lap	Polish	Ream	Grind	File	Scrape	Remachine	Test/Exercise	Lubricate	Align
															●											
																●										
	●				●	●									●			●								
	●	●		●				●			●	●									●			●			
	●	●																	●							
																								●		
															●									●	●	
			●																●						●	
	●											●														
●	●																									
	●		●				●				●															
															●											
			●				●								●											
	●													●												
														●												
														●												
	●																									
	●										●	●				●				●	●					
	●										●	●				●										
											●	●														
											●	●														
																●										
					●	●					●	●	●				●	●								
					●	●																				
											●	●			●										●	
																						●				
												●														●
												●			●											●
			●			●	●				●															
															●									●		
															●											●
	●								●	●					●											
															●									●	●	
															●									●		●
										●				●	●											
									●						●											
																								●		
																								●		
														●												
														●	●									●		
														●												
																										●

REMARKS

1.1 Inspect castings for crack formations-particularly on internal separating walls. If strainer cracked or broken, replace.

1.2 Watch for signs of fretting and binding.

1.4 Inspect stationary blading for damage by foreign bodies.

2.1 Inspect for cracks, blowholes & erosion.

2.4 Clean casing. If segments cracked or broken, replace.

3.1 Inspect for erosion, nicks, scratches. If coated, check boiler operation for carryover.

3.3 Perform run-out check. INSPECT cplg. hub/shaft fit.

3.4 Refer to chapter 7.1
Assure casing keys are free.

3.5 Check clearances.

4.1 Clean parts.

4.2 If badly worn, replace. If corroded, investigate oil system.

5.1 Refer to chapter 8.1
Refer to chapter 8.8

6.1 For piping refer to chapter 3.0.

Check condenser for leakage.

Table 9-4
Inspection Frequencies
Special Purpose Steam Turbines (Typical)

COMPONENTS	ACTION: Inspect	Replace	Clean	Change Oil	Test/Exercise	Lubricate	Align	INSPECTION FREQUENCY: 6M	3Y	5Y	I	II	III
1.1 VALVES & FITTINGS	●												●
Non-Return	●									●			
Trip & Throttle	●				●			●					
Governor Valve(s)	●								●				
Extraction Control	●									●			
Steam Strainers		●							●				
1.2 CONTROL VALVE GEAR	●		●						●				
1.3 STEAM CHEST	●												●
1.4 NOZZLE BOWL	●												●
2.1 CASING/CYLINDER	●												●
Drain Connections					●						●		
2.2 BLADE RINGS/COVERS	●												●
2.3 EXHAUST/EXH. HOOD	●												●
2.4 GLANDS/SEALS		●							●				
3.1 ROTOR BLADING	●												●
3.2 DISKS	●												●
3.3 SHAFT/DRUM	●												●
3.4 BEARINGS	●								●				
Pedestals	●								●				
Casing Keys	●								●				
3.5 ROTOR/CASING	●						●						●
4.1 OVERSPEED TRIP					●						●	●	
4.2 GOVERNOR	●			●				●					
4.3 TURNING GEAR	●											●	
5.1 LUBE OIL PUMPS	●				●							●	
5.2 L.O. FILTERS									●				●
5.3 L.O. RESERVOIR				●					●				●
5.4 L.O. CONDITIONER	●											●	
5.5 L.O. INSTR./CONTROL					●			●					
6.1 PIPING & CONDENSER	●												●

LEGEND: M=Months Y=YEARS
I=Each Trip II= Whenever Practical
III= Major Disassembly. Typical Frequency:
2Years 1st Run
5Years 2nd Run
10Years 3rd Run

Proper Break-in of Carbon Rings*

Incentives are:

- Long runs
- Higher turbine efficiency
- Protection of bearings and journals by keeping water out of the oil due to blowing steam past the seals into the bearing housing.
- In the winter the machine results in happy operators and a safer unit.
- Lower vibration levels

A common method of breaking in carbon rings involves mounting dial thermometers on the gland housing and observing its temperature rise at incremental speeds for about three hours. Stuffing box temperature rise is a function of carbon ring wear rate, heat transfer rate from the carbon rings through the gland housing, and steam conditions. Surface temperature monitoring procedures are highly questionable due to their poor time response to events happening at the sealing zone between the carbon rings and turbine shaft. Directly observing shaft vibration gives real time knowledge of the condition of the seals.

Factors affecting break-in. Figure 9-3 is a typical carbon ring gland housing assembly for a small steam turbine. The carbon rings that actually do the steam sealing are made of a special form of graphite that is self-lubricating. The seal is usually constructed of three or more segments bound together and against the rotor shaft by a garter spring. The carbon rings are prevented from rotating by a tang.

Mechanism of break-in. Assuming that the carbon ring packing clearances are within design specifications, the carbon rings are "broken in" when they acquire a slick glaze due to controlled rubbing action. Time required for the packings to wear in varies as a function of: steam temperature and pressure, clearances, pressure drop across the seals, sealing steam flow, shaft surface smoothness, shaft surface speed, seal casing configuration and carbon ring composition and design.

Break-in may take from 3 to 12 hours and occurs at about 2,500–3,500 rpm for 3–4 in. internal diameter carbon rings. Cold carbon ring to shaft clearance for 3–4 in. internal diameter rings is about 15–16 mils. Hot running clearance should be about 1–2 mils. Following wear-in, the car-

---

* Adapted from "Avoid Problems with Steam Turbine Carbon Ring Seals," by S. W. Mazlack, Amoco Oil Co., *Hydrocarbon Processing,* August 1981. By permission.

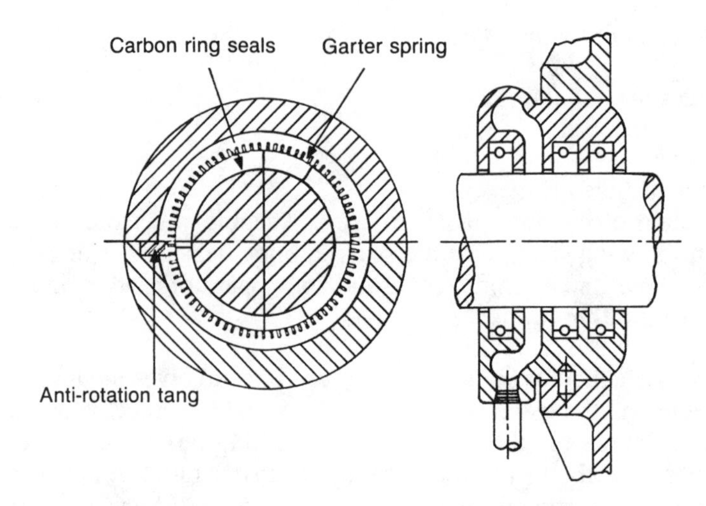

Figure 9-3. Carbon ring gland assembly.

bons fit rather loosely in the stuffing box and in the cold condition have a large clearance from the rotor shaft. In this condition, the seals are relatively immune to the sudden thermal changes that a turbine goes through during its normal duty cycle.

The differential coefficient of expansion between steel and carbon is .000004 in. per in. of diameter for each degree Fahrenheit increase in temperature. Since the thermal expansion of carbon is less than that of steel, too rapid of a wear-in often will result in broken rings, or excessive carbon ring to shaft clearances. Large clearances produce poor sealing and destructive steam "wire drawing" across the carbon ring faces. High levels of vibration, high gland box surface temperatures, noise and a big cloud of steam occurring shortly after turbine startup to full speed are sure signs that the carbon rings were inadequately broken in and are grabbing the shaft. If this happens, don't even ask if the carbon seal rings are "broken"—they are.

Warm up. Of the utmost importance for any turbine operation, including carbon ring break in, is proper warmup. The entire rotor case assembly must be allowed to heat up to its equilibrium temperature prior to starting slow roll. Even heating is required to avoid rotor rubs, high thermal stresses and unequal expansion of the seal rings. During heatup, if the steam plume starts at the case drain pipe outlet and is noisy, this means that water is flashing even if no liquid appears. Dry steam travels through a foot or more of clear space before a wet plume develops and there is much less noise. Again: do not start rolling until all case drains are blowing dry, without puffing. This heat-up may take several hours.

Sealing steam. For a condensing turbine, if possible, have the condenser vented and the sealing steam initially turned off. Sealing steam should not be turned on unless the rotor is turning. Cold air sucked across seals into a hot rotating or nonrotating turbine can distort the shaft as severely as hot sealing steam entering the seal area of a cold nonrotating turbine. Shaft distortion will cause a rotor bend or "bow" to form which can result in destructively high vibration levels. If the shaft develops or has a thermally induced rotor bow, a 1-hour 300–600 revolutions per minute slow roll usually will allow most of it to relax out.

The normal sealing steam pressure of a condensing turbine is about 3–4 psig. A higher sealing steam pressure is recommended at the outset to assist the outboard seals to begin break-in. This is important, for if the low pressure end is primarily sealed with the high pressure end seal leaking off steam, exhaust end packings may not get much steam until the unit actually is coupled up and running at normal speed. If this is the case, the low pressure seal area suddenly may get its first dose of hot steam preceded by a slug of water at full speed. The result may be a sudden seal "grab," carbon shattering, and violent failure. This is the cause of the mysterious severe turbine vibration that occurs shortly after the operator walks away from a machine that was just put on line.

Surface condenser use. The turbine is heated up and brought to minimum speed as a reduced back pressure machine. This maximizes heat input into the seal areas. Caution must be exercised to avoid overpressurizing the surface condenser expansion joint and the steam turbine exhaust casing. Steam flow to the condenser is minimal during an uncoupled slow speed run. As such, the exhaust pressure of the turbine, either positive or negative, can be controlled by balanced use of the surface condenser vent valve, cooling water flow to the condenser exchanger, proper hogging jet operation, and turbine case drain valve positions.

Use of vibration probes. For monitoring carbon ring break in, one temporary probe holder bracket mounted on the inboard face of each bearing housing, with a reasonably clean and nick-free shaft surface for the probe to monitor, will work. If the machine is to be permanently monitored with vibration probes, see API-670, "Non-Contacting Vibration and Axial Position Monitoring Systems" for additional details.

Normally, carbon ring break-in is performed with the turbine uncoupled from the driven unit. Having the turbine uncoupled eliminates most sources of external vibration, and has the turbine ready for its overspeed trip check immediately following completion of the carbon seal ring break in. Note, however, that a solo run turbine is quite different thermally from a coupled fully-loaded turbine at the same speed. This difference must be accounted for during carbon packing break-in.

Carbon Break-In Procedure

1. Heat the lubrication oil to a minimum of 100°F before beginning slow roll. Running oil temperature target is usually 110–120°F. Mount dial thermometers on the gland seal housing, mid-turbine case and exhaust casing. These temperatures are used to determine the steady state temperature point of the turbine prior to slow roll.
2. Open all case drains, trip and throttle valve and steam line drains leading to the turbine and begin slowly admitting warm-up steam. *Do not start slow roll until the turbine is hot.* Larger condensing turbines, particularly partial admission turbines, may require a special manufacturer's recommended startup procedure to avoid localized rotor bowing.
3. Slow roll at 500 rpm at least one hour. Open sealing steam line and establish 5–8 psig pressure.

   As the turbine gets hotter or the vacuum increases, it will speed up rapidly using the same steam flow due to the increased availability of energy.
4. Close case drains as appropriate.
5. Record vibration readings at both ends of the turbine.
6. Raise the speed to 1,000 rpm and immediately record vibration levels. Stay at 1,000 rpm for one hour minimum. At about 1,000–1,200 rpm, the bearing's oil film is carrying the rotor and the shaft has established a reasonably stable orbit in the bearing. Assuming that the rotor has relaxed its thermal bow, the "first" reading you will get at 1,000 rpm is primarily residual rotor unbalance. After about 15–30 minutes, you will observe an increase in vibration (about .25 to 1 mil). Gradually the vibration will drop nearly back to the first reading you took at 1,000 rpm. This is what you've been looking for; a slight carbon seal ring rub followed by a return to steady state.
7. Raise the speed to 3,500 rpm in 500 rpm increments, repeating the sequence of immediately taking "new speed" steady state vibration levels and watching for the vibration increase and decrease cycle caused by the carbon rings breaking in.

   At about 2,500–3,500 rpm, the new carbons are fairly well glazed and nearly run in. This is also the point where most people destroy their packings by assuming that the job is complete.

8. From 3,500 rpm, raise the speed by increments of 1,000 rpm up to running speed going rapidly through the criticals.
   If running smoothly, and a sudden severe jump in vibration occurs, immediately drop the speed to 2,000 rpm or less for about 1/2 hour. The carbon ring seals were grabbing the shaft. After a 1/2 hour cool down, return the turbine to operating speed.
9. Run at normal maximum running speed for one hour prior to checking the overspeed trip and coupling up.
   This procedure has produced consistently positive results with a variety of machines, some of which were considered to be characteristically bad performers. The key to a successful and long life carbon ring break-in is patience and the continuous presence of an operator through the entire procedure to handle any contingency.

Operation of Large Steam Turbines

Finally, if we adhere to the interface concept of large steam turbine maintenance and operation we must not forget to establish proper operating procedures. Here again, each plant will more often than not base its procedures on the manufacturer's general operating instructions. Typically these written procedures contain:

- General description of the turbine train.
- Emergency equipment and procedures.
- Initial startup.
- Normal startup.
- Emergency shut-down.
- Major components, care and feeding.

The following are a manufacturer's instructions for the operation of a special purpose condensing turbine.*

General. This procedure is recommended for starting and putting the turbine in operation. It is obvious that any such instructions can cover only the normal case and it will be recognized that under unusual circumstances, variations from this program will have to be adopted and the procedure to be followed will necessarily be determined by the best judgment of the operating engineers.

Before starting the turbine, clean off any dirt which may have accumulated during the installation or turnaround work and be sure that dirt has

* Courtesy Westinghouse Canada Ltd., Hamilton, Ontario, Canada.

not gotten into the bearing cavities or other internal parts. Be sure that the working parts of the governing mechanism are clean and in good working condition.

It is of utmost importance to see that the turbine casing and connecting pipe lines are drained properly at all times. During operation, any accumulation of water cools the adjacent metal and causes distortion which, if severe, may cause blade rubs or vibration. During shutdown periods, accumulation of water causes excessive corrosion that impairs the efficiency of the turbine.

The turbine casings are provided with built-in drains from each zone to the next lower pressure zone and finally to the exhaust. Orifices are provided for continuous drainage during normal operation, and hand-operated by-passes—where necessary—for use during starting and shutdown periods.

Similar drains must be provided from all connecting pipe lines. These include the steam inlet line and the atmospheric relief line. On condensing machines, all drains—except from the high-pressure steam inlet—should connect to the condenser or a vacuum trap because, when starting or operating at light load, vacuum may exist in the entire back end.

It is the duty of the operators to see that these drains function properly and to use those which are manually operated during starting and shutdown periods.

Check the overspeed trip mechanism by means of the hand-tripping device, and be sure it is working properly. Then reset it. It should be obvious that this tests only the trip mechanism and does not check the speed at which the overspeed trip weight actually functions.

Starting

1. Be sure that the oil supply to the turbine is operating. See that ample oil pressure is established at the bearings and in the control system.
2. See that the turbine casing drains, the extraction line drains, the gland leakoffs are open, and that the steam line is free of water.
3. Open the exhaust valve.
4. Establish water circulation through the condenser.
5. Open the throttle valve a sufficient amount to start the rotor immediately, then close it and open it again just enough to keep the rotor rolling 200 to 400 rpm. Listen for rubs or other unusual sounds, especially when the rotor is rolling with the steam shut off, for at this time a foreign noise can be heard most easily.
6. Start the condensate pump and operate intermittently, if necessary, to maintain level.

7. As soon as the rotor is in motion, turn on the water to gland condenser and steam to the ejector. Close all atmospheric casing drains to prevent drawing air through the drains when partial vacuum is established.
8. Start the second stage air ejector or the priming ejector if one is used.
9. Keep the turbine rolling at low speed—approximately 200 to 400 rpm—to allow the parts to become partly heated. Maintain this slow rolling about 20 minutes. The duration of the rolling period depends on the straightness of the rotor which, in turn, depends somewhat on the length of the previous shutdown. If the machine has been shut down long enough to become thoroughly cooled, the rotor should be straight. However, after shorter shutdowns—such as 4 to 8 hours—the machine is only partly cooled and the rotor may be distorted. In such cases, continued rolling at low speed will heat the rotor uniformly and straighten it.
10. At the end of the rolling period, bring the unit up to speed slowly, taking 10 to 15 minutes to reach full speed.

    After the unit comes up to speed, reduce the speed again and slow roll the turbine for a somewhat longer period of time.
11. Shut off the priming ejector, if one is used. When the maximum vacuum is obtained with the second stage ejector, start the first stage ejector.
12. Make sure that the governor properly controls the speed of the turbine with full steam pressure and vacuum.
13. Close the drains from pressure zones when it is assured that all water has been removed and condensation stopped.
14. Make certain that the temperature of the oil supply to the turbine is maintained between 110 and 120°F. The temperature of the oil leaving the bearings should not exceed 160°F.
15. Open throttle valve fully.
16. Make sure that the governor properly controls the speed of the turbine with full steam pressure.

Test of Overspeed Trip. When the turbine is first started after installation, it is very important to test the overspeed trip by actually overspeeding the machine. This may be done by use of the overspeed test device on the top of the governor. See the governor instructions for operation.

The overspeed trip should operate at approximately 10 percent above normal full speed. Where a speed changer is provided, the trip should operate at 10 percent above the upper limit of the speed changer. The proper tripping speed for each specific application will be found in the "special information" section of the instruction book and also stamped on the nameplate of the unit itself. A direct-reading, tachometer is preferred for reading the speed. A digital readout instrument is also satisfactory, provided the operator is familiar with its use and can read the speed correctly to a very close value.

During these tests, the speed should be increased slowly and the tachometer watched very carefully. An operator should stand by, ready to trip the mechanism by hand instantly if it does not trip automatically at about 5 percent above the specified overspeed. If the mechanism does not trip at the proper speed, it should be inspected and adjusted. The overspeed test should be made periodically, throughout the life of the machine, to ensure that this important safety device is kept in good working condition. Do not put the unit into service, or keep it in service, if it is known that the overspeed trip is not functioning properly.

Shutting Down

1. Decrease the load to about 20 percent of full load; except in an emergency shutdown, load should be removed gradually.
2. Then remove all load and shut down the turbine by tripping the overspeed trip mechanism manually.
3. Be sure that the oil supply is maintained until the machine becomes relatively cool. If this is not done, the heat conducted along the shaft from inside the turbine may injure the bearings.
4. When the turbine comes to rest, close the exhaust valve and open all drains between the throttle valve and the exhaust valve.
5. Shut off the air ejectors. Open the vacuum breaker if one is provided.
6. Shut off the water to gland condenser and steam to the ejector.
7. Shut down the condensate pump.
8. Open all blow-down drains.

How to Avoid Steam Turbine Distress

All steam turbines, like other turbomachinery, are prone to suffer from certain maintenance-causing problems. They are those triggered by:

- Oil contamination
- Foundation difficulties
- Alignment problems
- Piping loads
- Bearing difficulties
- Unbalance conditions
- Operating errors

We already dealt with all of these troubles in other pages of this book and elsewhere[5]. In the following we would like to review the problem of low steam purity because it is unique to steam turbines and particularly large special purpose steam turbines.

Steam contamination can cause stress-corrosion cracking, corrosion pitting, general corrosion and erosion, and can leave deposits in the turbine. A good deal of documentation exists describing the problems. One source[6] suggests the following preventive measures against stress-corrosion cracking:

- Keep contaminants in the steam at the lowest practical achievable level.
- Avoid caustic contamination of the turbine.
- Watch condensate and make-up demineralizers carefully.
- Maintain feedwater conductivity instrumentation.
- Permit only treated condensate in the steam path.
- Instrument the feedwater system to control steam chemistry.
- Do not use cutting fluids, with high concentrations of chlorine and sulfur, in machining operations during maintenance.
- Do not use cleaning fluids with unacceptable levels of caustic, chlorine, and sulfur.

The same source describes the mechanism of solid-particle erosion and corrosive pitting in large steam turbines, their effect and the repair methods used. We are referring our readers to this document.

Deposits on parts in the steam path from boiler carryover may have a considerable effect on capacity, efficiency, and reliability. The build-up of deposits can plug or partially plug turbine buckets, thus increasing thrust bearing load, which could lead to bearing failure with possible additional damage.

Deposits and other internal problems may be detected by monitoring of specific parameters, such as temperature, pressures, flows, and valve opening. An increase in temperature of thrust shoes as shown by embedded thermocouples would indicate increased thrust that could be the result of deposits.

The selection of a fouling detection system will be strongly influenced by the safety and complexity of a cleaning procedure. If the deposits are water soluble, internal washing is possible. In the simplest case this may involve injecting a quart of water into a single stage, mechanical drive turbine, with 30°F superheated inlet. On the other hand, the cleaning may involve removing 300° of superheat from 200,000 lbs/hour of steam entering an eight-stage turbine. This is a much more complex case.

The potential problems of water washing steam turbines are:

1. Misalignment due to piping stress as the temperature is reduced.
2. Water slugging.
3. Loss of clearance due to differential contraction between rotor and stator.
4. Vibration due to nonuniform deposit removal.
5. Thrust failure due to almost complete plugging of one stage.
6. Damage to blading if hit by water retained in the exhaust casing.

On most machines, the misalignment due to pipe stress will not be significant. After all, we are only disposing of the superheat, whereas during run up, the machine is exposed to a temperature change at least two times as large. However, if piping strains are a problem on startup, one must make sure all sliding supports are free before attempting to wash. We have had no problems other than an increase (doubling) of axial vibration due to misalignment.

We try to avoid water slugging by two measures. We always use a venturi nozzle for desuperheating. Also, we insist that the piping fall continuously between the desuperheater and the machine inlet. Even if the water is not broken up into droplets in the desuperheater, it will pass inoccuously through the turbine as a constant stream.

To prevent loss of clearance, we always limit the rate of temperature change to 180°F per hour. The greatest hazard would be failure of the injection pumps when at maximum injection rate. Such a failure would produce a very high rate of change of temperature and would most likely result in an axial rub. To guard against this, we try to use boiler feedwater, since these pumps are the most reliable in the plant.

We attempt to reduce the chances of nonuniform deposit removal by halting the increase of injection rate whenever a deposit is actually being removed. This condition is detected by measuring the conductivity of the exhaust condensate.

It is alleged that thrust bearing failures have occurred because of stage plugging when an upstream wheel has shedded its deposit before a downstream one. We have not found this to be the case.

One of the criteria we use to check a turbine design before purchase is "Can all the condensate be removed from the exhaust?" Some turbine designs are such that the blading is within about 1 in. of the bottom of the casing. Others don't have a casing drain at the lowest point. Others have a 1/2 in. or 3/4 in. casing drain. All of these designs are suspect unless the condensate can drain freely out of the exhaust.

We consider that a stage is washed adequately when the condensate conductivity falls to half its peak level. When this point is reached, the water injection rate is again increased. The wash is considered completed when the inlet steam is saturated and the exhaust conductivity is down to 200 micromhos.

After the wash is completed, the inlet temperature is raised to normal at a maximum rate of change of 180°F per hour.

Initiation of the normal steam flow path, bypassing the desuperheater, is the last hazard. We have found that water builds up in the line upstream of the valve, even when the bypass is left open. We now always leave the main valve cracked open to prevent the water buildup.

Over the past 13 years, we have successfully completed about 30 turbine washings. These involved six different machines, located in four plants. Based on this, we conclude that on-load washing is safe provided reasonable care is exercised. We have, however, observed that deposit solubilities vary considerably between subsequent washes on the same machine. This same variability has been observed on two machines supplied by steam from the same source for the same time period. Some machines can be successfully cleaned without making the inlet saturated but most have required a wet inlet.

During our earliest washes, we believed that condensate was essential as the desuperheating medium. We reasoned that any other water would leave salts behind during the temperature-increasing phase. Five of the machines have now been washed using boiler feedwater, without any observable problems or deterioration of the cleaning.

Also during our earliest washes we noted an apparent accelerated fouling rate during the first few days after a wash. We have no reasonable explanation of this phenomenon. It levels out quickly and does not appear to affect either the maximum mass flow or the efficiency. Currently, we merely warn the operators to disregard it if it is observed.*

Another form of steam turbine cleaning is by the use of chemical foam. Reference 7 describes this method.

General Purpose Steam Turbine Maintenance and Repair

Although steam turbine operation is not generally considered within the scope of this text, an overview is deemed appropriate here.

Turbine applications differ widely, therefore, operating and maintenance procedures must be tailored to each particular installation. The instructions here provide a recommended procedure for the initial startup

* Adapted from "On-Stream Cleaning of Turbomachinery," by B. Turner. Proceedings of Second Turbomachinery Symposium, Gas Turbine Laboratories Texas A&M University, October 1973. By permission.

and serve as a guide for establishing routine operating procedures for such general purpose turbines as Elliott Company's Type YR machines.*

In establishing the specific procedures applicable to a given turbine type or vendor's model, it is clearly advisable that operating and maintenance personnel review the technical material supplied with the equipment. Moreover, it is equally advisable that all appropriate persons familiarize themselves with the safety precautions and operating procedures for turbines. Particular attention should be directed to the warning and caution notes in this chapter.

Steam Supply

Steam should be free from moisture and preferably superheated. A receiver type separator with ample drains should be provided ahead of the shut-off valve to prevent water from entering the turbine. When a separator is not provided, a continuous drain must be connected to the lowest point of the steam inlet piping.

The steam strainer (2, Figure 9-4) protects the turbine from large particles of scale, welding beads, etc. This strainer does not guard against abrasive matter, boiler compound, acids, or alkaline substances, all of which may be carried over in the steam. These substances may corrode, erode, or form deposits on the internal turbine parts, thus reducing efficiency and power. It is necessary that feedwater treatment and boiler operation be carefully controlled to ensure a supply of clean steam, if prolonged satisfactory operation is desired.

Safety Precautions

1. Do not operate the turbine if inspection shows that the rotor shaft journals are corroded.
2. Be sure the rotor is not rubbing any stationary parts and rotates freely by hand before starting.
3. Check that all piping and electrical connections are made before operating the turbine.
4. Ensure that all valves, controls, trip mechanisms and safety devices are in good operating condition.

Under no circumstances should the trip valve be blocked or held open to render the trip system inoperative. Overriding the trip system, and allowing the turbine to exceed the rated (nameplate) trip speed may result in

---

* Source: Elliott Company, Jeannette, Pa. 15644. Adapted by permission.

fatal injury to personnel and extensive turbine damage. In the event the trip system malfunctions: immediately shut down the turbine and correct the cause.

5. If rubbing or vibration occurs during operation, immediately shut down the turbine; investigate and correct the cause.

Preparing the Turbine for Initial Startup

1. Disconnect the coupling between the turbine and driven machine (turbines driving through reduction gears can remain coupled to the gear and operated together).
2. Disconnect the steam inlet piping at the turbine and blow out the line with the supply steam to remove any foreign material from the pipe.
3. Check to be sure the steam strainer (2, Figure 9-4) is clean and properly installed in the steam chest inlet flange. Connect the pipe to the turbine with a permanent joint.
4. If operating condensing; clean rust preventative compound from internal turbine surfaces.

   Note: Rust preventative compound will foul surface condenser tubes if not removed before operating the turbine.

5. Remove bolting from the steam end bearing cap (21 or 53, Figure 9-5), and the exhaust-end bearing cap (12 or 54). Lift the caps approximately 1 in. (25 mm) and pry out the top bearing liners (16, Figure 9-5) to release the oil rings (49). Remove the bearing caps and roll out the bottom bearing liners (15), by rotating them away from their positioning lugs. Clean and inspect the bearing liners.
6. Clean the rotor shaft journals and the bearing housing oil reservoirs with clean, lint free rags. Flood the rotor locating bearing (50) and shaft journals with oil. (See Chapter 12 for proper oil levels and lubrication requirements.)
7. Lift the weight of the rotor and roll the bottom bearing liners into place. Make certain the positioning lugs on the liners are correctly seated in the bearing housing locating grooves.
8. Place the top bearing liners on the shaft journals and position oil rings in the slots in the top liners.
9. Replace the bearing caps, making sure that the positioning lugs on the top liners engage the grooves in the bearing caps. Insert the dowel pins and tighten all bolts.

(Text continued on page 366.)

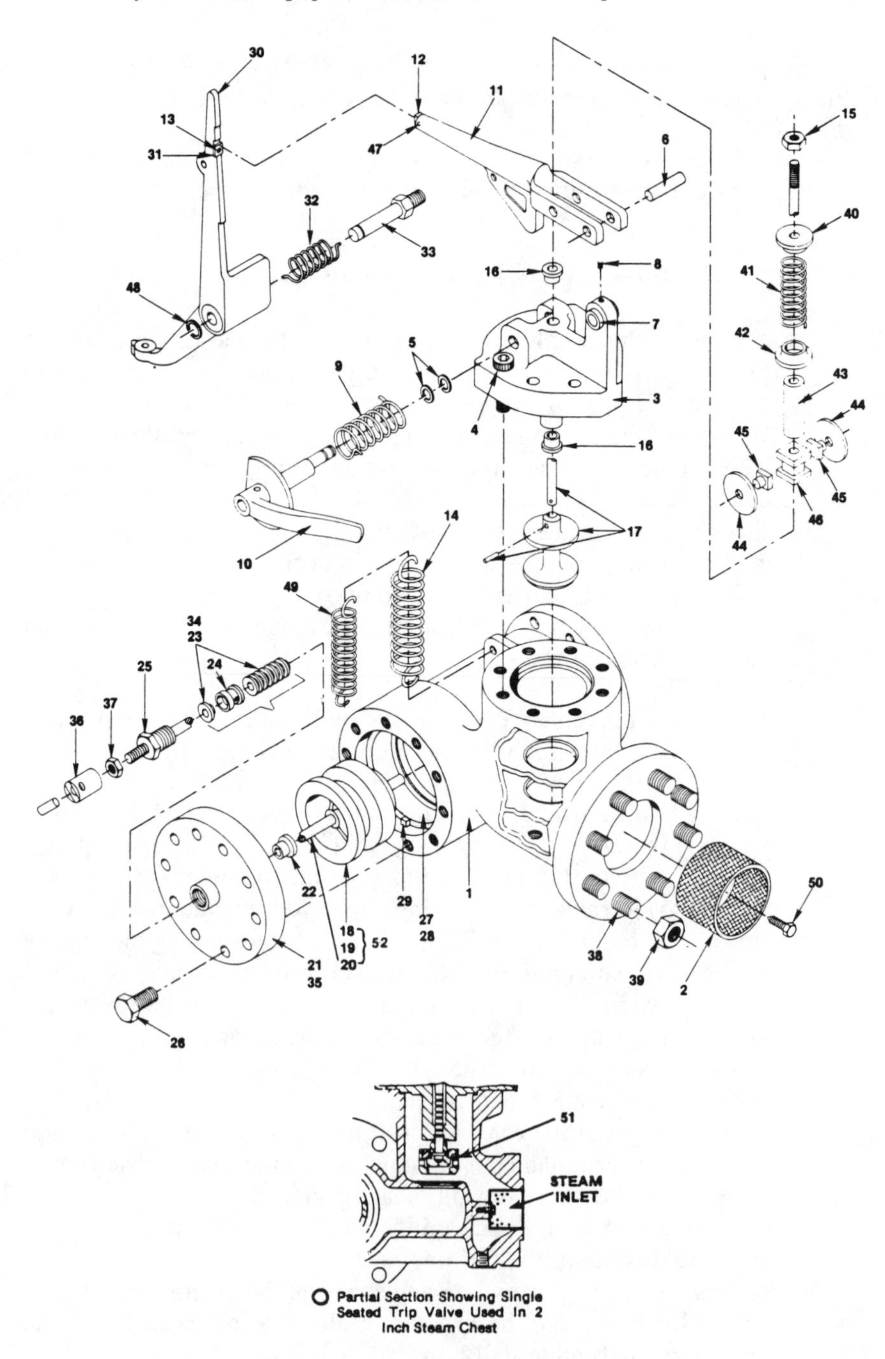

Figure 9-4. Steam chest assembly—typical general purpose turbine.

PARTS LIST

FIGURE ITEM NO.	DESCRIPTION	QUANTITY
4-12-1	Steam Chest Body	1
2	STEAM STRAINER	1
3	Trip Valve Cover	1
4	Cap Screw	8
5	RETAINER RING, AUX RESETTING LEVER	2
6	PIN, RESET LEVER	1
7	BUSHING, RESET LEVER	1
8	SET SCREW	1
9	SPRING, AUXILIARY RESET LEVER	1
10	AUXILIARY RESETTING LEVER	1
11	Resetting Lever	1
12	RESETTING LEVER KNIFE EDGE	1
13	Machine Screw	1
14	SPRING, CLOSING	1
15	Lock Nut, Trip Valve Stem	1
16	BUSHING, TRIP VALVE, LOWER	2
17	TRIP VALVE ASSEMBLY	1
18	Governor Valve	1
19	Pin, Governor Valve Stem	1
20	Governor Valve Stem	1
21	Governor Valve Cover (High Pressure)	1*
22	Bushing, Governor Valve Cover	1
*23	PACKING (HIGH PRESSURE)	1 Set*
24	Lantern Ring	1*
25	Follower	1
26	Machine Bolt, Governor Valve Cover	10
27	VALVE SEAT	1
28	BUSHING, VALVE SEAT (NOT SHOWN)	1
29	Weld Block	2
30	Hand Trip Lever	1
31	LATCH, KNIFE EDGE	1
32	SPRING, TORSION	1
33	Shoulder Stud	1
*34	PACKING (LOW PRESSURE)	1 Set*
*35	Governor Valve Cover (Low Pressure)	1*
*36	CONNECTION VALVE STEM	1*
37	Jam Nut, Valve Stem	1
38	Stud, Inlet	8
39	Nut, Inlet	8
40	Spring Seat (Top)	1
41	Spring, Backsetting	1
42	Spring Seat (Bottom)	1
43	Bushing, Trip Valve, Upper	1
44	Washer	2
45	Block, Resetting Lever	2
46	Connection, Backsetting	1
47	Roll Pin, Knife Edge	1
48	Retainer Ring	1
+49	Spring, Auxiliary Closing	1*
50	Machine Bolt, Strainer	1
51	Single Seated Trip Valve	1*
52	GOVERNOR VALVE & STEM ASSEMBLY	1

* Indicates part not used on all turbines or variable quantities.

+ Item 49 indicates the addition of a second spring used on turbines operating at 250 psig or higher maximum inlet steam pressure.

Figure 9-4. Continued.

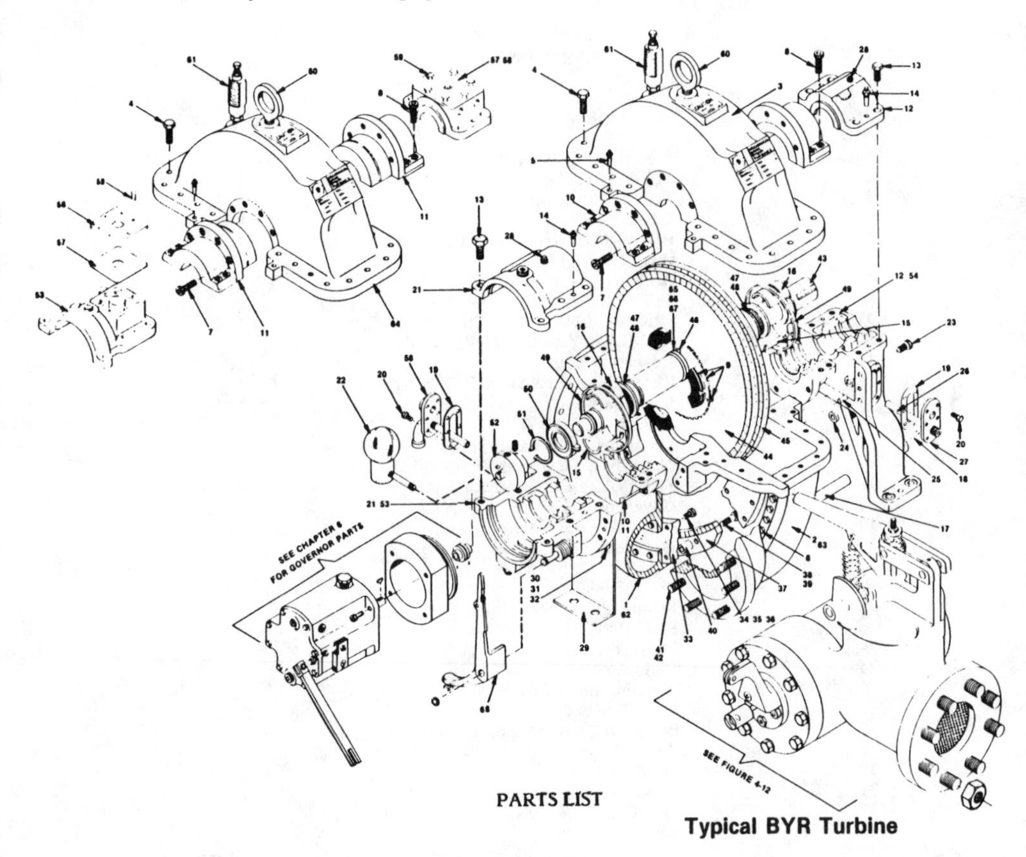

FIGURE ITEM NO.	DESCRIPTION	QUANTITY
4-11-1	Casing, Steam End, BYR	1
2	Casing, Exhaust End, BYR	1
3	Casing Cover, BYR	1
4	Machine Bolt (Cover)	22
5	Taper Dowel/Nut (Cover)	4
6	Cap Screw (Special	13
7	Cap Screw, Packing Case	4
8	Cap Screw, Packing Case	12
9	**CARBON RING ASSEMBLY, BYR**	8
	(CARBON RING ASSEMBLY, BYRIH)	12
10	Packing Case, BYR	2
11	Packing Case, BYRIH	2
12	Pedestal And Cap, BYR	1
13	Machine Bolt	10
14	Taper Dowel/Nut	4
+15	**BEARING LINER, BOTTOM**	2
⌓16	**BEARING LINER, TOP**	2
17	Combining Stud	2
18	Pin, Combining Stud	2
Δ 19	Gasket, Water Cooling Flange	2
20	Machine Bolt, Flange	12
21	Bearing Housing/Cap, BYR	1
Δ 22	Oiler	2

Figure 9-5. Typical general purpose steam turbine (Elliott Tyde BYR).

FIGURE ITEM NO.	DESCRIPTION	QUANTITY
23	Cap Screw Bearing Case to Casing	8
24	Spacer	8
25	Dowel	4
△26	Cooling Tube Assembly	2
△27	Flange (Exhaust End) Water Cooling	2
28	Pipe Plug, Oil Ring Inspection	6
29	Support, Steam Bearing Case	1
30	Machine Bolt, Support	2
31	Lockwasher	2
32	Nut, Support	2
33	Nozzle Ring	1
34	Cap Screw (Special)	13
35	Lockwasher	30
36	Cap Screw (Special)	17
*37	Reversing Blade Assembly	
*38	Cap Screw (Special) Reversing Blade	
*39	Lockwasher	
*40	Spacer, Reversing Blade	
41	Stud, Steam Chest To Casing	10
42	Nut, Steam Chest To Casing	10
43	Rotor Shaft	1
44	1st Disk Assembly	1
45	2nd Disk Assembly	1
46	Key, Disk	2
47	Sleeve Seal	3
48	Set Screw	6
O 49	**OIL RING**	2
50	**ROTOR LOCATING BEARING**	1
51	**RETAINING RING**	1
52	Trip Body	1
△53	Bearing Housing With Water Cooled Cap, BYRIH	1
△54	Bearing Pedestal With Water Cooled Cap, BYRIH	1
△55	3/4" Pipe Plug, Bearing Housing (Not Shown)	4
△56	Flange (Steam End)	1
△57	Gasket, Water Cooled Cover	2
△58	Cover	2
△59	Bolt	12
60	Eye Bolt, Cover, Lifting	1
61	Sentinel Valve	1
62	Casing, Steam End, BYRIH	1
63	Casing, Exhaust End, BYRIH	1
64	Casing Cover, BYRIH	1
65	Shrink Ring, Steam End	1
66	Shrink Ring, Center	1
67	Shrink Ring, Exhaust End	1
68	Trip Latch	1

* * Indicates variable quantity.

\+ + Bottom bearing liners used for oil ring lubrication, are not interchangeable with liners used for pressure lubrication.

⌓ Top liners used with Class 1 & 2 rotors are not interchangeable with top liners used with Class 3 rotors. (Rotor class designation on Page 409).

△ Not used on pressure lubricated turbines. Blank flange used in place of Items 27 & 56. Standpipe used in place of Item 55 when pressure lubricated turbine is equipped with Class 1 or 2 rotor. (Rotor class designation on Page 409).

O Not furnished with turbines equipped with Class 3 rotors. (Rotor class designation on Page 409).

Figure 9-5. Continued.

10. Inspect and lubricate the governor linkage. For specific details on preparing the governor for startup, see Governor Operation in Chapter 10.
11. Fabricate a clamp or other blocking device to secure the coupling sleeve (if applicable) to the hub while operating the turbine uncoupled.
12. Check for free movement and clearance between auxiliary resetting lever and cam mechanism.

Initial Startup, Noncondensing Turbines

1. Thoroughly drain the steam inlet piping, turbine steam chest and casing, and the exhaust piping of any accumulated water.
2. Open the turbine exhaust valve. If overload hand valves (1, Figure 9-6) are furnished, they must also be opened. Turn the governor speed adjustment to minimum speed (Governor Operation, Chapter 10).
3. Latch the resetting lever (11, Figure 9-4) and slowly open the steam shut-off valve until the turbine reaches approximately 500 r/min. Immediately check the operation of the trip valve by striking the trip lever (30, Figure 9-4). Close the steam shut-off valve as the turbine speed decreases.
4. Latch the resetting lever and slowly open the steam shut-off valve to bring the turbine back to 500 r/min. Remove the inspection plugs (28, Figure 9-5) from the bearing caps and check to be sure the oil rings are rotating. Monitor the speed carefully during the low speed operation.

   Caution: Steam should not be admitted to the turbine casing by partially opening the inlet steam shut-off valve while the rotor is stationary. This condition will cause uneven heating of the turbine rotor and casing which may result in a distorted casing, bowed rotor shaft or other related problems. Do not leave the turbine unattended at any time during the initial startup.

5. Introduce cooling water to bearing housing cooling chambers to prevent overheating. (See Table 9-5). Listen for any rubbing, unusual noises or other signs of distress in the turbine. Feel the bearing housings and oil lines, to detect overheating or vibration. Do not continue to operate if any of these conditions exist. Shut down the turbine; locate and correct the cause of the problem. See the

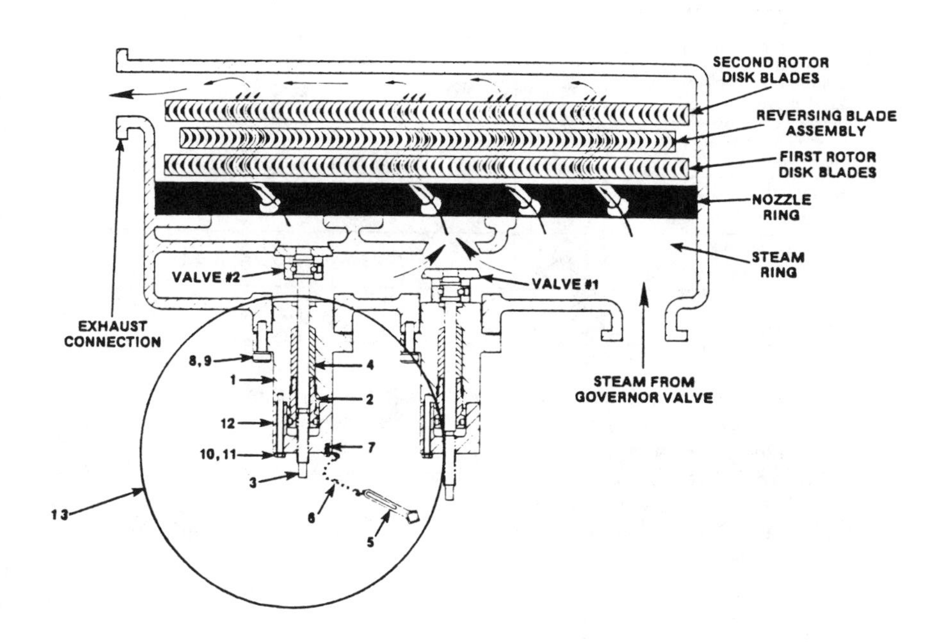

FIGURE ITEM NO.	DESCRIPTION	QTY.
4-10-1	HAND VALVE BODY	1
-2	FOLLOWER	1
-3	STEM & VALVE ASSEMBLY	1
-4	PACKING	1 SET
-5	WRENCH	1
-6	CHAIN	1
-7	SCREW	1
-8	CAP SCREW	6
-9	LOCKWASHER	6
-10	CAP SCREW	4
-11	LOCKWASHER	4
-12	COVER	1
-13	HAND VALVE ASSEMBLY	

Figure 9-6. Overload hand nozzle valve assembly.

Table 9-5
Normal Oil Pressure and Temperature Ranges for Condensing and Noncondensing Turbines

Method Of Lubrication	Oil Ring Lubrication		Pressure Lubrication			
			A (See Note 2)		B (See Note 3)	
Oil Reservoir Operating Temperature	130°F TO 160°F	54°C TO 71°C	130°F TO 160°F	54°C TO 71°C	130°F TO 160°F	54°C TO 71°C
Minimum Oil Temperature Before Starting	70°F	20°C	70°F	20°C	70°F	20°C
Oil Temperature From Bearings	140°F TO 190°F	60°C TO 88°C	140°F TO 190°F	60°C TO 88°C	140°F TO 190°F	60°C TO 88°C
Maximum Cooling Water Temperature (Nominal)	90°F	32°C	90°F	32°C	90°F	32°C
Oil Temperature From Cooler (To Bearings)	X	X	120°F	49°C	120°F	49°C
Normal Bearing Oil Pressure,	X	X	7 psi TO 9 psi	0.5 bar TO 0.6 bar	7 psi TO 9 psi	0.5 bar TO 0.6 bar

Notes

1. These guidelines are not intended to supersede the original manufacturer's recommendations. It is the intent to indicate the general service requirements and leave the particular recommendations to the OEM.
2. Column "A" provides the general guidelines for turbines lubricated by a turbine shaft driven pump or by the driven machine.
3. Column "B" shows the acceptable general guidelines for turbines lubricated from gear oil systems.

troubleshooting guide in Volume 2 for possible causes and corrective actions for abnormal conditions.

6. When the turbine is thoroughly warmed up and operation is determined to be satisfactory, check that all drain valves are open and gradually increase the speed. Increase the speed with the governor speed adjustment until the turbine is operating at the rated speed shown on the turbine nameplate. (Adjust the governor as outlined in Governor Operation, Chapter 11.) Continue to check the turbine for unusual noises, rubbing, vibration or other signs of distress. Do not continue to operate if any of these conditions exist. See the troubleshooting guide in Volume 2 for possible causes and corrective actions for abnormal operating conditions.

   Note: If the turbine is pressure lubricated, the oil pressure should be 7 to 9 psig (0.5 to 0.6 bar).

7. Check the overspeed trip by overcoming the governor to actuate the overspeed trip mechanism. Refer to Governor Operation for specific details on overspeeding the turbine.

Caution: Do not operate the turbine more than 2 percent above the rated trip speed listed on the turbine nameplate. If the overspeed trip does not operate within 2 percent of the designated speed, shut the turbine down and make trip adjustments as described in the section, "Overspeed Trip System."

8. Latch the resetting lever and bring the turbine up to speed. Operate the turbine for approximately one hour. Check the bearing temperatures and turbine speed. Listen for unusual noises, rubbing or vibration. After this period, the turbine can be shut down, doweled to the mounting surface and coupled to the driven machine. If the turbine is used with a speed reduction gear or other special equipment, follow all instructions pertaining to those particular items.

Initial Startup, Condensing Turbines

1. Thoroughly drain the steam inlet line, turbine casing, steam chest, and the exhaust line of any accumulated water. Close the drain valves when all water is drained from the system.
2. Adjust the governor speed setting to minimum speed. If overload hand valves (1, Figure 9-6) are furnished, they must be fully opened.
3. Latch the resetting lever (11, Figure 9-4) open the turbine exhaust valve and start the condensing equipment.
4. Open the steam inlet shut-off valve until the turbine speed reaches approximately 500 r/min.

   Caution: Steam should not be admitted to the turbine casing by partially opening the inlet steam shut-off valve while the rotor is stationary. This condition will cause uneven heating of the turbine rotor and casing which may result in a distorted casing, bowed rotor shaft, or other related problems.

5. Adjust the sealing steam valve so that a slight amount of steam is discharged from the leak-off drain lines.

   Note: 3 to 5 psig (0.20 to 0.35 bar) is usually sufficient sealing steam pressure. However, care must be taken to prevent steam from blowing out of the packing cases and along the turbine shaft.

Caution: If sealing steam is allowed to leak into the bearing housings, the lubricating oil may become contaminated and form sludge and foam. Adjust the sealing steam accordingly to prevent this condition.

6. Check the operation of the trip valve by striking the hand trip lever (30, Figure 9-4). Close the steam inlet shut-off valve as the turbine speed decreases.
7. Latch the resetting lever and slowly open the steam shut-off valve to bring the turbine back to 500 r/min. Remove the inspection plugs (28, Figure 9-5) from the bearing caps and check to be sure the oil rings are rotating. Monitor the speed carefully during the low speed operation.

Caution: Do not leave the turbine unattended at any time during the initial start-up.

8. Introduce cooling water to bearing housing cooling chambers to prevent overheating (See Table 9-5). Listen for any rubbing, unusual noises or other signs of distress in the turbine. Feel all bearing housings and oil lines to detect overheating or vibration. Do not continue to operate if any of these conditions exist. Shut down the turbine; locate and correct the cause of the problem. See the troubleshooting guide in Volume 2 for possible causes and corrective actions for abnormal conditions which might occur.
9. When the turbine is thoroughly warmed up and low speed operation is determined to be satisfactory, increase the speed with the governor speed adjustment until the turbine is operating at the rated speed shown on the turbine nameplate. (Adjust the governor as outlined in Governor Operation, Chapter 10.) If operational problems occur, shut the turbine down and locate and correct cause. (Refer to the troubleshooting guide in Volume 2 for possible causes and corrective actions for abnormal operating conditions.)

Note: If the turbine is pressure lubricated, the oil pressure should be 7 to 9 psig (0.5 to 0.6 bar).

10. Check the overspeed trip by overcoming the governor to actuate the overspeed trip mechanism. Refer to Governor Operation, later, for specific details on overspeeding the turbine.

Caution: Do not operate the turbine more than 2 percent above the rated trip speed shown on the turbine nameplate. If the overspeed trip does not operate within 2 percent of the designated speed, shut the turbine down and make necessary adjustments as described in the "Overspeed Trip System" section.

11. After the speed decreases by 15 percent to 20 percent of rated speed, latch the resetting lever and bring the turbine back up to speed. Operate the turbine for approximately one hour. Check the bearing temperatures and turbine speed. Listen for unusual noises, vibration or rubbing. After this period, the turbine can be shut down, doweled to the mounting surface and coupled to the driven machine. If the turbine is used with a speed reduction gear or other special equipment, follow all instructions pertaining to those particular items.

Initial Startup, Pressure Lubricated Turbines

1. Before startup, be sure that the oil pumps are primed and the oil reservoir is filled to the proper level. Start the auxiliary oil pump (if provided) and circulate the lubricating oil. Check the oil piping for leaks and be sure oil is being delivered to the bearings.
2. Check the oil temperature (See: Minimum Oil Temperature Before Starting, Table 9-5) then proceed with the applicable start-up procedure for Noncondensing or Condensing Turbines.
3. After the turbine is operating, closely observe oil pressures and temperatures. Introduce cooling water to the oil cooler as the system warms up. Refer to Table 9-5 for normal oil pressure and temperature ranges.

Routine Startup, Noncondensing Turbines

1. Check all oil levels. Fill lubricators as necessary.
2. Place any controls, trip mechanisms or other safety devices in their operative positions. Open hand nozzle valves (1, Figure 9-6), if furnished.
3. Open all drains from steam lines, turbine casing and steam chest.
4. Open the turbine exhaust valve.
5. Open the steam inlet shut-off valve and bring the turbine up to desired speed.
6. Close all drain valves when drain lines show the system is free of water.
7. Make necessary governor adjustments to attain desired speed as load is applied to the turbine. (See: Governor Operation, later.)

8. Introduce cooling water to bearing housing cooling chambers to prevent overheating. (Refer to Table 9-5.)
9. Observe bearing temperatures and overall operation for any abnormal conditions.

Routine Startup, Condensing Turbines

1. Check all oil levels. Fill lubricators as necessary.
2. Place all controls, trip mechanisms or other safety devices in their operative positions.
3. Drain steam lines, turbine casing and steam chest of all water, and fully open hand nozzle valves (1, Figure 9-6), if furnished. Close drain valves when the system is free of water.
4. Open the turbine exhaust valve, and start the condensing equipment.
5. Open the steam inlet shutoff valve.
6. When the shaft begins rotating, introduce sealing steam to the packing cases.
7. Bring the turbine up to speed and make any necessary governor adjustments.
8. Introduce cooling water to bearing housing cooling chambers to prevent overheating. (Refer to Table 9-5.)
9. Check bearing temperatures and overall conditions for smoothness of operation.

Routine Startup, Pressure Lubricated Turbines

1. Check the oil reservoir for proper oil level. Start the auxiliary oil pump (if provided) and circulate oil through the system.
2. Ensure that the oil temperature is 50°F to 70°F (10°C to 20°C) before operating the turbine.
3. Place all controls, trip mechanisms and other safety devices in their operative positions. Open hand nozzle valves (1, Figure 9-6), if furnished.
4. Open all drains from steam lines, turbine casing and steam chest.
5. If condensing, close all drain valves when drain lines indicate the system is free of water.
6. Open the turbine exhaust valve. If condensing, start the condensing equipment.
7. Open the inlet steam shutoff valve and bring the turbine up to rated speed. If noncondensing, close the drain valves when the system is free of water. If condensing, admit sealing steam to the packing cases when the rotor shaft begins to rotate.

8. Make necessary governor adjustments to attain desired speed as load is applied.
9. Observe bearing temperatures and introduce sufficient cooling water to the oil cooler to maintain bearing oil temperatures of 140°F to 190°F (60°C to 88°C).
10. Check the overall operation to determine all conditions to be satisfactory.

Overload Hand Nozzle Valves

Optional hand valves (Figure 9-6) are sometimes used to control the steam flow through an extra bank of nozzles. These valves can serve three functions:

1. When closed, the valves will provide more efficient turbine operation at reduced load with normal steam conditions by reducing the nozzle area and thereby reducing the steam flow.
2. In some applications, hand valves are used to develop the required power by opening the valves when steam conditions are less than normal (such as encountered during boiler startup).
3. Hand valves are sometimes used to develop increased power for meeting overload requirements with normal steam conditions.

(See also the section on hand valve positioning versus turbine power, speed, and operating steam conditions normally provided in the manufacturer's operating and maintenance manuals.)

Note: Hand valve must be positioned either fully open or fully closed. Turning the valve counterclockwise approximately 1½ turns will open the pilot valve. Turning the valve approximately nine additional turns will fully open the main valve disk. Open all hand valves during startups to ensure even heating of casing and prevention of valves binding in the casing.

Turbine Shutdown

1. Shut the turbine down by striking the top of the trip lever by hand.
2. Observe the action of the trip valve and linkage.
3. Close the inlet steam shutoff valve.

Note: Shutoff valves, located in the turbine inlet steam piping, must be closed after the trip valve has closed. Do not use the trip valve as a shutoff valve.

4. If noncondensing, close the exhaust valve and open turbine casing drains.
5. If condensing, shut down the condensing equipment, open the turbine casing drains and close the sealing steam shutoff valve.

   Note: Do not apply sealing steam to the packing cases while the turbine rotor is idle.

6. Allow the rotor to come to a complete stop and cool down before turning off the cooling water or stopping auxiliary oil pump, if furnished.
7. If the turbine is to be taken out of service for an extended period; follow the storage instructions given in Chapter 12.

Operation of Emergency and Standby Turbines

It is important that turbines used for emergency and standby services have drain lines open and isolating valves closed when the turbine is idle. Turbines not used for extended periods should be inspected and operated occasionally to make certain that they are in good working condition. Where impractical to operate the turbine, the rotor should be turned over by hand to introduce oil to the journal bearings (oil ring lubricated turbines). If an auxiliary oil pump is furnished (pressure lubricated turbines), oil can be supplied to the bearings by operating the pump. The introduction of dry nitrogen into the casing during shutdown periods is also advisable to prevent corrosion.

Emergency and standby turbines do not require a warmup period before applying the load. They may be placed in service as rapidly as desired. Steam should not be admitted to the turbine casing by partially opening the inlet steam shutoff valve while the rotor is stationary. This condition will cause uneven heating of the turbine rotor and casing which may result in a distorted casing, bowed rotor shaft or other related problems.

Only now are we ready to proceed to the topic of *Steam Turbine Maintenance.*

Maintenance Overview

Industrial steam turbines, like most quality machinery, require periodic maintenance and service. This section supplies turbine disassembly, assembly and adjustment procedures. These procedures should be a familiar subject to maintenance personnel to assure effective repair work and proper adjustments to components requiring service. Maintenance personnel should thoroughly understand, and at all times observe, all safety precautions related to turbine maintenance. It is of primary importance to ensure the turbine is isolated from all utilities to prevent the possibility of applying power or steam to the turbine when performing internal maintenance. Therefore, it is absolutely necessary to *close, lock and tag all isolating valves and open all drains to depressurize the turbine casing and steam chest before performing any internal turbine maintenance. Also, take necessary precautions to prevent possible turbine rotation due to reverse flow through the driven machinery.*

Nondestructive type testing is recommended for determining operational reliability of parts during turbine inspections. If major parts replacement (such as turbine shaft, disks, blading etc.) becomes necessary, it is advisable that the repair work be supervised by a vendor's representative or be done in a qualified repair shop.

Scheduled Maintenance

Scheduled maintenance inspections are necessary for safe and efficient turbine operation. Actual intervals between inspections cannot be specified rigorously because maintenance scheduling is dependent on factors best known by those directly involved with the turbine and its particular application. Table 9-6 serves as a general guideline for establishing a scheduled preventative maintenance program.

The actual frequency of required maintenance inspections can only be determined after carefully considering turbine performance records, maintenance history, corrosion/erosion rates, tests, observations and anticipated service demands. The established inspection schedule will usually be consistent with the availability of the turbine, necessary manpower and an adequate supply of repair parts. At the same time, scheduled inspections must be frequent enough to avoid unsafe operating conditions.

It is also necessary to test and adjust all safety devices on a definite schedule to ensure their operational reliability. These devices are designed to prevent injury to personnel and/or major equipment damage. If these devices are not operated at frequent intervals, they may not work when needed.

Turbine Casing and Miscellaneous Joints

The turbine steam joints are carefully made up and factory tested under pressure, to ensure steam tightness. Two types of sealing compounds are used on general purpose and moderate pressure (not to exceed 600 psig) turbines. For sealing of special purpose turbines, refer to Volume 3, Chapter 10 of this series. Of the sealing compounds used for general pur-

Table 9-6
Scheduled Maintenance Guidelines

Approx. Frequency	Maintenance Description
Daily	1. Check all oil levels and add oil as necessary. 2. Check bearing and lubricating oil temperatures. 3. Check turbine speed. 4. Check smoothness of operation, investigate sudden changes in operating conditions or unusual noises. 5. If daily shutdowns are made; test the trip valve by striking the hand trip lever.
Weekly	1. Exercise trip valve to prevent sticking due to deposits or corrosion. If on a continuous operating schedule; exercise the trip valve by striking the hand trip lever. Reset the lever when the turbine speed decreases to approximately 80% of rated speed.
Monthly	1. Sample lubricating oil and renew as necessary. 2. Check governor linkage for excessive play. Replace any worn parts. 3. Check the overspeed trip by overspeeding the turbine (if the driven machine permits).
Annually	1. Check all clearances and adjustments. 2. Remove and clean steam strainer. If strainer is exceptionally dirty, clean every six months. 3. Inspect governor valve and valve seat. Hand lap the valve if signs of uneven wear exists. Replace the governor valve stem packing if necessary. 4. Clean and inspect trip valve. Replace worn parts and hand lap if necessary. 5. Disassemble, clean and inspect overspeed trip and linkage. Inspect trip pin and check for ease of operation. Correct clearance to plunger. 6. Check journal bearings and rotor locating bearing for wear and replace if necessary. Adjust bearing case alignment for good liner bearing contact. 7. Inspect and clean bearing housing oil reservoirs and cooling chambers. 8. Lift turbine casing cover and inspect rotor shaft, disks, blades and shrouding. 9. Inspect carbon rings and replace as necessary. 10. Remove rotor assembly from casing and inspect reversing blades and nozzle ring. 11. Check operation of sentinel valve. 12. Adjust and check overspeed trip when turbine is put back in operation.

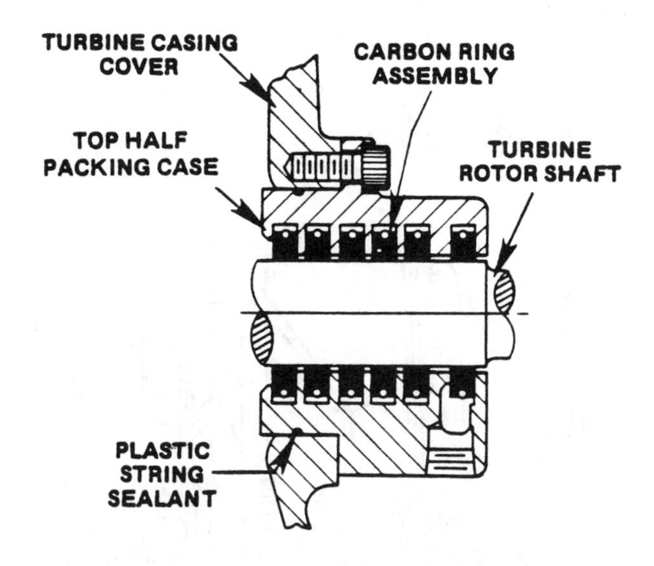

Figure 9-7. Packing case arrangement, including carbon rings.

pose and moderate pressure machines, one is a paste which is spread on the joints, the other is a plastic string type sealant.

A combination of 1/8 inch (3 mm) diameter string sealant and paste compound is used to seal the vertical joints between the packing cases and turbine casing on some models, and the joint between the outside radius of the packing cases and casing on turbines executed as shown in Figure 9-7. A combination of 1/16 in. (1.6 mm) diameter plastic string sealant, placed on a layer of paste compound is used to seal the following steam joints:

1. Steam chest body (1, Figure 9-4) to the turbine casing.
2. Governor valve cover (21, 35) to the steam chest body (1, Figure 9-4).
3. Trip valve cover (3) to the steam chest body (1).
4. Nozzle ring (33, Figure 9-5) to turbine casing.
5. The horizontal and vertical turbine casing joints. (Figure 9-8).

Screw threads subjected to high temperatures often gall during disassembly. It is recommended that anti-galling compound be applied to the threads of all studs, bolts, socket head cap screws, and other threads subjected to high temperatures.

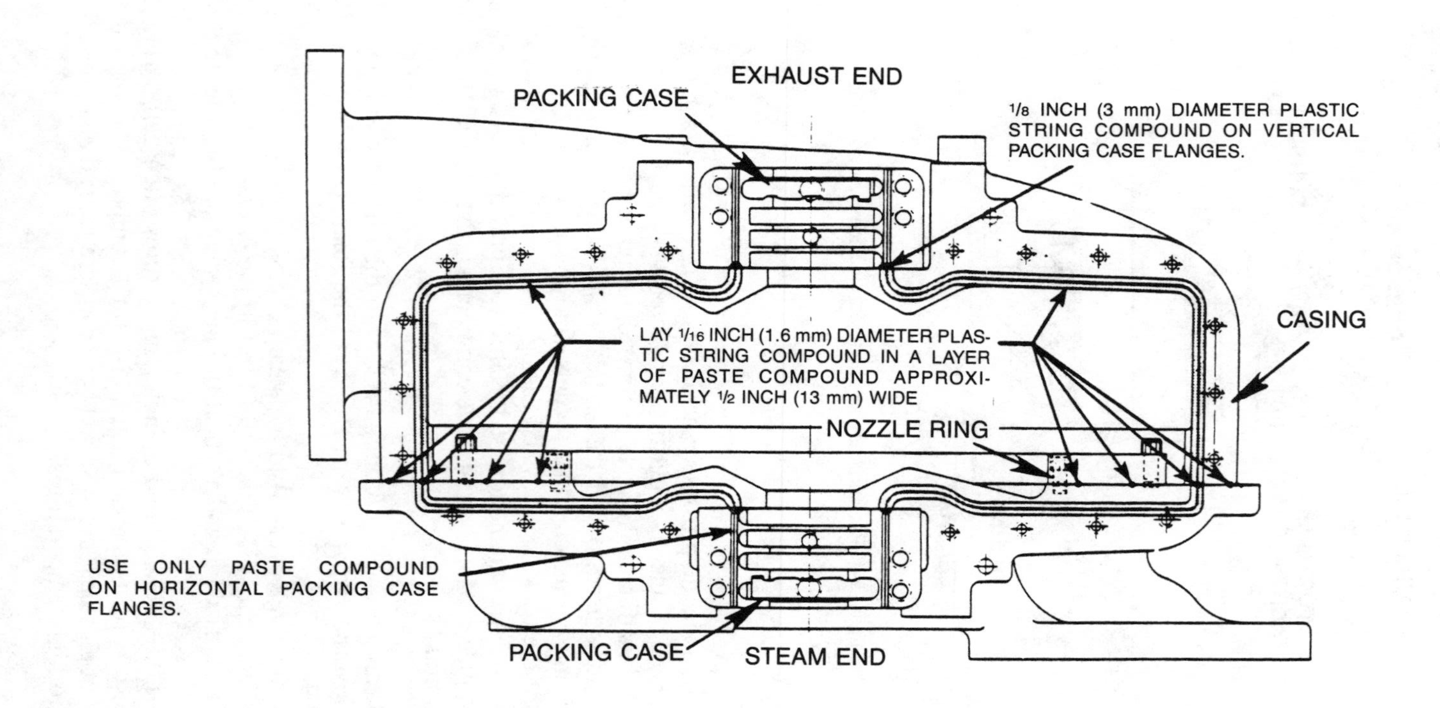

Figure 9-8. Horizontal joining detail on a typical general purpose turbine.

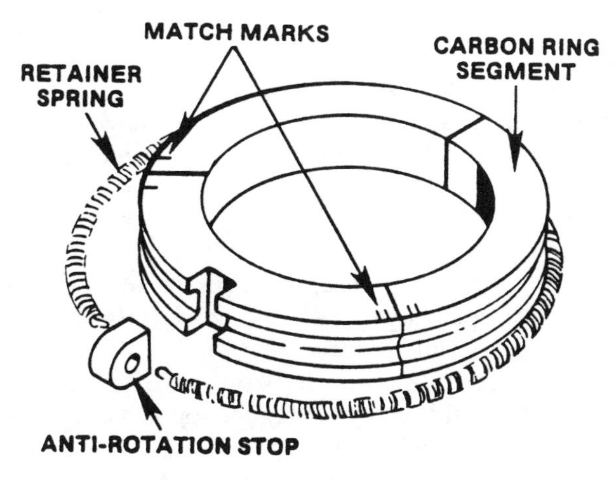

Figure 9-9. Carbon ring assembly.

The joints between the bearing caps and bearing housings may be made up with a thin coat of oil resistant sealant, such as "No. 2 Permatex," if desired.

Packing Case and Carbon Ring Service

The steam end and exhaust end packing cases furnished on small turbines (10, Figure 9-5), each house four carbon ring assemblies (Figure 9-9). Six carbon ring assemblies are housed in each packing case on larger turbines (11, Figure 9-5). Each carbon ring assembly consists of three carbon segments and an anti-rotation stop, which are held together by a retainer spring. Axial positioning of the carbon rings is maintained by machined grooves in the packing cases.

Packing Case Disassembly, Small Turbines

1. Remove cap screws (7 and 8, Figure 9-5) from the horizontal and vertical flanges on the top half packing cases.
2. Break the horizontal and vertical joints by prying the top half packing cases away from the bottom halves.
3. Carefully remove the top half packing cases by lifting straight up until they clear the carbon ring assemblies.

Packing Case Disassembly, Larger Turbines

The steam end and exhaust end packing cases (11, Figure 9-5) furnished on larger turbines, extend beneath the turbine casing cover. This design (shown in Figure 9-7) requires that the casing cover be removed before removing the top half packing cases.

1. Remove cap screws (7, Figure 9-5) from the vertical flanges on the top halves of the packing cases and remove turbine casing cover as outlined in Disassembly Section.
2. Remove cap screws (8) from horizontal packing case flange and break the horizontal joint by prying the top halves of packing cases away from the bottom halves.
3. Carefully remove the top halves of the packing cases by lifting straight up until they clear the carbon ring assemblies.

Carbon Ring Replacement

1. Unhook the retaining spring surrounding the carbon ring.
2. Remove the anti-rotation stop by sliding it off the retaining spring.
3. Remove the carbon ring segments by rotating them around the rotor shaft.
4. Pull the retaining spring from the packing case.

   Note: Do not mix carbon ring segments. Mark each ring so it can be returned to its original location.

5. Clean the packing cases, rotor shaft and all sealing surfaces on the packing case flanges. Blow out the packing cases with high pressure air.
6. Place the carbon ring retaining springs under and part way around the rotor shaft.
7. Roll the carbon ring segments around the shaft and into the packing case grooves. Align the match marks on the carbon ring segments to assure proper assembly.
8. Slide the anti-rotation stops onto the retainer springs and position the stops in notched carbon ring segments.
9. Hook the ends of each retaining spring together and rotate the carbon rings so the anti-rotation stops are seated in the anti-rotation notches in the bottom half packing cases.

Clearances

The inside diameters of new carbon rings are selected to match to the maximum expected turbine exhaust temperature (Refer to Table 9-7). The inside diameters of used carbon rings may be slightly larger than new rings. Measuring the carbon rings is difficult, however, an inside micrometer or snap gauges may be used with a fair degree of accuracy.

The cold clearances may be determined by measuring the inside diameter of the assembled rings and the diameter of the rotor shaft at sealing areas. The difference between the measurements is the cold diametral clearance.

Adjustments

The carbon rings are not adjustable. Replacement is recommended if excessive steam leaks from the packing cases. Packing case cleanliness is of the utmost importance in achieving proper carbon ring seating. If an air supply is available, blow out the packing cases before replacing the carbon rings. For best results, install new carbon rings in complete sets. Refer to pages 349 through 353 for break-in procedures.

Table 9-7
Minimum/Maximum Carbon Ring Dimensions*
For Operating Exhaust Temperatures to 750° (400°C)

OPERATING EXHAUST TEMP.	CARBON RING INSIDE DIA.	INCHES	MILLIMETRES
UNDER 300°F (150°C)	Min.	2.252	57.20
	Max.	2.253	57.23
301° TO 400°F (151° TO 204°C)	Min.	2.254	57.25
	Max.	2.255	57.28
401° TO 500°F (205° TO 260°C)	Min.	2.256	57.30
	Max.	2.257	57.33
501° TO 600°F (261° TO 315°C)	Min.	2.258	57.35
	Max.	2.259	57.38
601° TO 700°F (316° TO 370°C)	Min.	2.260	57.40
	Max.	2.261	57.43
701° TO 750°F (371° TO 400°)	Min.	2.262	57.45
	Max.	2.263	57.48

* Assumes a shaft diameter of 2.500 in.

Packing Case Assembly, Small Turbines

1. Clean packing case flange surfaces and mating turbine casing surfaces.
2. Blow out the packing cases with high pressure air.
3. Press 1/8 in. (3 mm) maximum diameter plastic string compound into the grooves provided in the packing case vertical flange faces. Cut the string to prevent it from extending beyond the horizontal flange.
4. Apply a thin coat of paste sealing compound to the horizontal flanges and inside bolt circles of the vertical flange faces (Refer to Figure 9-8).

   Note: Excessive paste sealant on packing case flanges may result in sealant entering the packing cases and adhering to carbon rings. This may prevent the carbon rings from seating properly. Keep paste sealant approximately 3/16 in. (5 mm) away from inside edges of flanges to prevent it from squeezing into carbon ring chambers.

5. Place top half packing cases in position and replace cap screws (7 and 8, Figure 9-5).

   Note: Turn cap screws (7) on vertical flange until snug. Tighten cap screws (8) on horizontal flange, then tighten cap screws (7) on vertical flange.

Turbine Casing

The turbine casing cover (3 and 64, Figure 9-5) must be lifted to inspect the rotor assembly, nozzle ring (33) or reversing blade assembly (37).

Disassembly

1. On small turbines, remove top half packing cases as outlined in Packing Case Disassembly. Remove only the vertical flange cap screws (7, Figure 9-5) from the packing cases on larger turbines.
2. Remove bolts (4) and dowels (5) from the horizontal casing flange.
3. Carefully lift the casing cover by the eyebolt (60) until it clears the rotor disks (44 and 45).
4. Remove the cover to a safe location. Take care to protect the machined surfaces of the cover.

Assembly

1. Clean all mating sealing surfaces between the bottom half turbine casing, casing cover and packing cases.
2. Apply sealing compounds to the sealing surfaces as shown in Figure 9-8.

   Note: Do not place plastic string sealant near turbine casing bolt holes. A poor seal may result if string sealant enters these holes.

3. Lower the casing cover onto the bottom half casing.
4. Seat the dowel pins (5).
5. Tighten bolts (4) at horizontal casing flange, starting with the bolts located closest to the packing cases.
6. On small turbines, replace packing cases (10) as described in Packing Case Assembly. On larger turbines, tighten cap screws (7) on vertical packing case flange.

Bearing Liners

Locating lugs on each bearing liner (15 and 16, Figure 9-5) engage grooves in the horizontal split of the steam end bearing housing (21 and 53) and exhaust end bearing pedestal (12 and 54). This arrangement retains the liners in the proper position.

Disassembly

1. Remove cooling water piping from bearing caps, if applicable.
2. Remove the dowels (14) and bolts (13, Figure 9-5) from the bearing cap joints.
3. Break the joints by prying the bearing caps away from the bearing housings.
4. Raise the caps approximately 1 in. (25 mm) and pry the top liners (16) (at the locating lugs) from the bearing caps with a screwdriver. This will release the oil rings (49) from the caps.

   Caution: Attempting to remove the bearing caps, without prying out the top bearing liners, can distort the oil rings. Distorted oil rings will not rotate to provide lubrication, thereby resulting in bearing failures.

5. Remove bearing caps and top journal bearing liners.

6. Lift the rotor slightly and remove the bottom bearing liners by rolling them away from the locating lugs. The rotor shaft will rest on the shaft sleeve seals (47) when the bottom liners are removed.

Clearances

Bearing liners used with rotors designated Class 1 and 2 provide a cold diametral clearance of .006 in. (0.15 mm) to .009 in. (0.23 mm). Diametral journal bearing clearances are .0035 to .0055 in. (0.09 to 0.14 mm) when turbine is equipped with a Class 3 rotor (refer to Figure 9-10).

To Check The Bearing Liner Clearances:

1. With the top liners (16, Figure 9-5) removed, place a piece of Plasti-gage® radially on the shaft journals.
2. Place the top bearing liners (16) over the shaft journals.
3. Place the oil rings (49) in the slotted guides in the top bearing liners.
4. Replace the bearing caps. Be sure the top bearing liner locating lugs engage the grooves in the bearing caps.
5. Insert dowels (14) and tighten bolts (13).
6. Remove the bolts (13). Lift the bearing caps approximately 1 in. (25 mm) and pry the top liners (16) from the caps with a screwdriver.
7. Remove bearing caps and top liners and measure the Plasti-gage® to determine the clearance between the shaft journals and top bearing liners.

Adjustments

Bearing liners are not adjustable. They should not be filed, scraped, shimmed, fitted or altered in any way. Worn bearing liners can lead to vibration and other operational problems. Replace worn bearing liners if the clearances exceed the maximum shown in Figure 9-10 by .002 in. (.051 mm). Bearing liners should also be replaced if inspection shows signs of scoring, wiping, cracking, flaking or loose bonding between the babbitt and the steel backing.

Assembly

1. Clean the flanges on the bearing caps and housings.
2. Drain and clean bearing housing reservoirs and refill with clean oil (refer to Table 9-8).
3. Lift the weight of the rotor and roll the bottom bearing liners (15) around the shaft journals and into the bearing housings. Be sure

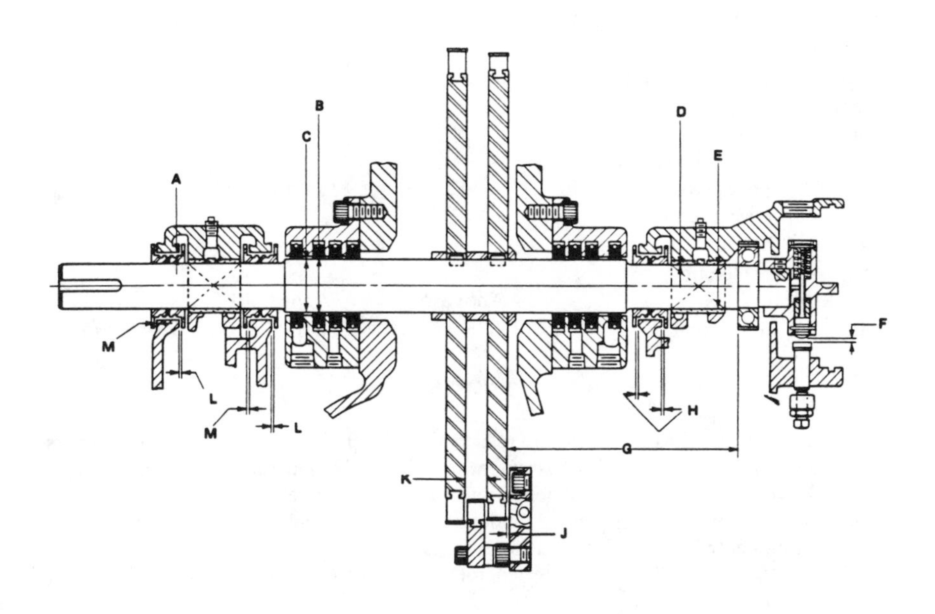

★NOTE:

SEE PAGE 408 FOR ROTOR CLASS DESIGNATION

			INCHES		MILLIMETRES	
			MIN.	MAX.	MIN.	MAX.
	A	Radial Clearance, Shaft Sleeve Seals	.010	.0145	0.25	0.37
	B	Carbon Ring Inside Diameters	SEE TABLE 9-7		SEE TABLE 9-7	
	C	Shaft Diameter	2.2500	2.2505	57.15	57.16
★	D	Diametral Clearance (Class 1 & 2 Rotors)	.006	.009	0.15	0.23
★	D	Diametral Clearance (Class 3 Rotors)	.0035	.0055	0.09	0.14
★	E	Shaft Diameter (Class 1 & 2 Rotors)	1.9320	1.9325	49.07	49.08
★	E	Shaft Diameter (Class 3 Rotors)	1.9350	1.9355	49.15	49.16
	F	Trip Pin/Plunger Clearance	.062		1.59	
	G	Axial Dimension	9.465	9.471	240.41	240.56
	H	Axial Clearance	.054	.064	1.37	1.62
	J	Axial Clearance	.042	.072	1.07	1.83
	K	Axial Dimension	.930	.950	23.62	24.13
	L	Axial Clearance	.068	.098	1.72	2.49
	M	Axial Clearance	.030	.040	0.76	1.01

Figure 9-10. Cold clearance diagram for typical general purpose steam turbine.

Table 9-8
Guidelines for Selecting Lubricating Oils (See Note 1)

Method of Lubrication	Oil Ring Lubrication	Pressure Lubrication A (See Note 2)	Pressure Lubrication B (See Note 3)
Viscosity, Saybolt Universal Seconds (Approx. SUS at 100°F)	300	150	300
Approx. Metric Viscosity (mm^2/s at 40°C)	60	30	60
Viscosity, Saybolt Universal Seconds (Approx. SUS at 210°F)	52	43	52
Approx. Metric Viscosity (mm^2/s at 99°C)	8	5	8
Viscosity at Operating Temperature, SUS	Above 90	Above 90	Above 90
Minimum Flash Point	350°F 175°C	350°F 175°C	350°F 175°C

NOTES

1. These guidelines are not intended to restrict the oil supplier to a definite set of numbers to which he must adhere. It is the intent to indicate the general service requirements and leave the particular recommendations to the oil supplier.
2. Column "A" provides the general guidelines for turbines lubricated by a turbine shaft driven pump or by the driven machine.
3. Column "B" shows the acceptable general guidelines for turbines lubricated from gear oil systems.

the liner locating lugs are firmly seated in the bearing housing locating grooves.

4. Place the top bearing liners (16) over the shaft journals.
5. Place the oil rings (49) in the slotted guides in the top half bearing liners.
6. A thin coat of oil resistant sealant may be applied to the bearing cap flanges, if desired.
7. Replace the bearing caps. Be sure that the top bearing liner locating lugs engage the corresponding locating grooves in the bearing caps.

Caution: Bearing caps must seat firmly on the bearing housings. Do not force the caps down by tightening the bolts. Forcing the caps down will damage the bearing liners.

8. When the locating lugs are properly seated, replace the dowels (14).
9. Tighten all bolts (13).
10. Connect cooling water piping to bearing caps, if applicable.

Water-Cooled Bearings

Provisions for cooling the bearing oil are supplied as standard equipment on oil ring lubricated turbines. The lubricating oil is cooled by water flow through chambers in the bottom halves of the steam end bearing housing and exhaust end bearing pedestal. Water-cooled bearing caps are supplied when additional cooling is required on oil ring lubricated turbines. Water-cooled bearing caps are furnished as standard equipment on many larger turbines when oil ring lubricated.

> Caution: If the turbine is idle during cold weather, the cooling water chambers must be drained to prevent damage from freezing water.

Disassembly

1. Disconnect cooling water piping from the cooling chamber flanges (27 and 56, Figure 9-5) and bearing caps (53 and 54) as applicable.
2. Remove machine bolts (20) from the cooling chamber flanges.
3. Remove the flanges (27 and 56), gaskets (19) and cooling tube assemblies from the bearing housings.
4. Remove the bolts (59) from the bearing cap covers (58), if applicable.

Adjustments

1. During operation, adjust the water flow through the chambers to approximately 2 gpm (7.5 l/min). Cooling water pressure should not exceed 75 psig (5 bar).
2. Annually clean and inspect the cooling water chambers. (See Table 9-6).

Assembly

1. Install new gaskets (19 and 57, Figure 9-5).
2. Replace the flanges (27 and 56) on the cooling chambers.
3. Replace the cooling chamber flange bolts (20) and connect the cooling water piping.

4. Replace the bearing cap covers (58) and tighten bolts (59), if applicable.

Rotor Assembly

The rotor assembly must be removed from the turbine casing before removing or replacing the following (Refer to Figure 9-5):

1. Oil Rings (49)
2. Shaft Sleeve Seals (47)
3. Trip Body (52)
4. Rotor Locating Bearing (50)
5. Nozzle Ring (33)
6. Reversing Blade Assembly (37)
7. Bearing Housings (12, 21, 53 and 54)
8. Packing Cases (10 and 11)

Removal

1. Disconnect the coupling between the turbine and driven machine.
2. Remove the turbine casing cover (3 and 64, Figure 9-5), top half packing cases (10 and 11) and carbon rings (9) as described earlier.
3. Remove the journal bearing liners (15 and 16) as described in Disassembly Section.
4. Disconnect the governor linkage and remove the governor as outlined in the Governor Maintenance Section (page 405).
5. Place a sling between the rotor disks (44 and 45), and slowly lift the rotor approximately 1 in. (25 mm).

   Caution: Keep the rotor level when lifting, to prevent binding it in the casing or damaging machined surfaces.

6. Lift the oil rings (49) from the bearing housings. Move the rings to the side so that they are free of the bearing housing support castings, then lift the rotor assembly out of the turbine casing.

   Caution: Chock the rotor assembly with blocks to prevent it from rolling when removed from the casing. Also protect the rotor journals and carbon ring sealing areas by wrapping them with clean rags or other suitable covering.

Clearances

Refer to the Cold Clearance Diagram, Figure 9-10, for rotor dimensions.

Replacement

1. Lower the rotor assembly to within 1 in. (25 mm) of full replacement in the casing. Carefully guide the rotor while lowering it into the casing to prevent the disks (44 and 45, Figure 9-5) from contacting the reversing blade assembly (37).
2. Position the oil rings (49) so they fall into the openings between the bearing liner supports located in the bottom of the bearing housings.
3. Position the anti-rotation tab on the rotor locating bearing (50) to engage the groove in the steam end bearing housing (21).
4. Slowly lower the rotor into the casing.
5. Check bearing housing alignment.
6. Replace the journal bearing liners and caps.
7. Replace the governor and connect the governor linkage (Refer to Governor Maintenance Section).
8. Replace the carbon rings (9) top half packing cases (10 and 11) and casing cover (3 and 64) as outlined in Carbon Ring Replacement Section, and Assembly Section, earlier.

Exhaust End Bearing Pedestal Replacement

The exhaust end bearing pedestal (12 and 54) is attached to the turbine casing by four socket head cap screws (23) and two combining studs (17). The combining studs are threaded into the bottom half turbine casing and pinned to the pedestal.

Two dowel pins (25), pressed into the exhaust end of the turbine casing, position and hold the pedestal in correct horizontal and vertical parallel alignment with the steam end bearing housing (21 and 53). Spacers (24), located between the pedestal and turbine casing, are used to adjust and maintain proper angular bearing alignment.

Disassembly

1. Remove the rotor assembly as outlined in Removal Section, earlier.
2. Remove the hold-down bolts and dowel pins from the pedestal support feet.
3. Support the weight of the turbine exhaust end casing with a jack, wooden blocks, or other adequate means.
4. Remove the tapered pins (18, Figure 9-5) from the combining studs (17).
5. Loosen the four cap screws (23) three or four turns and pry the pedestal away from the casing until the spacers (24) are free to move.

6. Remove the cap screws and spacers. Mark each spacer so it can be returned to the location from which it was removed.

   Caution: If spacers (24, Figure 9-5) are not returned to their original locations, bearing misalignment may occur. This will cause uneven bearing wear or possible failure.

7. Slide the pedestal off the combining studs and dowel pins (25).

Replacement

Note: Bearing anti-rotation locating grooves must be provided at the horizontal split on replacement bearing pedestals. These grooves may be made by hand filing. Hold the liner so the tab is on the upstream end for clockwise rotation (looking in direction of steam flow) and on the downstream end for counterclockwise rotation.

1. Slide the pedestal onto the combining studs (17, Figure 9-5) and dowel pins (25).
2. Replace the spacers (24) and cap screws (23). Spacers must be returned to the same locations from which they were removed.
3. Tighten the cap screws (23) and insert the taper pins (18) in the pedestal and combining studs (17).
4. Replace the rotor assembly as outlined in Replacement Section.
5. Replace the bottom half journal bearing liners (15). Check the bearing alignment and adjust as necessary.

Steam End Bearing Housing Replacement

The steam end bearing housing (21 and 53, Figure 9-5) is attached to the turbine casing by four socket head cap screws (23). Two dowel pins (24), pressed into the steam end of the turbine casing, maintain the bearing housing in correct horizontal and vertical parallel alignment with the exhaust end bearing pedestal (12 and 54). Spacers (24), located between the housing and turbine casing, are used to correct any angular misalignment and also to adjust the axial position of the turbine rotor in the casing.

Disassembly

1. Remove the rotor assembly as outlined in Removal Section.
2. Remove the hold-down bolts and dowel pins from the steam end bearing support (29, Figure 9-5).

3. Place a jack, wooden blocks or other adequate support under the steam end of the turbine casing and steam chest.
4. Remove the bolts (30) securing the support (29) to the bearing housing (21 and 53).
5. Loosen the socket head cap screws (23) and pry the bearing housing away from the turbine casing until the spacers (24) are free to move.
6. Remove the cap screws and spacers. Mark the spacers so they can be returned to their original locations.

   Caution: If the spacers (24, Figure 9-5) are not replaced in their original locations, bearing misalignment may result. This can cause uneven bearing wear or possible bearing failure. Nozzle ring to rotating blade clearance may also be affected, resulting in poor performance or mechanical failure.

7. Pull the bearing housing off the dowel pins (25).

Replacement

Note: Bearing liner anti-rotation grooves must be provided at the horizontal split on replacement bearing housings.

1. Push the bearing housing onto the dowel pins (25, Figure 9-5) which are pressed into the steam end turbine casing.
2. Replace the cap screws (23) and spacers (24). Ensure that the spacers are returned to the same location from which they were removed.
3. Bolt the support (29) to the bearing housing.
4. Replace the rotor assembly as outlined in Replacement Section.
5. Replace the bottom half journal bearing liners (15), check the bearing alignment and adjust as necessary.

Exhaust End Bearing Pedestal and Steam End Bearing Housing Alignment

To obtain the correct bearing and rotor shaft journal contact, the bores of the exhaust end bearing pedestal and the steam end bearing housing must be in parallel and angular alignment. Dowel pins (25, Figure 9-5), pressed into the turbine casing, position the pedestal and bearing housing in horizontal and vertical parallel alignment. Spacers (24), located between the pedestal and turbine casing and between the steam end bearing housing and turbine casing, are used to correct any angular misalignment and to position the turbine rotor axially in the turbine casing.

To Check The Alignment

1. Remove the rotor assembly from the turbine casing and clean the shaft journals.
2. Apply a light coating of soft blue to both shaft journals.
3. Install bottom half journal bearing liners (15, Figure 9-5) in the bearing pedestal and steam end bearing housing. Be sure the liners are properly seated.
4. Lower the rotor assembly until the full weight of the rotor is supported by the journal bearing liners.
5. Rotate the rotor assembly one quarter turn in each direction.

   Note: Ensure the rotor shaft is seated on the bottom of the bearing liners and not moving sideways or upward while being rotated.

6. Remove the rotor assembly from the turbine casing and check the bearing contact.

   Note: The exhaust end bearing pedestal and steam end bearing housing are considered to be in alignment when bearing contact with the shaft journals is no less than 85 percent along the bottom of the bearing liners and when the contact along the sides of the liners is parallel with the bearing bore and equal on each side (See Figure 9-11).

To Correct Any Misalignment

1. Place shim stock, in increments of .002 in. (0.05 mm), behind the spacers (24, Figure 9-5) to correct the misalignment.
2. Recheck the bearing contact and continue to add shims to achieve proper alignment.
3. After the correct bearing contact is obtained, the shims must be removed from each spacer, and the thickness of the opposite spacer altered accordingly. (Surface grinding is the preferred method.)

   Example: It is necessary to add .004 in. (0.10 mm) shim thickness to the two bottom spacers, to achieve correct alignment; .004 in. (0.10 mm) must be ground from the two top spacers to maintain the alignment after the shims are removed.

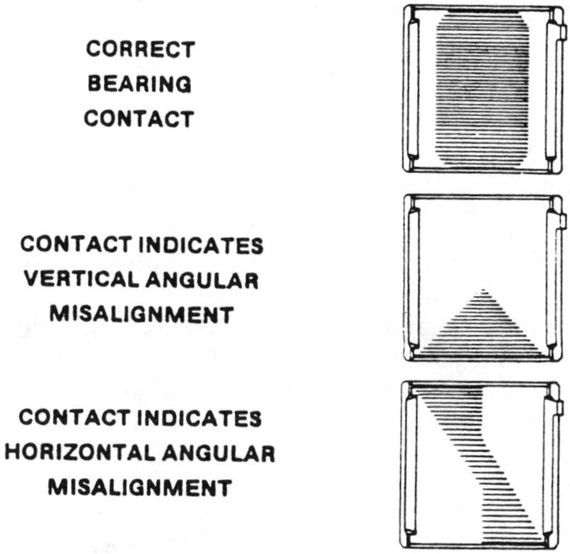

CORRECT BEARING CONTACT

CONTACT INDICATES VERTICAL ANGULAR MISALIGNMENT

CONTACT INDICATES HORIZONTAL ANGULAR MISALIGNMENT

Figure 9-11. Journal bearing and rotor shaft contact.

4. Recheck the bearing contact, after the ground spacers have been installed.

Rotor Locating Bearing

The rotor locating bearing (50, Figure 9-5) maintains the correct axial position of the rotor assembly to the nozzle ring and steam end bearing casing. The bearing is mounted on the rotor shaft with the shielded side of the bearing toward the trip body. A beveled retainer ring (51) holds the bearing in place on the rotor shaft. The outer bearing race fits into a groove in the steam end bearing housing (21 and 53), and is prevented from rotating by an anti-rotation tab which is permanently attached to the outer race. The anti-rotation tab fits into a slot at the horizontal split of the bearing housing.

Disassembly

1. With the rotor removed from the turbine casing; disassemble and remove the trip body as described in Trip Body Removal Section, later.

2. Remove the retainer ring (51, Figure 9-5) with ring expanding pliers.
3. Remove the locating bearing (50) with a bearing puller.

Clearances

To check the axial bearing clearance, an axial rotor float check must be made:

1. Mount a dial indicator perpendicular to a vertical shaft face (such as the coupling hub or a rotor disk).
2. Shift the rotor as far as possible in both axial directions while observing the dial indicator. The normal axial rotor float is from .010 in. (0.25 mm) to .018 in. (0.46 mm). In no case should the total indicator reading exceed .025 in. (0.64 mm).

Adjustments

The rotor locating bearing is not adjustable. Worn bearings must be replaced when the axial rotor float reaches .025 in. (0.64 mm).

Assembly

1. Install the bearing on the shaft by using a sleeve type bearing driver which contacts the inner bearing race. Seat the bearing solidly against the machined shoulder on the shaft (43, Figure 9-5).

   Note: Be sure the shielded side of the bearing is positioned toward the trip body (52).

2. Replace the retainer ring (51). Seat the ring firmly in the groove on the rotor shaft, with the beveled edge of the ring positioned toward the trip body.
3. Replace the trip body as outlined in Trip Body Replacement Section on page 400.
4. Flush the locating bearing with oil before replacing the bearing cap.

Nozzle Ring and Reversing Blade Assembly

The relative locations of nozzle rings and reversing blade assembly are shown in Figure 9-11. The nozzle ring (33, Figure 9-5) directs the steam flow from the steam chest to the blades of the first rotor disk (44). Steam exits the blades of the first disk and passes through the reversing blade assembly (37) which directs it into the blades on the second rotor disk

(45). The reversing blade assembly is positioned between the two rotor disks and is bolted to the nozzle ring. The reversing blade assembly is positioned by spacers and requires no further adjustment.

Disassembly

1. Remove the rotor assembly as outlined in Removal Section.
2. Remove the bolts (38, Figure 9-5), lockwashers (39), and spacers (40), and lift out the reversing blade assembly (37). Mark each spacer (40) so that it may be returned to its original location.
3. Remove the nozzle ring bolts (34 and 36), lockwashers (35) and nozzle ring (33) from the casing.

Clearances

The clearance between the nozzle ring (33, Figure 9-5) and the shroud on the first rotor disk (44) must be checked whenever the rotor assembly, nozzle ring or reversing blade assembly is replaced. This clearance is a minimum of .042 in. (1.07 mm) and a maximum of .072 in. (1.83 mm). The clearance can be measured with a feeler gauge.

Adjustments

Inspect the nozzle ring and reversing blade assembly annually. Clean scale or boiler compound deposits as necessary. Replace eroded parts.

Assembly and Replacement

1. Clean the casing and nozzle ring sealing surfaces.
2. Apply a thin coat of paste type sealer and plastic string compound to the nozzle ring sealing surface on the steam end turbine casing (Reference Figure 9-8).
3. Apply anti-galling compound to the threads of the nozzle ring bolts (34 and 36, Figure 9-5).
4. Bolt the nozzle ring to the turbine casing. Be sure that lockwashers (35) are used with all bolts.
5. Place lockwashers (39) on the reversing blade assembly bolts (38) and apply anti-galling compound to the bolt threads. Put the bolts through the holes in the reversing blade assembly and slip the spacers (40) over the bolts.
6. Position the reversing blade assembly (37) in the turbine casing and bolt it to the nozzle ring (33).

   Note: Be sure the reversing blade assembly is installed in the same location from which it was removed so that it covers all the nozzles and overlaps the end nozzles by a minimum of five blades. The reversing blade trailing edges must point in the same direction as the nozzles.

7. Return the rotor assembly to the turbine casing.

Shaft Sleeve Seals

Three seals (47, Figure 9-5) mounted on the rotor shaft, prevent oil leakage from the steam end bearing housing (21 and 53) and exhaust end bearing pedestal (12 and 54). The seals also restrict the entry of steam, dust and dirt into bearing housings.

Disassembly

1. Remove the rotor assembly from the turbine casing as described in the Removal Section.
2. Remove the drive coupling from the rotor shaft.
3. Remove the trip body (52, Figure 9-5) and rotor locating bearing (50) from the rotor shaft.
4. Remove the two set screws (48) from each sleeve seal (47) and slide the sleeve off the rotor shaft (43).

Replacement And Adjustment

1. Place sleeve seals (47, Figure 9-5) on the rotor shaft (43).
2. Replace the rotor locating bearing (50) and trip body as outlined in the Assembly Section.
3. Install the drive coupling on the rotor shaft.
4. Return the rotor assembly to the turbine casing.
5. Position the shaft sleeve seals to provide the axial clearances shown in Figure 9-10.
6. Tighten set screws (48) to lock the sleeves (47) in position on the shaft.

   Note: The tops of the set screws must be below the outside diameter of the sleeve seals.

7. Lock the set screws in place by staking the sleeve seals.
8. Replace the casing cover (3) and bearing caps.

Clearances

1. Disassembly is not required to check axial clearances (H, L & M, Figure 9-10).
2. Remove bearing caps to check radial clearances (A, Figure 9-10) with feeler gauges.

Overspeed Trip System

Warning: Under no circumstances should the trip valve be blocked or held open to render the trip system inoperative. Overriding the trip system, and allowing the turbine to exceed the rated (nameplate) trip speed, may result in fatal injury to personnel and extensive turbine damage. Always close all isolating valves and open drains to depressurize the turbine casing and steam chest before performing maintenance on the overspeed trip system.

The overspeed trip pin assembly is contained in the trip body mounted on the turbine rotor shaft. When the turbine speed increases above the rated operating speed, centrifugal force exerted on the trip pin (1, Figure 9-12) increases. When the centrifugal force overcomes the force of the trip pin spring (2), the weighted end of the pin protrudes from the trip body. The pin strikes the plunger assembly (7), forcing it against the adjustable jack screw (8) in the bottom of the hand trip lever. The lever pivots on a shoulder stud; causing the top of the lever to move away from the resetting lever. This movement disengages the latch from the resetting lever knife edge and allows the closing spring to pull the trip valve closed. This stops the steam flow through the turbine.

Disassembly, Overspeed Trip Mechanism

Note: To check the trip pin for cracks, it is recommended that either the zyglo or dye check method be used. The "U" lock staples should also be examined for bends or cracks. The overspeed trip pin assembly can be checked by monitoring the frequency of overspeed trips. Check the assembly at least every 30 overspeed trips and at two year periods.

1. Remove the steam end bearing cap as outlined earlier.
2. Remove the "U" lock staple (3, Figure 9-12), surrounding the adjusting nut (4), by prying it out of the trip body.
3. Remove the adjusting nut, trip spring (2) and washers (5), if provided.

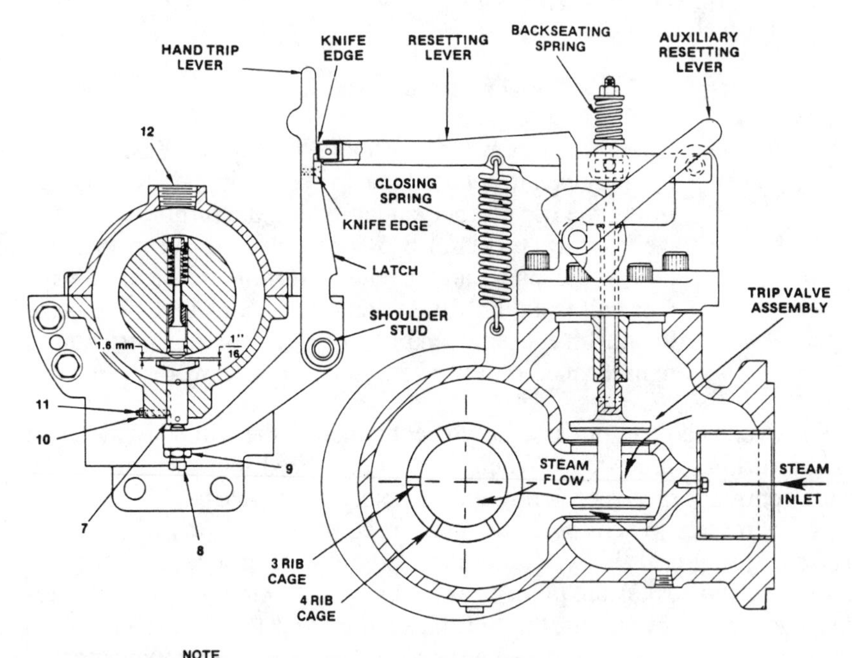

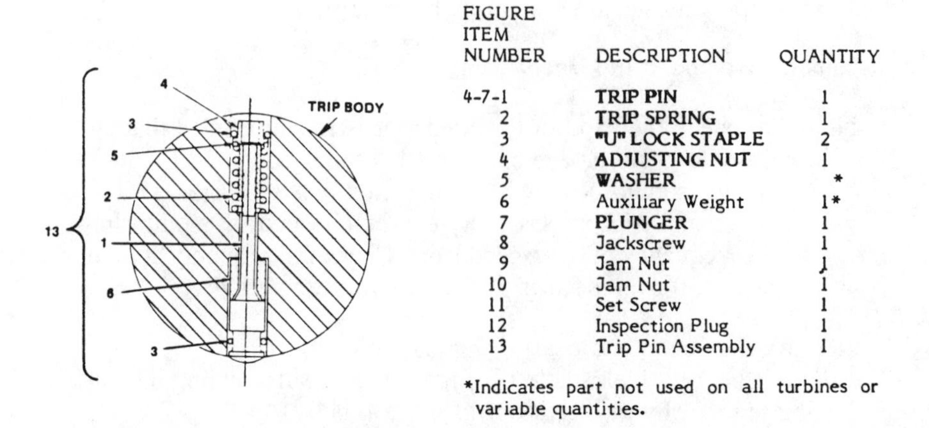

FIGURE ITEM NUMBER	DESCRIPTION	QUANTITY
4-7-1	**TRIP PIN**	1
2	**TRIP SPRING**	1
3	**"U" LOCK STAPLE**	2
4	**ADJUSTING NUT**	1
5	**WASHER**	*
6	Auxiliary Weight	1*
7	**PLUNGER**	1
8	Jackscrew	1
9	Jam Nut	1
10	Jam Nut	1
11	Set Screw	1
12	Inspection Plug	1
13	Trip Pin Assembly	1

*Indicates part not used on all turbines or variable quantities.

Figure 9-12. Overspeed trip system.

Note: Record the number of turns required to remove the adjusting nut (4) so it can be returned to its original setting during assembly.

4. Rotate the rotor shaft 180° and remove the "U" lock staple surrounding the weighted end of the trip pin (1).
5. Remove the trip pin from the trip body. (Remove the auxiliary weight (6), if furnished.)

Trip Body Removal

1. Remove the rotor assembly from the turbine casing.
2. Remove the set screw from the trip body (52, Figure 9-5).
3. Heat the trip body evenly with a torch. Apply heat as rapidly as possible, then pull the trip body from the rotor shaft.

Caution: Care must be exercised to prevent heating the rotor locating bearing and the rotor shaft when heating the trip body. Protect both by wrapping in asbestos cloth.

Plunger Assembly Replacement

The plunger assembly (7, Figure 9-12) can be removed by lifting it out of the bearing housing while the rotor assembly is out of the turbine casing. (Except turbines equipped with bearing cases having a flanged governor fit such as PG, UG and O Governors.) Turbines equipped with PG, UG and O Governors employ an "Umbrella" plunger. A retainer ring, washer and spring arrangement must be removed from the bottom of the "Umbrella" plunger before lifting it from the bearing housing.

If necessary, worn plunger assemblies can be replaced without removing the rotor assembly. (See applicable details for "Umbrella" plunger removal, later.) To remove the plunger assembly shown in Figure 9-12:

1. Remove the steam end bearing cap.
2. Remove the governor and adapter piece from the steam end bearing housing.
3. Loosen the jam nut (10, Figure 9-12) and remove set screw (11) from the side of the bearing housing.
4. Remove the set screw from the plunger assembly and separate the two halves of the plunger to remove them from the bearing housing.
5. Assemble in reverse order.

Note: When replacing the plunger assembly in this manner, both parts of the new plunger assembly must be installed.

Trip Body Clearance

1. The trip body to the shaft diametral interference fit is .001–.002 in. (.025–.051 mm).

   Caution: Consult the manufacturer if interference is less than .001 in. (.025 mm).

Trip Body Replacement

1. Heat the trip body in hot oil or an oven. Do not exceed 500°F (260°C).
2. Place the heated trip body on the rotor shaft and align the set screw holes in the trip body and shaft.
3. Tighten the set screw to ensure proper positioning on the shaft, then back the set screw out of the body one or two turns.
4. Tighten the set screw when the trip body has cooled to ambient temperature.
5. Check the trip body runout. Runout should not exceed .003 in. (0.07 mm) on the outboard end of the trip body.
6. Gradually overspeed the turbine by overcoming the governor (See Governor Operation, Chapter 10).
7. Check that the plunger assembly (7) is properly positioned in the bearing housing, and return the rotor to the turbine casing.

Assembly, Overspeed Trip Mechanism

1. If furnished, place the auxiliary weight (6, Figure 9-12) on the trip pin (1).
2. Insert trip pin (1) into the trip body. Position the weighted end of the pin on the opposite side of the trip body set screw.
3. Press the "U" lock staple (3) into the trip body to secure the weighted end of the trip pin. Be sure the staple is fully seated in the circular groove in the trip body.
4. Place the trip spring (2) in the trip body. (Install washers (5), if furnished.)
5. Return the adjusting nut (4) to its original setting, by tightening the nut the same number of turns recorded during disassembly.
6. Press the "U" lock staple (3) into the trip body to lock the adjusting nut (4). Be sure the staple is fully seated in the circular groove in the trip body.

   Note: The overspeed trip should be tested after performing any maintenance on the trip system.

Adjusting The Trip Pin And Plunger Clearance

1. Remove the inspection plug (12, Figure 9-12) from the steam end bearing cap:
2. Rotate the rotor shaft, by hand, until the adjusting nut (4) can be observed through the inspection hole. This will position the weighted end of the trip pin (1) directly above the plunger assembly (7).
3. Latch the resetting lever and loosen the jam nut (9) on the trip lever jackscrew (8).
4. Push the plunger assembly (7) upward and into the bearing housing, until it is in solid contact with the trip pin.
5. Adjust the jackscrew to obtain 1/16 in. (1.6 mm) clearance between the base of the plunger (7) and the jackscrew (8).
6. Tighten the jam nut (9) and recheck the clearance.

Caution: The jam nut (9, Figure 9-12) must be locked, at all times, to prevent the jackscrew from vibrating loose during operation. A loose jackscrew can render the trip system inoperative.

Adjusting The Turbine Trip Speed

1. Remove the inspection plug (12, Figure 9-12) from the steam end bearing cap.
2. Rotate the rotor shaft, by hand, until the adjusting nut (4, Figure 9-12) can be viewed through the inspection hole.
3. Latch the resetting lever (11, Figure 9-4).
4. Place a non-ferrous drift pin on the adjusting nut and strike the drift pin sharply to ensure that the trip pin (1, Figure 9-12), trip valve and trip linkage function properly.
5. Latch the resetting lever and start the turbine. Closely monitor the turbine speed during operation.
6. Gradually overspeed the turbine by overcoming the governor (See Governor Operation, Chapter 11).

   Caution: Do not allow turbine to exceed 2 percent above the rated (nameplate) trip speed.

7. If the overspeed trip does not function within 2 percent of the rated trip speed, manually trip the turbine, by striking the top of the hand trip lever. Then close the steam inlet shutoff valve.
8. When the rotor shaft stops rotating, turn the shaft, by hand, until the adjusting nut is visible through the bearing cap inspection hole.

9. Partially pry the "U" lock staple away from the trip body until the adjusting nut is free to turn.
10. Turn the adjusting nut to change the trip speed. Turning the nut counterclockwise will decrease the trip speed. Turning the nut clockwise will increase the trip speed.
11. Push the "U" lock staple into the trip body, and check that the trip pin (1, Figure 9-12) moves freely. (Be sure staple is firmly seated.)
12. Start the turbine and check the trip speed. Continue to make trip adjustments until the turbine trips at the rated (nameplate) trip speed.

Disassembly, Trip Valve

1. Place the trip valve in the tripped position and disconnect the closing spring (14 and 49, Figure 9-4) from the resetting lever (11).
2. Remove the cap screws (4) from the valve cover (3) and lift the trip valve assembly (17) and cover from the steam chest body (1).
3. Remove nut (15), spring (41), bushing (43) and spring seats (40 and 42) from the valve stem.
4. Turn the valve stem out of connection (46) and remove the valve assembly (17) from the cover (3).

Guide Bushing Replacement

1. Disassemble the trip valve as described under "Disassembly, Trip Valve" Section.
2. Drive the bushings (16, Figure 9-4) out of the valve cover (3) with a non-ferrous drift pin.
3. Clean and de-burr the valve cover.
4. Press new bushings into the valve cover and lock them in place by staking the cover.
5. Assemble the valve as outlined under "Assembly, Trip Valve" Section.

Assembly, Trip Valve

1. Clean the sealing surfaces on the valve cover flange and steam chest body.
2. Insert the valve stem into the lower guide bushing (16, Figure 9-4) and push the valve stem through the valve cover.
3. Turn the valve stem into and through connection (46) and replace spring seats (40 and 42), bushing (43), spring (41) and locknut (15).

4. Apply a combination of paste and plastic string sealants to the sealing surfaces of the steam chest valve cover flange.
5. Return the valve assembly and cover to the steam chest body and tighten the cap screws (4).
6. Backseat the trip valve per Backseating the Trip Valve Section, and connect the closing spring (14 and 49) to the resetting lever (11).

Backseating The Trip Valve (Refer to Figure 9-13)

1. Disassemble and inspect the trip valve and linkage to ensure cleanliness of all parts. Replace worn linkage pins, guide bushings, valve stem, knife edge, latch, etc.
2. Reassemble the trip valve and linkage per Assembly, Trip Valve Section.
3. Disconnect closing spring (1) from resetting lever (2).
4. Remove locknut (3) from trip valve stem (4).

   Note: Firmly grasp spring (10) to prevent rapid decompression while removing locknut (3).

5. Raise connection (5) to backseat the valve (12) against the lower guide bushing (9) by prying against the bottom of connection (5) and the valve cover (6) with a long screwdriver as shown in Figure 9-13.
6. Slightly release the pressure on the screwdriver and turn valve stem (4) in or out of connection (5) to provide .12 in. (3 mm) overlap between the bottom of the resetting lever (2) knife edge and the top of the hand trip lever (8) latch. Turning the valve stem (4) clockwise (rotation viewed from top of trip valve) decreases the overlap; counterclockwise increases.

   Note: Turning the valve stem in small increments will have great effect on the overlap adjustment. Care must be taken to prevent over adjusting.

7. Replace and fully tighten locknut (3) until the upper spring seat (11) is firmly seated against bushing (13). Prevent the valve stem from turning by placing a wrench on the valve stem flats (4) located below the connection (5).
8. Raise the resetting lever (2) until the valve (12) backseats against bushing (9) and check that the bottom of the resetting lever (2) knife edge is still .12 in. (3 mm) below the top of the hand trip lever (8) latch.
9. Latch the resetting lever and verify that spring (10) compresses.

Warning: If spring (10) does not compress, readjustment is required. If backseating occurs with knife edge more than .12 inches below latch, the "play" in connection (5) will be exceeded. This may cause the lever (2) to bend and result in excessive forces holding the knife edges, so that the trip system is rendered inoperative or damage to the trip valve stem pin occurs.

10. Reconnect closing spring and check trip valve operation.

Trip System Linkage

Note: The backseating adjustment has a direct effect on steam leakage along the trip valve stem. Check trip valve operation after every adjustment.

1. Frequently inspect the trip system linkage for cleanliness and freedom of movement.
2. Replace pin (6, Figure 9-4), bushing (7), blocks (45) or shoulder stud (33) if the linkage develops excessive play.
3. Lubricate the linkage pins, shoulder stud and auxiliary resetting lever (10) with a high temperature water resistant silicone grease.

Trip System Clearances (Refer To Figure 9-4)

1. With the resetting lever latched, maintain 1/16 in. (1.6 mm) clearance between the weighted end of the trip pin (1, Figure 9-12) and plunger assembly (7, Figure 9-12).
2. The diametral clearance between the valve cover bushings (16) and the trip valve stem should be .008 to .010 in. (0.20 to 0.25 mm).
3. To ensure positive backseating, adjust the trip valve to provide .12 in. (3 mm) overlap between the resetting lever knife edge (12) and the hand trip lever latch (31).
4. The resetting lever knife edge (12) and latch knife edge (31) must overlap approximately 1/8 in. (3 mm) when the resetting lever (11) is latched.

Note: Latch knife edge (31) can be rotated in 90° increments to provide a new latching surface for the resetting lever knife edge (12). The resetting lever knife edge (12) can be rotated 180° to provide a new mating surface for the latch knife edge (31). Replace the knife edge and latch when adjustment can no longer be made to compensate for worn latching surfaces.

Caution: The clearance between the resetting lever (11) and the auxiliary resetting lever cam (10) should be 1/16 to 1/8 in. (1.5 to 3 mm) with the trip valve in the closed position. This clearance will prevent the cam shaft from bending and hampering the trip operation.

Governor Valve

Warning: Close all isolating valves and open drains to depressurize turbine casing and steam chest before performing maintenance on the governor valve or its linkage.

The governor valve (18, Figure 9-4), located in the steam chest body (1), regulates the steam flow through the turbine. The valve is positioned through mechanical linkage, by the speed governor.

Governor Valve Disassembly (Refer to Figure 9-4)

1. Remove the linkage connecting the governor valve to the governor.
2. Remove bolts (26, Figure 9-4) from the valve cover (21, 35) and pull the cover and valve (18) away from the steam chest body (1).
3. Remove the valve stem connection (36) and jam nut (37) from the valve stem (20) and remove the stem from the cover assembly.
4. The valve seat (27) has a shrink fit in the steam chest. Welded blocks in the steam chest prevent valve seat movement. These blocks (29) must be removed by chipping or grinding before removing the valve seat.
5. Chill the valve seat (27), by packing with dry ice (CO_2), and pull the seat from the steam chest with a puller.
6. Remove the valve seat bushing (28) by driving it from the valve seat with a nonferrous rod. Replace the bushing by pressing it into the valve seat. Stake the valve seat to lock the bushing in place.

Clearances

1. The governor valve must move freely at all times. A smooth sliding fit is necessary between the valve stem (20, Figure 9-4) and the packing (23, 34), and between the stem and guide bushings (22, 28).
2. The governor lever and linkage should also be smooth sliding fits. The governor valve travel should be set in accordance with the value shown on Page ii. The governor valve has a maximum travel of 1/2 in. (13 mm).

Adjustments

1. Inspect the governor linkage, valve stem and bushings for loose fittings or excessive play before attempting any adjustments. Replace as necessary.
2. With the governor valve, governor servo motor or actuator, and linkage fully assembled:
   a. Turn the valve stem (20, Figure 9-4) from the connection (36) until the valve is fully seated.
   b. Adjust the jam nut (37) so the distance between the jam nut and connection (36) is equal to the valve travel dimension shown on Page
   c. Screw the valve stem into the connection until the jam nut contacts the face of the connection.
   d. Lock the jam nut by tightening it against the connection.

Inspect the governor valve stem (20) and guide bushings (22, 28) for wear and replace as necessary.

Remove the packing follower (25, Figure 9-4) and replace the valve stem packing (23, 34) if excessive steam leakage is evident. (See details A and B, Figure 9-14.)

> Note: Do not overtighten the packing follower (25, Figure 9-4). The governor valve stem can bind in the valve cover and result in erratic speed control.

Lubricate the governor linkage pins with high temperature, water resistant silicone grease.

Assembly (Refer to Figure 9-4)

1. Chill the valve seat (27) with dry ice (CO_2) and press it into the steam chest body (1).

   > Note: The number of ribs (either 3 or 4) found on the valve cage body depends on the size of the steam chest and governor valve. Position these ribs to allow the inlet steam flow between them (Figure 9-12). Do not confuse these ribs with guide bushing ribs. Do not weld blocks (29) to the governor valve seat (27). This may distort seating surfaces.

2. Weld blocks to the steam chest (180° apart) to secure the valve seat.
3. Place the governor valve stem (20) in the valve cover (21, 35).
4. Replace connection (36) and jam nut (37) on the valve stem (20).

5. Clean the joint between the valve cover (21, 35) and the steam chest body (1). Apply a combination of paste and plastic string sealing compounds on the sealing surfaces.
6. Replace the cover and tighten bolts (26).
7. Connect the governor valve linkage and adjust the valve travel.

Overload Hand Nozzle Valve

Note: Close all isolating valves and open drains to depressurize turbine casing and steam chest before performing maintenance on the hand valve.

Disassembly (Refer to Figure 9-6)

The hand valve assembly is bolted to the bottom half steam and turbine casing. Remove cap screws (8) to remove hand valve from casing.

Adjustments

1. Keep the valve stem packing (4) tight by adjusting the packing follower (2).
2. Replace the packing when follower adjustment no longer prevents steam leakage along the valve stem.

Assembly

1. Apply a thin coat of paste type sealing compound on the valve body flange.
2. Coat the cap screw threads (8) with an anti-galling compound.
3. Bolt the valve body to the turbine casing and tighten.

Steam Turbine Lubrication

Proper lubrication is a primary factor in achieving maximum trouble-free operation. Only the best grades of oil should be used for turbine lubrication. Using the best oils will help eliminate costly downtime due to bearing failures and other lubrication related problems.

Basic Oil Requirements

Turbine manufacturers sometimes do not recommend specific brands of oil; they ask equipment owners to consult reliable oil suppliers regarding their lubrication requirements. The oil should be a premium quality mineral lubricant which will readily separate from water and have minimum tendency to emulsify or foam when agitated. It should have high rust and oxidation resistance and minimum sludge, lacquer, varnish or resin forming tendencies. In addition to these requirements, Tables 9-5 and 9-8 contain other necessary information to aid in selecting the proper lubricating oil for your turbine.

Turbines driving through speed reduction or increasing gears are often pressure lubricated by the gear lubrication system. Refer to the gear manufacturer's instructions for gear oil requirements.

Care of Oil

Lubricating oil should be maintained in first class condition by preventing contamination from moisture, dust, dirt or other impurities. An oil maintenance analysis program is recommended for determining the frequency of oil changes. Consult your oil supplier for assistance in establishing a program that will meet your specific lubrication maintenance requirements. Refer also to Chapter 12.

Methods of Lubrication

Most steam turbines are furnished with either an oil ring lubrication system or a pressure lubrication system. Pressure lubricated turbines

Table 9-9
Rotor Class Designations and Construction Features for Single-Stage Turbines

	CLASS 1	CLASS 2	CLASS 3
Service	LOW PRESSURE, MEDIUM	MEDIUM PRESSURE MEDIUM	MEDIUM PRESSURE HIGH
Component	TEMPERATURE	TEMPERATURE	TEMPERATURE
Shaft	Medium carbon steel	Chrome-molybdenum steel	Chrome-molybdenum steel
Disk	Low alloy high strength steel	Medium alloy high strength steel	Medium alloy high strength steel
Trip Device	Standard	Standard	Special

equipped with a Class 1 or Class 2 rotor (rotor class designated on Table 9-9), are furnished with oil rings. The oil rings provide bearing lubrication during turbine startup and shutdown. This arrangement employs a standpipe, installed in the bearing housing oil drain connection, to maintain the proper oil level for oil ring operation.

Turbines equipped with a Class 3 rotor are pressure lubricated and use an auxiliary oil pump for bearing lubrication during turbine startup and shutdown. Standpipes and cooling tubes are also used with this arrangement but oil rings are not.

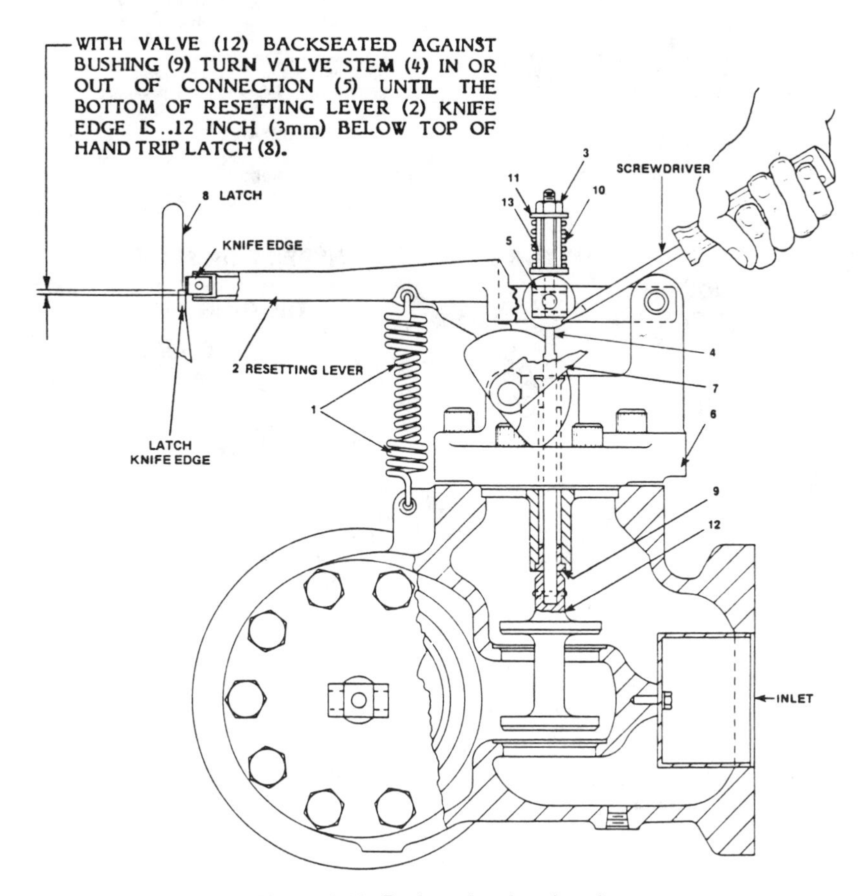

Figure 9-13. Backseating the trip valve.

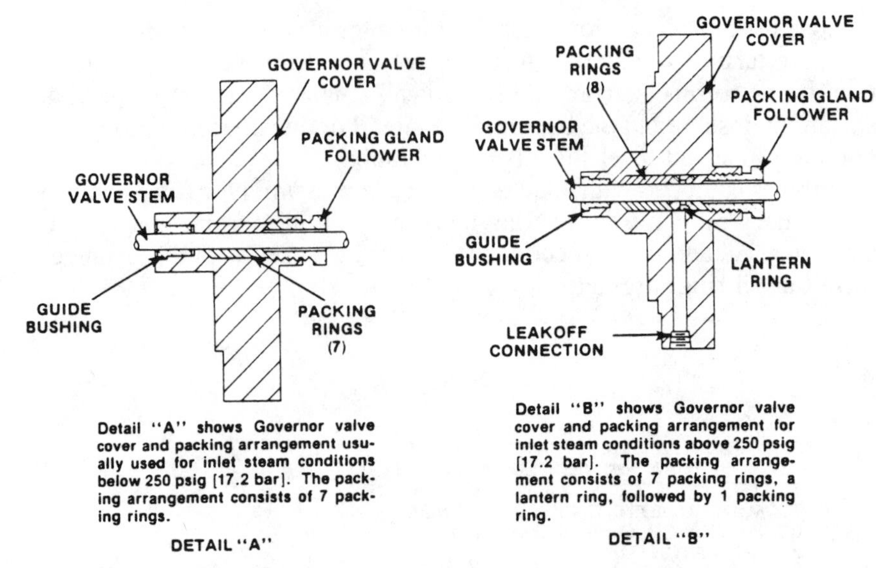

Figure 9-14. Governor valve packing arrangement.

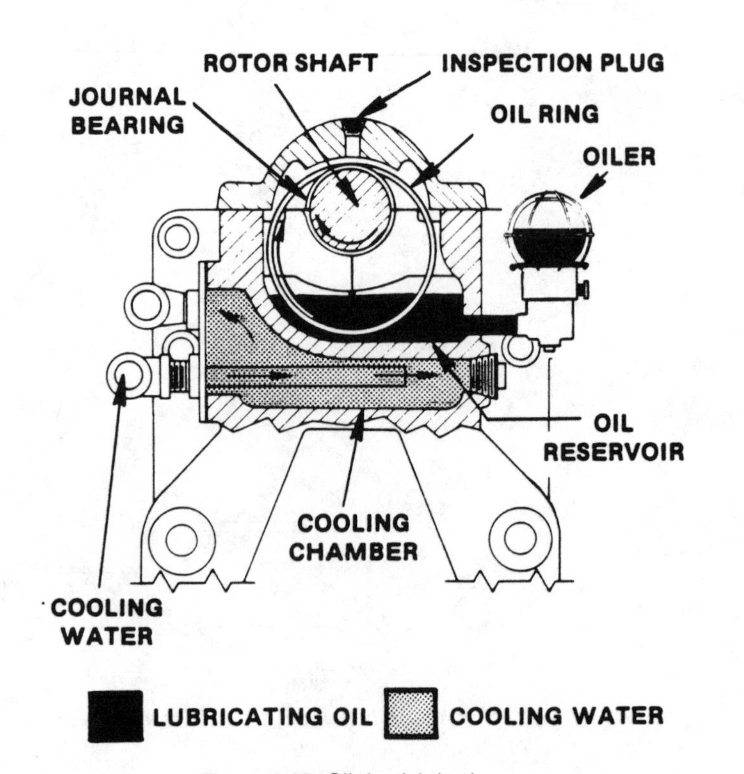

Figure 9-15. Oil ring lubrication.

Oil Ring Lubrication (Refer to Figure 9-15)

The oil ring lubrication system is a simple system which employs oil rings to deliver oil to the turbine bearings. The rings, approximately twice the diameter of the shaft journals, are rotated by the journals and carry oil from the bearing housing reservoirs to the top half bearing liners. The oil flows axially along the top liners through grooves. Rotating journals carry the oil to the clearance between the bearing liners and the shaft journals. Oil drains from the ends of each bearing liner and returns to the bearing housing reservoirs to be cooled and recycled. Oil discharged from the steam end journal bearing floods the rotor locating bearing before draining into the bearing housing reservoir.

A cooling water tube arrangement, as shown in Figure 9-15, is used to cool the oil in the bearing housing reservoirs of oil ring lubricated turbines. Turbines having additional cooling requirements are furnished with water cooled bearing caps (Figure 9-16).

Shielded glass oilers (Figure 9-17), installed on the sides of the bearing housings, maintain a constant reservoir oil level when the turbine is oil ring lubricated. To assure the oil level is correct, the level adjuster crossarms must be adjusted to provide the 15/16 in. (23.8 mm) dimension (between the bottom of the lower crossarm and the bottom of the level adjuster base) shown on Figure 9-17. Lock the crossarms in place by tight-

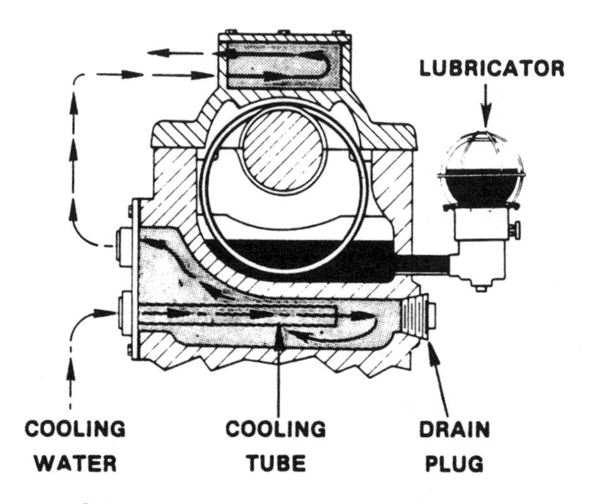

Figure 9-16. Schematic view, water cooled bearing housing and cap.

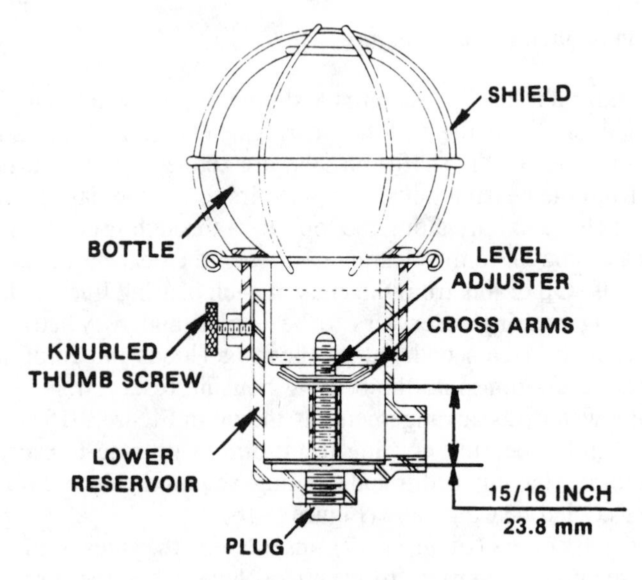

Figure 9-17. Glass oiler.

ening the upper crossarm down against the lower crossarm. Oil must be visible in the oiler bottles at all times during operation. Empty bottles indicate possible low reservoir oil levels. Care must also be taken to avoid overfilling the bearing housing reservoirs. High oil levels will restrict oil ring rotation and may also cause inadequate bearing lubrication.

Note: Correct bearing casing oil is maintained by filling reservoir to 1/2 in. above inside diameter of bottom half of oil ring. This level is maintained by the crossarm adjustment.

To Change Oil in Bearing Housing Reservoir:

1. Remove 3/4 in. drain plug (item 55, Figure 9-5) from bottom of bearing housing reservoir.
2. Replace plug after draining bearing housing.
3. Remove oiler bottle from lower reservoir (Figure 9-17).
4. Fill bearing housing reservoir by pouring oil into the lower reservoir of the oiler until the oil level reaches the bottom of the level adjuster crossarms.
5. Fill oiler bottle with oil and install in the lower reservoir.

Note: Ensure bottle is seated on level adjuster crossarms.

6. Tighten knurled thumb screw to secure bottle in the lower reservoir.

Pressure Lubrication

Details of pressure lubrication systems vary widely. Each system is designed to meet the turbine application requirements. Factors such as the type of driven equipment, operational and environmental conditions and individual preferences in the selection of various components, may affect the specific design of the system.

Pressure lubrication systems are generally similar in that each employs a pump to draw oil from a reservoir and deliver it under pressure to the bearings and other parts requiring lubrication. The oil then drains by gravity flow back to the reservoir to be recirculated.

In most cases, the oil supply piping contains an oil filter, oil cooler and a pressure control device. Twin oil coolers, filters, and transfer valves are sometimes used in the pressure lubrication system. This arrangement allows the cooler or filter to be isolated for maintenance or repairs without shutting down the system.

Various types of optional monitoring, control and safety devices can be used with pressure lubrication systems. Among these devices are pressure and temperature indicators, auxiliary pumps, alarms and emergency shutdown devices.

Governor Lubrication

This segment provides lubrication, operation and maintenance instructions for a typical small turbine governor. This governor system employs a Woodward, Model TG-10, Mechanical Hydraulic Speed Control Governor as shown in Figure 9-18.

The governor is bolted to an adapter which is mounted in the steam end bearing housing. The governor driveshaft is coupled to the trip body by a flexible coupling. The governor valve lever, adjustable connecting rod, and terminal shaft lever connect the governor rotary terminal (output) shaft to the governor valve.

The TG-10 governor has a self contained oil reservoir with a 1.75 quart (1.7 liter) capacity. A breather, located on the top of the governor, vents the reservoir and also serves as a plug for the oil filler hole. An oil level sight glass on the side of the governor indicates the operating oil level. A reservoir drain plug is located in the governor end cover.

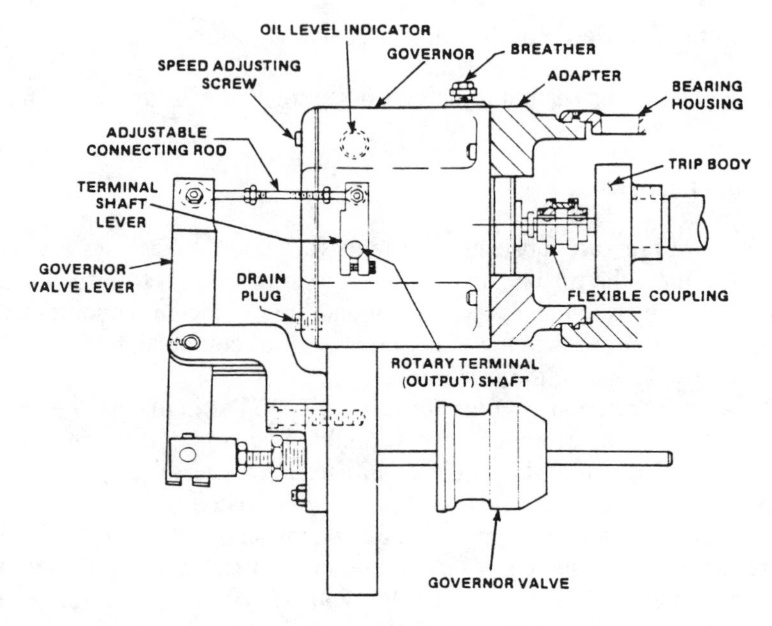

Figure 9-18. Governor system arrangement.

To replenish the oil supply; remove the breather and fill the governor until oil is visible in the sight glass. A quality turbine oil having a viscosity of 100 SUS at 100°F (14 mm^2/s at 50°C) to 220 SUS at 100°F (26 mm^2/s at 50°C) is recommended. Most SAE oils, without a tendency to foam, are acceptable.

It is advisable to change the oil at regular intervals to ensure trouble-free operation and long governor life. Operating the governor with dirty oil or with a low oil level can cause the governor to malfunction and result in possible damage to the turbine and/or the governor.

Principle Of Governor Operation (Refer to Figure 9-19)

The TG-10 governor is coupled to the trip body, and driven by the turbine rotor shaft. The governor uses mechanical force to sense the turbine speed, and hydraulic force to correct the speed. The hydraulic force is generated by an internal oil pump. The pump draws oil from the governor reservoir and discharges it to a spring loaded accumulator which stores high pressure oil to help maintain the full work capacity of the governor. A relief valve, built into the accumulator, maintains 150 psig (10 bar) operating oil pressure in the governor oil passages.

Spring loaded flyweights sense the turbine speed. When the speed changes, centrifugal force causes the flyweights to pivot outward or inward. The flyweights actuate the pilot valve plunger which opens or closes a control port in the pilot valve bushing. The control port directs the control oil to or from the bottom side of the power piston.

The power piston is connected through linkage to the rotary terminal (output) shaft. Control oil moves the power piston. The motion is transmitted mechanically through the terminal shaft and connecting governor valve linkage, to position the turbine governor valve.

Initial Start-Up (Refer To Figure 9-20)

1. Fill the governor with oil and check for signs of leakage.
2. Check the governor linkage for ease of movement.

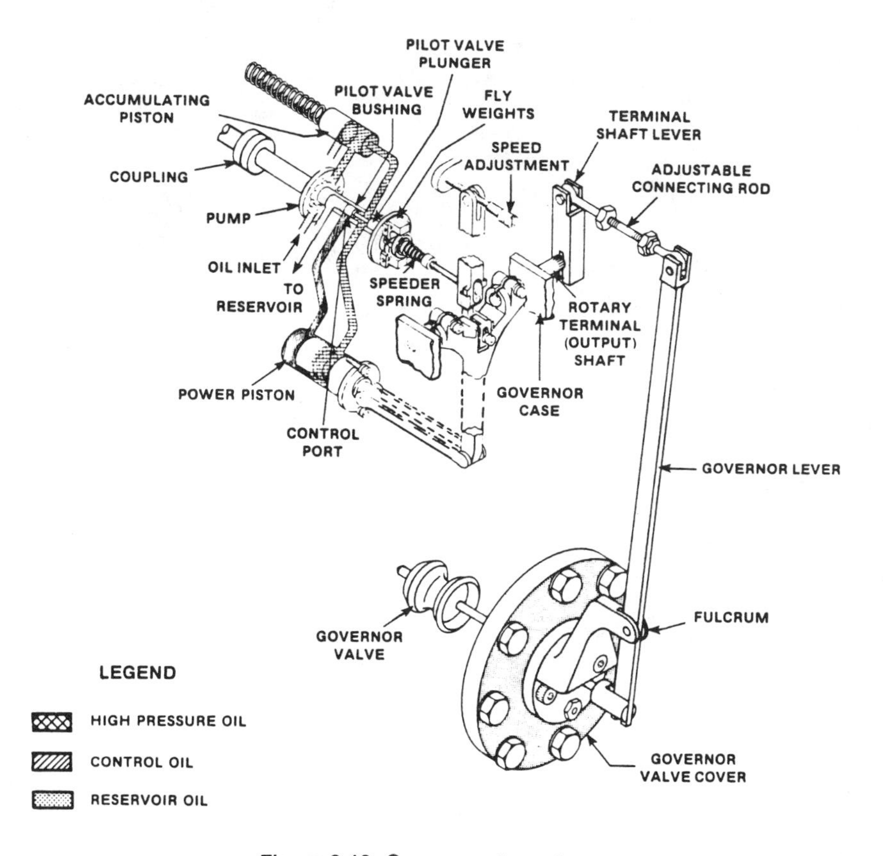

Figure 9-19. Governor schematic.

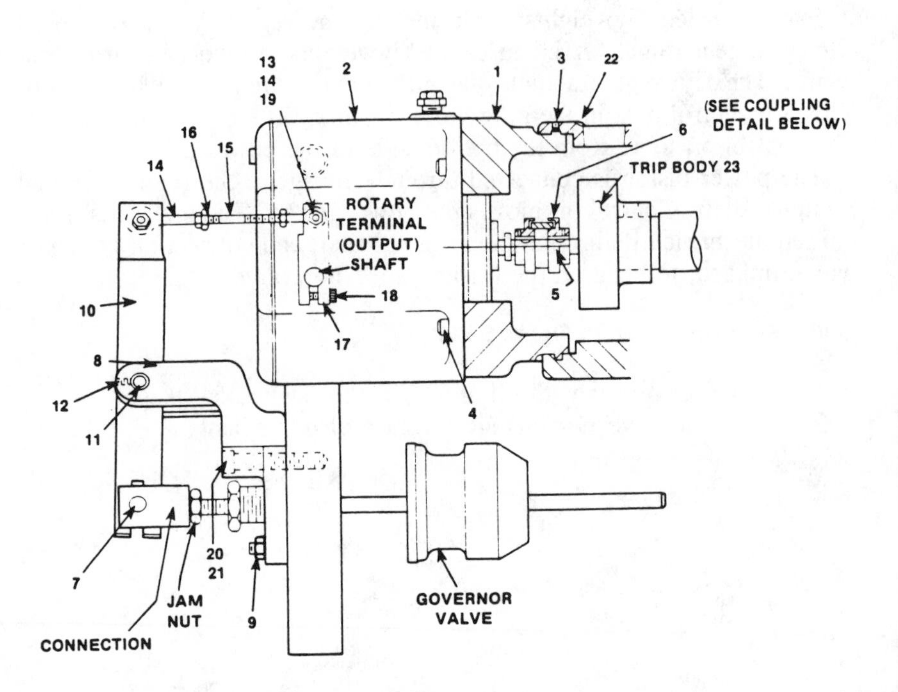

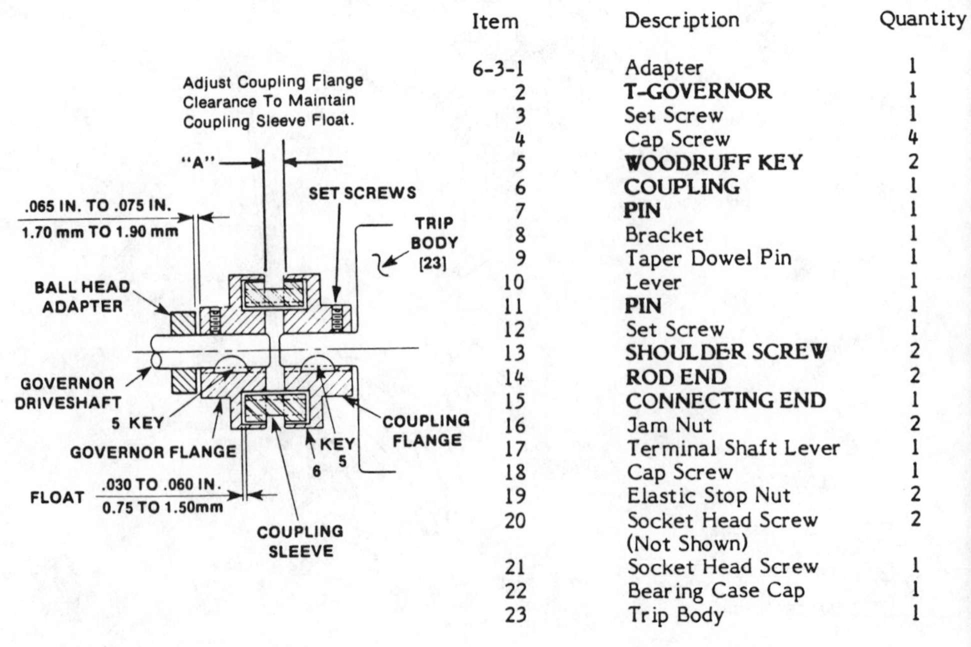

Item	Description	Quantity
6-3-1	Adapter	1
2	T-GOVERNOR	1
3	Set Screw	1
4	Cap Screw	4
5	WOODRUFF KEY	2
6	COUPLING	1
7	PIN	1
8	Bracket	1
9	Taper Dowel Pin	1
10	Lever	1
11	PIN	1
12	Set Screw	1
13	SHOULDER SCREW	2
14	ROD END	2
15	CONNECTING END	1
16	Jam Nut	2
17	Terminal Shaft Lever	1
18	Cap Screw	1
19	Elastic Stop Nut	2
20	Socket Head Screw (Not Shown)	2
21	Socket Head Screw	1
22	Bearing Case Cap	1
23	Trip Body	1

Figure 9-20. T—Governor system.

3. Check the linkage jam nuts, set screws and shoulder screws to ensure that they are tight.
4. Turn the speed adjusting screw to minimum speed (counterclockwise).
5. Start the turbine in accordance with the initial startup procedures, given earlier, and check the governor for oil leakage, vibration and ease of linkage operation.
6. After the turbine is thoroughly warmed up and operating satisfactorily at minimum speed; slowly increase speed, by turning the speed adjusting screw clockwise, until the turbine is operating at rated speed.
7. Continue to monitor the overall governor operation for any abnormal conditions.

   Note: Due to the design characteristics of the TG-10 governor the operating temperature may often exceed 200°F (94°C). Continuous operation at this temperature is acceptable and does not exceed the design limits. The governor may feel unusually hot to the touch, but this is not an indication of overheating.

8. After the turbine has been operated satisfactorily at rated speed (for approximately one hour), check the overspeed trip by turning the speed adjusting screw full travel in the clockwise direction while closely monitoring the turbine speed. If the rated trip speed cannot be obtained by turning the adjusting screw, it will be necessary to overcome the governor, by placing a bar between the governor valve lever and the valve cover and slowly prying the governor valve open.

   Caution: Pry the valve evenly and squarely from the valve body. Uneven force on the valve stem can bend the stem and cause binding. Also, monitor the turbine speed closely. If the turbine does not trip within 2 percent of the rated trip speed, strike the hand trip lever and adjust the overspeed trip as discussed earlier.

Routine Start-Up

Start the turbine as outlined in Routine Startup procedures, earlier. Monitor turbine speed and turn the speed adjusting screw as required to bring the turbine to the desired speed.

Governor Maintenance (Refer to Figure 9-20)

Routine Maintenance

Because of the simplicity of the TG-10 Governor System, a minimum of maintenance is required. The following checks should be made:

1. Check the governor oil level daily.
2. Frequently sample the governor oil. Change oil if the sample shows signs of contamination.
3. Check the governor linkage for binding, excessive play and loose bolts, jam nuts or set screws.
4. Keep the linkage clean and well lubricated with a high temperature, water resistant silicone grease.

Governor Disassembly

Internal governor maintenance is not recommended. It is advisable to replace the governor if defective. If the governor must be dismantled in the field, Woodward Bulletin 04041 should be requested. This Bulletin lists all special tools and replacement parts necessary for making repairs to the TG-10 governor.

Governor Removal (Refer To Figure 9-20)

1. Disconnect the adjustable connecting rod (15) from the terminal shaft lever (17) by removing elastic stop nut (19) from shoulder screw (13).
2. Loosen socket head cap screw (18) on the terminal shaft lever (17) and slide the lever off the terminal shaft.
3. Remove the four cap screws (4) which secure the governor (2) to the adapter (1) and remove the governor from the adapter.
4. Loosen the coupling flange set screws and remove the flange and key (5) from the governor driveshaft. (See Coupling Detail, Figure 9-20).
5. Remove the steam end bearing cap as outlined earlier, and remove the adapter (1) by lifting it out of the groove in the steam end bearing housing.
6. Remove the coupling sleeve from the trip body coupling flange.
7. Loosen the coupling flange set screws to remove the flange and key (5) from the trip body (See Coupling Detail, Figure 9-20).

Clearances (Refer To Coupling Detail Figure 9-20)

1. Maintain .065 in. to .075 in. (1.7 mm to 1.90 mm) gap between the coupling (6) and the ball head adapter.
2. Adjust gap "A" to maintain the coupling sleeve axial clearance .030 to .060 in. (.76 to 1.5 mm).
3. Maintain the coupling sleeve axial clearance float of .030 to .060 in. (.76 to 1.5 mm).

Note: Proper float of .030 to .060 in. (.76 to 1.52 mm) is to be maintained to protect the TG-10 governor from any turbine shaft thrust and thermal expansion loads, and to ensure sufficient coupling tooth engagement.

Governor Installation

1. Assemble with the steam end bearing cap (22) removed. (The coupling can be inspected with this cap removed and "A" clearance can be checked.)
2. Place the key (5, Figure 9-20) and coupling flange (6) on the trip body (23).
3. Install the coupling flange (6) on the trip body (23) allowing .080–.090 in. (2.03–2.29 mm) clearance between their vertical surfaces. Snug up set screws since the flange will be repositioned later to obtain "A" clearance.
4. Install coupling sleeve in the coupling flange.
5. Mount governor adapter (1) into machined groove on bottom half steam end bearing casing.
6. Place the key (5, Figure 9-20) and governor flange on the governor drive shaft.
7. Adjust the coupling flange to provide the proper clearance between the flange and the ball head adapter and lock the flange in place by tightening the set screws. (Coupling detail, Figure 9-20).
8. Mount the governor (2) on the adapter (1) making sure the coupling sleeve engages the governor coupling flange.
9. Replace the four cap screws (4) to secure the governor to governor adapter.
10. Loosen the set screw and slide the coupling flange up tight against the coupling sleeve and the locked governor flange. Measure the gap between the coupling flanges, "A" dimension.
11. Add the float .030 to .060 in. (.76 to 1.52 mm) to this gap (measured in previous step) and then reset "A" to this clearance. Tighten the flange setscrew and recheck the float.

Linkage Adjustments

1. Rotate the terminal output shaft to the middle position of its total travel. Place the terminal shaft lever (17, Figure 9-20) on the shaft so that it is in the vertical position.
2. Tighten cap screw (18) to secure lever (17) to the terminal shaft.
3. Move terminal shaft lever (17) to the full open position (clockwise direction).
4. Connect adjustable connecting rod (15) to the terminal shaft lever (17) by placing shoulder screw (13) through rod end (14).
5. Replace elastic stop nut (19) on shoulder screw (13).
6. Loosen the governor valve stem jam nut, and turn the valve stem out of the governor valve connection until the valve is firmly-seated.
7. Back the valve off of the seat, by turning the stem into the connection to provide design valve travel.
8. Lock the valve stem jam nut.

Operation and Maintenance of Cryogenic Plant Turboexpanders*

Turboexpanders have been used for many years in cryogenic processing plants to generate the deep, low-temperature refrigeration required by industry for gas separation and liquefaction or purification, the recovery of power from waste heat, and for other related processes. Figure 9-21 shows one such machine.

The turboexpander is a specialized, high-efficiency turbine, developing the required low temperature by removing heat from the process stream as power, thus chilling the gas. The power developed is a by-product of the gas expansion, and the amount of chilling is equal to the power generated. To absorb this energy, various loading devices are used. Machines developing less than 50 hp normally dissipate the energy in an oil turbulence device (dynamometer). For the recovery of higher amounts of power, integral compressor loads, electric generators, or pumps are usually used.

Today's high-speed turboexpanders are of rugged construction, suitable for years of trouble-free service with minimum attention to maintenance after installation and startup. Provisions are built into the systems to resist numerous abuses to which they may be subjected, such as ice deposits, solids passing through from plant lines, pressure surges, sudden cooldown, etc.

Figure 9-21. Expander-compressor.

The turboexpander is usually mounted on a single rigid steel baseplate with support systems for simplicity of installation, requiring only connections to the plant process, seal gas supply, cooling water, and power supply lines.

Proper checkout at installation and regular routine maintenance of the system will ensure years of trouble-free operation with a minimum of downtime.

Pre-Checkout

A simple "Three-Step" precheck of the system provides an adequate inspection of all components:

First, disassemble, inspect, and clean the rotating parts of the machine which are incorporated in the "mechanical center section" (Figure 9-22) to ensure that the unit is totally free of dirt and moisture, and that all critical parts are free of damage. After reassembly, provide adequate protection to prevent recontamination prior to reinstallation on the skid.

Figure 9-22. Mechanical center section of expander-compressor.

Second, with the mechanical center section removed, connect the oil supply piping to the oil drain line and flush the lubrication system for 6 to 24 hours. Use the same grade of oil that is to be used during operation, and heat it to 150°F, or as hot as possible without endangering critical pump clearances. Heat will decrease the viscosity of the lube oil to effect better cleaning. Circulate the oil with both pumps operating to help heat the oil and to increase its velocity and turbulence. On completion, discard or purify the flushing oil and filter cartridges, and fill the system with clean oil, installing new filter cartridges throughout the lube system (for details on oil lube purification or reclamation, see Volumes 1 and 2 of this series). Third, verify that all gauges, controls, and safety devices are properly connected, set and operative.

Fourth, check the electrical and hydraulic systems with the rotating parts reinstalled on the unit(s), seal gas flow established,* and the oil system connected and operating. Prepare a table similar to Table 9-10

* New seals may not leak at the specified rate but will soon break in. Meantime, maintenance of a seal gas pressure above the pressure being sealed is sufficient.

showing desired vs. actual values of operating parameters. Settings are made at the factory but may drift during shipment and must be checked and reset as necessary. Today's control/annunciation systems are largely trouble-free, using easily replaceable solid-state printed circuit cards without the complicated wiring connections used in earlier systems. This facilitates maintenance and checking of the electrical system.

All alarms and shutdowns should be checked under simulated conditions, and all components tested through the plant annunciator control system. Verify that the inlet trip closes within the specified time (normally one-half* second) to provide maximum expander protection. Activate all shut-down controls, both for the system and the plant, to test the shutdown valve. Experience has shown that precise settings and checkout of the instrumentation at the site are key factors in smooth startup and operation of all systems.

Pre-Startup

The first component of the auxiliary system to be started is the seal gas. It is also the last to be shut off to prevent contamination of the process zone by the lubricating oil. An optional interlock feature is often used to prevent startup of the lube oil system until an adequate flow of seal gas is established. The seal gas is filtered and regulated to a given supply pressure. A graduated needle valve or other flow control** regulates the flow and pressure of the seal gas to the labyrinth seal, providing positive containment of the process gas and preventing its loss or contamination by the lube oil and possible dilution of the lube oil by the process gas.

To start the lubrication system, open the pump and filter bypass valves, close the accumulator inlet valve, and then turn on the pump(s). The filter selector switch should be so positioned as to allow the purging of trapped gas or air in both filters. If two pumps are provided, both should be started.

After several minutes, when the system is thoroughly purged, close the pump and filter bypass valves. Once the system is brought up to pressure, the second pump should be stopped and placed in the auxiliary "standby" mode. The oil system is normally operated with one pump and one filter.

Failure to purge the system of gas at low pressure in this manner creates the risk of collapsing the filter cartridges when the pumps are turned on.

---

* Some generator systems require more rapid response; and for some compressor-loaded systems a slower response is specified—of the order of several seconds.

** In some systems a differential pressure control across the seal is used.

Table 9-10
Typical Field Service Running Report

JOB #_______ FIELD SERVICE RUNNING REPORT

	Design Cond.	Running Conditions													
Unit #															
RPM															
Exp. P_1															
Exp. P_2															
Exp. Wheel Press.															
Load P_1															
Load P_2															
Comp. Wheel Press.															
Front Bearing Thrust Press.															
Back Bearing Thrust Press.															
Seal Gas Supply															
Seal Gas Exp.															
Seal Gas Comp.															
Res. Press.															
Bearing Oil Press.															
H.P.D. Press.															
Oil B/Filter Press.															
Oil A/Filter Press.															
Res. Temp.															
H.P.D. Temp.															
Oil in Temp.															
Drain Temp.															
Front Bearing Temp.															
Back Bearing Temp.															
Vibration															
TIME															
DATE															

The accumulator system requires careful attention, since it maintains oil pressure to the bearings during the first critical seconds of coast-down in case of power failure (both pumps stopped). Ambient temperatures below 50°F require heat tracing and insulation of the accumulator.

The accumulator should be flushed and checked periodically to ensure that the oil in the holding tank is clean and uncontaminated. First, place the seal gas and lube oil systems in operation with the accumulator inlet valve open so that the accumulator will be filled. Then close the accumulator inlet valve, and open the drain valve. Connect a pressure gauge to the top of the accumulator and adjust the pressure, as necessary. If the oil is contaminated, drain and refill as often as necessary to obtain a clean flow of oil.

Pre-Start Mode

Place the system in the prestart mode by bringing the plant to startup pressure. This is accomplished by starting the seal gas system and the lube oil system (to protect the bearings in case pressurization rotates the expander). Finally, pressurize the unit by opening first the expander discharge block valve, the compressor inlet block valve, and the compressor discharge block valve. Verify that all atmospheric vents are closed. Maintain the prestart mode long enough for the oil pressure to cease fluctuating, and the temperature to stabilize at prescribed levels.

Startup

Note 1: Prior to startup, verify that the range of the nozzle actuator rod responds proportionally to the specified air pressure signal.

Note 2: Purge compressor (and if applicable, expander) casing for possible accumulated liquid.

Remember that turboexpanders come in different sizes and that expander, compressor and auxiliaries may differ from model to model. Our procedures are typical, but may have to be modified or adapted to a specific situation. Always consult with the manufacturer if discrepancies need to be resolved. On typical machines, with the seal gas flow established and oil temperature and pressure stabilized, proceed as follows:

1. Clear all alarms and verify that the "expander ready" light is illuminated.
2. Verify that the expander inlet block valve is closed.
3. Set the actuator inlet vane control to a low power position.
4. Open the inlet trip valve by pressing the start button on the panel (if provided).

5. Slowly turn the inlet block valve to full open to feed gas to the expander. This should cause it to rotate at reduced speed.*
6. Use the variable inlet nozzle control to increase the speed of the expander in 10 percent speed increments, at intervals of 20 to 30 minutes.
7. As the expander flow is increased, the flow passing through the Joule-Thomson bypass valve should be correspondingly decreased and, finally, flow control switched to automatic nozzle control.

A minimum flow of approximately 65 percent of design must be maintained through the compressor to avoid surge.

Check the expander and compressor inlet screens during startup, and repeatedly during the first few hours of operation, for blockage by material from the plant lines.

Adjust the seal gas flow to the minimum found to be required to prevent perceptible cooling of the oil drain line when the unit is operating on cold gas.

Normal Operation and Periodic Maintenance

Normal operation requires only routine monitoring of system controls and gauges.

The following should be checked during each shift during initial operations, and daily later on, and the readings logged:

1. Reservoir lube oil level
2. Seal gas flow
3. Lube oil pressures
4. Lube oil temperatures
5. Thrust meter readings

The lube oil should be changed every three to six months, or whenever there are indications of oil contamination, as evidenced by reduction of the oil viscosity. Frequent oil checks are necessary after startup to establish the best timing and frequency for oil inspections.

Filters do not require frequent cartridge change, except to remove contaminated oil.

Annual inspection of the expander-compressor is not essential, unless damage is suspected.

* Some systems attain reduced speed and pressurization prior to opening the trip valve by slowly opening a bypass valve around the trip valve. If such a bypass valve is used, great care must be exercised to be sure the bypass valve is closed and kept closed while the trip valve is open. Otherwise, the trip valve loses effectiveness.

Shutdown

The protective system for the expander includes alarm points for most or all shutdown conditions. This is intended to provide time for the operator to correct a problem before it reaches emergency proportions and the unit goes into shutdown condition. Alarm and shutdown signals are normally provided for overspeed, high bearing temperature, vibration, excessive bearing thrust load, low seal gas pressure, and low bearing oil pressure. (The alarm on the latter also starts the auxiliary oil pump and locks the circuitry so that restoration of oil pressure will not stop the auxiliary pump, which could result in cycling.)

If a condition reaches emergency proportions, the unit is shut down by the automatic closing of the inlet trip valve, allowing the machine to coast to a stop. Alarm and shutdown signals may also occur from other sensors or plant protection circuits sensing low inlet gas temperatures, high liquid level, low process flow, etc. If any one of these emergency conditions is not critical then the shutdown circuitry may utilize a built-in time-delay relay to prevent unnecessary shutdown.

Special Designs

In the past, uncontrolled problems often caused extreme wear and premature failure of mechanical parts in turboexpanders. With the feedback from hundreds of units operating in the field for decades, continued research and development has resulted in solutions to practically all operating problems. Improved, field-tested designs now offer increased reliability and ease of maintenance to reduce costly downtime and product loss.

One key problem area was overcome with the development of the dust-free seal design (Figure 9-23) which has successfully eliminated erosion of the seal behind the expander rotor. Previously, this was a frequent problem during the startup period. Fine particles in the process stream are now quickly removed from the seal and wheel labyrinth and directed out through the balance holes in the rotor to the discharge gas stream.

The development of the automatic bearing thrust load control system (Figure 9-24) solved another problem. Previously, machines were designed for an equalized thrust load, but had no provision for thrust load monitoring or load correction. A recent development in automatic thrust load monitoring and control uses the thrust bearing oil film pressure to indicate the actual thrust load on the meters, and to activate a control piston which operates a control valve, maintaining a proper balance of thrust forces. The control valve causes the gas pressure behind the compressor impeller or a thrust balancing drum to increase or decrease as

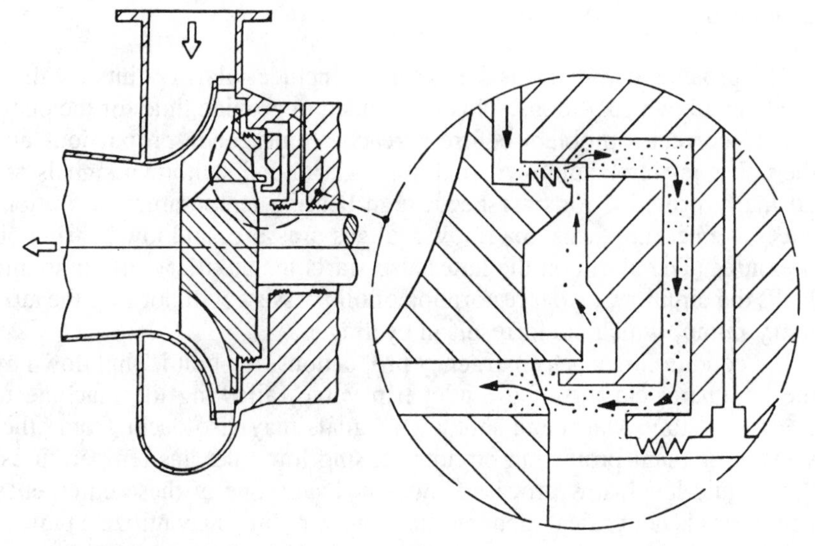

Figure 9-23. Dust-free seal design (source: Rotoflow Corporation).

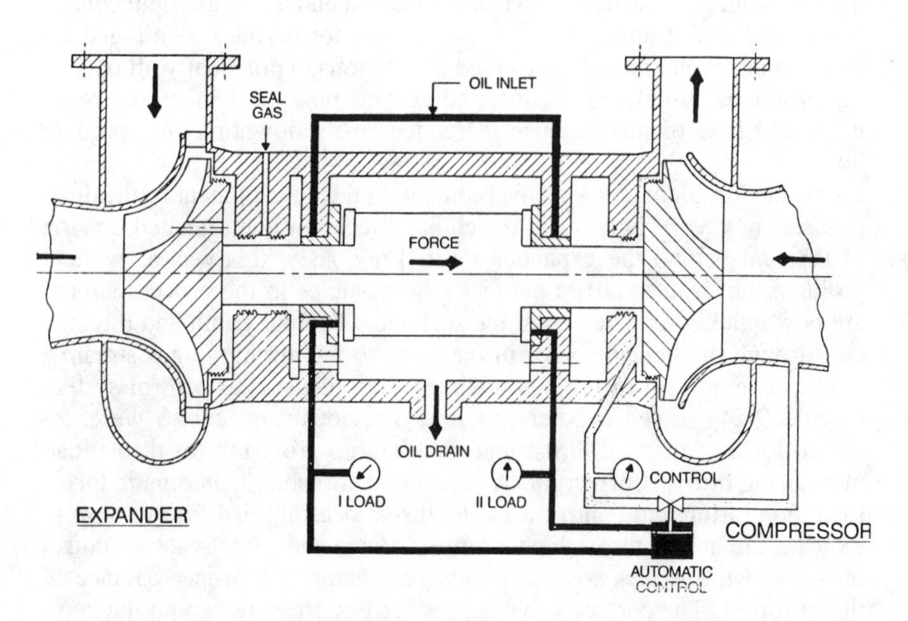

Figure 9-24. Thrust load control scheme (source: Rotoflow Corporation).

required to maintain thrust balance and prevent thrust bearing overload and possible failure.

Special rotor blading arrangements permit unlimited condensation through radial inflow expanders, or operation on flashing liquid streams with no loss of efficiency or liquid erosion of the rotor blading.

Wheel resonance has been exhaustively studied. The results have been computerized to provide wheels that are free of critical resonance conditions within the normal operating range, and usually free of critical speeds below trip speed. When a minor resonance occurs below design speed, systems are available to bring the expander up to speed, automatically passing rapidly up or down through the critical zone to prevent wheel damage.

A solid state control system for the startup and speed control of multiple, series-connected turboexpanders is available when required.

Periodic Checks

Maintaining accurate running records for a machine is essential to analyze any problems that may occur. Operating data should be recorded at least twice daily, and should include specific data on oil reservoir level, thrust bearing pressures, oil temperatures and pressures. Running time for the oil pumps (normally operated on a 30-day, equal-time basis) should be included. An oil sample should be analyzed monthly to determine viscosity and change-out or purification schedules for the oil. Accumulator charge pressure should be checked at least every two months to ensure adequate protection for the bearings in the event of a shutdown. (A typical periodic check sheet was reproduced earlier in Table 9-10.)

Troubleshooting

Troubleshooting guidelines for expander operating problems follow:

Low Oil Pressure to the Bearings

This condition will actuate the alarm and start the auxiliary pump and, if it reaches emergency proportions, will result in unit shutdown.

Check the lube system completely. Check filters for excessive pressure drop, and verify the settings for all relief valves and pressure regulators. Be sure to check the pressure-controlling bypass valve for excessive leakage. Sample the oil and test the viscosity. Reset the instruments and gauges in the system as necessary. If the problem persists, the unit should be disassembled and inspected for excessive bearing clearance.

Cold Front Bearing

The indications of this condition are a low oil discharge temperature, a low temperature reading on the front bearing thermocouple, or frost on the bearing housing or drain line near the expander bearing. This condition must be corrected to prevent loss of process gas and refrigeration, and to prevent freezing of the oil within the bearing during shutdown. Probable causes and corrective measures are as follows:

Cause	Correction
Seal gas shutdown while the expander is cold and pressurized.	If the unit is running, simply increase the seal gas flow rate until freezing has disappeared and further increase does not change the oil or bearing temperature. If the unit is shut down, the bearing oil film probably is frozen and must be thawed before startup. To accomplish this, the seal gas flow should be increased and the bearing housing warmed with steam or hot water until the bearing temperature is in a safe range.
Heat barrier wall leaking or damaged.	If the first treatment does not correct the problem, again check seal gas pressure and flow. Then disassemble the unit and inspect the heat barrier wall for cracks and correct clearance, replacing it or adjusting its position, as necessary.
Expander shutdown during extremely cold weather.	Thaw the unit at the front bearing with the application of steam or hot water on the bearing housing.

A good rule of thumb when there are no temperature indicators on front bearings is that if the housing is warm to the touch at the front bearing area, the bearing is probably warm enough to start.

Excessive Vibration

An accumulation of ice or frozen material in the rotor will cause an imbalance condition indicated by excessive radial vibration. A vibration monitor will provide an alarm signal before the unit shuts down. This condition may be corrected by passing clean, warm, dry gas through the expander or by injecting methanol into the case.

Another possible cause is the application of excessive pipe loads on the unit flanges, causing possible internal rubs or bearing malfunction, or shaft coupling misalignment.

Higher than normal expected readings may also be produced by the inherent resonant vibration of a stationary component, often aggravated by low oil viscosity.

Excessive Thrust Load

Excessive pressure on the bearing thrust face could exceed the design thrust force, causing damage and rapid deterioration. Protection against this is normally provided by the use of a differential pressure switch for alarm or shutdown.

The thrust control system should be checked to eliminate errors in indications and in the control system. The oil lines to the meters should be checked and purged of dirt and gas pockets. Excessive thrust loading can be caused by a higher than normal pressure behind the expander rotor which the balance system is unable to control. This suggests rotor back seal deterioration or icing.

The plant pressures and flows should also be checked. If the pressures and flows are not within normal limits, the machine may not be able to run properly. This situation should be checked with the manufacturer.

Filter Collapse

Failure to open the pump and filter bypass valves during the prestart check can result in trapped gas crushing the filter cartridge when the oil pumps are started. The filter cartridge should be checked for the correct part number and micron size for its usage. After a long period of downtime, the cartridge could become contaminated with water, causing weakening and reduced permeability.

Oil Loss in the System

Oil loss can be determined by the reservoir level. One condition that can cause oil loss is high oil level in the reservoir with a high seal gas flow rate. This causes flooding of the mist eliminator in the reservoir vent line, sending the oil into the compressor. To correct the problem, the oil level should be maintained between the specified minimum and maximum levels, and the seal gas should be maintained at the minimum required flow rate.

With hundreds of turboexpanders in service worldwide, an increasing demand for a working knowledge of the machines and supporting systems has developed. The "Three-Step" precheck described earlier is intended to eliminate errors and omissions in the system checkout prior to startup.

In review, the expander is protected from unexpected plant problems by the use of special designs, such as the dust-free seal, the automatic thrust balance control system, special rotor blading to accept condensation, rotor and nozzle design to avoid rotor resonance damage, special instrumentation to protect against freezing, etc. These systems in conjunction with the protective annunciator system are designed to prevent premature failure of the expander.

Proper operation and periodic maintenance will ensure a reliable, dependable system.

The need for complete running records must be emphasized. When an operator understands the readings and knows what to do when a problem first occurs, he can often make corrections and prevent downtime for the machine; or assist the manufacturer's service personnel in making the necessary analysis and repairs quickly and efficiently.

Installation of Turboexpanders

Arrangements for the installation of turboexpanders must take into account the specific guidelines issued by a given manufacturer for a given model or type of machinery. The following generalized installation procedure is offered as a typical example.

Equipment Foundations

The skid dimensions for mounting the system on its foundation are, or should always be given on the machinery arrangement drawing.

Handling

A suitable hoist or crane should be provided for installation purposes. Most models have provisions for the attachment of lifting yokes on the side of the skid.

Plant Process Connections

Connection sizes and locations for plant process inlet and discharge piping are usually shown on the machinery arrangement drawing.

A 60-mesh cone-type inlet screen and a 30-mesh cone-type screen need to be installed in the expander and compressor inlets of the unit, respectively. The screens must be installed in the process piping with the screen side upstream. We will discuss cleanliness later, but keep it in mind during installation.

Note: The expander inlet screen should be positioned where it can be conveniently removed for cleaning and still not interfere with the operation of the safety shutdown valve.

Refer, also, to the maximum allowable pipe nozzle load drawing, which experienced vendors will furnish.

Seal Gas System

Connect the plant's seal gas supply to the filters on the control panel. Refer to vendor-supplied data for requirements and to the system schematic and control panel drawings for location.

Lube Oil Cooler

If the oil system incorporates air cooling, connect proper voltage to each air-fan motor. Install, set and test vibration switch.

Variable Nozzles

After installing the nozzle actuator, connect the plant instrument air supply and air signal line to the variable nozzle positioner. The vendor should have provided data on pertinent pressures.

Safety Shutdown Valve

The expander shutdown valve (with solenoid) should be placed in the process piping at the expander inlet. See that it does not interfere with the screen. An adequate air supply must be connected to the solenoid valve to enable the safety trip valve to close within one-half second or less.

Electrical Connections

Refer to the system schematic and/or electrical schematic (or wiring diagram) for wiring, and any applicable component manufacturer's data. On all components, care must be taken to use the terminal combinations for the voltage to be used.

Verify that all switching power feeders and circuit breakers are locked out in the off position when connecting the power supply.

NOTE: To avoid electrical abuse, make certain that sensor lines to the various switches are properly connected to the plug-in circuit boards inside the junction box and *not* shorted to ground.

Junction Box

Junction boxes should always remain closed except during the actual installation period. In no case should it be allowed to stand open overnight or for extended periods. This will protect the electrical circuitry from dirt, rain, moisture and other sources of potential electrical abuse.

Electrical Conduit Seals

If the process gas is combustible, electrical conduit seals are to be poured only after verifying that all electrical components and circuitry are functioning correctly.

Control Panel

Install and check instrument air lines and electrical sensor lines as indicated on control panel junction box and system schematic drawings furnished by vendor. Before applying power, check that there are *no* shorts to ground.

Operating Instructions

Precautionary Instructions

If, after installation, it is believed there is significant possibility of injurious foreign matter having entered the unit, it must be disassembled and all parts cleaned and reassembled before initial startup. Verify that the cold section of the expander is free of oil or grease. Assembly and disassembly instructions must be carefully observed.

This may be a good time to have the manufacturer's service person instruct your operating personnel in the assembly and disassembly of the unit.

It should be noted that there is a danger of overspeed until the compressor suction reaches design pressure. The expander could reach full speed with less than full expander power until the compressor suction design pressure has been established.

Procedure for Filling Reservoir

Refer to the system schematic drawing furnished by the expander manufacturer.

Fill the lube oil reservoir with lubricant in accordance with specifications given by the manufacturer.

When the system is not under pressure nor in service, the reservoir may be filled using the inspection hole on either end of the reservoir. When the reservoir is pressurized, oil may be added by using a filler pump.

After filling all components, such as cooler or filters, ascertain that the lube oil reservoir is filled to a point between maximum and minimum levels, as specified by the equipment manufacturer.

If the reservoir needs flushing, do not use a detergent-containing oil. The regular lube oil is suitable for flushing.

Prestart Instructions

The following steps must be taken before putting the system into operation:

Check the shaft alignment between oil pumps and drivers. These are usually aligned at the factory but may have become misaligned during transit.

Verify that the accumulator bladder is charged to the correct pressure level specified by the manufacturer. To charge the accumulator bladder before putting the machine in service:

1. Isolate the accumulator from the oil system by closing the block valve.
2. Open drain valve.
3. Connect a suitable vendor-provided charge kit to the filling valve on top of the accumulator.
4. Fill with nitrogen gas to the required pressure level.
5. Close the drain valve.
6. Slowly reopen the block valve.

Note: If the lube system is pressurized, follow special instructions given by the expander manufacturer.

- Verify that all electrical circuits are properly connected and that motors operate in the proper rotation.
- Verify that there are no shorts to ground in any sensor line.
- Verify that all valves are in the correct open or closed position.
- Again verify that all gauges, controls and safety devices are properly connected, set and operative. Don't forget to ascertain correct lube oil operating temperature and pressure readings, and monitor all other control settings. Note that although settings were made at the factory, they may have drifted during transit.

- Verify that a suitable seal gas supply is properly connected. Pressure and flow requirements should have been specified by the equipment manufacturer.
- Verify that the range of the nozzle actuator rod responds proportionally to the specified air pressure signal. The range capability of the air actuator is usually greater than the nozzle actuator travel, which is set near the middle of the actuator range. The positioner on the actuator is adjusted to make the nozzle position respond proportionally to the air pressure signal.
- Verify that the safety inlet trip system is functional. The solenoid valve should be sized to close the trip valve in one-half second or less.

Starting Procedure

Refer to the manufacturer's data sheets and prepare a table for seal gas data, correct lube oil operating temperatures and pressures, and all recommended control settings. Use Table 9-10 for general guidance.

Caution: Verify that expander and compressor inlet valves are closed before proceeding. Verify installation of an expander inlet screen in the expander inlet process line with the screen side upstream. Likewise, in the compressor suction line.

Startup is as follows:

1. Admit seal gas to expander-compressor seals at the specified rate of flow. Next, verify the existence of a free passage for the seal gas leaking from the seals, especially for startup and shutdown conditions. New seals sometimes require a brief "break-in" period before reaching the specified flow rate. Meantime, hold the seal gas supply at a pressure approximately 5 percent higher than the pressure level to be sealed (e.g., expander rotor back pressure or compressor impeller back pressure).
2. Open the pump bypass valve and the filter bleed valve. Then start both lube oil pumps to sweep the lines free of gas. When the pressure readings on the oil pressure gauges have stabilized, then slowly close the pump bypass valve, close the filter bleeder valve, and stop the auxiliary pump.
3. With manual control unit, regulate diaphragm actuator to set variable nozzle at a low power position. Refer to manufacturer's data for nozzle positioner requirements.

4. Verify that the process block valve upstream of the expander inlet valve is closed before opening the expander discharge valve.
5. Open safety trip valve in expander inlet line. (In most installations it is not permissible to use the trip valve for starting.)
6. Open compressor inlet valve, open compressor discharge valve, and monitor control panel pressure gauges for correct pressures.

Note: *Do not make abrupt changes in seal gas flow, expander pressure, or compressor pressure.*

7. Slightly open expander inlet valve allowing the process gas to flow sufficiently for the expander to rotate at approximately 25 percent of design speed. Until the speed stabilizes, monitor all gauges. This should provide enough time to "break-in" new seals. Verify that lube oil to the bearings is at the correct operating pressure.
8. Using the manual control unit to operate the nozzle actuator, increase the process flow through expander inlet valve in steps of about 10 percent of full pressure every two to five minutes until full process pressure to the expander inlet is reached.
9. Meantime, monitor thrust meters and verify operating pressures.
10. Over a period of 10 to 30 minutes, gradually increase flow to desired capacity by adjustment of nozzle actuators.

Note: *Care must be taken to avoid compressor surge by maintaining minimum flow (approximately 65 percent of design) through the compressor, even if gas must be recycled back to the compressor suction.*

11. Expander and compressor inlet screens should be checked at the earliest opportunity and then repeatedly until no foreign matter is found in the screens. *Blocked inlet screens reduce flow.*

Seal Gas Flow Rate

Seal gas flow rate is usually controlled in normal operation by a differential pressure regulator.

Thrust Meter Gauges

Three meters are mounted on the turboexpander base support. The difference between the readings of the outer two of the three meters gives the measurement of the thrust force loading on the respective thrust bearings. See manufacturer's data for maximum permissible thrust meter difference.

For reference purposes a third meter, the center one, reads the controlled pressure behind the compressor impeller.

A thrust balance valve is generally furnished. The adjustment of this valve is set at the factory and readjustment is not needed unless:

1. The operating conditions change.
2. There is a major deterioration of one of the seals, as by erosion from dust particles entrained in the gas.
3. Deposits of ice or the like collect in the vent passages. An unexplained, significant change should be investigated.

On modern turboexpanders, adjustments are made automatically at all speeds. An alarm switch is installed to warn the operator, should pressure reach a preselected pressure on either thrust bearing. The unit will shut down automatically when the pressure reaches a higher, preselected, shutdown pressure on either thrust bearing. If activated, an investigation of wear on seals or wheels should be made immediately.

Normal Operation and Periodic Service

Normal operation requires only routine monitoring of system controls and gauges.

1. The following should be checked frequently:
   a. Reservoir oil level
   b. Seal gas flow
   c. Lube oil pressures
   d. Lube oil temperature
   e. Thrust meters

2. The lube oil should be changed as often as found necessary by indications of lube oil contamination. However, frequent checks, especially of oil viscosity, should be made.
3. Filter elements should be changed with each lube oil change.

Filter or Heat Exchanger Servicing While in Operation (*Caution!*)

If a filter or heat exchanger is taken out of service and drained, it must be refilled manually before being placed back in service. Otherwise, the filling of it may take excessive oil from the bearing supply. Also, the displaced air may be forced through the bearings.

To replace a filter element, turn the switching valve to opposite filter from the one to be serviced.

1. Close equalizing valve.
2. Carefully remove drain plug of the filter housing to drain the oil.
3. Open the filter housing to remove the used element.
4. Insert new filter element.
5. Replace drain plug in the filter housing.
6. Manually fill the filter housing with clean lube oil.
7. Reassemble the filter housing. Make certain it is closed securely.
8. Open the equalizer valve.
9. Open the bleed valve.

After it is purged for ten minutes, the bleed valve should be closed and the switching valve cautiously and gradually returned to a desired position. *Be careful not to significantly reduce the pressure of the oil to the bearings during the switching process.*

Note: The equalizing valve always remains open except when changing filter elements.

Annual inspections of the expander-compressor are not essential, unless damage is suspected.

On units furnished with bladder-type accumulators, periodically check accumulator bladder pressure. To do this, isolate the accumulator from the system pressure by closing the block valve. Then drain the accumulator by carefully opening drain valve.

Caution: The accumulator is under high pressure and some method must be devised for receiving the high pressure of the oil to be drained.

Connect charge kit to the filling valve located on top of the accumulator, and add nitrogen gas to achieve the pressure recommended by the manufacturer.

To place the accumulator back in service while the machine is in operation proceed as follows:

1. Close the drain valve.
2. Open the block (isolating) valve *very gradually* to fill the accumulator to its capacity. *Make sure this step is done gradually to avoid a sudden pressure drop in the lube system which could trigger a shutdown and perhaps cause damage to the bearings.*

Normal Shutdown

1. Close the expander inlet block or safety trip valves.
2. Observe the expander to determine that rotation has stopped.

Note: It is preferable not to close the nozzles on shutdown, because slight leakage from the shutdown valve sometimes causes "jet streaming" which can make the expander rotate when the nozzles are closed.

3. Shut off compressor gas flow.

Note: There must be a provision, such as a check valve, to prevent reverse flow through the compressor.

4. Shut down the lube oil system.
5. When the oil pressure reads zero, shut off the seal gas.

Emergency Shutdown

In the event of system malfunction, the safety devices will automatically close the expander safety trip valve to shut off the process stream to the expander. Under normal conditions, the seal gas and lube systems will continue to function unless those systems are the cause of the emergency shutdown.

Turboexpander Maintenance

In these procedures, every question obviously cannot be answered nor every problem anticipated. The maintenance person must use judgment in handling the parts, and be careful to follow the manufacturer's drawings. If maintenance must include disassembly of the turboexpander, disconnect all lube oil and seal gas piping and all instrumentation lines. Plug all lines to prevent the entry of any foreign matter.

The manufacturer's reference drawings should clearly indicate the attachment of all assembled parts.

Field Removal

Refer to the manufacturer's service assembly drawing and parts list. Some of the components are heavy and a hoist should be made available for support during field removal. Extreme caution should be exercised in the handling of all internal parts. These have been precision machined and dynamically balanced. Careless injury to parts may result in irreparable damage. Avoid marring any flange face or gasket face.

Disassembling Turboexpanders

The modular design approach followed by many manufacturers allows us to disassemble turboexpanders into major subassemblies or modules. With this concept the machine separates into three major components: expander case, rotating section, and compressor case. This eliminates the necessity of total disassembly when the main concern is isolated in one of the three primary areas. To separate the system into major components refer to Figure 9-25 and proceed as follows:*

1. Support the compressor inlet spacer (196) with a hoist or forklift. Remove screws (401). Remove compressor inlet spacer (196).
2. Support bearing housing (147). Remove screws (423). Slide away compressor section (188). Take care not to bump the compressor impeller (170). Remove O-ring (430). Remove retaining screw (179) and retaining washer (180). Use a jack screw to remove compressor impeller (170). *Do not use any other tool or device to remove this wheel.* Remove keys (421).
3. With the bearing housing (147) still under support with a hoist or crane, remove screws (440). Carefully withdraw the entire rotating section. Remove seal (966). Take care not to bump the expander rotor (107). Remove the rotating section to a clean work area.

Note: The machine is now separated into three major components. In this mode the variable nozzle assembly remains attached to the expander case. For further disassembly of the major components see the appropriate section in the following procedures.

These instructions describe the total disassembly of the machine.

Compressor Case and Impeller

1. Disconnect compressor process piping and instrumentation lines.
2. Support compressor inlet spacer (196) with a hoist. Remove socket head cap screws (401), compressor inlet spacer (196), and O-ring (405).
3. If the assembly includes a transition piece between parts (196) and (188), remove all fasteners. Also, if there is a shear ring or split-ring spacer, remove screws and remove this part, or parts.

* Figure 9-25 represents a typical Rotoflow expander and may differ from other manufacturers' equipment.

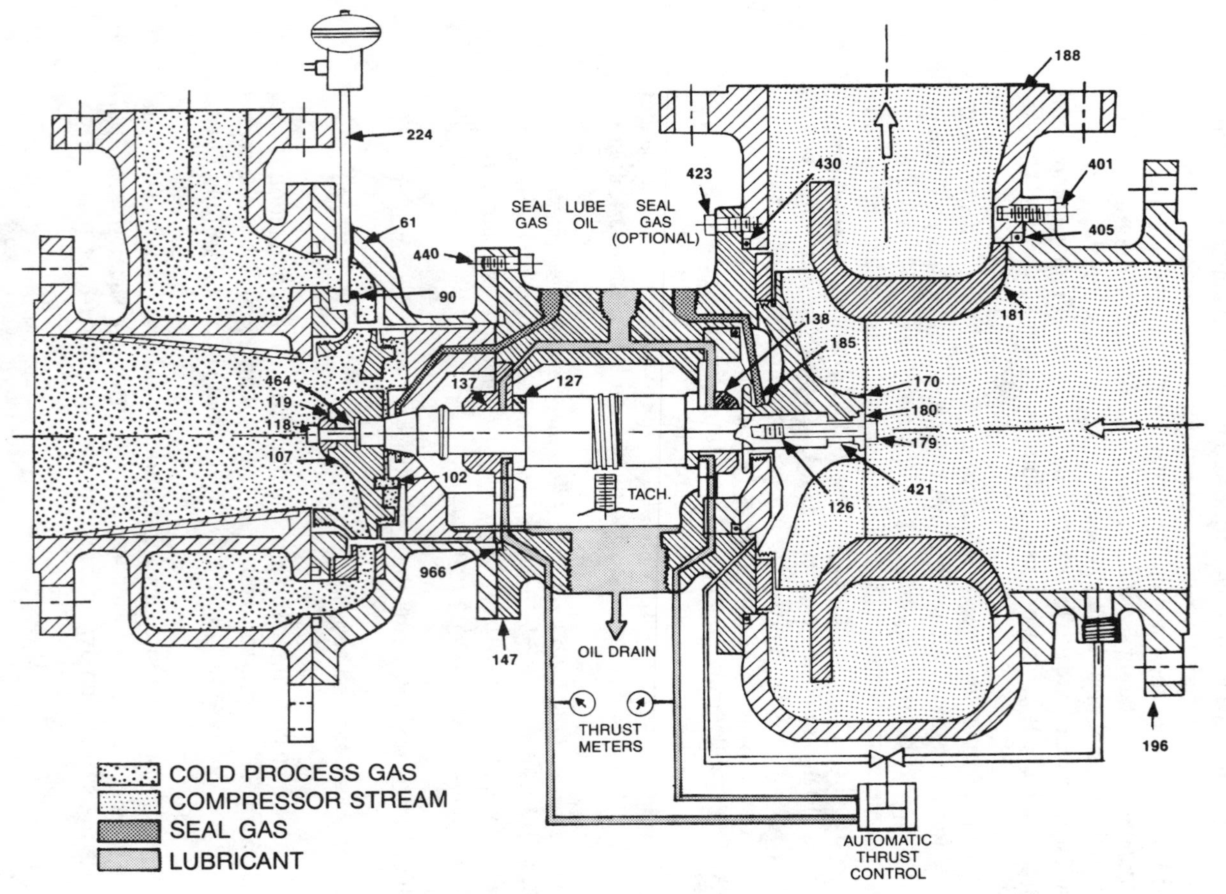

Figure 9-25. Turboexpander disassembly.

4. After removing the split-ring spacer, reinsert the screws to remove the compressor inlet transition piece. The screws usually serve as jack screws. Remove any O-rings not previously removed.
5. Remove any screws attaching impeller follower (181) to other parts. Remove impeller follower (181). Threaded holes are usually provided for jack screws to aid removal. Remove shims which axially locate (181) relative to the impeller vanes.
6. Compressor inlet diffuser (not shown) may be removed by removing appropriate screws. Two threaded holes are usually provided for jack screws.
7. Remove retaining screw (179), retaining washer (180), and compressor impeller (170) from shaft (126). The impeller usually has a threaded hole provided for a jack screw. *Do not use the wrong tools to remove this wheel.* Remove keys (421).
8. Steadying bearing housing (147) with a hoist, remove screws (423) and slide compressor case (188) away from housing, taking care not to strike compressor case against shaft (126) or the back impeller seal (185). The temporary use of two long threaded rods placed on opposite sides of bearing housing in compressor case will aid in this step. Remove O-ring (430).

Rotating Assembly

1. Steadying unit with a hoist, remove screws (440) which secure bearing housing (147) to expander case (61). The complete rotating assembly may now be withdrawn. Care must be exercised not to strike expander rotor (107) while withdrawing assembly. The temporary use of two long threaded rods placed on opposite sides of the bearing housing in expander case will aid in this step, also. Remove seals (966).
2. The entire rotating assembly should now be taken to a clean location for disassembly. The unit should be positioned horizontally on a workbench for ease in removal of components. *Care must be taken not to mar back impeller seal (185) or shaft.*

Expander Rotor and Heat Barrier Wall

1. Carefully remove expander rotor (107) by removing retaining screw (118), and retaining washer (119). Remove rotor (107) utilizing the threaded hole provided for pulling purposes. *Use jack screw only to remove wheel.* Remove keys (464). When reinstalling the rotor, the required torque for retaining screw (118) must be observed.

2. Remove exposed screws and dust free seal from heat barrier wall assembly (102). Threaded holes are usually provided for jack screws in the dust free seal.
3. Remove screws and washers and heat barrier wall (102). Again, two threaded holes are usually provided for jack screws.
4. Remove front oil retainer and O-rings which will be accessible after the heat barrier has been removed. Two threaded holes are provided for jack screws.

Back and Front Journal Bearing

1. Remove accessible screws and back impeller seal (185). Remove exposed screws to remove compressor seal and seal O-rings. Two threaded holes are usually provided for jack screws in both the back compressor seal and back impeller seal.
2. Remove back oil retainer and associated O-ring. Again, two threaded holes are provided for jack screws.
3. Break exposed lockwires and remove screws and washers. Using the two threaded holes provided for jack screws, remove back grooving bearing (138). Remove O-ring associated with part (138).
4. Withdraw shaft (126), together with back and front bearing thrust washers and pins (127). Exercise care to keep shaft centered in order to avoid damaging the journal bearing. Do not drop either thrust washer.
5. Break exposed lockwires and remove screws and washers. Remove front grooving bearing (137). Two threaded holes are provided for jack screws. Remove O-ring.

Nozzle Actuator Assembly

Refer to the manufacturer's nozzle actuator assembly drawing.

1. Remove locking nut for air motor while there is still about 6 psi of signal pressure applied. This is to relieve the spring pressure for easy removal of locking nut. After its removal, disconnect all instrument air lines. Remove self-locking pin in actuator rod, making certain there is *no burr* in pin hole.
2. To remove bonnet assembly, loosen gland follower and remove screws at base of assembly. To remove upper section of actuator rod, lift assembly and remove jam nut normally supplied with this assembly.

Variable Nozzle Assembly

The actuator rod (224) should be positioned to relieve the force on the pin in the nozzle adjusting ring (90) in order to ease removal of nozzle assembly. Care must be exercised in removal to ensure that actuator rod or pin does not become bent. Support nozzle assembly with a hoist.

1. To remove nozzle assembly from expander case (61), release shoulder screws. Threaded holes are provided for jack screws. Take component to a clean work area for further disassembly. Be aware of piston ring inserted in part (90). If it does not fall off, remove it.
2. Position housing horizontally on a workbench with nozzle assembly end up. Remove shoulder screws and nozzle assembly clamp. Lift off nozzle adjusting ring (90) and nozzle segments with pins from nozzle cover.
3. Nozzle fixed ring and shims may be removed from expander case by releasing exposed screws. Two threaded holes are usually provided for jack screws. To remove the expander case diffuser you will have to release a mounting screw.

Note: To reinstall nozzle assembly, the following procedures are carried out in a clean work area:

1. Place nozzle segments on pins of nozzle cover in closed position. Place adjusting ring (90) over nozzle segments and engage segment pins in slots of adjusting ring. Install shoulder screws and nozzle assembly clamp to hold adjusting ring and nozzle segments to nozzle cover.
2. The nozzle assembly may now be installed in expander case by carefully engaging adjusting ring pin in actuator rod (224). This may require some manipulation of adjusting rod. Care must be taken to avoid bending the pin. When the pin is fully engaged, replace shoulder screws.

Cleaning and Reassembly

After disassembly, the parts should be thoroughly cleaned. Parts other than those of expander housing should be coated with a thin film of lubricating oil to protect them against rusting.

Reassembly of the unit is in the reverse order of disassembly. Ensure that the expander end of the unit is not tilted below a horizontal position, which could allow coating oil to leak into the labyrinth seal housing.

If prevention of gas seepage is imperative, all gaskets should be coated with a viscous sealant.

The torque to be applied to the screws used in the reassembly of the machine can be found in the following torque table issued by the manufacturer.

Inspection and Allowable Wear Data

Check journal, thrust bearings, and shaft.

The maximum allowable clearances are given in Table 9-11. Carefully examine bearings and thrust washer faces for score marks. If scoring is excessive, replace them with new parts. It is recommended that all seals and seal inserts be replaced.

The dynamic balance of shaft, expander rotor and compressor impeller should also be checked in accordance with the tabulated figures. The rotating assembly components have been accurately machined and dynamically balanced; to check the balance of the assembly or to follow alignment marks between its components is unnecessary, unless suspicion of unbalance exists.

The following table specifies allowable wear and dynamic balance figures on the expander rotating assembly components. If inspection reveals that clearance or balance figures exceed those specified, the condition should be corrected, if possible, or the parts replaced.

Table 9-11
Allowable Clearance and Dynamic Balance

Component	Allowable Clearance
Expander Side (front) Journal Bearing & Shaft	.0040–.0046 in. (dia)
Compressor Side (back) Journal Bearing & Shaft	.0040–.0046 in. (dia)
Shaft End Play	.007–.011 in. (axial)
Heat Barrier Labyrinth (with thrust clearance in direction of seal)	.003–.015 in. (axial)
Component	**Dynamic Balance Required**
Expander Rotor Assembly	.002 in.-oz
Shaft only	.014 in.-oz
Compressor Impeller only	.005 in.-oz

Labyrinth Seal Clearance

Should it become necessary to ascertain if the heat barrier labyrinth bore is worn, proceed as follows:

1. Remove heat barrier wall (102) with dust free seal. Measure and record shaft axial end play.
2. Loosen expander side bearing screws and pull the bearing (137) away from the housing. Remove compressor impeller (170). Replace heat barrier wall and dust-free seal. Measure and record shaft end play to heat barrier wall.
3. If the reading from Step 2 exceeds the reading from Step 1 by more than .030 in., replace the heat barrier wall.

Similarly, check the labyrinth bore of the compressor seal and replace if the reading exceeds .030 in.

To set the clearance between the expander rotor and the nozzle fixed ring, and the impeller follower and impeller, pull the shaft toward minimum clearance and check with clay on the face of the blades. The clearance should be a nominal .020 to .025 in., measured between the contour surfaces. Adjustment is made with shims.

References

1. *De Laval Engineering Handbook,* edited by H. Gartmann, McGraw-Hill, New York, third edition, page 5–6.
2. *Handbook of Loss Prevention,* Allianz Versicherungs-AG, Berlin/Munchen, Germany. Springer Verlag, New York, 1978, page 164.
3. Bloch, H. P. and Geitner, F. K. *Machinery Failure Analysis and Troubleshooting,* Gulf Publishing Co., Houston, Texas, 1983, pages 289–476.
4. Bloch, H. P. *Improving Machinery Reliability,* Gulf Publishing Co., Houston, Texas, 1982, pages 38–47.
5. Reference 3, pages 268–275.
6. Cowgill, T., and Robbins, K., "Understanding the Observed Effects of Erosion and Corrosion in Steam Turbines," *Power,* September 1976, pages 66–72.
7. LaLena, P. and Glaser, F., "Consider Foam Cleaning for Turbines," *Power,* March 1980, pages 66–68.

Chapter 10

Gas Turbines and Hydraulic Governors*

Gas Turbine Maintenance Philosophy and Objectives

There are probably as many different approaches to gas turbine maintenance as there are users. Each philosophy must be the right one for a given user, in a given location, for a given type of gas turbine. The most important thing is to decide on an approach or philosophy that will optimize the resources of the user's company and provide an acceptable level of risk for the business in which the gas turbine is applied. This decision should be made as early as possible for it will affect which type of turbine is right for the application. Once a gas turbine is purchased the available options in maintenance practice are narrowed considerably.

It is difficult to approach gas turbine maintenance as a single subject because there are two distinct groups of turbines which require greatly different approaches. These are the *heavy duty industrials* and the *aircraft derivatives.* Primarily we will be discussing the maintenance of the heavy duty industrial turbines. (Figure 10-1.) Aircraft derivative units, as the name suggests, are derived from turbines designed for aircraft use and thus their maintenance has to a great extent followed the practices of the aircraft industry.

Proper selection of a gas turbine should focus on the relative strengths and weaknesses of each design.[1] The jet engine (Figure 10-2) was designed to be a very compact, lightweight powerplant for aircraft propulsion and is far superior to any other prime mover for this application. In the design of any component, or system of components, it is important to

* Compiled by Dean H. Jacobson, Esso Chemical Canada, Ltd., Redwater, Alberta, Canada.

Figure 10-1. Heavy duty industrial installation (courtesy R. Pellefier).

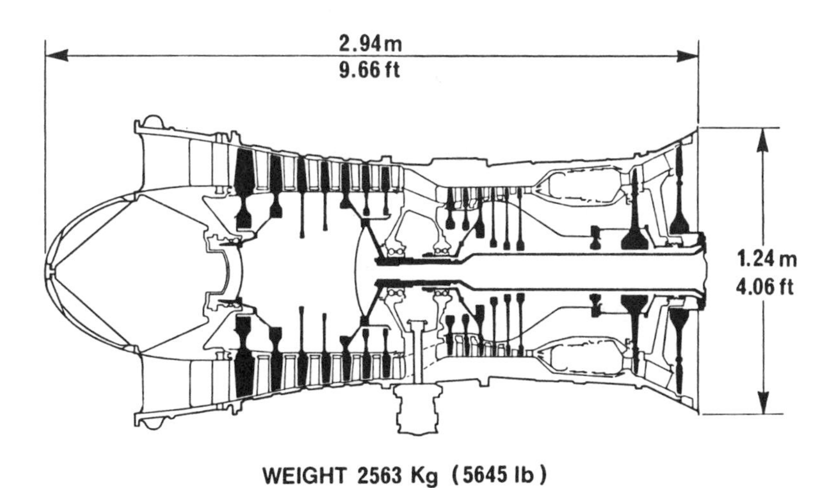

Figure 10-2. Typical aircraft derivative gas generator (courtesy Rolls Royce).

address the basic parameters for that design. For a jet engine these are to minimize frontal area through the use of long blades and a small hub diameter, minimize combustor length and diameter to minimize drag and overall engine length, and to ensure dimensional symmetry upon rapid heat up and cool down. A jet utilizes turbine metallurgy and advanced cooling techniques to withstand the relatively short periods of time (takeoffs) during which the engine is fired hard. It must be recognized that an aircraft engine typically spends 98 percent of its life operating in a clean, cool atmosphere, burning a good clean fuel, and at a fraction of its rated firing temperature. These factors all contribute to the high reliability we have come to expect from jet engines.

The design priorities of an industrial-type turbine are quite different from the jet. Weight is not one of the most significant factors, although industrial turbines such as those manufactured by Solar (Figure 10-3) incorporate many of the lightweight features of the jets. Larger rotor hub

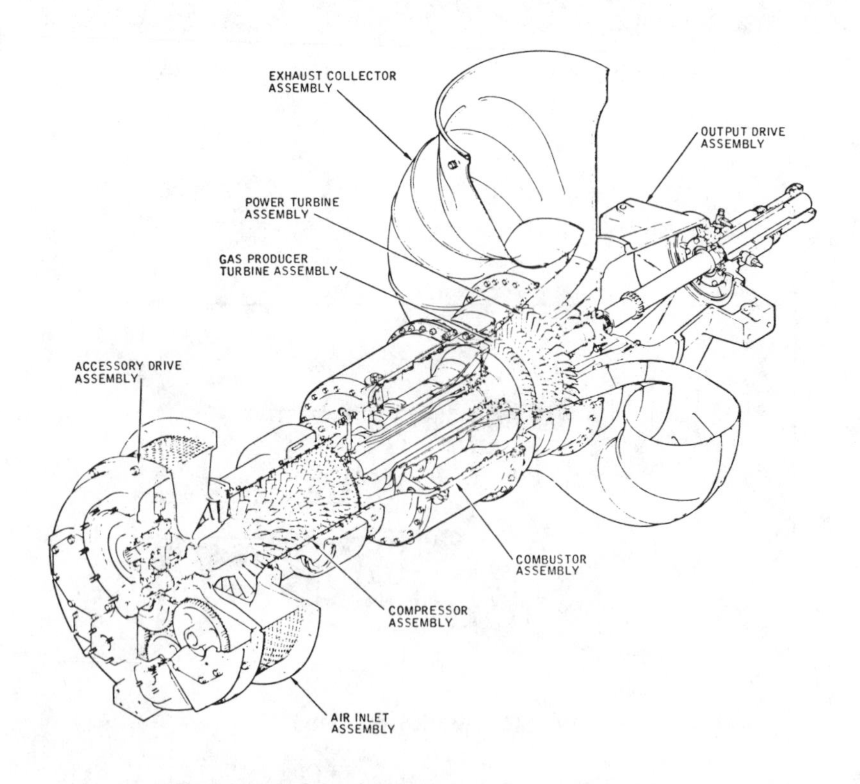

Figure 10-3. Solar Saturn gas turbine (courtesy Solar).

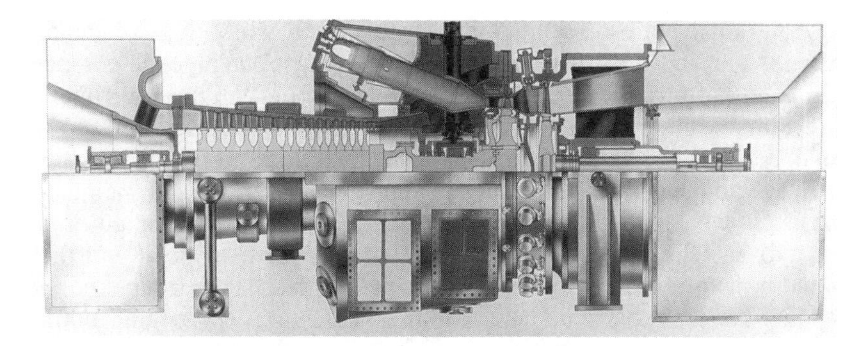

Figure 10-4. Heavy duty industrial turbine (courtesy Westinghouse Canada).

diameters along with thicker and shorter blades characterize the industrial turbines (Figure 10-4). Long, uninterrupted runs at rated firing temperatures are desired from many units. This has led to a much more conservative application of metallurgical and mechanical design.

It is not fair to say that the designers of each of these types of turbines are not learning from each other. Newer industrial designs are incorporating some of the advanced blade and vane cooling systems to allow higher firing temperatures and greater efficiency. Aircraft derivative manufacturers and their users are "industrializing" their units more as time progresses to achieve longer uninterrupted run times and longer intervals between overhauls.

Aircraft derivative users have two primary styles of maintenance:

1. Remove gas generators on a regular basis as determined by the manufacturer and their experience and send them back to the manufacturer's facility for overhaul.
2. Operate a repair facility of their own for most of the routine overhaul procedures. Procedures which they were unable to perform could still be contracted to the Original Equipment Manufacturer (OEM) on a component basis.

There are advantages and disadvantages to both of these approaches. Clearly if you only operate a few turbines and they are of different types the second approach would not be very attractive. If, however, you operate many identical units, the savings of the second approach could easily cut your maintenance budget in half.

Operational philosophy also has a bearing on which type of maintenance is most cost effective for a given company. One pipeline company may elect to run all of its units hard for 8,000 hours and then overhaul on a regular basis. If they operate their own maintenance facility and the volume of work is within their capability, then this approach could minimize the amount of unscheduled downtime experienced and maintain a consistent workload in their shop. If the company was sending all of this work to an OEM's shop, their return would not be so great and there would be more incentive in extending run lengths. Another pipeline, depending on their particular situation, could run every engine until it showed some sign of distress before doing an overhaul and do a lot less overhauls. This type of approach would necessitate more conscientious monitoring and a good routine inspection program so as not to experience catastrophic failures.

Heavy duty industrial turbines are more frequently installed in facilities which have some on-site maintenance forces and in many cases extensive engineering staffs. These types of facilities have a few more options in terms of maintenance philosophy than do the remote locations. As was stated previously, there are as many different approaches to gas turbine maintenance as there are users. Which one is right for each user depends upon the application which his turbine or turbines are in, the value the user places on downtime, his proximity to the OEM or a non-OEM maintenance facility. The relationship between his company and the OEM, and the relative strength or weakness of the people on site also influence the type of maintenance which is practical.

Four approaches to heavy duty industrial turbine maintenance follow, beginning with the approach requiring the least owner involvement:

1. A service contract with the OEM for all maintenance, including routine inspections and overhauls.
2. In-house maintenance supervision with engineering and service support from the OEM and a contract maintenance crew to perform the work.
3. In-house maintenance supervision and repair crew, with engineering from a third party (either a consultant or another part of the user's company) and assistance from the OEM's service engineering.
4. A multidiscipline team approach by the user, drawing from internal company resources (mechanical, metallurgical, instrumentation, maintenance planning and supervision, etc.) with minimal assistance by the OEM.

As was the case for the aircraft derivative engines, each approach has certain advantages and disadvantages. The approach taken is highly dependent upon the specific circumstances. As a general rule, the more units operated and the more time in operation the further down the list the user progresses. Although there are still users who find an OEM maintenance contract attractive, as they gain more experience users tend to seek more direct involvement in their gas turbine maintenance.

In refineries and chemical process plants, unscheduled shutdowns can be extremely expensive. Therefore, turbines are generally maintained at a high level by experienced personnel employing sophisticated condition monitoring and corrective maintenance techniques. By contrast, some standby electrical generator and crude oil transmission installations are maintained at a very low level and have effectively no maintenance procedures beyond contacting the manufacturer's representative when the equipment fails to start or operate. The majority of the gas turbine equipment operating is maintained at a level somewhere between these extremes. The operating availability and reliability is directly related to the level and quality of maintenance performed.[2]

A maintenance policy which realistically addresses the costs involved in outages, spare parts, and engineering services should be developed as early as possible in order to select the level of maintenance which is suited to the application. Ideally this policy should be adopted before a gas turbine is selected and should form an integral part of the driver selection. Most maintenance managers unless specifically experienced in gas turbines are not capable of realistically setting this type of policy. It should determine the level of support to be provided to ensure that corporate objectives can be fulfilled. It should establish goals and provide guidelines as to what is practical, reliable, and economical based on the operating and availability requirements of the turbine.[2]

The setting of such a policy may be facilitated by a competent outside consultant if such an individual is not available within the user's company. Primary emphasis should be placed on ensuring that personnel with the right skills are available, or that the people who are available receive the necessary training.

Along with a maintenance policy or a part of it the objectives of planned maintenance should be defined and their relative importance determined. It is difficult to define the objectives of planned maintenance, but in general they can be summarized as follows:

1. Minimize capital investment.
2. Maximize turbine reliability and availability.
3. Minimize operating and maintenance costs.
4. Maintain original design performance.
5. Incorporate appropriate product design improvements.[7]

Since the multidiscipline approach is the most user-intensive it is worth exploring in greater detail. Recently C. D. Hall presented one major petrochemical company's approach to gas turbine maintenance.[3] Problems were identified with the more traditional approaches to gas turbine maintenance:

1. Adequate records are not kept on mechanical condition before repairs; specific repairs made; and clearances, fits, and finishes when new, before repairs, and after repairs.
2. Knowledge gained from failure analysis walks out the gate with the consultant and/or repair crew, and it never gets back to the people who own the equipment.
3. The personnel actually responsible for day-to-day operation of machinery often do not learn the characteristics of their machines, the things to watch for, and how to improve the machine performance.
4. Often, engineers or specialists sitting in remote offices make decisions that affect a machine's performance. However, these decision makers have never experienced living with their decisions.

The objectives of the approach taken at Hall's facility were:

1. To create in-house technology, so repairs can be made intelligently and without delay.
2. To operate machines at the maximum efficiency allowable by construction materials, instruments, and operating parameters.
3. Not to shut down a machine for overhaul without a scientific reason for doing so.

To accomplish these objectives the responsibility for gas turbine maintenance was distributed among various plant functions and specialists. These included alignment, lubrication, metallurgical engineering, welding, inspection/NDT, and mechanical engineering. In essence, what has been done is to achieve managerial control of the engineering factors that influence the reliability of the machines. This has enormous advantages if managed correctly, but is only as strong as the individual specialists in the team.

Mr. Hall lists the results of this approach at his site as:

1. Gas Turbine turnaround time has been reduced approximately 50 percent.
2. Manpower usage has been reduced approximately 50 percent.

3. Machine failures have been reduced drastically, with only one gas turbine operating failure in eight years. This failure was caused by operator error—the operator didn't believe the instrumentation.
4. Component life has been extended by careful inspection, metallurgical analysis, rewelding, and heat-treating.

Spare Parts/Special Tools

A maintenance policy for gas turbines must address the level of spare parts required to achieve the reliability and availability objectives of the user organization. There is no absolute guide to the purchasing of these parts. The OEM will normally recommend a list of spare parts or in some cases a number of lists such as commissioning spares, normal operating spares, major replacement spares, and in some cases contingency spares. In certain instances users have been known to purchase entire engines as spares, depending on the design, number, and critical nature of the units.

A typical recommended spare parts list for a two-shaft, simple cycle industrial turbine is shown in Table 10-1. This list does not include any of the bolting materials which will certainly be required on any major overhaul or the spare parts required for the control system.

Table 10-1
Typical Recommended Spare Parts List for Industrial Gas Turbines

Item	Quantity	Description
1	1	Set of gas turbine and auxiliary gearbox bearing shoes and gas turbine thrust plate and thrust bearing pads.
2	1	Set of primary combustor baskets
3	1	Set of secondary combustor baskets
4	1	Set of combustor swirl plates
5	1	Set of fuel nozzles
6	1	Set of cross flame tubes and clamps
7	2	Flexible fuel gas hoses
8	1	Set of gas turbine thermocouples
9	1	Speed probe and extension cable
10	1	#2 Bearing vibration probe extension cable and gland assembly and proximitor.
11	2	Sets of lube oil filter cartridges and gaskets
12	2	Ignitor plugs
13	1	Ignitor extension cable

For an application where downtime is costly, the purchasing of spare rotors is imperative. Consideration should also be given to stocking of stationary vanes for the hot gas path, axial compressor inlet guide vanes, transition pieces, and other components along the hot gas path.

Considerable savings can be achieved by ordering these parts with the original purchase order for the machine. This is due to the added leverage enjoyed by the customer before he places an order, and by allowing the vendor to minimize his cost of production by manufacturing the parts in parallel with the machine.

Once a level of spares is established adequate enough to protect the operation, the decision can be made to either replenish this inventory with new OEM parts when they are consumed, or to try some of the repair techniques currently available. In recent years techniques for the repair of alloys once unrepairable have advanced and will continue to advance as long as there is a market for this type of service.[4, 5] The lead time required for the purchase of new hot parts can at times be unacceptably long. In this case remanufacturing or repairing parts may be the only acceptable alternative.

The level of spares required for a gas turbine control system varies with the type of system and the complexity of the design. Currently systems vary from older relay-based systems, through analog and hybrid analog/digital, to fully redundant digital systems. In the redundant systems the spares are on-line. In solid state analog systems (which are generally not redundant) a decision has to be made whether to stock individual cards, or a whole spare system and whether it should be kpet powered-up or wrapped on a warehouse shelf. Generally a powered-up complete spare is the most reliable approach, but is also the most costly. Again, an assessment must be made as to the relative costs versus the reliability objectives for the unit.

Spare parts for aircraft derivative units are generally kept as complete gas generator assemblies. Other external assemblies associated with the engine are stocked as well. These include fuel manifolds, thermocouples, bleed valves, speed pickups, and guide vane actuators. Each operator must assess how many spare engines should be stocked to service his operation. This may mean none, or a share of a pooled spare engine or in the case of a major Canadian pipeline one spare for every four running engines.[6]

Special tools are normally required to perform routine maintenance on both industrial and aircraft derivative engines. They are usually identified in the maintenance manual for fabrication by the user, or supplied with the engine by the manufacturer. The responsibility for ensuring that these tools are available for an overhaul rests with the user.

Maintenance Organization/Planning

The key to all satisfactory maintenance activity is planning.[7] This is especially true when dealing with gas turbines. Even the simplest of maintenance tasks, whether it is recognized or not, requires some measure of preplanning. A maintenance policy for gas turbines must address the increased importance of inspection and overhaul planning if unplanned outages are to be avoided. Each user regardless of the number or type of the engines must institute the following basic types of planning to be successful in gas turbine maintenance:

1. Material/Spare Parts Planning
2. Manpower/Training
3. Technical Analysis/Historical Record keeping
4. Contingency Planning

Each of these basic types come together into a formal turbine overhaul and inspection plan. Without any one of them, the overall effort will never achieve maximum effectiveness. This type of planning is fundamentally no different from other types of rotating machinery planning.

Condition Monitoring

Advances in condition monitoring over the last few years along with the high cost of inspections and overhauls are gradually shifting the emphasis in gas turbine maintenance from a preventive to a more predictive approach. More and more users of both industrial and aircraft derivative engines are challenging the manufacturer's recommended inspection schedule in favor of a "more scientific approach." In process plants, the manufacturer's schedule often just does not mesh very well with the manufacturing plan for the process unit. In other cases units are shut down due to a problem in another part of the plant. This gives an opportunity for some minor inspection, such as a borescope inspection, and based on this data run lengths are extended. It must be remembered that many parts of a gas turbine are designed for a finite life, typically 100,000 hrs., and that there is a statistical probability that the part will not last that long. This is one of the main factors which sets gas turbines apart from other types of rotating equipment. Whereas many companies are extremely successful in operating compressors, motors, and to a certain extent steam turbines based on observed condition only, this is not yet feasible with gas turbines. Most users rely on a "home-brewed" mixture of fired hours, historical data, and condition monitoring to determine gas turbine run lengths.

What is meant by condition monitoring? Quite simply it is the observation and analysis of externally observable parameters to estimate the internal condition of the machine. As the word "estimate" implies, this is not an exact science, but one that requires a great deal of skill and experience. The technology of condition monitoring is advancing rapidly as lower cost computers make sensing, trending, and analyzing turbine data easier and more accessible. The parameters most closely monitored are performance and vibration.

Monitoring may be as simple as the analysis of a few key variables which are manually recorded on log sheets. It may also be as extensive as the computerized collection and analysis of 150 variables for a single gas turbine train. Process plants which place a high value on outage time and operate unspared single train units can easily justify the monitoring of every conceivable variable around a gas turbine on the basis of proper identification of the cause of trips alone (Figure 10-5).

Vibration monitoring of gas turbines presents a few difficulties not normally encountered on other types of turbomachinery. High temperatures in and around the engine require that special precautions be taken to ensure probe wires remain cool. On industrial turbines with noncontact type probes many times this means running the probe leads in the bearing oil drain piping, since this path is cooled by the flow of oil. Monitoring with seismic probes is another option employed by some industrial users and is the only practical method for aircraft derivative units. On large industrials establishing good locations for these probes in close proximity to the bearing is sometimes difficult. Many users have chosen to retrofit noncontact probes with good success when unavailable through the manufacturer.

Performance monitoring is carried out in basically the same fashion on each engine, although the analysis of the data taken is quite different depending on the type of turbine and control scheme employed. Basically one tries to determine the engine's ability to produce power for a given amount of fuel. If carried out properly this analysis should not only indicate the deterioration of thermodynamic performance, but give an indication of the mechanical health of the engine. The simplest form of performance monitoring involves the trending of key variables (e.g., pressure ratio, compressor speed, turbine speed, fuel flow) associated with the engine and comparing them with a baseline which has been developed. This method is commonly used on fixed geometry engines, both industrial and derivative. Variable geometry engines, however, do not lend themselves easily to this type of analysis making trending extremely difficult in comparison to the fixed geometry engines. Thermodynamic gas path analysis is really the only good way to establish the performance of a variable[8] geometry turbine. This involves an iterative process of matching the per-

Figure 10-5. Computerized machinery monitoring system (courtesy Zonic Corp.).

formance of the compressor with the turbine section for a given output speed and exhaust gas temperature.

Inspection/Overhaul/Repair

This section deals primarily with one of the two basic types of maintenance practiced on gas turbines, that is scheduled maintenance. The other type, unscheduled maintenance, is the type which we seek to avoid, or at least minimize. Unscheduled gas turbine maintenance in a process plant is always extremely expensive and when carried out well is characterized by rapid identification of the problem and implementation of the solution. Time is not usually spent thinking about "what else should be done." It was once said that, "emergency shutdowns lead to hasty decisions which in turn lead to costly mistakes."

Scheduled maintenance in its broadest sense includes many of the concepts dealt with earlier in this chapter. Scheduled maintenance is a continuous activity which incorporates the condition monitoring program, routine inspections by operating personnel who identify problems which can be rectified while the unit is running, and such activities as online compressor cleaning. Although there are many different types of maintenance and inspection activities which are commonly discussed, we will simplify them into two major classifications: *combustion inspection* and *major overhaul.* Not all comments will apply to every turbine. Primarily we will focus on industrial type turbines commonly used for mechanical drive applications in process plants. These are generally between 15–35,000 HP and are either one- or two-shaft designs.

A combustion inspection for our purposes is defined as an inspection and/or replacement of the major components of the combustion system. These include fuel nozzles, combustors, transition pieces, cross-flame tubes, and in some cases even first stage nozzles! Various other work is also carried out with this inspection (Table 10-2), but the work stops short of the removal of any of the upper casings. Borescope inspections are carried out into the various portions of the hot gas path which is not directly visible except by this method. Stationary vane surfaces should be inspected for:

- Foreign object damage
- Corrosion
- Cracks
- Trailing edge bowing

Table 10-2
Inspection/Maintenance Task List

Item	Action Performed	Combustor Inspection	Major Overhaul
1	Inspect fuel nozzles	x	x
2	Inspect ignitors	x	x
3	Inspect combustors and cross flame tubes	x	x
4	Inspect combustion transition pieces	x	x
5	Inspect 1st stage nozzles and blades	x	x
6	Check all field transducers	x	x
7	Replace faulty thermocouples	x	x
8	Check control calibration	x	x
9	Check back-up systems	x	x
	—UPS system		
	—Lube/control oil system		
	—Fire protection system		
10	Inspect compressor IGV's	x	x
11	Inspect all compressor stages		x
12	Inspect all turbine stages		x
13	Check bearings and seals		x
14	Inspect and check alignment		x

- Erosion
- Burning

Similarly the rotating blade airfoil surface may exhibit the following defects:

- Foreign object damage
- Corrosion
- Cracks
- Erosion
- Burning

In addition to the above, a visual check may be made of the rotating blade tip clearances. The root areas are not visible when using the borescope. To view all rotating blades the rotor must be turned manually. Normally there is some facility provided for this purpose.

The downtime associated with combustion inspections can be somewhat reduced by utilizing a spare set of combustors, transition pieces, and fuel nozzles. Inspection and repairs can be made on the removed parts without holding up the reassembly.

A major overhaul includes all of the tasks of a combustion inspection along with the removal of the upper half of the casing for inspection of the entire gas path. A major overhaul may include the removal of the rotor(s) from the casing for blade inspection, the removal of blades from the rotor for blade root and disk inspection, or an inspection "in-situ." Some may call the latter a hot gas path inspection rather than a major overhaul, but generally if the casing has to be removed we would consider this major work.

Crack detection is the primary thrust of inspection activities on gas turbine overhauls.[9] A number of techniques are used both in the field and the shop. These include:

- Visual inspection
- Dye penetrant inspection
- Magnetic particle inspection
- Fluorescent (Zyglo) inspection
- Ultrasonic inspection
- Eddy current inspection

Visual Inspection

Visual inspection in its broadest sense includes observations made with the unaided eye, borescope inspections, and observations of measuring instruments. Visual inspection is usually the first method employed and is useful in assessing the macro condition of a component. It normally indicates that one of the other types of inspection should be performed.

Dye Penetrant/Fluorescent (Zyglo) Inspection

These are perhaps the most widely used methods of crack detection, especially in the field. They are both limited to the detection of surface imperfections since they depend on the liquid being drawn into the imperfection by capillary action. Dye penetrant is usually easier to use in the field since it is very portable and can be done in a variety of light conditions. Zyglo is generally preferred, where possible, since it is more sensitive, but it does require darkness and the use of a black light. Small parts such as blades which can be removed to a darkened booth, are almost always inspected by the Zyglo method (Figure 10-6).

Figure 10-6. Zyglo inspection (courtesy Canadian Westinghouse).

Magnetic Particle Inspection

This method is also widely used in industry, although it is losing some of its popularity with machinery engineers. It relies on the attraction of iron particles to discontinuities in a magnetized piece of steel. There are a number of disadvantages with this method. First, the material to be tested has to be capable of being magnetized. This leaves out most of the high nickel and cobalt alloys used in gas turbine blading, along with austenitic stainless steels, copper alloys, aluminum, etc. Materials associated with rotating machinery which have been inspected in this fashion normally retain some residual magnetism. There is evidence that this is the cause of shaft currents, which leads to premature bearing failure. Any parts which have been inspected in this manner should be checked for residual magnetism with a gaussmeter. The sensitivity of this type of inspection is also greatly dependent upon the skill of the operator and the relative orientation of the discontinuity to the applied field.

Eddy Current Inspection

Eddy current inspection utilizes the same principle as the eddy current probe (sometimes referred to as the noncontact probe, or the Bently probe). This is a relatively new technology which requires a significant amount of operator skill. It can be extremely sensitive and is most commonly applied to the detection of fine cracks in the fir-tree roots of austenitic blade materials.

Ultrasonic Inspection

Both surface and subsurface flaws can be detected by ultrasonic inspection. This technique utilizes high frequency sound waves which are introduced into the material being tested and are sensed by a probe held on the surface of the material. Ultrasonics are very versatile, but like eddy current inspection require a very skilled operator, and frequent calibration of instrumentation.

Air Inlet and Compression

Both the heavy duty industrial and the aircraft derivative gas turbine use large volumes of air. In the case of a 17,700 KW gas turbine the amount of inlet air approaches 1 million lbs per hour. If only 10 ppm of dust is carried in that air, it would be equivalent to 240 lbs of particulate passing through the machine every day.

Filtration systems are employed to provide protection from erosion, corrosion and fouling.[10, 11]

Erosion

Both the axial compressor and the hot gas path parts can be affected by erosion from hard, abrasive particles, such as sand and mineral dusts. As they impact upon the compressor blades, they cut away a small amount of metal. The net rate of erosion, although not precisely quantifiable, depends on the kinetic energy of the particles, the number of particles impinging per unit time, and the mechanical properties of both the particles and the material being eroded. In general, the experience of one major gas turbine manufacturer indicates that particles below 10 microns do not cause erosion, whereas particles 20 microns and above normally cause erosion when present in sufficient quantities.

Corrosion

Two types of corrosion are generally to be avoided. These are compressor corrosion and hot section corrosion. Compressor corrosion is typically caused by moisture in combination with sea salt and in some cases acids. They cause the compressor blades and vanes to become pitted, adding drag, increasing their susceptibility to fouling and decreasing their aerodynamic performance. Hot section corrosion is far more serious and occurs when certain metals combine with sulfur and/or oxygen during the combustion process, and deposit on the surfaces of the hot gas path parts. Of primary concern are sodium, potassium, vanadium, and lead. These metals, either as sulfates or oxides, cause the normally protective oxide film on hot gas path parts to be disrupted so that the parts oxidize several times faster than in the presence of exhaust gases free of these metals.

Fouling

Compressor fouling in the short term is the most noticeable effect of poor inlet air filtration. The type of performance monitoring outlined earlier addresses this problem. With increasing cost of fuel the savings to be realized from operating at maximum efficiency have become significant.

The efficiency of an axial compressor is dependent on, among other considerations, the smoothness of the rotating and stationary blade surfaces. These surfaces can be roughened by erosion, but more frequently roughening is caused by the ingestion of substances which adhere to the surfaces. These include oil vapors, smoke, and sea salt. The output of a turbine can be reduced as much as 10 percent by compressor fouling.

There are various methods of removing deposits from blading in use today. Some can be accomplished with the unit running and others require the unit to be shut down. The two basic types are the dry media cleaning and the solvent/water (wet) method. Dry media cleaning employs either some type of ground nutshells (usually walnuts or pecans) or a fine alumina catalyst such as is used in most Catalytic Cracking units. This method is always carried out on the run and is usually quite effective. There are some vendors who do not approve of this method and it has proven ineffective if deposits are very sticky such as those mixed with a lot of oil. The wet methods are carried out in a number of ways usually specific to the individual machine involved. Some are done while the unit is shut down, some at cranking or ignition speed and some while the engine is being fired (although normally at minimum load). One method which has been successfully applied to a two-shaft industrial turbine heavily fouled with oil and dirt was to coat the axial compressor blades with a strong industrial detergent while at ignition speed. The unit was allowed to sit for a time (rather like cleaning an oven) and then the mixture was washed off with water, again while the unit was at ignition speed.

All turbine manufacturers are interested in seeking the best quality air possible for their turbines. This usually leads to the purchase of one or two types of filtration systems. The older system includes an inertial separator, which removes coarse particles, and a high efficiency media filter (Figure 10-7). The inertial separator is usually adequate to protect against erosion, while the high efficiency media removes fine contaminants which could lead to hot path corrosion. A prefilter is sometimes

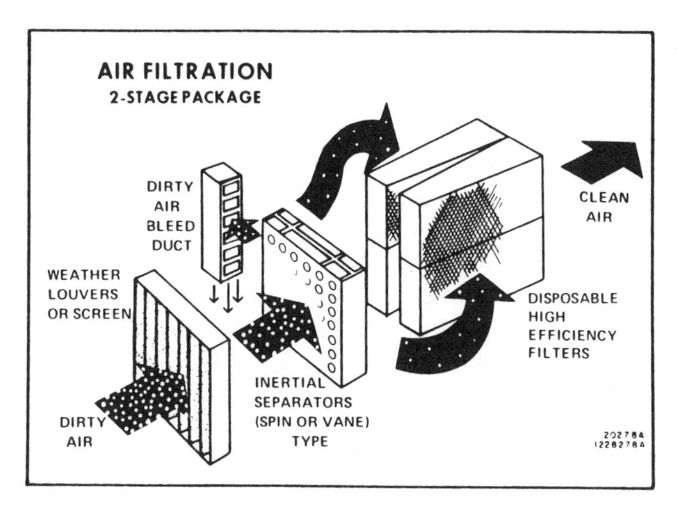

Figure 10-7. Inertial/High efficiency filter (Reference 10).

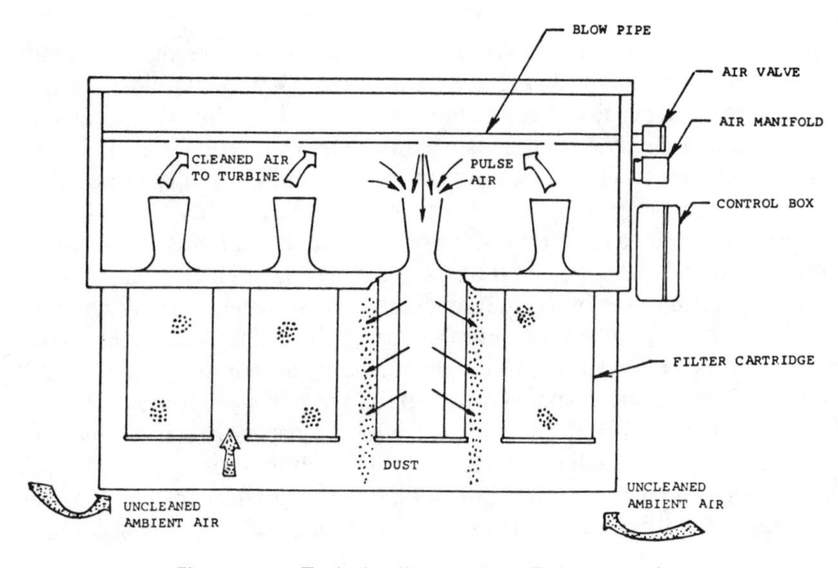

Figure 10-8. Typical pulse type filter (Reference 11).

installed to extend the life of the more costly high efficiency media. In recent years the development of the automatic self-cleaning filter (Figure 10-8) has greatly decreased the complexity of standard filtration systems and reduced the amount of maintenance required. They are designed for an element life of approximately two years at a fairly constant pressure drop of two to three inches of water column. Self-cleaning filters are especially attractive in cold climates where other filter types would require costly preheat systems for anti-icing.

Mechanically, axial compressors require very little routine maintenance. Problems normally encountered include tip rubs, sealing strip rubs, and foreign object damage (FOD). In the design of the axial, extreme care must be taken to avoid blade resonances; however, these problems are rare in the field on equipment other than prototype machines. Figure 10-9 illustrates the axial compressor rotor from a two-shaft industrial gas turbine in the balance stand. With all of the inlet precautions described it would seem unlikely that FOD would occur, but sometimes it does. Often when a shallow nick is discovered in the upper portion of the blade, it may be left "as-is" or "blended," that is ground to minimize the stress concentration factor. If it is too deep or too low on the blade profile, blade replacement may be necessary. This can be either very easy or very difficult depending on the design of the specific axial compressor.

Some axials have wheel spaces which are wider than the width of a blade and facilitate the individual replacement of blades. Other models require that the rotor disks be completely disassembled to get at a single blade. The maintenance savings in the first instance are obvious.

Fuel System and Combustion

Gaseous or liquid fuels can be burned in gas turbines. Since fuels vary widely in calorific value, hydrocarbon composition, contaminants present and viscosity, each fuel must be considered separately in terms of engine performance, reliability, operation, and maintenance.[7]

The use of natural gas fuel for industrial gas turbines is wide-spread. Natural gas is generally considered to be the optimum fuel when supplied to the unit in clean, dry and sweet form. Machines fired on good quality natural gas will invariably exhibit superior operation compared to similar

Figure 10-9. Axial compressor rotor in balance stand (courtesy Canadian Westinghouse).

units operating on liquid fuels. In line with most turbine manufacturers' recommendations, the gas supplied should be free of particulate, contain the minimum amount of hydrogen sulfide, and should be at least dry and preferably have 20°F superheat at entry to the turbine fuel system.

In general, problems associated with gas fuel systems are either caused by variations in the calorific value of the gas or by the condensation of hydrocarbon liquids. Due care and attention should be paid to the design and operation of the knock-out facilities associated with the gas turbine. A secondary knock-out drum or inline cyclone separator should be installed as close to the turbine as possible, to eliminate any condensation in the line from the primary knock-out facility.

Operation on distillate fuels such as diesel fuel generally does not present a major problem. This type of fuel is usually free of the type of contaminants which cause hot corrosion of the turbine blades. Crude oil, on the other hand, represents a serious risk of hot gas path corrosion if not properly treated to remove the heavy metal contaminants which are usually present.

Life expectancies for combustion hardware operated on liquid fuels are generally less than those which are operated on gaseous fuels. Gas turbines operated on dual fuel, that is liquid and natural gas, suffer from coke build-up in the liquid fuel nozzles if the liquid nozzles are not adequately purged when not in use. Purging is usually done with air or natural gas.

The basic components of the combustion system are the fuel nozzles, combustor baskets, transition pieces, cross flame tubes, and associated seals, clips, etc. A typical primary combustor basket from a heavy duty industrial turbine is shown in Figure 10-10.

Combustion system hardware is generally the portion of the gas turbine that has the shortest life and requires the most maintenance. This is due to the extreme turbulence around the combustors, and the high temperatures associated with the combustion process. The key to extending the maintenance interval on most gas turbines lies in extending combustion system component life.

Both fretting and fatigue cracking are common on combustion system components. The combustors must be designed to accommodate thermal expansion, but yet be rigid enough to handle the turbulence in the area. This normally involves sliding fits, which are prone to fretting. Numerous holes and spot welds provide stress concentration points to initiate fatigue cracks, which are propagated through the hot metal by vibratory forces caused by turbulence. Design improvements seek to minimize these problems, but to date have not eliminated them. Combustor inspections are hopefully timed to catch failures of these components before they cause serious downstream damage to the machine.

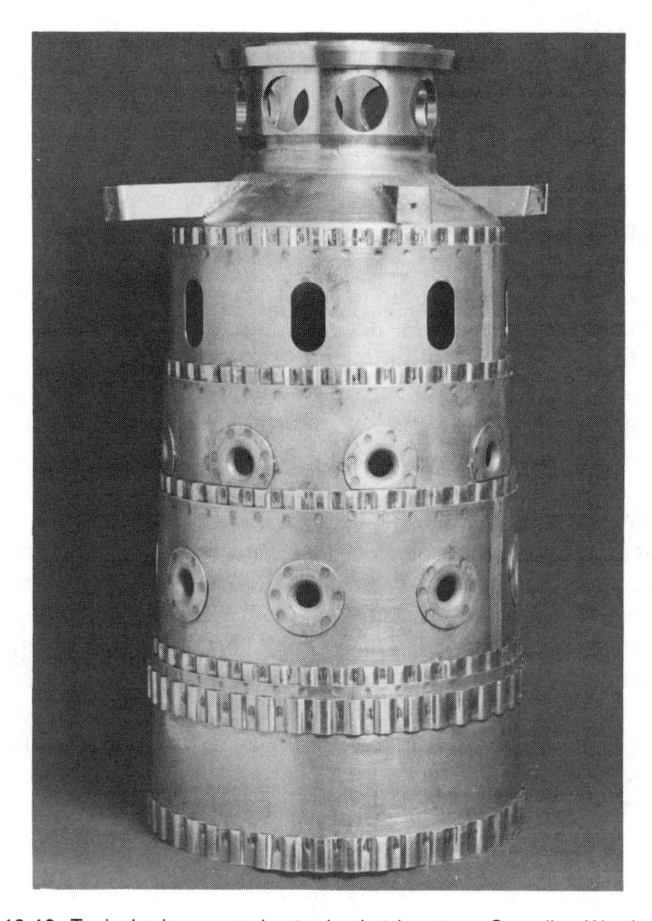

Figure 10-10. Typical primary combustor basket (courtesy Canadian Westinghouse).

Turbine and Exhaust

Whereas the combustion system of a gas turbine requires most of our routine maintenance, the downstream hot parts are where maintenance becomes most expensive. Annual expenditures for routine maintenance on a gas turbine may average ten to twenty percent of the initial purchase price.[4]

In order to achieve the highest possible operating efficiency, gas turbine combustion temperatures are kept high. Thus, turbine inlet temperatures are held rather high. This, in turn, generates severe thermal stressing in both rotating and stationary hot components, stressing which taxes the properties of even the most up-to-date superalloys used as materials of construction.

As the operating temperature for gas turbine components has increased, the development of suitable alloys has progressed. Each new development has resulted in increased high temperature strength, improved creep and fatigue resistance, or enhanced corrosion resistance. However, minor deviations in temperature control, burner condition, cooling air flow or fuel quality can cause sudden and major damage to the hot components. Component deterioration due to tensile fracture, thermal fatigue, or hot corrosion results in varying degrees of metal disintegration and further damage from erosion or more severe mechanical impact. Progressive corrosion in a typical superalloy and foreign object damage can result when upstream components break up and strike downstream components such as turbine blades.

In most heavy duty industrial engines the life of the first stage turbine nozzles limits the interval between major overhauls. Typical damage includes sidewall and vane cracking, vane bulging, impact damage and, to a lesser extent, hot corrosion. These problems tend to be worsened by the collapse of cooling core inserts and plugging of vane section cooling air holes. A trend seems to have been established to group these vanes in smaller segments. This minimizes the thermal variations within the segment. The probability of cracking is reduced and replacement of a smaller piece is made possible when cracks do develop. One design has gone so far as to use single vanes which are replaceable without removing the upper half of the casing.

Since vane segments are stationary parts not subjected to extremely high stress levels, it is possible to carry out considerable weld repair of cracks and impact damage. Whole sections of the vane may be cut out and new pieces welded in place.

Rotating turbine blading, especially the first stage rotor blades, are some of the most vital components of any gas turbine. First stage blades see the most severe environment in terms of stress levels and metal temperatures. The most common defects in turbine blading result from impact damage, hot corrosion, thermal fatigue, and creep void growth due to long term exposure to stress and temperature.

Weld repairs to these blades are generally restricted to the restoration of blade tips damaged because of rubs, cracking, or minor foreign object damage. Creep damage may be reversed through hot isostatic pressing (HIP). This technique may also be used to enhance the as-cast properties of new blades prior to putting them into service.[5]

It is typically claimed that parts can be recovered in "as new" condition for between 30 and 60 percent of the cost of a new part. Care should be exercised to ensure that the part recovery procedure is proven and economic in terms of cost versus additional life expected. Newer and better repair techniques are being developed rapidly. As confidence in these remanufacturing procedures increases, it is expected that recycling of hot gas path components will become a more accepted practice in the process industries.

Controls

Gas turbine control systems are without doubt the most complex controls on any form of rotating machinery. Driven by advances in digital technology, these systems are undergoing rapid changes. To generalize about controls is almost impossible, since nearly every gas turbine manufacturer uses his own control system. Moreover, even these are not always consistent from one model to another.

Reliability is the key to all control systems. Reliable starting, stopping, and governing are the basic objectives. The type of execution used to achieve this objective has developed from pneumatic and relay-based systems, through solid state analog systems to fully redundant digital systems. Figure 10-11 shows the type of control panel which was commonly

Figure 10-11. W191 Control panel (courtesy Westinghouse Canada).

Figure 10-12. CW352 Control Panel (courtesy Westinghouse Canada).

designed for machines through the 60's and early 70's. They attempted to achieve reliability through simplicity. In the 70's, solid state electronics became popular for gas turbines. Digital and digital/analog hybrid systems (Figure 10-12) were developed. These systems allow much more flexibility at the expense of simplicity. Currently manufacturers and users alike debate whether added complexity and redundancy provide reliability.

Any control system faces its most severe test during the start cycle.[2] Consequently more problems are encountered at this time and these problems are the most difficult to troubleshoot. It is critical to understand the sequence of events which the control system is trying to execute.

The normal sequence of events in the starting of a gas turbine is:

1. The starter is energized and the gas turbine accelerates to ignition speed.
2. The fuel valve is opened and the ignition system energized. Some period of time is alotted to confirm that lightoff has occurred.
3. The turbine is accelerated using an acceleration fuel schedule along with help from the starting device.
4. At some point the gas turbine achieves a self sustaining speed and the starting device drops out. The turbine continues to accelerate until it reaches the speed setting of the governer.
5. After a suitable warm-up period, load is applied to the turbine.

All control systems, no matter how complex, receive information about the turbine from devices mounted on the machine such as thermocouples, transmitters, probes, UV flame detectors etc. These devices are critical and should be checked at every service outage. Sensors which have failed should be replaced and those which can be adjusted recalibrated.

Lube Oil System

As in other process plant machinery trains, the quality of the lube oil system which is supplied with the train has a definite impact on the service factor of the unit. Gas turbines are normally supplied with a lube oil package mounted in the bedplate of the turbine. These packages do not normally meet the type of standards (such as API 614) which have been developed to ensure the reliability desired for refinery and process plant application. The gas turbine purchaser may have a number of options from which to choose.

1. Accept the vendor's standard. In some cases this is all that can be done and as long as vendor and user agree on its adequacy, this will be the least expensive and easiest option.
2. Review and propose changes to the vendor's standard package. This method is the most common. Care should be taken to avoid making the vendor's design less reliable and unnecessarily expensive.

3. Purchase an API system. This will normally mean that the vendor's standard system in the bedplate must be removed and a separate system installed. This type of system is not practical with many gas turbines, since the lube oil system is heavily integrated into the vendor's design. In a process plant, reliability improvements can nevertheless be made by this approach. A steam turbine driven main lube oil pump can eliminate the auxiliary drive gearbox and its associated couplings, etc., making the gas turbine train itself less complex.

Regardless of the type of lube oil system supplied, the calibration of instruments associated with the system is critical in maintaining a high service factor. These instruments should be located in such a way as to ensure they are not physically abused, and are readily available for calibration during maintenance outages. A cabinet containing all of the pressure switches and gauges associated with the gas turbine such as shown in Figure 10-13 facilitates maintenance. This cabinet can be located as

Figure 10-13. Pressure switch and gauge cabinet (courtesy Canadian Westinghouse).

Figure 10-14. Gas turbine lube oil system showing gauge cabinet (courtesy Canadian Westinghouse).

shown in Figure 10-14, away from heat and vibration, yet close enough for easy access.

Oil leaks present a greater hazard on gas turbines than on most other types of machinery. Care must be taken to minimize the number of places where leaks can occur, especially in hot areas. However, oil leaks will occur and one should ensure that provision is made to collect the oil in a safe location.

Maintenance Records

Once again, gas turbines represent a slightly different problem in terms of record keeping. Generally much more information is supplied with a gas turbine than any other type of machine. As the gas turbine is operated, inspected, and overhauled, a great deal of valuable information becomes available. If collected and properly organized, these data will form the cornerstone of an effective maintenance strategy.

Maintenance records should include a maintenance log of work performed on the turbine and operating data from the machine as well. This does not mean that the operating log sheets should be kept in the maintenance file, but that operating parameters around the engine should be recorded before and after mechanical work to provide a baseline operating record.

Effective gas turbine maintenance requires documenting the condition of every part on an inspection or overhaul. Most parts have serial numbers, which makes it much easier to keep track of the history of a single component. Pictures and sketches are imperative to keep track of accumulated fretting, cracking, foreign object damage and weld repairs on combustion and hot gas path parts.

Control systems on gas turbines which employ programmable logic represent a special problem in terms of maintenance records. It is imperative that all changes, no matter how small, be recorded and logic diagrams modified to reflect these changes. The flexibility to easily modify control logic is a tremendous asset, but a much worse liability if record keeping is not kept up.

Keeping maintenance statistics is also useful in order to assess how the gas turbine is performing compared to others. To achieve a fair comparison, reliability and availability statistics should only be compared on units with similar use factors.[7] In addition to indices which are normally kept on other major machinery trains (reliability and availability), two other indices are useful in comparing gas turbines. Starting reliability is defined as the ratio of successful starts to the total number of starts attempted. Since starting of the turbine so heavily involves the control system, this index is primarily a measure of the performance of the controls.

"Use Factor" is important in comparing gas turbines, since a significant number of turbines are not required to operate continuously. This is generally not the case in process plants, but turbines can record remarkable reliabilities and availabilities if they are only run 10 to 20% of the year.

It is equally important to record the number of starts which the turbine undergoes each year or per maintenance interval. This record should be closely monitored to establish what type of thermal cycles the unit has experienced.

Maintenance record-keeping and indices close the loop in the maintenance strategy which has been developed for a unit. Merely keeping records for records sake is not appropriate. For best results, much effort must go into the analysis of those records. It is important to determine whether the maintenance strategy as initially defined is optimum for the unit in question. This means a continuous reassessment of spare parts stocking levels, machine design, repair procedures, control strategy, and overall operational philosophy to ensure that corporate objectives are being met.

Maintenance of Hydraulic Governors*

This segment will provide typical adjustment and maintenance information for mechanical speed governors. The mechanical governor as described here is the Woodward PG-PL (pressure compensated governor) type with a pneumatic speed setting mechanism. The PG-PL shown in Figure 10-15 is representative of most mechanical governors found in the petrochemical industry. All PG governors have the same basic components regardless of how simple or complex the complete control may be.

The standard oil pressure for PG governors is 100 psig. However, with appropriate modifications the oil pressure may be increased, thus increasing the work capacity of the power cylinder assembly.

Control air pressure must be three psig minimum and 100 psig maximum. Typical pneumatic ranges are 3 to 15 psig and 10 to 60 psig.

Linkage Adjustment is Important

The linkage from the governor to the fuel or steam control should be properly aligned. Any friction or lost motion should be eliminated. Unless the engine or turbine manufacturer has given special instructions, the linkage should be adjusted so that when the governor power piston is at the end of its stroke in the "OFF" direction, the gas or steam valve, or diesel fuel pumps will just be closed.

Oil Specifications Must be Observed

Use SAE 20 or 30 oil for ordinary temperature conditions. If governor operating temperatures are extremely hot, use SAE 40 to 50; if extremely cold, use SAE 10. In most cases, the same oil that is used in the engine or turbine may be used in the governor.

Keep the governor oil level between the lines on the glass of the oil level gauge when the engine or turbine is running. The oil should never be above the line where the case and column castings meet. Oil above this level will be churned into foam by rotation of the flyweight head. The governor can run safely with the oil level quite low in the gauge glass.

It is important to always check governor oil level during operation once the governor has reached normal operating temperature. Add oil, if necessary, to the required level.

---

* Compiled by R. S. Adamski, using copyrighted material provided by and with permission of the Woodward Governor Company, Fort Collins, Colorado

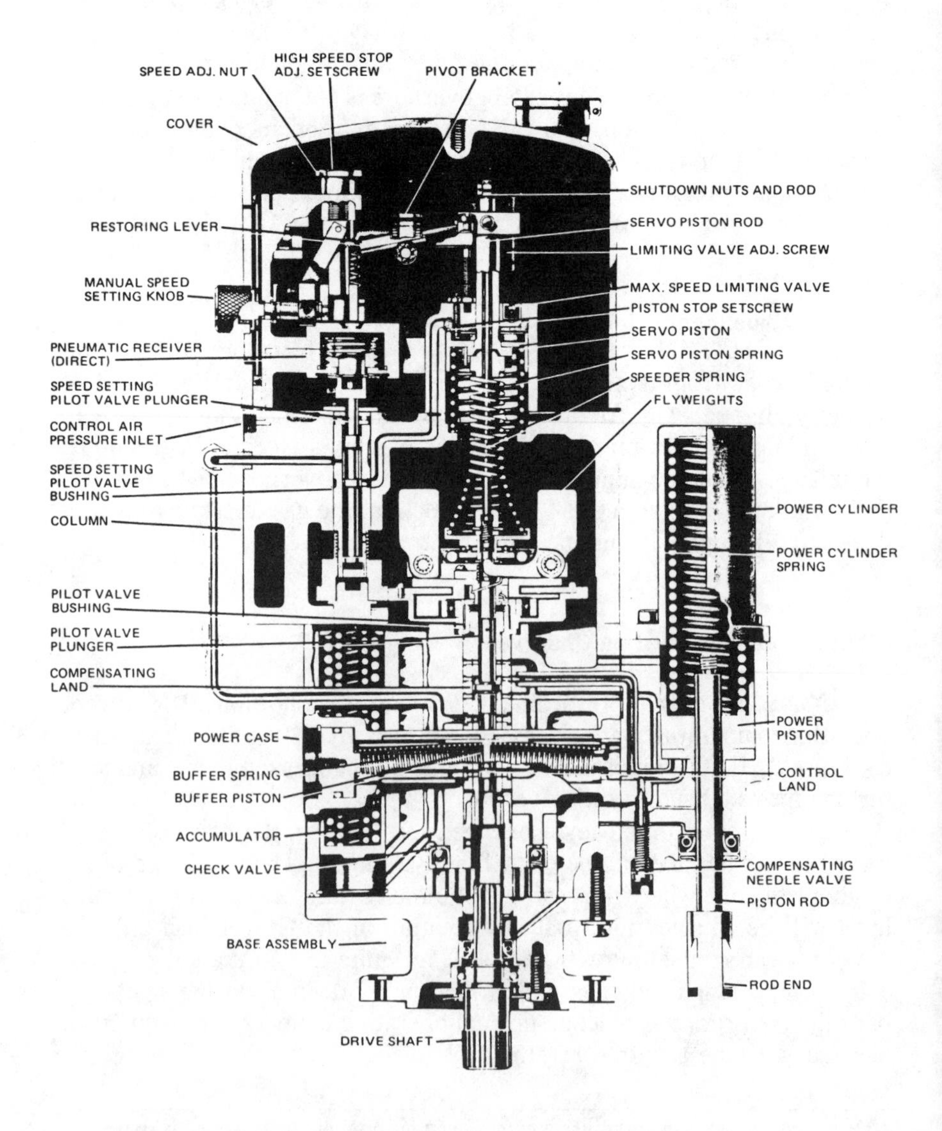

Figure 10-15. Cutaway view of PG-PL Governor.

Air Must Be Purged and Needle Valve Adjustment Made

When the engine or turbine is started for the first time, or after the governor has been drained of oil, cleaned and refilled with oil, any trapped air in the governor must be removed using the following steps:

1. Start the engine or turbine and run it at idle for at least 15 minutes to allow the governor and engine to warm up.
2. Add oil to the governor to maintain the oil level between oil level lines on the gauge.
3. Back off needle valve (ccw) several turns to allow the governor to hunt. Close the needle valve (cw) until hunting just stops. If the needle valve closes completely before hunting stops, proceed with the next step.
4. If hunting has not stopped, open the needle valve several turns to allow the governor to hunt for about five minutes using as much terminal shaft travel as possible.
5. Repeat Step 3. If the governor continues to hunt proceed with the following.
   Caution: The vent screw is under pressure. Do not remove while operating the governor.
6. Loosen the vent screw on the side of the governor case, enough to establish an oil leak. Bleed until the air bubbles stop, about one-half cup of oil.
7. Tighten the vent screw and refill the governor with oil. If the vent screw leaks after tightening, shut down the engine and remove the plug. Coat the plug with a good grade of pipe sealant, replace the plug and tighten it.
8. Repeat Step 3.
9. Run at normal maximum operating speed and check governor stability. The needle valve may have to be closed slightly to achieve stability.

With preloaded buffer springs, the needle valve should be no more than 1/16 turn open for smooth operation. The needle valve must never be closed tight, as the governor cannot operate satisfactorily when this condition exists.

On some installations, opening the needle valve will not cause the engine or turbine to hunt. In such cases, bleed the air from the governor by disturbing engine or turbine speed to cause the governor to move through full stroke in both directions a sufficient number of times to force out all trapped air.

After the needle valve is adjusted correctly for the engine, it should not be necessary to change the setting except for a large permanent temperature change affecting the viscosity of the governor oil.

Speed Adjustment

The pneumatic speed setting mechanism furnished with the governor is either a direct type which increases the governor speed setting as the control air pressure signal increases, or a reverse type which increases governor speed setting as the control air pressure signal decreases. Perform the following procedures as applicable to set the maximum and minimum operating speed of the governor. See Figures 10-15 and 10-16.

Direct Speed Setting Mechanism

1. Set the manual speed adjusting knob to the minimum speed position (fully counterclockwise until clutch slips).
2. Adjust the high speed adjusting setscrew as required until upper end of screw is flush with top of speed setting screw.
3. Apply specified minimum control air pressure signal to the unit; adjust the speed adjusting nut as required to obtain corresponding specified minimum speed (clockwise to decrease); be sure the pneumatic low speed adjusting screw does not touch the restoring lever at this time.
4. Adjust the limiting valve adjusting screw as required so that it does not unseat the maximum speed limiting valve as speed is increased. Set governor speed range to control air pressure range as follows:

   a. Slowly increase control air pressure signal to maximum. Be sure engine does not exceed specified maximum speed.
   b. If specified maximum speed is obtained before control air pressure signal is increased to maximum, adjust the pivot bracket to move the ball bearing pivot toward the speed setting servo.
   c. If specified maximum speed is not obtained with maximum control air pressure signal, adjust the pivot bracket to move the ball bearing pivot away from the speed setting servo.
   d. Adjust the pivot bracket as follows: Loosen the socket head screw in top of the pivot bracket; loosen knurled nut on appropriate side of bracket and turn opposite knurled nut to move bracket; tighten screw and knurled nuts.

5. Repeat Steps 3 and 4 until specified minimum speed is obtained with minimum control air pressure and specified maximum speed is

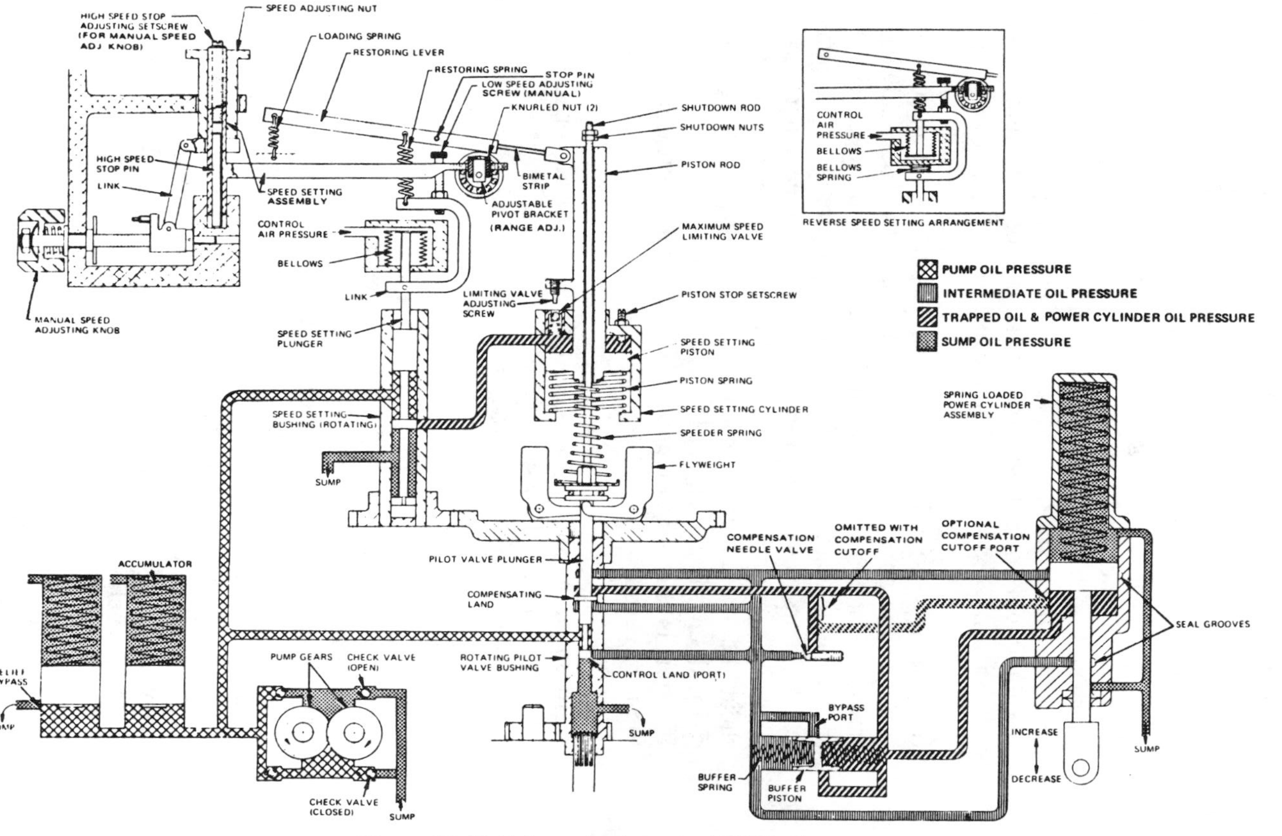

Figure 10-16. Schematic diagram of PG-PL Governor.

obtained with maximum control air pressure. Speed should begin to increase as the control air pressure begins to increase from minimum.

6. Apply maximum control air pressure to maximum speed. Adjust the limiting valve adjusting screw so that it just contacts the ball in the maximum speed limiting valve. Increase control air pressure slightly above specified maximum; the maximum speed limiting valve should open prior to engine reaching five rpm above specified maximum speed. Readjust screw as necessary.
7. Apply minimum control air pressure signal for minimum engine speed. Perform Step a or b as applicable.

   a. If engine is to go to low speed upon loss of control air pressure signal to the governor, set the pneumatic low speed adjusting screw to just contact the stop pin in the restoring lever with the engine running at low speed. Shutdown nuts are usually omitted on governors which are arranged to go to low speed upon loss of control air pressure. If nuts are included but not used, lower nut should be a minimum of 1/32-in. above the speed setting piston rod with engine running at low speed.
   b. If engine is to shut down upon loss of control air pressure signal to the governor:

      - Lift up on the shutdown rod to take out any slack or lost motion; do not lift the rod so far as to cause the engine speed to drop. While holding the rod up, position the lower shutdown nut 1/32-in. above the top of the speed setting servo piston rod and lock in position with upper nut.
      - Turn the piston stop setscrew down until it touches the speed setting piston, then turn the screw counterclockwise two turns and lock in position with nut. This adjustment limits the upper movement of the piston when the engine is shut down, and it minimizes the cranking required when the engine is restarted.
      - Adjust the pneumatic low speed adjusting screw so that it is .040–.050 in. below the stop pin in the restoring lever. Turn off control air pressure signal to the governor (engine will shut down). Adjust the adjusting screw so that it is from .002 to .005 in. below the stop pin in the restoring lever.

8. With control air pressure signal removed (engine does not go to shutdown with loss of control air pressure signal), turn the manual speed adjusting knob clockwise to increase engine speed to maxi-

mum. Turn the high speed adjusting setscrew in until it touches the high speed stop pin (this adjustment stops the downward movement of the speed adjusting nut at high speed).

Reverse Speed Setting Mechanism

1. Set the manual speed adjusting knob to the minimum speed position (fully counterclockwise until clutch slips).
2. Adjust the speed adjusting nut so that the speed setting assembly protrudes approximately 1/4-in. above the nut.
3. Adjust the high speed adjusting setscrew as required until screw is flush with the top of speed setting screw.
4. Adjust the limiting valve adjusting screw as required so that it does not unseat the maximum speed limiting valve as speed is increased. Apply minimum control air pressure signal to the governor (pressure at which specified maximum engine speed is to be obtained). Be careful that engine does not exceed specified maximum speed.
5. Turn the manual speed adjusting knob clockwise to increase engine speed to specified maximum. Turn the high speed adjusting setscrew in until it just touches the high speed stop pin. If screw is turned down too far, speed will decrease.

   If the specified maximum speed is not obtained with the manual speed adjusting knob fully clockwise, turn the knob approximately two turns counterclockwise, back out high speed stop adjusting nut counterclockwise until specified maximum speed is obtained. Turn high speed adjusting setscrew down until it touches the high speed stop pin (if the screw is turned down too far, speed will decrease). Turning the speed adjusting knob fully clockwise should not increase speed beyond the specified maximum.
6. Slowly increase control air pressure signal until specified minimum speed is obtained. The pneumatic low speed adjusting screw should not touch the stop pin in the restoring lever and the piston stop setscrew should not stop the speed setting piston as it moves up to decrease speed.

   If specified minimum speed is obtained before the control air pressure signal is increased to specified maximum, adjust the pivot bracket to move the ball bearing pivot toward the speed setting cylinder.

   Adjust the adjustable pivot bracket as follows: Loosen the socket head screw in top of pivot bracket; loosen knurled nut on appropriate side of pivot bracket and turn opposite knurled nut to move the pivot bracket; tighten screw and knurled nuts.

7. Repeat Steps 4, 5, and 6 until specified minimum speed is obtained with maximum control air pressure signal and specified maximum speed is obtained with specified minimum control air pressure signal. Ensure engine speed begins to increase as the control air pressure signal begins to decrease from maximum.
8. After setting speeds pneumatically, apply minimum control air pressure signal (governor will go to maximum speed setting). Turn manual speed adjusting knob counterclockwise until specified minimum speed is obtained. Alternately turn speed adjusting nut 1/2 turn counterclockwise (increasing speed) and adjusting knob counterclockwise (decreasing speed) until adjusting knob is fully counterclockwise. Turn off control air supply (speed will rise slightly). Adjust speed adjusting nut to obtain specified minimum speed. If adjusting nut is turned fully counterclockwise without reaching the specified minimum speed, turn off control air supply (speed will rise slightly). Adjust speed adjusting nut to obtain specified minimum speed.
9. With the engine operating at specified minimum speed, turn the piston stop set screw down until it just touches the top of the speed setting piston; then turn the screw two turns counterclockwise; lock in position with lock nut. This adjustment limits the upward movement of the piston when the engine is shut down, and it minimizes the cranking required when the engine is restarted.
10. If shutdown nuts are used, lift up on the shutdown rod to take out any slack or lost motion; do not lift the rod so far as to cause the engine speed to drop. While holding the rod up, position the lower shutdown nut 1/32-in. above the top of the speed setting servo piston rod and lock in position with upper nut.
11. With the control air pressure signal turned off, turn the manual speed adjusting knob clockwise to increase engine speed to maximum. Adjust the limiting valve adjusting screw so that it just contacts the ball in the maximum speed limiting valve. Increase engine speed slightly above the specified maximum; the maximum speed limiting valve should open prior to the engine reaching five rpm above maximum speed. Readjust screw as necessary.
12. Turn the manual speed adjusting knob fully counterclockwise and apply maximum control air pressure to the governor. Adjust the pneumatic low speed adjusting screw to just contact the stop pin in the restoring lever with the engine running at low speed.
13. Turn the manual speed setting knob fully clockwise for normal pneumatic speed control operation.

Malfunctions

The source of most troubles in any governor stems from dirty oil. Grit and other impurities can be introduced into the governor with the oil, or from foam when the oil begins to break down (oxidize) or become sludgy. The moving parts within the governor are continually lubricated by the oil within the governor. Thus, grit and other impurities will cause excessive wear of valves, pistons, and plungers, and can cause these parts to stick and even "freeze" in their bores.

In many instances erratic operation and poor readability can be corrected by flushing the unit with fuel oil or kerosene while cycling the governor. The use of commercial solvents is not recommended as they may damage seals or gaskets.

If the speed variations of the governor are erratic but small, excessive backlash or a tight meshing of the gears driving the governor may be the cause. If the speed variation is erratic and large and cannot be corrected by adjustments, the governor should be repaired and/or replaced.

Lubricant Quality

The oil used in the governor should be clean and free of foreign particles to obtain maximum performance from the governor. Under favorable conditions, the oil may be used for six months or longer without changing. Change oil immediately when it starts to break down or darken.

Disassembly

Disassemble the governor following the sequence of index numbers assigned to Figures 10-17 and 10-18, giving special attention to the following. Circled index numbers do not require further disassembly unless replacement parts are required.

1. Clean exterior surfaces of governor with clean cloth moistened with cleaning solvent.
2. Discard all gaskets, O-rings, seals, retaining rings, cotter pins, clips, etc., removed in the process of disassembly.
3. Do not remove press fit components unless replacement is required.
4. Disassemble power cylinder assembly as applicable.
5. Disassemble base assembly as applicable.
6. To remove accumulator springs and pistons from the power case, place the power case (260 Figure 10-18) in an arbor or drill press

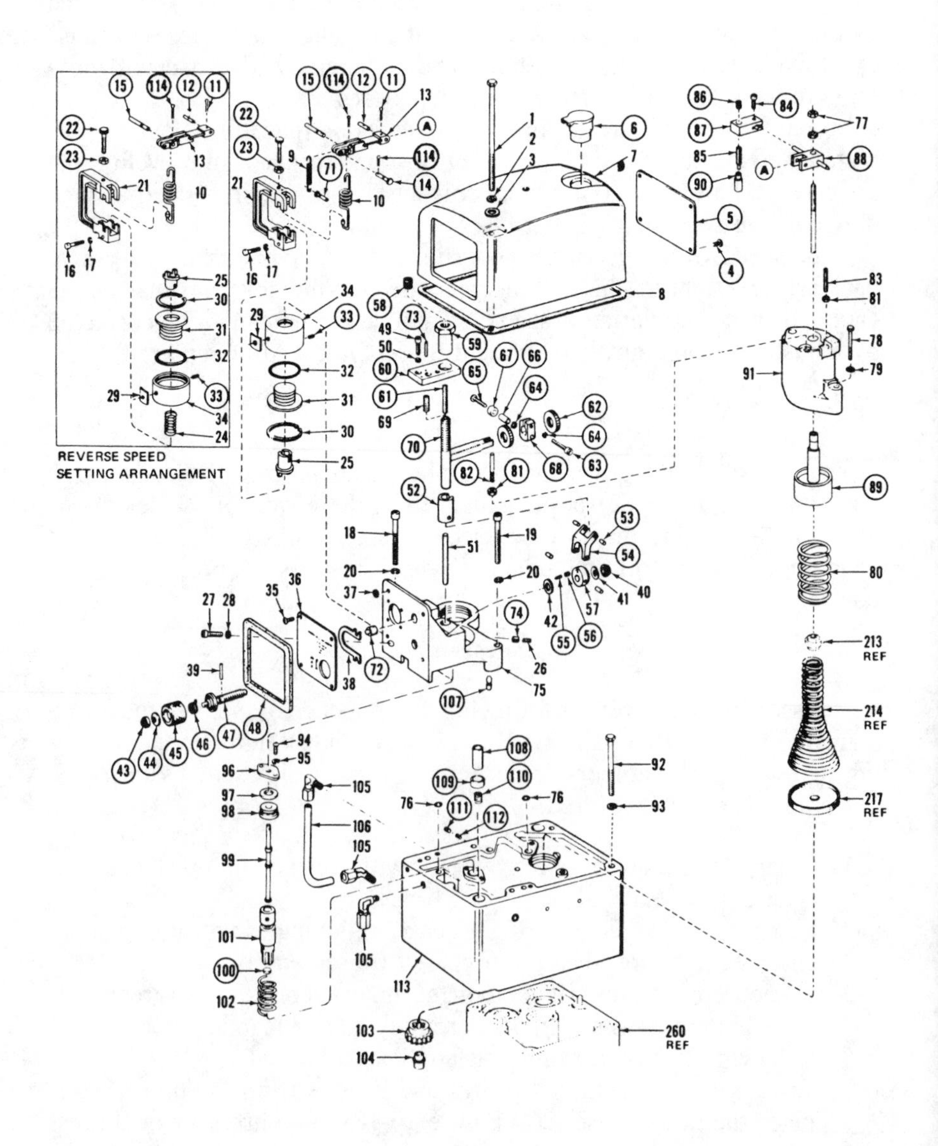

Figure 10-17. Exploded view of column.

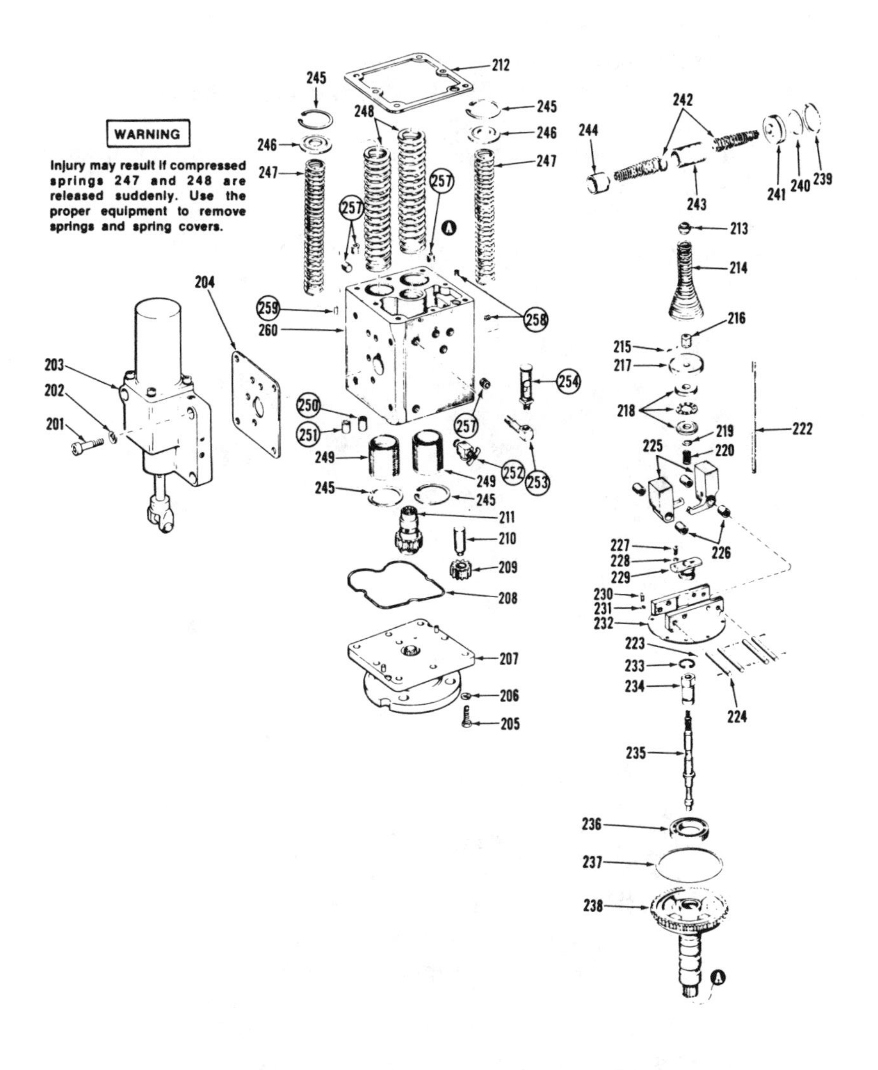

Figure 10-18. Exploded view of case.

with the bottom down. With a rod against the spring seat (246), compress accumulator springs (247 and 248) to permit removal of upper retaining ring (245). Remove spring seat and springs (see Figure 10-19). Invert the power case and remove lower retaining ring and accumulator piston (249).

7. If necessary to remove check valve assemblies (250 and 251), proceed as follows:

   a. To remove inner check valves (250), pry the retainer plate from the check valve assembly and remove springs and check balls.
   b. To remove outer check valves (251), press the check valves through and out of the valve case.
   c. Then tap all four check valve cases with 1/4-28 in. tap. Using a 1/4-28 in. bolt with a small plate as a jack, pull the four valve cases.
   d. Remove two balls from the lower case.

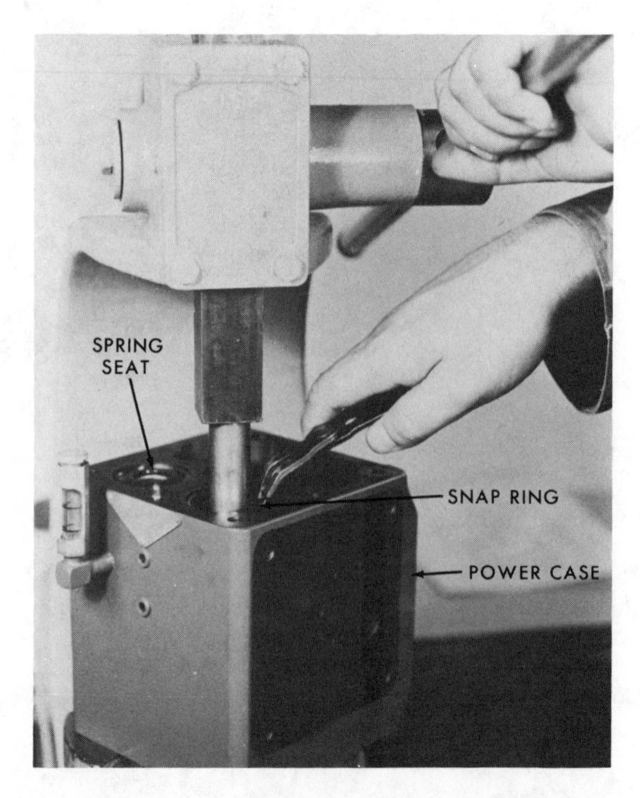

Figure 10-19. Removal of spring seat and springs.

Cleaning

1. Wash all parts ultrasonically or by agitation while immersed in cleaning solvent.
2. Use a nonmetallic brush or jet of compressed air to clean slots, holes, or apertures.
3. Dry all parts after cleaning with a jet of clean, dry compressed air.

Inspection

1. Visually inspect all parts for wear and damage.
2. Inspect bearings in accordance with standard shop practice. Replace bearings when there is any detectable roughness.
3. All pistons, valves, plungers, rods, and gears should move freely without excessive play. Do not lap in parts if possible to free by other means.
4. Mating surfaces must be free of nicks, burrs, cracks, or other damage.
5. Inspect flyweight toes for wear. Replace flyweights if any detectable wear or doubtful areas are found.
6. It is recommended that speeder spring be replaced at time of overhaul if it is pitted or damaged in any way.

Repair or Replacement

1. Repair of small parts of the governor is impractical and should generally be limited to removal of nicks and burrs from mating flanges and light burnishing of mating parts.
2. Replace damaged thread inserts in accordance with standard shop practice.
3. Polish slightly corroded areas with fine grit (600 grit) abrasive cloth, or paper and oil.

Assembly

Assemble governor assembly in reverse order of index numbers assigned to Figures 10-17 and 10-18, following the special instructions given here.

Note: A dust-free area is recommended for assembly if acceptable results are to be obtained.

During assembly ensure no lint or other foreign matter is present on the parts. The governor may be assembled dry or a small amount of clean lubricating oil may be applied to the parts as they are assembled into the governor. When the governor is assembled, apply a liberal amount of clean lubricating oil over all moving parts to ensure initial lubrication. Apply a small amount of joint compound to pipe plug threads as plugs are installed. Ensure compound does not enter cavity.

Obtain new gaskets, O-rings, seals, retaining rings, cotter pins, etc., to replace those discarded during disassembly.

1. Press spring loaded check valve (250, Figure 10-18) into power case (260). Press plain check valve into power case.
2. Install accumulator piston (249) and lower retaining ring (245) into power case. Place power case in an arbor or drill press with bottom down, (see Figure 10-19) install springs (247 and 248) and spring seat (246); compress springs, using a rod on spring seat, and install upper retaining ring.
3. Attach base assembly (207) to power case loosely, rotate drive shaft until splined and engages with splines in pump drive gear. Continue turning drive shaft to check for alignment and free rotation of the drive gear and idler gear while tightening base screws.
4. Attach power cylinder assembly (203) to power case in the proper plan and quadrant; ensure holes in gasket (204) are aligned with holes in power case.
5. When assembling the flyweight head pilot valve bushing assembly (238), align the missing tooth in the bushing with the corresponding missing tooth in the spring coupling assembly (229).
6. Install three piece thrust bearing (218) onto stem of pilot valve plunger (235)—bearing race with the larger hole must be against the flyweight toes.
7. When Items 216 through 238 have been assembled, center pilot valve plunger as follows (Figure 10-20): Apply a slight pressure to speeder spring seat (217), adjust pilot valve plunger nut (216) until flyweights (225) move from their extreme inward to their extreme outward position and there is the same amount of control land showing in the control port at such extreme. The control ports are the bottom holes in the pilot valve bushing.
8. When assembling speed setting mechanism, ensure retaining ring (30, Figure 11-3), is positioned with opening in line with set screw (33).
9. Assemble manual speed setting shaft assembly (43 through 47), tighten nut (43) approximately seven turns; insert roll pin (39) to protrude through shaft (43) approximately .090-in.

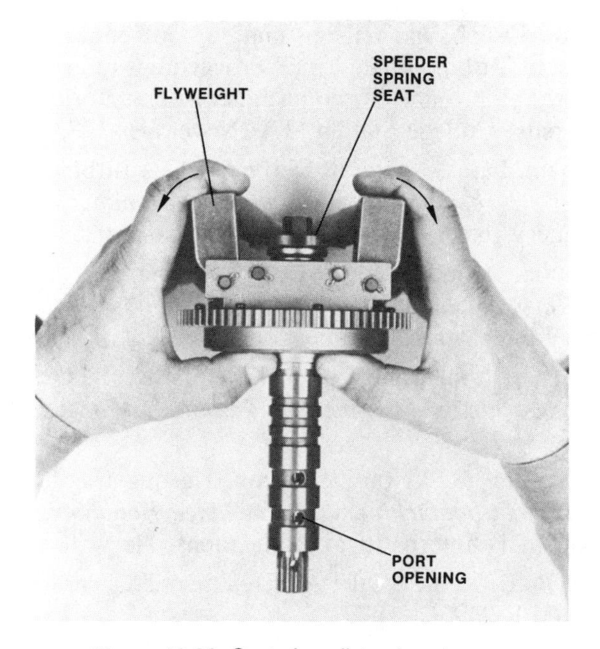

Figure 10-20. Centering pilot valve plunger.

References

1. "Comparison of Jet Derivative Gas Turbines with Heavy Duty Industrial Gas Turbines." Canadian Westinghouse Technical Data 09-100, December 1975
2. Tushinski, A. H., "Small to Intermediate Horsepower Industrial Gas Turbines." *Sawyer's Turbomachinery Maintenance Handbook, Vol. 1.*, Turbomachinery International Publications, Norwalk, Conn (1980)
3. Hall, Chester D., "A Technical Approach to the Maintenance and Overhaul of Gas Turbines." Proceedings of the 12th Turbomachinery Symposium, Turbomachinery Laboratories, Texas A&M University, College Station, TX (November 1983)
4. Smit, Jan, and Capstaff, Arthur E. Jr., "Hot Component Repairs—for Heavy Duty Gas Turbines." *Sawyer's Turbomachinery Maintenance Handbook, Vol. 3.*, Turbomachinery International Publications, Norwalk Conn (1980)

5. Liburdi, Joseph, and Wilson, James, "Guidelines for Reliable Extension of Turbine Blade Life." Proceedings of the 12th Turbomachinery Symposium, Turbomachinery Laboratories, Texas A&M University, College Station, TX (November 1983)
6. Langham, Ian J., "Aircraft Derivative Gas Turbines in Pipelines." *Sawyer's Turbomachinery Maintenance Handbook, Vol 1.*, Turbomachinery International Publications, Norwalk, Conn (1980)
7. Anderson, A. W., "Heavy-Duty Industrial Gas Turbines." *Sawyer's Turbomachinery Maintenance Handbook, Vol. 1.*, Turbomachinery International Publications, Norwalk, Conn (1980)
8. Saravanamuttoo, H. I. H., and MacIsaac, B. D., "Thermodynamic Models for Pipeline Gas Turbine Diagnostics." ASME paper 83-GT-235
9. Hoppe, Paul J., "Non-Destructive Testing Gas Turbine Parts." *Sawyer's Turbomachinery Maintenance Handbook, Vol. 3.*, Turbomachinery International Publications, Norwalk Conn (1980)
10. Tatge, R. B., et al., "Inlet Air Treatment." General Electric Company, GER-2490C (1979)
11. Ostrand, Gary G., "Gas Turbine Inlet Air Filtration." *Sawyer's Turbomachinery Maintenance Handbook, Vol. 3.*, Turbomachinery International Publications, Norwalk Conn (1980)

Chapter 11

Maintenance of Electrical Motors and Associated Apparatus

Electric Motor Maintenance*

Electric motors, particularly those using alternating current, have become highly sophisticated devices that are now the workhorses of our industrialized society. Induction motors serve by far the majority of petrochemical process applications. Motors operating at or above 4,000 volts, or sized at 1,000 horsepower or more, are usually custom built for a specific application. Figure 11-1 shows a representative distribution of electric motors in North American plants by voltages and horsepower ranges, for total plant loads of (a) 5 megawatts and more, and (b) less than 5 megawatts.

Types of Motors

The choice of an induction motor versus a synchronous alternating or a direct current motor is often made on the basis of cost, but other considerations also play a role. For example, spare parts availability, repair capabilities, and similar factors usually favor the induction motor. Synchronous motors offer "nonslip" performance and power factor adjustments by over- or under-exciting the field.

* Reviewed by Delta Enterprises (Sarnia) Ltd., Sarnia, Ontario, Canada.

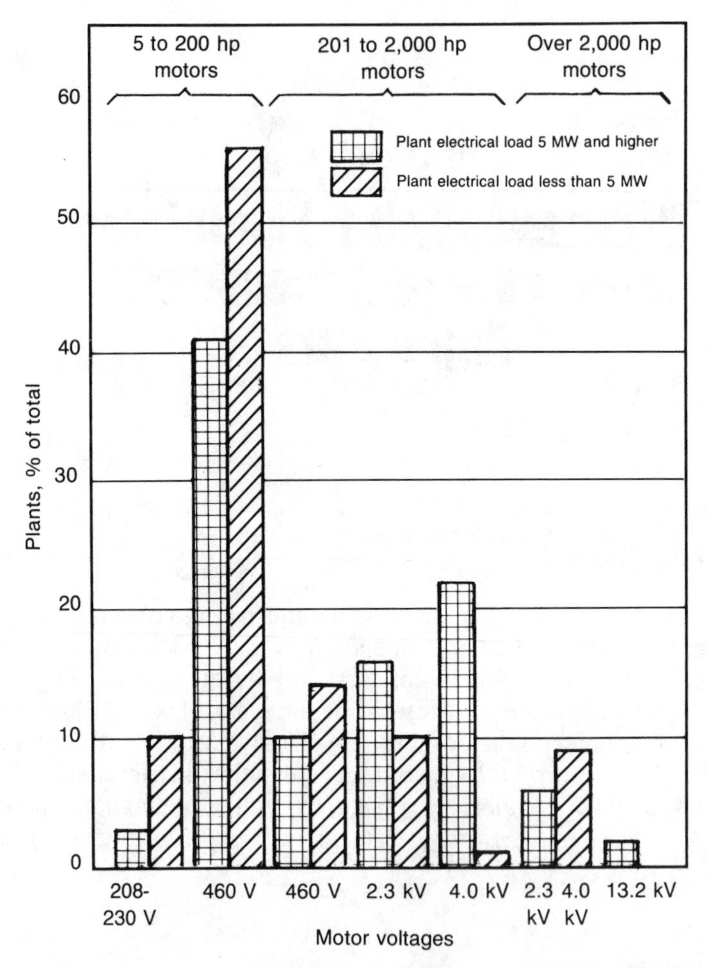

Figure 11-1. Distribution of electric motors by voltages and horsepower ranges.[1]

Usage of motor types, mainly driven by economic considerations, is as follows:[1]

Type	Size, HP
AC induction motors	1/8 to 1,500
AC synchronous motors	200 to 15,000
DC motors	1/8 to 2,000

Induction or squirrel cage motors have many standard design features such as oversized bearings, balanced rotors and a greater number of leads to give versatility in starting methods as well as various rotor designs to match torque and efficiency to particular applications. The induction motor is the most commonly used prime mover in the petrochemical process industry because it is simple, low cost, easily maintainable, efficient, and available in standard sizes. Recent developments in variable frequency controllers will further enhance the use of the squirrel cage motor where variable speed drives are required.

The main components of a squirrel cage induction motor are the two bearings, the rotor, a stationary winding and a case. Figure 11-2 shows a sectional view of a typical three-phase squirrel cage motor.

Types of Enclosures

Motors in petrochemical plants work under a variety of adverse conditions, ranging from wet and corrosive environments to hazardous areas containing combustible liquids, vapors, gases, and dusts. Because of this wide range of environmental conditions, a number of different types of enclosures have been developed. The two basic types of motor enclosures are the open-type and the totally-enclosed type as shown in Figure 11-3.

Open Motors. An open motor is one in which a free exchange of air is permitted between the surrounding atmosphere and the interior of the motor. Cooling air is drawn into the motor through ventilating openings at each end and exhausted through similar openings at the bottom of the motor. The air is circulated by means of fans formed on each end of the

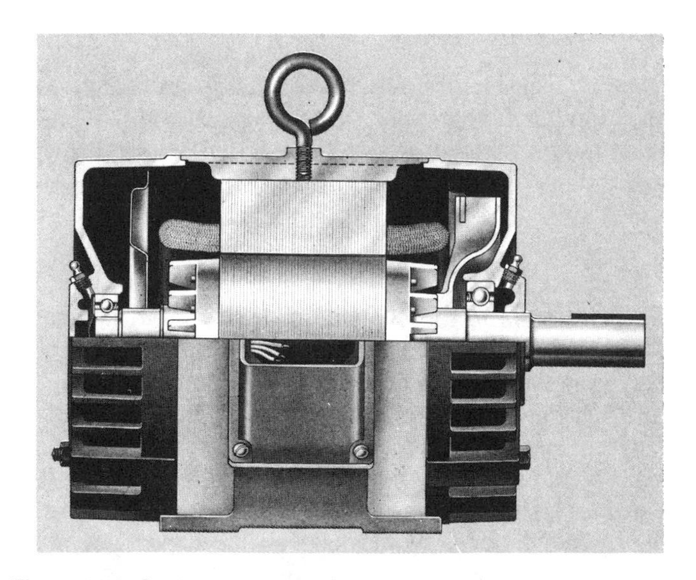

Figure 11-2. Sectional view of a typical squirrel cage (induction) motor.

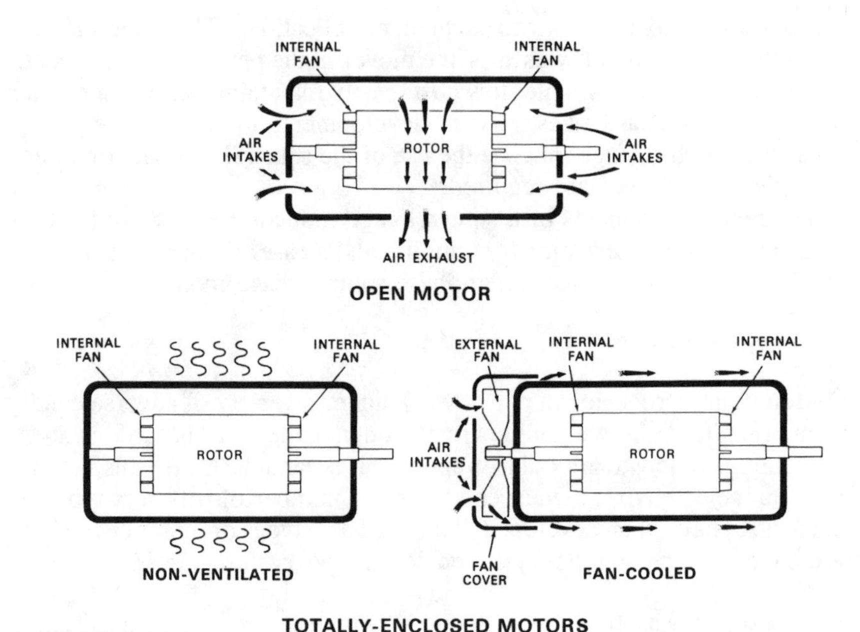

Figure 11-3. Basic types of motor enclosures.[2]

rotor assembly. Open motors may be either drip-proof, splashproof, or floodproof. Another type popular in the petrochemical industry is the weatherproof motor. Weatherproof I has one baffle before incoming air passes the windings. Weatherproof II has two baffles and filters in the incoming air flow.

Totally-Enclosed Motors. A totally-enclosed motor is one in which there is no free exchange of air between the surrounding atmosphere and the interior of the motor. The interior of the motor is completely covered by the stator frame and the end covers, but not sufficiently enclosed to be airtight. Totally-enclosed motors may be either nonventilated or fan-cooled. Totally-enclosed nonventilated motors are cooled solely by heat radiation from the surface of the motor. Totally-enclosed fan-cooled motors are cooled both by radiation and by means of a shaft-driven fan mounted in a housing external to the main motor enclosure. The fan draws air into its housing and directs it along the outer surface of the main enclosure between cooling fins cast into the stator frame. In both the totally-enclosed nonventilated and totally-enclosed fan-cooled motor types, the radiation of heat is helped by fans formed on each end of the rotor assembly. These fans keep the air within the main motor enclosure circulating to improve the transfer of heat to the exterior.

In addition to the two basic types of totally-enclosed motors, nonventilated and fan-cooled, other types of totally-enclosed motors are in use. These are explosion-proof, dust-explosion-proof and pipe-ventilated motors.

Explosion-Proof Motors are similar in appearance to standard totally-enclosed motors. They are, however, constructed with wide metal-to-metal joints having close clearances, nonsparking fans, and thicker enclosure material. This enables them to withstand an explosion inside the motor and also to prevent any sparks, flashes, or explosions generated within the enclosure from igniting the surrounding atmosphere. Explosion-proof motors must be used in all locations where hazardous gases or vapors surround the machine. Class II dust-explosion-proof motors are constructed similarly to Class I explosion-proof motors. The enclosure is designed to exclude ignitable amounts of dust and to prevent a dust explosion inside the motor. The metal-to-metal joints of the enclosure prevent any sparks or hot gases inside the motor from igniting exterior accumulations or atmospheric suspensions of dust on or near the motor. Dust-explosion-proof motors must be used in all locations where hazardous dusts surround the machine.

Pipe Ventilated Motors find a typical application on large process compressor trains. Their construction is similar to standard drip-proof motors with the exception of openings that are provided at each end of the motor enclosure for the connection of cooling ducts or pipes. The cooling ducts enable cooling air to be drawn from a nonhazardous environment remote from the operating area of the motor. Air is circulated either by a fan on the motor shaft inside the enclosure or by a separate remotely located fan or blower connected to the ducting.

This concludes our discussion of motor enclosures. In the following section we are going to discuss induction motor nameplate information and its significance.

Motor Nameplate Data

The nameplate is a motor's I.D. card, and maintenance people should be familiar with what it wants to tell us. Here is a quick overview:

Horsepower (HP) is the power the motor is capable of putting out continuously. Continuously in this context means that at the correct operating load and voltage specified, under standard ambient conditions, the motor will run indefinitely.

Phase Data (Ph) indicate whether the motor is a single or polyphase machine.

Hertz (Hz) is the frequency of the electrical source. In the United States and Canada this frequency is 60 Hz or cycles per second.

Frame Size (Frame) is a number that defines the physical dimensions of the motor. It relates to a standard number defined by the National Electrical Manufacturers Association (NEMA)*. A motor of a particular manufacturer will interchange physically with a motor produced by any other manufacturer if they have the same frame number and suffix. The motors may differ in overall appearance, but all critical dimensions such as shaft size and shaft extension, mounting bolt hole locations, etc., will be the same.

Most motors in the petrochemical process industry are of the "T-frame" type, i.e., the "T" letter suffix is shown together with the numerical frame designation on the nameplate. The T-frame motor was introduced as a NEMA standard in 1964. Until then the prevailing NEMA motor frame was the U-frame, which was introduced in 1953.

The most important and apparent difference between the U-frame and the T-frame motor is that the T-frame allows more horsepower to be put into a smaller package.

Voltage (Volts) is the voltage rating at the motor terminals. Usually satisfactory operation can be expected at a 10 percent variation from the indicated voltage.

Full Load Current (Amps) is the current draw of the motor connected to the nameplate voltage, loaded at nameplate horsepower and running at nameplate speed.

NEMA/CEMA Design Letter (Design) governs motor torque and slip characteristics. Design letters A through D (see box) determine these characteristics for various types of application.

Letter Code (Code) applies to starting conditions in kilovolt/amps per horsepower (kVA/HP) when starting the motor on full voltage. Table 11-1 shows NEMA locked rotor code letters and kVA/HP values.

Service Factor (SF) indicates the ability of the motor to deliver more than the nameplate horsepower. To arrive at the increased rating, multiply the nameplate horsepower by the service factor. The same also applies to the current. With the exception of 1-hp, 3,600 revolutions per minute (which have a service factor of 1.25), all standard NEMA general purpose drip-proof integral horsepower T-frame motors through 200 hp have a service factor of 1.15. T-frame motors above 200 hp, and totally enclosed T-frames have a service factor of 1.0.

* CEMA in Canada

Table 11-1
NEMA Locked Rotor Code*
Letters and kVA/HP Values

Code	kVA/hp	Code	kVA/hp	Code	kVA/hp
A	0.00–3.14	F	5.00–5.59	L	9.00–9.99
B	3.15–3.54	G	5.60–6.29	M	10.00–11.19
C	3.55–3.99	H	6.30–7.09	N	11.20–12.49
D	4.00–4.49	J	7.10–7.99	P	12.50–13.99
E	4.50–4.99	K	8.00–8.99	R	14.00–15.99

* *NEMA Standards MG 1-10.36*

Speed (Motor RPM) is the speed at which the motor rotor shaft will rotate when loaded with the nameplate horsepower. For induction motors, "slip" means the difference between synchronous speed—the speed of the revolving magnetic field—and operating speed. Percentage slip can be calculated by subtracting the operating speed from the synchronous speed, dividing the difference by the synchronous speed, and then multiplying the result by 100.

Motor Receiving, Handling and Storage

After an electric motor shipment has arrived at the plant site, it should be carefully examined. Any damage to either packaging or contents should be recorded and reported to the carrier and to the motor manufacturer or repair shop. This receiving inspection should include examination of any items that may have been shipped separately from the main assembly. The owner or his contractor should proceed with either installation or proper storage as fast as possible. This really means that if the motor is not to be installed immediately, it should be stored in a clean, dry place at a temperature not less than 60°F or 15°C in its original packaging.

Handling. Lifting positions for packaged machines are usually marked on the shipping boxes or skids. Consult the approved outline drawings for lifting instructions, center of gravity, and lift points. Consider the use of spreaders to avoid damaging any machine parts. The handling of components shipped separately from the main assembly may require the use of steadying slings, since the components cannot always be lifted at their exact center of gravity. Avoid slinging on or around polished surfaces such as bearing journals or electrical parts. Always move complete machines or major components with all assembly bolts in place and secured. Packed or unpacked, move the machine in its "as shipped" position, including the shaft clamp, when one is supplied. Machines with oil-lubricated bearings are usually shipped without oil.

Storage. Outdoor storage of electric motors without special precautions is not recommended because variations in temperature and humidity can cause condensation, producing rust upon unprotected metal parts and a deterioration of the electrical insulation. The following remarks, therefore, cover the minimum acceptable storage arrangements in an unheated but protected environment. It would be preferable to use a heated facility in Northern climates, which generally would simplify meeting these conditions. However, when outdoor storage cannot be avoided, the reader should refer to our section on machinery storage and "mothballing," Chapter 12.

Storage Facility Requirements. The storage facility for electric motors must provide protection from contact with rain, hail, snow, blowing sand or dirt, accumulations of ground water, corrosive fumes, and infestation by vermin or insects. There should be no continuous nor severe intermittent floor vibrations. There should be adequate power service for illumination and for space heaters. There should be fire detection facilities and a fire fighting plan. The machines must not be stored where they are liable to accidental damage, or exposed to weld spatter, exhaust fumes or dirt and stones kicked up by passing vehicles. If necessary, suitable guards or separating walls should be erected to provide adequate protection. Avoid storage in an atmosphere containing corrosive gases, particularly chlorine, sulphur dioxide and nitrous oxides.

Temperature Control. Whenever the motor temperature is equal to or below room temperature, and relative humidity is above 60 percent, water vapor can condense on and within the machine to promote rapid deterioration. In order to prevent condensation, the machine's space heaters—see nameplate for voltage rating—must be energized. The goal here is to keep the machine temperature above room temperature by at least five degrees F or three degrees C. However, during periods of extreme cold or rapid temperature drops, the space heaters may not be adequate to maintain this temperature differential, and supplementary space heating may be required. It would of course make sense to provide supplementary space heating if the motor is not equipped with space heaters. A note of caution: If a motor is boxed or covered in any way while space heaters are energized, there should be thermostatic control and sufficient surveillance to quickly detect an over-temperature condition. Ensure that temporary packaging does not come into contact with the space heaters.

When electrical windings are sound and their insulation resistance is well above the formula minimum discussed earlier, low temperature is not a problem. However, if the insulation resistance drops, windings can be permanently damaged by freezing. Therefore, the machine temperature should be kept above the freezing point.

Storage Record. It is recommended that a storage record be kept regardless of the extent of the storage period. The following information should be recorded:

1. Location and description of the storage facility.
2. Identification of the stored motor, including serial and model numbers.
3. Date and condition of the motor when first placed in storage.
4. Date and results of in-storage inspections and tests.
5. Date and description of storage maintenance operations.
6. Machine and room temperature, and relative humidity records.
7. Date and condition of the machine when taken out of storage.

Preparation For Storage

One large motor manufacturer recommends the following procedure[3]:

1. Remove the shipping box or crate and the polyethylene shroud, but leave the machine mounted on its skid.
   Unnecessary shipping timbers, boxes and crates are a fire hazard and a source of moisture.
2. Ensure that all exposed machined surfaces—such as shaft extensions and mounting feet—have rust preventives in place.
3. Ensure that there is a moisture barrier between the machined feet and the skid timbers.
4. Ensure that protective covers, gaskets, caps, and blind flanges are in place and secured.
5. Land the skidded machine on firm supports in its normal mounting position. Vertical machines must have their vertical axis plumb within one degree or 1.75 in. per 100 in. of motor diameter.
6. For machines supplied with water coolers, it is desirable to drain or pump out any water in the coolers and fill them 90 percent full with a proper mixture of antifreeze.
   If there is any chance of freezing temperatures, ensure that the flanges have a protective coating and are covered with gasketed blind flanges. Remove the drain plug from the leak detector housing.
7. For machines with collector rings and brushes, ensure that there is a strip of protective film—such as Mylar—between the brushes and the rings.

8. Provide an overhead cover of polyethylene to protect against settling dust, dirt or debris, but ensure that there is adequate clearance for good ventilation around the machine.
9. Provide sufficient access around each machine in storage to carry out all the specified tests and inspections.
10. For long term storage of electrical machines with oil-lubricated bearings, it may be desirable to consider removal of the rotor, if it was not shipped separately.

    In this case, the manufacturer should be contacted for further instructions.

Protective Coatings, Separators, Cleaners, and Solvents.

1. Select a suitable protective coating for machined surfaces.
   a. Before applying, remove all rust with fine emery papers and polish with crocus cloth. A wire brush should be used on threads.
   b. While wearing rubber gloves, swab with Varsol or an equivalent cleaning agent. Apply a finger-print suppressor. Remove any excess suppressor with a clean dry cloth, taking care not to touch the machined surfaces with bare hands.
   c. Apply an even, heavy layer of the protective coating and allow to dry for at least four hours.
   d. Check with a deposit-measuring gauge to assure the coating is at least .003 in. or .08 mm thick. Alternatively, visual inspection should show a uniform dark brown color.
   e. If the protected surface is to be in contact with a wooden surface, apply two layers of a moisture barrier between them.
2. If fire insurance and other ordinances require fire proofing of timbers, ensure that the materials used do not give off corrosive fumes.
3. Use a suitable moisture barrier.
4. For a cleaning agent use Varsol or an equivalent petroleum solvent.

Periodic Tests and Inspections of Stored Motors. The objective of tests and inspections is to reveal deterioration or failure of protective systems such as shelter, coatings and temperature control or of the machine itself. Any deterioration or failures discovered must be corrected without delay. The storage area should be inspected to the criteria described earlier. The stored machine should be inspected for:

1. Physical damage.
2. Cleanliness.
3. Signs of condensation.
4. Integrity of protective coatings.

5. Condition of paint, i.e., discoloration.
6. Signs of vermin or insect activity.
7. Integrity of all closures.
8. Satisfactory space heater operation. It is recommended that an alarm system be in place to annunciate space heater power interruption. Alarms should be responded to immediately.
9. Measure and record the ambient temperature and relative humidity adjacent to the machine, the winding temperatures—by its resistance temperature detectors (RTD's), if supplied, with space heaters on—and the insulation resistance described as follows. All readings should be recorded.
10. Turn motor shafts once in three months in the case of long duration storage.

Winding Insulation Resistance Measurement

These steps are recommended:

1. Use a 500 volt insulation-resistance meter and apply for one minute. Record this reading and correct it to a 40°C basis using the graph on Figure 11-4. If this corrected value is less than rated kilovolt plus one megohms, check first for possible causes of erroneous reading. Consult Reference 4 for further details. If surge capacitors are connected to the power leads, disconnect their ground cables and take resistance readings again. If the winding insulation resistance is still unsatisfactory, contact the manufacturer.
2. Treat the rotor windings of wound-rotor motors in the same way as the stator windings, but note that their rated voltage is the value given for open-circuit secondary voltage.
3. On windings rated 2,300 volts and above and when the insulation resistance is not significantly higher than the formula minimum, a polarization index test is a good indication of the winding condition. Apply the insulation-resistance meter—500 or 1000 volts—for ten minutes, recording the readings at one minute and at ten minutes. For a sound winding, the ratio of the ten minute value to the one minute value should be greater than two.

Motor Installation

Location. An electric motor should be installed in a location consistent with the type of protection that is indicated by the motor enclosure. Care should be taken to allow enough space around the motor to permit the

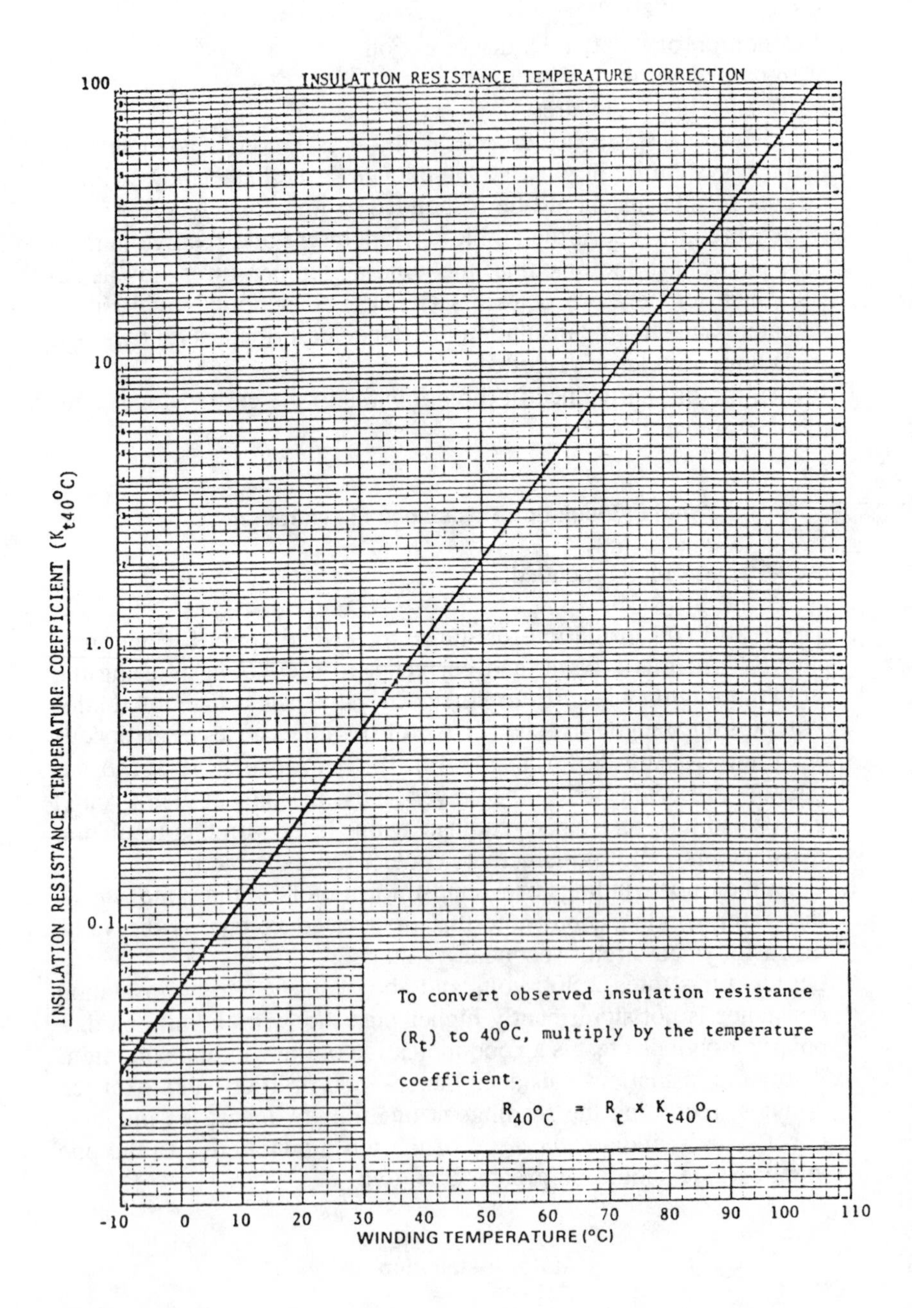

Figure 11-4. Motor winding insulation resistance temperature correction. From Reference 3.

free flow of ventilating air. If dust, moisture and splashing liquids can be avoided in areas where open motors are installed, reduced motor maintenance and increased life will be achieved. Further, electric machinery should never be placed in a room with a hazardous process, or where flammable gases or combustible material may be present unless it is specifically designed for this type of service.

Bases. All bases, except those meant to be "self-supporting" are usually designed to act only as a spacer between the foundation and the machine and must not be trusted to carry any weight when they are unevenly supported. Care should be taken during handling and lifting to prevent warping or distortion of the base.

Mounting. Motors should be mounted on machined sole plates or pads on fabricated bases or skids. These in turn should be bolted to a firm foundation and grouted according to guidelines described in our section on grouting procedures. Sole or bed plates, bases, skids, or platforms must be rigid enough to prevent any vibrations of the unit and also prevent transfer of vibrations from other sources. These structures must not impose bending or twisting strains on the housing. When mounting, we recommend the use of slotted shims as it may be necessary to remove or add shims during the alignment procedure. The use of proper shims under each mounting foot will prevent distortion of the housing when the mounting bolts are tightened. The following procedure is recommended when mounting a motor:

1. Add shims under the lowest mounting foot and tighten mounting foot bolt.
2. Insert feeler gauge under the other mounting feet to determine the amount of shims required.
3. Insert required shim under each mounting foot and tighten mounting bolt.
4. Use the smallest number of shims possible.
5. Measure the alignment between motor and driven machine shaft using the procedures described in Chapter 5, Volume 3 of this series.

Pre-Installation Checks. The following precautionary steps should be taken before finishing an electric motor installation:

1. If dampness in a standard insulated motor is suspected as the result of shipment and/or storage conditions, the insulation resistance of the windings should be measured as mentioned earlier. The value of the insulation resistance should not be less than shown in Table

Table 11-2
Insulation Resistance Values for Electric Motors

INSULATION RESISTANCE	
Rated Voltage	**Insulation Resistance**
600 volts and below	½ megohm
2300 volts	2 megohms
4000 volts	3 megohms

11-2. If the insulation resistance is lower than shown, the windings should be dried out by heating in a well-ventilated area.

2. Check that the motor shaft turns freely by hand. Motor antifriction bearings are usually greased at the factory and need no attention prior to operation. Sleeve bearing equipped motors are usually shipped with their oil wells drained. Before filling with oil, it is advisable to flush the bearing thoroughly with a suitable flushing material in order to remove any foreign matter. Turning the motor shaft by hand should be repeated after the motor is coupled to the load to make sure the entire drive train is mechanically free before applying power.
3. Check that the voltage and frequency stamped on the motor and control nameplates correspond with those of the power supply.
4. Totally enclosed motors for nonhazardous locations are equipped with plugged vent holes. To achieve self-venting of these motors when they are operated under severe moisture conditions, the plugs should be removed.
5. Standard practice dictates that the conduit box be located on the right hand side of the motor when viewed from the opposite-to-drive end. The conduit box may be generally changed to the opposite side by removing the end shields and rotor and turning the stator frame around. Relocating the position of the secondary leads on wound-rotor motors may require drilling and threading of the end shields at the desired point.

Pre-Operational and Operational Checks. Before coupling up to the driven machine it would be advisable to start up the motor for a "solo" run. The following checks should be performed prior to start-up:

1. Motor and control wiring, overload protection and grounding should be in accordance with national electrical codes and local requirements.

2. In excessive moisture conditions the conduit system must be provided with adequate drains so that water will not accumulate in the motor conduit box.
3. On wound-rotor motors, the brushes should be positioned on the slip rings before starting.
4. Motors with more than three leads should be connected in accordance with the diagram furnished with the motor.
5. Direction of rotation should be checked immediately after start-up. To reverse the rotation of three-phase motors, interchange any two line leads. Note that some large high speed motors are designed for rotation in one direction only, which is indicated by an arrow on the directional nameplate.
6. Check location of magnetic center and verify the manufacturer's magnetic center pointer location on sleeve-bearing motors during "solo" run.

Connecting the Motor to Its Load. There are several variations of motor to load connections. Direct connection by way of a flexible coupling should always be considered, but in a large number of applications, belt and chain drives, for instance have to be used. Flexible couplings are preferred because they allow for a slight amount of misalignment between the motor and the driven machine and prevent transmission of thrust to the motor bearings in many cases. Belt, chain and certain gear drives must be carefully selected and installed to ensure that motor bearing loads are within their design capacities.

Coupling Sleeve Bearing Motors. Motor sleeve bearings are in most cases not designed to carry axial thrust loads. If the driven machine exerts thrust it must be equipped with its own thrust bearings. Motor sleeve bearings have a specified end play as indicated in Table 11-3. During op-

Table 11-3
Electric Motor Rotor End Float and Coupling End Float*

Motor Horsepower	Synchronous Speed	Minimum Rotor End Float Inches	Maximum Coupling End Float Inches
125–200	3600 and 3000	1/4	3/32
250–450	1800 and below	1/4	3/32
250–450	3600 and 3000	1/2	3/16
500 and above	all speeds	1/2	3/16

* *NEMA MG1-6.11*

eration the motor rotor will seek its magnetic center which will fall between the end play limits established in the manufacture of the motor. Additionally, the magnetic center and the limits of end play are scribed on the shaft at the factory.

In order to prevent thrust from being transmitted to the bearings, a limited end float coupling should be used. The limited end float coupling is designed to have less end play than the total end play of the motor. Thrust that tends to separate the coupling is restrained by lips or coupling shoulders. Thrust that tends to push the coupling together is restrained by buttons on the shaft ends or by Micarta discs in the gap between shaft ends. The difference between the motor end play and the limited end float of the coupling represents the allowable tolerance in setting the motor. A motor built with 1/2 in. end play requires that the end float be limited to 3/16 in. Medium size motors having 1/4 in. nominal end play require that free floating type couplings have their end play limited to 3/32 in. Small motors having 1/4 in. end play or less seldom require limited end float couplings.

Gear type as well as disc type couplings find their application when limited end float couplings are required. The assumptions are that both type of couplings exert a minimum of thrust and can be used on applications of any rating and end play. However, there has been sufficient evidence supplied[5] that gear couplings—when worn and poorly maintained—can exert considerable amounts of thrust onto driver or driven equipment. When limited end float couplings are being installed, it is important to understand Figure 11-5. It tells us that the motor should be located so its shaft is in the center of rotor float. This means that dimension "Y" should equal dimension "Z."

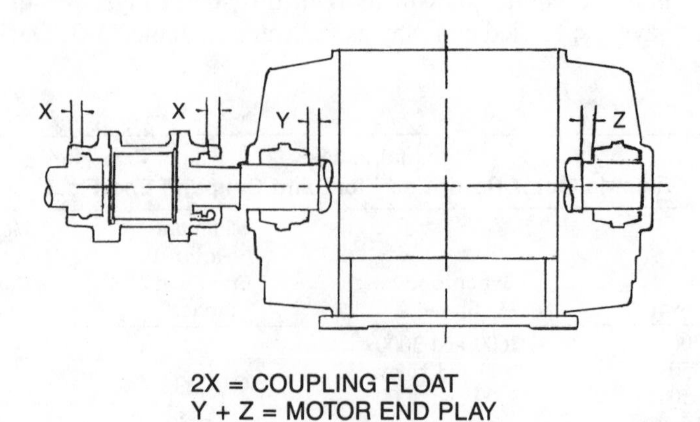

Figure 11-5. Determination of end play on limited end float couplings.

When installing the coupling, sufficient clearance, as specified by the coupling manufacturer, must be provided between shaft ends on the hubs of the flexible coupling. This is to allow for adequate thermal expansion of the shafts.

Electric Motor Shop Repair. It should be obvious now that electric motor problems appear in two forms: mechanical rotor bearing difficulties and electric winding and supply-system problems. A recent review of motor repairs by severity and kind in a petrochemical plant showed that 80 percent of those motors had bearing problems. Consequently, one should always pay a good deal of attention to the mechanical aspects of the motor repair business. Fortunately, electric motors are unique in that they are capable of being repaired much more economically than most other prime movers. Also, the chances of restoring the original quality during a motor repair are excellent—in fact, there are *many* examples of successful motor upgrading and uprating projects. Electric motor repairs are frequently limited only by the quality and experience of the persons in charge.

In our experience common motor problems have been caused by mechanical difficulties such as coupling misalignment, lubrication problems—under or overgreasing, excessive shaft-loading from a belt or chain drive, snow and ice, or dirt in the windings. All these cases have never been able to present a major challenge to a capable motor repair shop.

When it comes to the selection of a motor repair shop we should apply the same criteria put forth in an earlier section in Volume 3. Most petrochemical process plants will have their motor repair specification ranging from the simple one page motor repair report form—Appendix 11A—to an elaborate document for large critical motors. Usually the EASA* standards for the electrical apparatus sales and service industry—Appendix 11B—will have been consulted or made part of the document. If petrochemical companies now and then embark on rewriting their standards for smaller motors—1 to, say 250 HP range—it is because motor shop quality problems are quite frequent.

Usually one-to-one communication between the machinery maintenance person and the motor repair shop manager will yield the best results. It goes without saying that adequate documentation should exist in these quality control cases so both parties can look at the facts.

A capable electric apparatus repair and service shop will take on a variety of petrochemical process work. Figures 11-6, 11-7 and 11-8 show an appropriate cross section of electrical motor shop repair activity in progress.

---

* Electrical Apparatus Service Association

Figure 11-6. Large motor repair floor in an electrical apparatus repair and service shop.

Motor Preventive Maintenance

Any electric motor PM program should be designed with these four elements in mind:

- Inspection and test
- Lubrication and impregnation
- Cleaning
- Protection and safety measures

The four element approach contains the following steps:

1. Inspection. We need to determine what is to be inspected, the equipment has to be examined, visible defects have to be identified, and we have to make sure that minor adjustments are made and major defects are being reported. General motor inspection should be directed toward these components:

Figure 11-7. Rewinding synchronous motor stator.

a. Air passages.

- Inspect rotor and stator air passages for obstructions on explosion-proof motors and totally enclosed fan cooled motors.
- Air filters on weather-protected (WP II) motors.
- Inertia filters, filter media and remote fans on pipe ventilated motors.

b. Bearings.

- Inspect for overgreasing or undergreasing.
- Inspect for dirt and foreign matter in bearing housings.
- Check periodically for proper alignment to avoid bearing overloads.

Figure 11-8. 1500 horsepower, 1800 rpm, 4160 volt TEFC induction motor ready to have bearings rebuilt and rotor balanced.

c. Motor windings.

- Keep them cool.
- Keep them clean.
- Inspect stator winding air gaps.

d. Squirrel cage rotors.

- Inspect for heat checked end rings.
- Inspect for fractured rotor bars.

e. Winding insulation values.
f. Vibration monitoring.

- Set limits for small motors as a function of speed.
- Set limits for large motors (see Volume 2 of this series).

2. Lubrication. Lubrication practices on antifriction bearing motors vary widely. We should determine proper lubricants, quantities, and equipment. After establishing a lubrication schedule, it has to be implemented. Refer to Chapter 12 for details.
3. Cleaning. We need to determine cleaning standards and requirements, establish a cleaning schedule and implement it.
4. Protection and Safety. Safety requirements have to be established, safety equipment installed, and its effectiveness monitored.

The task that follows now is the building of appropriate check lists for each PM element. It should contain the following information:

1. Motor name and "yard" number.
2. Location.
3. Frequency designation of PM activity.
4. Operation sequence number.
5. A simple description of the PM operation using verbs at the beginning of a sentence.
6. Required tools and/or special equipment and instruments.
7. Required replacement parts.
8. Required time to perform PM activity.
9. Space for the inspector's comments or any other special comments.

After the various electric motor PM checklists have been developed, thought should be given to the implementation of the program. Two main considerations will help in the implementation of the motor PM program. They are motor criticality and the cost of the program.

To determine whether an electrical motor is critical, one has to look at the cost implications of a failure. For instance, the breakdown of an unspared 4,000 horsepower ethylene refrigeration compressor drive motor would have a different financial impact than a small motor driving a fully spared water pump in a batch type cleaning operation. As to the case of the 4,000 HP pipe-ventilated induction motor, there would be no question as to whether a PM program should be considered. Frequently, cleaning and inspection of this critical motor becomes an integral part of the process plant's turnaround preparations. Figure 11-9 illustrates the elaborate fixture necessary to accomplish the task on site in an economical way.

However, there are a multitude of cases where a decision to do or not to do PM on motors does not appear that straightforward. Questions that

Figure 11-9. Field removal of rotor for winding cleaning and inspection of a large pipe ventilated induction motor.

should be answered during the development of the implementation schedule for an electric motor PM program are:

1. How critical is the motor?
2. What will operation downtime cost?
3. Is similar or alternate equipment available?
4. Are spare motors available or easy to obtain?
5. Are major spare parts such as winding coils and bearings available?
6. What is the motor failure experience?
7. What is its age?
8. What is the motor operating severity?

In conclusion, it is necessary to determine the benefits and costs of an electrical motor PM program. Only after its cost factors have been compared to current maintenance practices and other related downtime costs can it be determined if a motor PM program will truly support itself.

Preventive Maintenance of Nonrotating Electrical Apparatus

Power transformers and electrical switch gear are another link in the chain that ends in a reliable motor-driven process machine. As on other occasions, we are defining preventive maintenance in this context as the planned periodic testing, cleaning, adjusting, and lubricating of each component.

A preventive maintenance program of nonrotating electrical apparatus should above all be done by experienced personnel. Quite often such a program can be truly predictive because it allows detection of incipient failures. Consequently, action can be taken before serious and extensive damage will occur. It is essential for a successful preventive maintenance program that accurate records be kept of all work performed. The records must also show a forecast of future work.

Power Transformers. Transformers generally require less care and attention than most other kinds of electrical power apparatus. Often they are the most vital link in an electrical distribution system.

Oil preservation is the most important aspect of transformer maintenance. It requires an understanding of oil deterioration and how it can be prevented.

The main factors determining the chemical action in transformer oils are:

- The moisture concentration
- The oxygen concentration
- The presence of copper
- Temperature

Each of these factors contributes to oil deterioration. The result is a large number of possible chemical products which can affect the oil. One of the products is acid, which can attack the insulation and metals of the transformer and reduce the dielectric strength of the insulation. Other products are sludges which impede the transformer cooling.

As a consequence the oil has to be cleaned or replaced. Selection of proper limits at which oil should be filtered or replaced is partly a matter of judgment and experience. Oil preservation depends primarily on keeping oxygen and moisture away from the oil. This fact in turn requires that the case and the openings in the case be made pressure tight.

The following are oil defect limits:

1. The dielectric strength of transformer oil should not be allowed to fall below 22,000 volts.
2. The moisture content should not be higher than 20 parts per million.
3. New transformer oil has usually an acid No. of .040. If the acidity rises to an acid No. of .10, the transformer oil should be replaced or purified via vacuum dehydration (see Volumes 1 and 2 of this series).

4. Transformer oil samples should be taken at least every two years. The samples may be taken from the sampling valves at the bottom of the transformer tank. Drain off and discard at least two quarts of oil before saving a test specimen. Water and condensate in oil will collect in the bottom of tanks and in drain lines and will drain off with the initial oil withdrawn from the tank.

Power transformers above 1,000 kVA are provided with a dial type thermometer calibrated in degrees centrigrade. This thermometer is used to indicate both top oil temperature and winding temperature. The dial has two pointers—the white indicates the temperature at any given moment—the red pointer is carried along by the white to show the maximum temperature reached since it was last reset. Name plates on power transformers will indicate the maximum allowable temperature rise above ambient. The thermometer may be equipped with up to three switches which may be used for alarms or other protective devices.

The temperature of main substation transformers should be checked at least once a month. A visual check of the tanks for rust spots and discoloration should be made at the same time.

Usually the tanks have level gauges. These gauges are calibrated to indicate the level of oil in the main tank. On ratings above 7,500 kVA, a low level alarm contact is provided. The oil level should be checked at least once a month.

Large power transformers are usually supplied with a fault pressure relay. It is also referred to as a gas detector relay and is used to give an indication of faults of a major nature resulting in a sudden increase in internal pressure. It consists basically of a bellows which operates a microswitch. The contacts of the microswitch are used to operate circuit breakers. The fault pressure relay can be tested by applying a small amount of air pressure through the test check valve. Preferably, this test should be done when the transformer is not in service. It can be done under load by opening the tripping circuits to the circuit breakers involved.

On a transformer with a gas-oil seal, a two-compartment expansion tank provides an oil seal between the gas above the oil in the transformer tank and the outside air. The transformers are filled to the correct level at 25°C, sealed and shipped with all accessories in place. If the temperature rises above 25°C, a positive pressure will be indicated. If the temperature falls below 25°C, the pressure gauge will indicate a vacuum.

The pressure-vacuum gauge on this type of transformer should be checked regularly. If it is approximately zero at different liquid temperatures, a leak is indicated. All gaskets, radiators, and tanks should be thoroughly checked for leaks.

Power transformers are frequently provided with external cooling fans to give additional load capacity above the natural circulation cooling system. The cooling fans are operated by thermostatic control. To check their operation a manual test position bypasses the thermostatic control.

Dry-Type Transformers. This type of transformer is not generally used as a main power transformer, but mainly for lighting service, distribution centers and control power. Because its case must have openings to allow for air circulation, the windings are subject to dust infiltration over a period of time, reducing the cooling air through the windings.

An inspection period of every two years is recommended at which time the windings should be cleaned and checked for loose connections.

High Voltage Insulators. The buildup of foreign materials on high voltage insulators will lead eventually to flashovers. A flashover will generally result in the loss of an incoming line because of its similarity to a ground fault. The rate at which buildup will occur will depend upon the following environmental conditions.

- High humidity
- Corrosive gases
- Salt laden atmosphere
- Heavy dust

Under normal operating conditions, it is not usually possible or practical to check the insulation value of a high voltage insulator unless the complete structure is deenergized. However, a visual inspection is sufficient to detect contamination buildup.

Experience has shown that in areas where there is a lot of contamination in the atmosphere, insulators should be washed once a year. This period may be extended to two years if the contamination is not too severe. Special equipment is required to wash insulators while maintaining a plant in operation. Utilities are usually well equipped to do this work. For the short time required to do this work, engaging the local utilities has proved to be the most economical method of accomplishing this work.

Power Circuit Breakers. A circuit breaker must be kept in good condition if the expected performance is to be obtained. It has only one chance to operate each time a fault occurs and if it fails to trip, serious damage may result. The more common types of circuit breakers are *oil circuit breakers and air circuit breakers.*

The frequency at which oil circuit breakers must be inspected depends upon the severity of the service. The following outline may be used as a guide.

The breaker should be taken out of service and the following work carried out:

- Inspection of oil—carbonization of the oil is caused by arcing. The same limits as for transformer oil should be used.
- Inspection of contacts—if the contacts are burnt or badly pitted, they must be replaced.
- Inspection of spring pressure on contacts—evidence of uneven burning or pitting may be due to uneven contact pressure. If so, contact pressure should be checked.
- Inspection of tank for bulging or rust—if the tank is bulged due to excessive pressure, it should be replaced or a spare breaker used.
- Inspection for broken insulators—replace if necessary.
- Alignment of limit switches and control contacts—realign as required.
- Test closing and tripping of the breaker in the test position.

The maintenance of air circuit breakers is similar to oil circuit breakers. In an oil circuit breaker, the arc is broken under oil. In an air circuit breaker, the arc is extinguished by the elongation of the arc through arc chutes. In some breakers the arc chutes are augmented by blow out coils.

The frequency at which air circuit breakers must be maintained again depends upon the number of operations per year. The following outline may be used as a guide.

- Inspection of main contacts—replace if burnt or badly pitted.
- Inspection of arcing horns—remove spatter from insulation material.
- Inspection of arc chutes and phase barriers—replace if cracked or broken.
- Alignment of limit switch and control contacts as required.
- Test for closing and opening in Test Position (or with main fuses removed).

Relay Protection. Speed and selectivity are two demands placed on protective relaying. Speed in disconnecting the faulty portion from the overall system and selectivity in that the disturbance to the remainder of the system is held to a minimum.

There are many different types of relays used in a petrochemical plant. It is, therefore, important that the manufacturer's literature be consulted before any adjustments are made. Each relay in service is calibrated for a specific operating current and operating time. Both functions must fit into the overall coordination plan of the distribution system.

When a relay receives a signal to operate, the response must go through a number of components before reaching the trip coil on a circuit breaker. Each component must be functional at the time the relay receives the signal to trip. Normally, a relay is not required to operate frequently. There are long periods of inactivity which permit the accumulation of dust and in some instances, rust. These factors will greatly decrease the effectiveness of a relay.

In most installations, relays can be removed from their cases without interruption of service. When this is possible, it is recommended that a relay be checked and calibrated at least every three years. The following brief description highlights the maintenance of an induction type overcurrent relay.

- Remove relay from case.
- Manually check the rotating disc for dust, freedom to rotate and iron filings between the disc and the magnets.
- Note the current tap and time dial setting.
- From the manufacturer's literature, determine the tripping time at twice and four times the current tap and time dial setting.
- Using alternating current, check the "as found" tripping time against the manufacturer's curves.
- Minor adjustments may be made by changing the position of the magnets or by changing the time dial setting. In most cases the relays will be slow due to the accumulation of foreign matter.

Literature supplied by the manufacturer will indicate the internal connections, i.e., current coils, voltage coils, contact terminations. These diagrams should be used in order to avoid damage to the relay.

Because of the number of components in a tripping and closing circuit, it is important that the circuits be proven from the relay and control switch contacts. Tripping checks can be performed by manually rotating the disc on the relay until the tripping contacts close to open the breaker. To prove the closing circuit, the control switch is turned to the "close" position and released.

It goes without saying that operational checks should be done during turnarounds or at times when service to the operating units can be interrupted.

Finally, once preventive maintenance frequencies have been established it is a good idea to communicate them to the interested parties. Table 11-4 shows a typical example of this communication effort.

Table 11-4
Possible Preventive Maintenance Schedule for Electrical Apparatus in a Petrochemical Process Plant

Electrical Apparatus	Preventive Maintenance	PM Frequency					
		Weekly	Monthly	Annually	Every 2 Years	Every 3 Years	Change Only If Required
Battery Systems	Check Voltage		●				
	Check Voltage Alarm		●				
	Specific Gravity and Plates			●			
Emergency Transfer Schemes	Standby Generators—run up	●					
	Complete Transfer		●				
Main Circuit Breakers	Operational Check			●			
	Oil Inspection				●		
Main Substation Transformers	Temperature & Load Check		●				
	Oil Inspection				●		
Motors	Bearings						●
	Grease*						●
	Ventilation						●
Protection	Cleaned and Checked for Calibration					●	
	Kilowatt-hour Meters					●	
Substation High Voltage	Incoming Lines—Dirty Atmos			●			
	Incoming Lines—Clean Areas				●		
Unit Substations	Ground Indicators		●				
	Sump Pump		●				
	Housekeeping		●				
	Temperature & Load Check			●			
	Oil Inspection—Transformers				●		

* *Refer to Chapter 12 for details.*

References

1. Brown, T. and Cadick, J. L., "Electric Motors Are The Basic CPI Prime Movers," *Chemical Engineering,* March 12, 1979, pp. 85–91.
2. Canadian General Electric, Industrial Apparatus Dept., "Industrial Induction Motor Manual," Filing No. 300, 1967.
3. Canadian General Electric, "Receiving, Handling and Storage Of Induction Machines," Bulletin PGET-5070A.
4. IEEE, "Recommended Guide For Testing Insulation Resistance Of Rotating Machinery" IEEE, publication No. 43.
5. Bloch, H. P. and Geitner, F. K., *Machinery Failure Analysis and Troubleshooting,* Gulf Publishing Company, Houston, Texas, 1983, pp. 606–611.

Appendix 11-A

Electrical Machines Maintenance Report

ELECTRICAL MACHINES MAINTENANCE REPORT

(Maintenance Dept.)

Make ________ Type ________ H.P. ________ Voltage (V) ________

M. No. ________ Service ________ Location ________

Signed ________ Date ________

REASON FOR REPAIR

☐ Excessive Vibration ☐ Noisy ☐ High Current

☐ Runs Hot ☐ Seized ☐ Brg. Failure

☐ Burned Out ☐ Dirty ☐ Preventative Maint.

☐ Other (explain)

INSPECTION REPORT

Lubrication ☐ Clean ☐ Dirty ☐ Lacking

Windings ☐ Clean ☐ Dirty ☐ Burned Out

Windings Epoxy Coated ☐ Yes ☐ No ☐ Type

Lead Condition: ________

Shaft Runout ____ × ____ in.

Shaft Endfloat ____ in. Before Repair ________ in.

Bearing No. ________ D.E. ________ O.D.E. ________

WORK DONE

☐ Rewind ☐ Cleaned

☐ Changed O.D.E. Brg. Bearing No. ________

☐ Changed DE Br. Bearing No. ________

☐ Rotor Balanced

☐ Replaced Leads

☐ Thermal Protection Type

☐ Other (Explain below)

Oil Used ☐ Ter. 52 ☐ Ter. 43 Other ________

Grease Used ☐ Unirex Other ________

TEST

___ Shaft End Float ________ in. Insulation (2 × V + 1000) ________

___ Inspect & Run in ________ No Load Amps. ________

Thrust Direction

Vibration - DE ________ ODE ________

COMMENTS ________

Appendix 11-B

EASA* Standards for the Electrical Apparatus Sales and Service Industry

AC MOTORS

1.0 - *General - All Electrical Apparatus*

1.1 - *Purpose:* The purpose of this set of standards is to promote integrity and good faith between members, suppliers and customers by maintaining a level of workmanship that will produce a quality product.

1.2 - *Scope:* The basic work will include tests and inspection of windings and of all parts for defects, wear, and condition to determine if repairs are justifiable. Repairs shall meet the conditions outlined in this standard. Apparatus shall be reassembled, tested, and made ready for shipment.

1.3 - *Nameplate:* All electrical apparatus must have a permanent nameplate containing the principal information needed to put it into service. The necessary information is the HP, KW, or KVA rating, primary, secondary, field and/or armature voltage and amperes, power factor, RPM, frequency and phase if applicable, temperature rise or ambient temperature, time rating and class of insulation. If known, the manufacturer's name, serial number, model or style number, service factor, type and frame size should be shown. Should a winding be redesigned the original nameplate, whenever possible, should remain on the apparatus and a new nameplate mounted adjacent to it with the words "Re-

* Electrical Apparatus Service Association, St. Louis, Mo. By permission.

designed" or "Changed" and the new ratings shown. The new nameplate should contain the date of change and the name of the company who made the changes.

1.4 - *Cleaning:* All windings and parts must be cleaned free from dirt, grit, grease and oil, then properly dried. Cleaning agents must be completely removed.

1.5 - *Stripping:* Defective windings shall be removed in such a manner than no mechanical damage is done to the laminations or frame. Heating of laminations shall be controlled to avoid impairment of the magnetic qualities of the core and distortion of any parts.

1.6 - *Laminations:* Shorted or damaged laminations shall be repaired or replaced.

1.7 - *Leads:* All apparatus will be equipped with lead wire of rated temperature and voltage insulation and of sufficient current carrying capacity. All leads shall be suitably marked or colored where necessary to indicate correct connection and be of sufficient length for ease of connecting to power supply at terminal box or to terminal blocks. A print or plate must be furnished, where necessary, indicating correct connections. All lead markings should conform to NEMA standards and be of sufficient durability to withstand the environment involved. Leads on TEFC and explosion-proof apparatus shall be properly sealed to meet environmental operating conditions.

1.8 - *Varnishing and Coating of Windings:* The windings of all rewound apparatus shall be varnish-treated or coated using a material and method of application of sufficient quality to withstand the normal application of the apparatus. The varnish or coating shall be compatible with the entire insulation system and be suitable for the environment of the apparatus.

1.9 - *Terminal Box:* Terminal boxes must be large enough to accommodate the connections without crowding. Knockout holes must be large enough to accommodate the proper size conduit and not have a sharp edge.

1.10 - *Exterior Finish:* Apparatus shall be externally cleaned and refinished with a good quality oil resistant coating.

1.11 - *Crating and Shipping:* All apparatus must be boxed, crated, or skidded to protect it during shipment.

2.0 - *Mechanical Portion - Rotating Apparatus*

2.1 - *Shafts:*

2.1.1 - *Inspection:* The shafts shall be checked for undue wear, scoring and straightness.

2.1.2 - *Bearing Journals and Shaft Extension:* The bearing journals and shaft extension must be concentric with the shaft centers, smooth, polished, and of proper size and fit. If a sale is involved the diameter and the length of the shaft extension or extensions shall conform to NEMA standards for the frame size invovled or the customer be notified of the shaft dimensions. The tolerance for permissible shaft runout for standard length shafts, when measured at the end of the shaft extension, shall be 0.002 inch indicator reading for 5/8 to 1 5/8 inch diameter shafts inclusive and 0.003 inch indicator reading for over 1 5/8 to 5 inch diameter shafts inclusive.

2.1.3 - *Keyways:* Keyways should be true and accommodate keys to a tap fit. If a sale is involved keyways must be NEMA standards in dimensions, if not, offset keys of standard size must be furnished with the apparatus.

2.1.4 - *Oil Rings:* Oil rings must be true and rotate freely and be of proper size to carry sufficient lubrication to prevent bearing wear. Retainers, when provided, shall be inspected and replaced if necessary.

2.2 - *Bearings:*

2.2.1 - *Ball and Roller Bearings:* Ball and roller bearings must be free from defects and surface wear. Bearings must be properly installed, fitted, and aligned.

2.2.2 - *Sleeve Bearings:* Sleeve bearings must be uniform in diameter, be of proper fit in the housing, be smooth internally, and suitably grooved for proper distribution of lubricant. For A.C. apparatus the maximum clearance between the bore of the bearing and the diameter of the shaft must not be more than 0.002″ for the first inch of shaft diameter plus 0.001″ for each additional inch of shaft diameter. For D.C. apparatus the maximum clearance shall be 0.005″ per inch of shaft diameter, air gap permitting.

2.3 - *Lubrications:*

2.3.1 - *Grease:* Ball and roller bearings will be properly greased with a lubricant or proper grade and recognized quality. The grease pipes should be clean and have a proper cover.

2.3.2 - *Oil:* Oil must be of proper weight and quality. There must be a covered overflow cup of sufficient height to maintain the proper oil level in the bearing housing. There should also be a drain plug.

2.4 - *Frame and End Shields:* The motor frames and end shields shall be examined for defects and fits. All cracks or breaks must be repaired and fits brought to standard.

2.5 - *End Thrust and End Play:* All rotors and armatures of horizontal apparatus must be positioned on the shaft to eliminate end thrust against either bearing. Total end play in sleeve bearing apparatus should be approximately 1/16″ per inch diameter of shaft journal. Total end play in ball bearing apparatus should be sufficient to allow for shaft expansion caused by temperature rise.

2.6 - *Balancing:* Recommendations of vibration limits for special AC and DC motors and generators are established by each manufacturer. The standard balance for integral horsepower AC and DC motors built in frame 143 and larger when made in accordance with NEMA procedure is as follows:

Speed, RPM	Maximum Amplitude
3000–4000 incl.	0.001 inches
1500–2999 incl.	0.0015 inches
1000–1499 incl.	0.002 inches
999–and below	0.0025 inches

2.7 - *Rotors and Armatures:*

2.7.1 - *Laminations:* The armature and rotor laminations must be firm and secure on their shafts or support sleeves and the sleeves secured on the shafts.

2.7.2 - *Slip Rings:* Slip rings shall be checked to ensure that they are concentric, smooth and polished, properly insulated and have sufficient stock to be mechanically strong.

2.7.3 - *Commutators:* For commutators refer to paragraph 3.5.2.

3.0 - *Electrical Portion - Rotating Apparatus*

3.1 - *Windings:*

3.1.1 - *General:* All rewound equipment must produce essentially the same torque, speed, horsepower, efficiency, power factor, and temperature characteristics as the original winding unless a design change has been made for a specific application. Any change in the electrical rating must comply with paragraph 1.3.

3.1.2 - *Insulation System:* The entire insulation system shall constitute materials and methods of application that will be equal to or better than used by the manufacturer. It shall be sufficient to withstand the normal operation of the apparatus.

All the components that constitute an insulation system must be compatible with each other and the varnish or coating applied.

All insulation materials used must be of the proper NEMA temperature class to meet the required temperature rise of the apparatus.

All insulation shall be installed in a neat and workmanlike manner and as recommended by the insulation manufacturer.

3.1.3 - *Magnet Wire:* The current carrying capacity, insulation, and mechanical qualities of the magnet wire shall be suitable to the operational environment and shall be adequate to withstand the normal life of the apparatus. When the metal in the magnet wire has been changed it shall be equal to or better than the original material in all aspects of performance and application.

3.1.4 - *Lacing and Shaping:* The coil ends shall be particularly laced and shaped to maintain all necessary clearance such as rotor, end shields and through bolts and be able to endure starting and running currents with a minimum of distortion. On larger motors where coil support (surge) rings are used, they shall be suitably insulated, accurately fitted and laced to insure adequate support for the winding. All lacing and shaping should be done in a neat and workmanlike manner.

3.1.5 - *Connections:* All connections shall be properly soldered, brazed, or welded with materials that will be mechanically strong enough to withstand the normal operating conditions. Materials such as solder paste, fluxes, inhibitors and compounds, where employed, shall be neutralized after using. They shall be suitable for the use and of a type that will not adversely affect the conductors. All connections and splices shall be so constructed as to have a resistance equal to or less than the conductors of the winding. Connections to terminal shall be of the type approved by the National Electrical Code that will ensure a good electrical and mechanical contact without injury to the conductors.

3.1.6 - *Insulating Connections:* All connections are to be adequately insulated with materials that will withstand the temperature, voltage and frequency rating of the apparatus and be mechanically adequate to withstand normal operation. All connections and leads shall be laced, tied, or otherwise securely fastened to prevent movement and subsequent failure. All connections shall be made and insulated in a neat and workmanlike manner.

3.1.7 - *Banding:* Glass banding tape may be applied directly to the winding. It should be applied at proper tension and be of sufficient thickness and width to support the coils to minimize distortion.

When banding wire is used, it shall have a tensile strength of no less than 200,000 psi. The area to be banded should be properly shaped and insulated. The banding wire should be applied at the proper tension.

All types of bands should be sufficiently secured, tied, or laced and be mechanically strong enough to withstand the centrifugal force, current surges, and vibrations of normal operation.

Caution: Replacing wire banding with resin filled glass banding may change the magnetic circuit, effect commutation, and speed.

3.1.8 - *Thermal Protectors:* Thermal protectors shall be checked for electrical and physical defects. Replacement of protectors shall be identical with original in tripping characteristics. Removal of protectors should be done only with customer consent or manufacturer approval.

3.2 - *Random Wound Coils:* The coils should be wound and inserted in the slots with a minimum of crosses. Care should be taken not to damage the insulation or the magnet wire during winding and insertion of the coils. All coils should be properly wedged to hold the magnet wire secure in the slots.

3.3 - *Form Wound Coils:* Loops shall be made and spread to form coils without damage to the magnet wire insulation. Application of layer insulation shall be uniformly and tightly applied to eliminate stress points and air voids.

Insulated coils shall be placed in slots with no damage to the coil insulation. Coils shall tightly fit slots. Coils shall be secured to surge ring and laced to one another as necessary to prevent distortion.

3.4 - *A.C. Rotating Apparatus*

3.4.1 - *General:* The condition of the winding and the extent of repairs shall be determined by visual inspection and as necessary, an ohmmeter reading, phase unbalance or surge comparison, growler test, and/or appropriate high potential ground test.

Winding data shall be accurately taken and recorded.

3.4.2 - *Stators:* Stators shall be properly stripped (reference paragraph 1.5) and prepared for winding. The slots shall be clean and free of sharp edges or particles. Laminations shall be inspected for damage (reference paragraph 1.6). Stator bore shall be true and concentric. New windings shall comply with paragraphs 3.1, 3.2 or 3.3.

3.4.3 - *Squirrel Cage Rotors:* All slot conductors must be free of open circuits and have essentially the same resistance and conductivity as the original. The end rings should be welded, brazed, soldered or otherwise joined to the slot conductors to maintain relatively the same end ring resistance and conductivity. The entire squirrel cage winding shall maintain the same electrical characteristics as the original.

3.4.4 - *Wound Rotors:* Rotors shall be properly stripped (reference paragraph 1.5) and prepared for winding. The slots shall be clean and free of sharp edges or particles. Laminations shall be inspected for damage (reference paragraph 1.6).

New windings shall comply with paragraphs 3.1, 3.2 or 3.3. Windings shall be treated with a high bond strength varnish and banded.

For banding reference paragraph 3.1.7.

For slip rings reference paragraph 2.7.2.

For balancing reference paragraph 2.6.

3.4.6 - *Single Phase Motors:*

3.4.6.1 - *Stators:* Rewinding and repairing of single phase motors shall comply with paragraph 3.1, 3.2 and 3.4.

3.4.6.2 - *Squirrel Cage Rotors:* Squirrel cage rotors shall comply with paragraph 3.4.3.

3.4.6.3 - *Armatures:* Armature windings shall be tested for shorts, opens and grounds. Inspection of solder joints in commutators shall be made. New armature winding shall comply with paragraph 3.1. The commutator should be turned to concentricity of the shaft center or bearing surface. No flat spots, high, low or loose segments shall exist. The commutator shall be undercut as required. The surface of the finished commutator must be a smooth, polished finish to give full brush current carrying capacity.

3.4.6.4 - *Brush Holders and Brushes:* Brush holders shall be clean to prevent contamination of the commutator or brushes. All holders must be free working. Brush springs shall be in condition to give proper brush tension. Brushes must be the correct size and grade. All pigtails must be tight and connections to the holder must be clean and tight.

3.4.6.5 - *Accessories and Mechanisms:* Capacitors shall be checked for proper value and condition. Short circuit devices, centrifugal mechanisms, switches, and starting relays shall be checked for proper electrical and mechanical operation at correct speed and voltage. Terminal boards shall be replaced if damaged.

3.4.6.6 - *Leads:* In general, leads shall apply to paragraph 1.7. When single phase motors use lead colors instead of letter markings to identify the leads, the color assignment shall be determined from the following:

T-1 Blue	T-5 Black
T-2 White	T-8 Red
T-3 Orange*	P-1 No color assigned
T-4 Yellow	P-2 Brown

3.4.7 - *Final Tests:*

3.4.7.1 - *General:* The condition of new windings shall be tested by the following methods as applicable.

3.4.7.2 - *Ground Tests:* Ohmmeter reading of windings should be made and recorded.
Overpotential ground tests shall comply with section 8.0.

3.4.7.3 - *Winding Defect Tests:* Tests for winding defects shall be made by phase balance, surge comparison, growler, resistance, voltage drop and/or reverse coil tests.

3.4.7.4 - *No-Load Tests:* When possible no-load running tests shall be made at rated voltage. The no-load current and speed recorded and the mechanical operation checked.

3.4.7.5 - *Full-load Tests:* Full-load tests may be made as arranged with customer or as necessary to check the operating characteristics of the apparatus.

* Formerly green

3.5 - *D.C. Rotating Apparatus*

3.5.1 - *Mechanical*

3.5.2 - *Commutators*

3.5.3 - *Brush Holders*

3.5.4 - *Windings*

3.5.4.1 - *General*

3.5.4.2 - *Armatures*

3.5.4.3 - *Field Coils and Interpoles*

3.5.4.4 - *Wire Size*

3.5.4.5 - *Testing*

3.5.5 - *Testing Existing Windings*

3.5.5.1 - *Armature*

3.5.5.2 - *Fields*

3.5.5.3 - *Ground Test*

3.5.6 - *Leads*

3.5.7 - *Banding*

3.5.8 - *Varnish Impregnation and Coating*

3.5.9 - *Air Gaps*

3.5.10 - *Polarity Testing*

3.5.11 - *Brush Yoke*

3.5.12 - *Running Test*

4.0 - *Transformers - Liquid Filled*

4.1 - *General*

4.2 - *Preliminary Tests*

4.3 - *Untanking*

4.4 - *Disassembly*

4.5 - *Accessories and Auxiliary Items*

4.6 - *Core and Coils - Rewinding*

4.6.1 - *Laminations*

4.6.2 - *Coils*

4.6.3 - *Varnish and Coatings*

4.7 - *Care of Tanks*

4.8 - *Assembly*

4.8.1 - *Core and Coils*

4.8.2 - *Leads*

4.8.3 - *Bushing, Gaskets, and Accessories*

4.8.4 - *Insulating Liquid and Filling*

4.9 - *Final Tests*

4.10 - *Exterior Finish*

5.0 - *Transformer - Dry type, Distribution*

5.1 - *General*

5.2 - *Preliminary Tests*

5.3 - *Core and Coils - Rewinding*

5.3.1 - *Laminations*

5.3.2 - *Coils*

5.3.3 - *Insulation*

- 5.3.4 - *Varnishing and Coating*
- 5.4 - *Assembly*
  - 5.4.1 - *Core and Coils*
  - 5.4.2 - *Leads*
  - 5.4.3 - *Bushings and Accessories*
  - 5.4.4 - *Enclosure*
- 5.5 - *Final Tests*
- 5.6 - *Exterior Finish*

6.0 - *Hand Power Tools*

- 6.1 - *Cords*
- 6.2 - *Switches*
- 6.3 - *Brushes and Brush Holders*
- 6.4 - *Armatures*
- 6.5 - *Commutators*
- 6.6 - *Fields*
- 6.7 - *Gears*
- 6.8 - *Bearings*
- 6.9 - *Seals*
- 6.10 - *Grease and Oil*
- 6.11 - *Housings*
- 6.12 - *Chucks*
- 6.13 - *Tests*
  - 6.13.1 - *125 V Tools*

6.13.1.1 - *Ground Test - Blades to Metal*

6.13.1.2 - *Ground Test - Ground Pin to Metal*

6.13.2 - *125 V Double Insulated Tools*

6.13.2.1 - *Insulation Strength Test #1*

6.13.2.2 - *Insulation Strength Test #2*

6.13.2.3 - *Ground Test*

6.13.3 - *230 V Tools*

6.13.4 - *Speed Tests*

6.13.5 - *Final Run*

6.14 - *Finished Appearance*

7.0 - *Hermetic Stators*

7.1 - *Scope of Work*

7.2 - *Stripping*

7.3 - *Cleanliness*

7.4 - *Insulation System*

7.5 - *Windings*

7.6 - *Varnish Dipping and Baking*

7.7 - *Tests*

8.0 - *Testing*

8.1 - *Over-Potential Tests*

8.1.1 - *Induction and Direct Current Motors:* The high potential test for new windings of induction and direct current motors rated 1/2 horsepower and larger shall be made by applying 1000 volts A.C. 25–60 hertz plus twice the rated voltage of the motor for one minute.

The high potential test for new windings of induction and direct current motors rated less than 1/2 horsepower and for operation upon circuits not exceeding 250 volts shall be made by applying 1000 volt A.C. 25–60 hertz for one minute.

8.1.2 - *D.C. Field Windings:* New D.C. field windings should be high potential tested at ten times the excitation voltage.

8.1.3 - *Equivalent Test:* An alternating current test voltage of 1.2 times the one (1) minute test voltage specified may be applied for one (1) second as an alternative if desired.

8.1.4 - *Maintenance Proof Tests:* Overpotential tests for windings other than new are performed at sixty (60) percent of new winding test values.

8.1.5 - *D.C. High Potential Testing:* The testing of electrical apparatus by the use of D.C. high potential test equipment is recommended up to 34.5 KV. The standard ratio for converting A.C. test potential to equivalent D.C. values is 1.6. D.C. testing must be for one (1) minute duration.

Caution: After a D.C. over-potential test, the winding must be grounded to the frame until the charge has decayed to zero.

Electric Motor Winding Test Table

	New Windings			Old Windings		
Rated Voltage	**1 Min. A.C.**	**1 Sec. A.C.**	**1 Min. D.C.**	**1 Min. A.C.**	**1 Sec. A.C.**	**1 Min. D.C.**
600 and less	2200	2640	3520	1320	1585	2110
2400	5800	6960	9280	3480	4175	5570
4160	9320	11180	14900	5590	6710	8950

8.2 - *D.C. Field Testing:*

8.2.1 - *D.C. Voltage Drop Test:* Rated D.C. voltage should be applied to the field pole circuit. The variation in voltage drop should not be greater than 5% between coils.

8.2.2 - *A.C. Voltage Drop Test:* Same as 8.2.1 except the applied A.C. voltage shall be double the D.C. rating.

8.2.3 - *Resistance Tests:* Shunt field coils shall be checked individually with an ohmmeter or wheatstone bridge. The variation in resistance shall not exceed 2%.

9.0 - *Electrical Safety Standards*

9.1 - *Test Area*

9.1.1 - *Enclosure:* Test area shall be enclosed with a fence or colored rope - preferably yellow.

9.1.2 - *Grounding and Gates:* When a metallic fence or cage is used it must be grounded. Gates provided for entry of equipment and personnel must be equipped with interlocks so power to test area is interrupted if gate is opened.

9.1.3 - *Signs:* Approved signs must be posted warning unauthorized personnel to stay out of area.

9.1.4 - *Units in Area:* Only the unit under test should be in the test area.

9.1.5 - *Lighting:* The area must be well lighted.

9.2 - *Test Panels*

9.2.1 - *Construction:* Construction shall be of the "dead front" type.

9.2.2 - *Voltages:* Output voltages must be clearly marked.

9.2.3 - *Warning Lights:* An approved warning light shall indicate when the panel is energized. An additional approved light shall indicate when power leads to unit under test are energized.

9.2.4 - *Disconnect:* Panel shall have an approved main disconnecting means on the line side.

9.2.5 - *Switch:* An emergency hand or foot operated switch to disconnect the power source shall be in a convenient location.

9.2.6 - *Leads:* Test leads should be of adequate size and voltage class. Clips shall be insulated.

9.2.7 - *Mat:* Operating personnel shall stand on an approved insulated mat.

9.3 - *Grounding:* An equipment ground shall be installed on all test apparatus during test as approved by the National Electric Code.

Part III

General Preventive and Predictive Maintenance

Chapter 12

Storage Protection and Lubrication Management

Storage Protection

Preservation or corrosion inhibiting of inactive process machinery depends on the type of equipment, expected length of inactivity, and the amount of time required to restore the equipment to service.

Petrochemical companies will usually develop their standards to take these criteria into account. One recent typical mothballing program for indefinite storage in a northern temperature climate zone was planned and executed as follows and forms the basis for our recommendations.

Centrifugal and Rotary Pumps

1. Flush pumps and drain casing.
2. Neutralizing step required on acid or caustic pumps.
3. Fresh water flush and air dry all cooling jackets.
4. Fill pump casing with mineral oil containing 5 percent rust preventive concentrate.
5. Plug cooling water jackets—bearing and stuffing box—but keep low point drain valve cracked open slightly.
6. Coat space where shaft protrudes through bearing or stuffing box housings with Product 1 (see Table 12-1) and cover with tape.
7. Coat all coupling parts except elastomers with Product 1.
8. Coat all exposed machined surfaces with Product 1.
9. Fill bearing housing completely with mineral oil containing 5 percent rust preventive concentrate.
10. Pumps do not require rotation.
11. Close pump suction and discharge block valves.

Table 12-1
Corrosion Inhibiting Materials for Machinery Protection

PRODUCT	TYPE	APPLICATION	TRADE NAME
1	Solid Film Corrosion Inhibitor	Hot Dip Hot Brush	RUST-BAN 326 * or equal
2	Solvent Cutback Rust Preventive	Spray After Thinning	RUST-BAN 392 * or equal
3	Solvent Cutback Rust Preventive	Spray Brush	RUST-BAN 394 * or equal
4	Rust Preventive Concentrate	Mix or Full Strength	PROCON * or equal
5	Barrier Material-Grade C Waxed Paper	Wrap	US Govt. Spec. MIL-B121-D or equal
6	Oil and Moisture Resistant Coating (Aluminum Paint)	Spray	RUST-BAN PH6297 Aluminum Phenolic
7	Petrolatum—(Neutral Unctuous)	Hand Apply	Vaseline or equal

* *EXXON/IMPERIAL OIL Products*

Reciprocating Pumps

1. Flush and drain pump casing.
2. Neutralizing step required—if caustic or acid.
3. Blind suction and discharge nozzles of pump.
4. Fill liquid end with mineral oil containing 5 percent rust preventive concentrate. Bar piston to coat all surfaces. Allow some space for thermal expansion.
5. Fill steam end with mineral oil containing 5 percent rust preventive concentrate. Bar piston to coat all surfaces.
6. Close inlet and outlet valves.
7. Coat all joints where shaft protrudes from casings with Product 1. Cover with tape.
8. Coat exposed piston rod, shafts, and machined parts with Product 1.
9. Fill bearing housing and gearbox with mineral oil containing 5 percent rust preventive concentrate.
10. Fill packing lubricator with mineral oil containing 5 percent rust preventive concentrate.

Turbines

1. Isolate from steam system.
2. Seal shaft openings with silicone rubber caulking* and tape.
3. Dry out with air.
4. Fill turbine casing with oil containing 5 percent rust preventive concentrate including steam chest. Hold governor valve open as necessary to ensure chest is completely full. Vent casing, as required, to remove trapped air. Fill trip and throttle valve completely with oil.
5. Install a valved pipe on casing which can serve as filler pipe for adding oil to fill casing. Allow space for thermal expansion of oil in pipe.
6. Coat all external machined surfaces, cams, shafts, levers, and valve stems with Product 1.
7. Coat space between case and protrusion of shaft with Product 1. Cover space with tape.
8. Fill bearing housing completely with oil.
9. Coat casing bolts with Product 1.

Large Fans

1. Coat coupling and all external machined surfaces with Product 1.
2. Spray Product 2 on fan wheel.
3. Crack open casing low point, drain valve.

Gearboxes

1. Fill gearbox and piping completely with oil containing 5 percent Product 1.
2. Plug all vents. Allow space for thermal expansion.
3. Install a valved pipe on casing which can serve as filler pipe for adding oil to fill casing.

Large motors

1. Blank oil return line.
2. Seal shaft openings with silicone rubber caulking and tape.
3. Fill bearing housing with oil containing 5 percent rust preventive concentrate.

---

* Sealastic® or equal—black, to prevent pilfering.

4. Install a valved standpipe such that the inlet is higher than the bearing housing.
5. Coat all exposed machined parts with Product 1.
6. Do not rotate motor.

Centrifugal Process Compressors

1. Purge compressor casing of hydrocarbons.
2. Flush internals with solvent to remove heavy polymers.
3. Pressurize casing with nitrogen.
4. Mix 5 percent rust preventive concentrate to existing lube and seal oil. Circulate oil through the entire system for one hour.
5. Blank oil return header.
6. Seal shaft openings with silicone rubber caulking and tape.
7. Fill bearing housing with oil containing 5 percent rust preventive concentrate by running turbine-driven pump at reduced speed.
8. Fill oil console with mineral oil containing 5 percent rust preventive concentrate.
9. Filling should be done when compressor is at ambient temperature. Turn off all heat tracers.
10. Coat all exposed machined parts, including couplings, with Product 1.

Lube and Seal Oil System

1. Add 5 percent rust preventive concentrate to lube and seal oil.
2. Circulate oil throughout piping system. Open and close control and bypass valves so that all piping and components will receive oil circulation and become coated. Circulate for one hour. Vent trapped air from all components and high points.
3. Block in filters and coolers. Fill completely with oil containing 5 percent rust preventive concentrate but allow small space for thermal expansion. Water side of coolers should be drained and air dry. Plug all vents. Lock drain connections in slightly open position.
4. Fill reservoir with oil containing 5 percent rust preventive concentrate. Blind or plug all connections to tank including vent stack.
5. Coat exposed shaft surfaces and couplings of oil pumps with Product 1.

Reciprocating Compressors (see also page 552).

1. Purge compressor casing of hydrocarbons.
2. Blank compressor suction and discharge.

3. Fill crankcase, connecting rod and valves with oil containing 5 percent rust preventive concentrate. Install a valved standpipe. Allow space for thermal expansion.
4. Coat all exposed machined parts with Product 1.
5. Top up oil level in the cooling jacket.

Another occasion would be the three to twelve months' storage of machinery at a construction site. Usually termed a preventive maintenance program, storage protection plans would look like this, again in a northern, dry climate:

Rotation

Rotate all motors, turbines, compressors, pumps, excluding deep well pumps with rubber bushings, fin fans, blowers, aerators, mixers and feeders every two weeks.

Visual Inspection

Check when rotating exposed machined surfaces, shafts and couplings to see that protective coating has been applied and has not been removed. Reapply if needed.

Check all lubricating lines to see if any tubing, piping, tank, or sump covers have been removed. Retape ends and cover. Do this when discovered. If flanges are open on machinery, notify pipefitter general foreman or other designated responsible person in unit.

Inspect the interior of lube oil consoles on a six week schedule. Check to see if clean, and rust and condensate-free. Clean and dry out if needed, then fog with rust preventive concentrate.

Draining of Condensate

Drain condensation from all bearing housings, sumps, and oil reservoirs on a once a month schedule. If an excessive amount of condensation is found, recheck once a week, or at two week intervals depending on amount of condensate present.

Bearings

Fill all bearing housings that are oil lubricated but not force-fed with rust preventive concentrate, bringing the oil level up to the bottom of the shaft. For bearings that are force-fed the upper bearing cap and bearing will be removed. A heavy coat of STP® can be applied to the journal and bearing surfaces. This should be reapplied as needed.

Turbines

Turbines should be spot checked by removing the upper half of the turbine case and visually inspected. Plan to open a sampling of these turbines, selecting from the first preserved and in the worst condition. This should be done on a three month schedule. Other turbines may be inspected by the manufacturer's field service engineer on his periodic (monthly) visits. Small, general purpose turbines should be fogged with rust preventive concentrate through the opening in the top case as the rotor is being rotated. This should be done on a three month schedule.

Compressors

Manufacturers' representatives should inspect the compressors on a monthly visit basis. Preservative needed can be applied under their supervision. Centrifugal process compressors should be fogged and consideration be given to placing dessiccant bags in these machines. They should be inspected on a two month schedule. High speed air compressors should be inspected on a three month schedule. Axial compressors should be inspected and fogged on a three month schedule.

Pumps

Reciprocating pumps should be opened and inspected on a two month schedule. Centrifugal and inline pumps should be fogged with rust preventive concentrate. Volute cases need not be filled unless it is anticipated that they will remain out of service for over one year.

Electric Motors

Electric motors having grease type bearings need not be lubricated. If received with a grease fitting it should be removed and plugged or capped. For other lubrication type bearings see bearings.

Reducing or Speed Increasing Gears

The interior of the housing should be fogged with rust preventive concentrate. Tooth contact points should be coated with STP®. Gears and interiors should be visually inspected on a three month schedule by removing inspection plates.

Blowers

Blowers should be inspected on a three month schedule for rust.

Mixers

Mixers should be filled with rust preventive concentrate.

Fin Fans

Drive belts should stay on. Run several minutes at least every two weeks or whenever snow load dictates.

Miscellaneous Equipment

Miscellaneous equipment should be lubed as applicable and should be rotated on a two week schedule.

Other Considerations

In a warm, high precipitation climate—see Figure 12-1—it would be wise to look for alternate solutions to the problem of field storage during construction and prior to startup. If oil mist lubrication is not already part of the original design, it should be installed with top priority and activated as soon as possible. Figure 12-2 shows temporary field tubing to supply oil mist to the bearing points of a turbine drive pump row. Figure 12-3 shows a similar installation, feeding oil mist to pump and motor bearings, and Figure 12-4 illustrates construction site storage oil mist supply lines to a vertical mechanical drive turbine as well as to a large feed pump motor.

The third and last case of machinery storage protection arises when stand-by capability of laid-up equipment is desired. Reference 1 describes such a case. It appears as though there are no limits to the ingenuity displayed by operators—as long as a "do nothing and take your chances" stance is not taken. By the way, all preventive maintenance applied should be logged by item number in the maintenance log.

While the case of extended stand-by protection does not seem to present a problem for process pumps and other general purpose equipment—especially where oil mist lubrication is installed and operating—it might well be a challenge to operators of steam and gas turbines as well as reciprocating engines and compressors. One company[1] has had excellent success with their in-house developed program for stand-by storage of critical machinery, particularly gas turbines. One manufacturer[2] recommends the following procedure for the stand-by protection of gas engines or gas engine driven compressors:

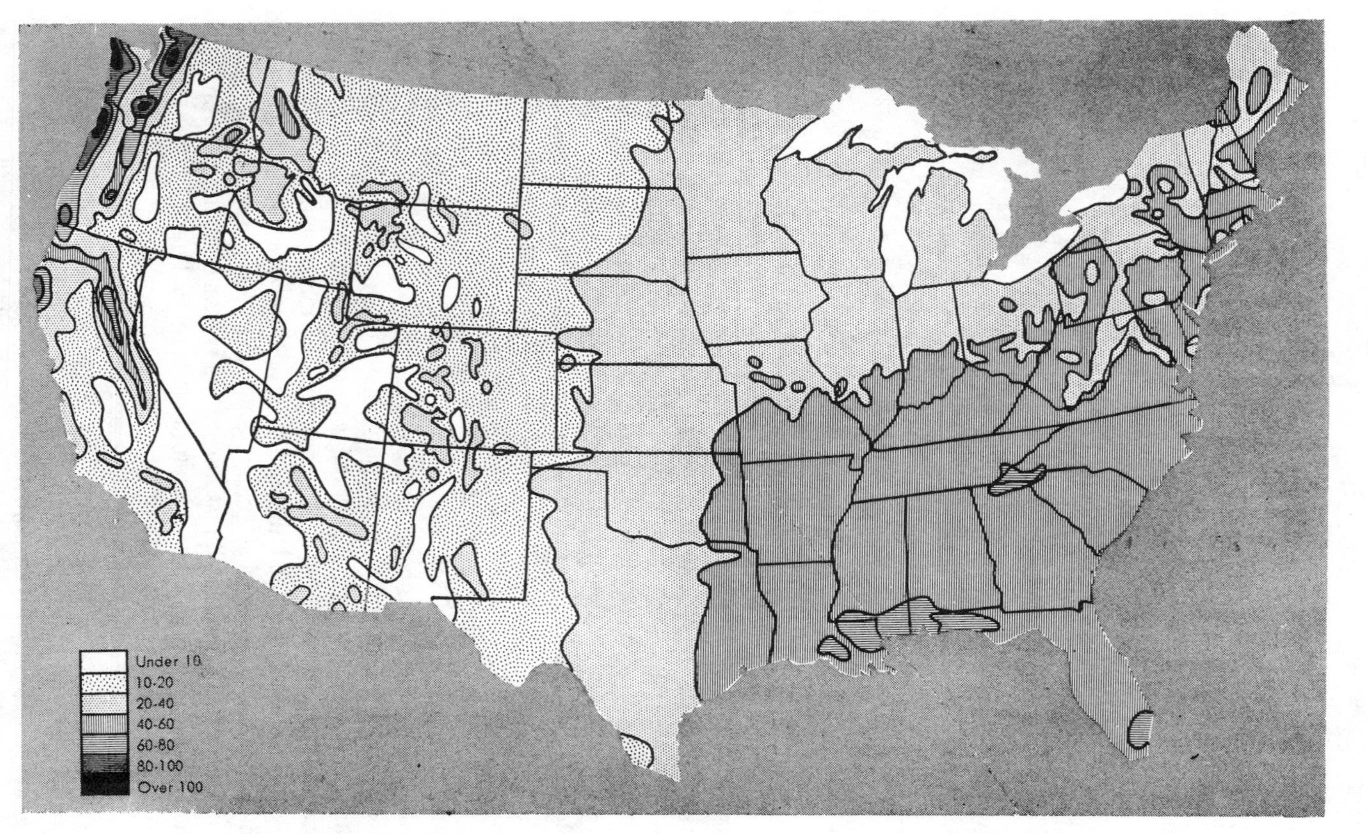

Figure 12-1. Precipitation (inches) map of the United States (*Collier's Encyclopedia*).

Figure 12-2. Temporary oil mist installation for protection of machinery during construction phase—pump row.

Figure 12-3. Temporary oil mist installation for protection of machinery during construction phase—motor/process pump set.

Figure 12-4. Temporary oil mist installation for protection of machinery during construction phase—electric motor and vertical turbine with speed governor.

Drain the water jackets and then circulate the proper compound* through the jackets making sure that all surfaces in the jacket are reached. Drain the system and plug all openings.

Lubrication System

On engine lubrication systems, proceed as follows:

1. Drain the lubricating oil system, including filters, coolers, governors, and mechanical lubricators. Flush the complete system with standard petroleum solvent that will take the oil off the surfaces. Use an external pump to force the solvent through the system. Spray the interior of the crankcase thoroughly. Then drain.
2. Refill to the minimum level—just enough to ensure pump suction at all times—in each case with the proper compound.* Crank the mechanical lubricator by hand until all lines are purged. Where compressors are used be sure to flood the compressor rod packing.
3. Using air pressure or any other convenient means turn the engine at sufficient speed and for a sufficient length of time to thoroughly circulate the compound through the engine.

---

* Product 4, Table 12-1

4. Stop and drain the engine sump, filters, coolers, governor, lubricators, etc. Plug all openings.
5. Remove the spark plugs or gas injection valves and spray with Product 2 inside the cylinders so as to cover all surfaces. While doing this rotate the engine by hand so that each piston is on bottom dead center when that particular cylinder is being sprayed.
6. After this operation the engine should not be turned or barred over until it is ready to be placed in service. Tag the engine in several prominent places with warning tags.
7. Where compressors are involved—including scavenging air compressors—remove the valves and spray inside the cylinder so as to cover all surfaces. Dip the compressor valves in Product 1 and drain off the excess. Reassemble valves in place.

Lubrication Management

General Plant Lubrication Concepts*

The correct lubrication of plant equipment is an important factor in sustaining production with reduced equipment outage and lower maintenance costs. When a well planned and coordinated lubrication program has been established, the production plant has an effective preventive maintenance program. A complete program will encompass all phases of plant lubrication.

The need for protecting operating equipment against excessive wear and for keeping plant delays to the absolute minimum brings lubrication into a very prominent position. All moving parts of the equipment require lubrication to permit it to function successfully and contribute toward high plant production. This can only be achieved when a complete "Lubrication Program" has been established and is being maintained.

The success of any plant-wide lubrication program depends upon close cooperation among certain individuals and groups of individuals within the various units and divisions of the organizational structure, including operating, maintenance, engineering, and lubrication personnel. In addition, valuable technical assistance can be obtained from the equipment manufacturer and the lubricant supplier.

* Partial source: Allen M. Clapp, as published in "Proceedings of Seventh Texas A&M Turbomachinery Symposium," December 1978. Adapted by permission of Texas A&M Turbomachinery Laboratories.

Lubrication Survey of the Plant

If the plant has not recently been surveyed from a lubrication standpoint, a careful survey should be conducted to determine the current lubrication practices.

The main part of this survey will consist of a detailed lubrication inspection of all plant equipment. Each machine must be studied and its lubrication-related characteristics tabulated. These characteristics must include the items listed in Figure 12-5.

The listing would thus be comprised of electric motor bearings, flexible couplings, reduction gears and bearings, etc. The name of the lubricant being used, together with the means and frequency of application should also be listed for each lubrication point. Likewise, any abnormal conditions encountered, such as excessive oil leakage, water entering the oil, frequent bearing changes and the like should be recorded.

Obtaining this information is time consuming and several days may be required to survey the plant. However, such a survey is the only way of obtaining an accurate picture of current lubrication practices and it is the basis upon which future steps to improve lubrication will be made. Therefore, the time spent in conducting a thorough survey is well justified. Since a general knowledge of the design of a machine is required for making any critical analysis of its lubrication requirements, it may be necessary to make frequent reference to machine drawings to clarify points that cannot be determined by an external inspection. When lubrication or other maintenance problems are being encountered on a machine, every effort should be made to inspect the machine when it is opened for repair to determine the points where excessive wear or other problems exist.

Lubricant Consolidation

For best performance results, buy and use only carefully selected lubricants. As shown in Figure 12-6, establish a list of lubricant types which is sufficiently extensive to cover essentially all the lubricant requirements for the entire plant. Each type should have characteristics which make it particularly suitable for applications under certain operating conditions. A common name or number should be assigned each type for use on lubrication schedules, stores records, etc. Although the total number of different lubricants required, including the various viscosities and consistencies, may amount to 15 to 25 for a large plant, some major plants have been able to reduce the number of lube types to as few as 7 without in any way compromising equipment reliability.

1. Inventory Machinery
2. Machine Characteristics
3. Points of Lubrication
4. Methods of Lubrication
5. Frequency of Lubrication

Figure 12-5. Lubrication-related characteristics to be used in a plant lubrication survey.

1. List Lubricant Types
   . . . Common Name
   . . . Identity Code
2. Determine General Characteristics
   . . . Supplier
   . . . Laboratory Testing
3. Final Selection
   . . . Field Testing
   . . . Past Experience

Figure 12-6. Lube oil selection parameters.

The general characteristics of any particular lubricant may be determined from information received from the supplier, or from testing a sample of the lubricant to determine its measurable physical and chemical properties. The final selection of a particular lubricant should be made only after a careful observation of the lubricant in one or more typical applications in the plant.

Master Lubrication Schedules

After the detailed plant survey has been conducted, the lubricant types established, and the necessary consolidation of lubricants effected, the next step is the preparation of the master lubrication schedule as shown in Figure 12-7.

1. Identification of Equipment
2. Points of Lubrication
3. Lubricants
4. Methods
5. Frequencies
6. Name of Worker

Figure 12-7. What lubrication schedules should contain.

Lubrication schedules are then printed by the computer and placed in the hands of the personnel who apply the lubricants. Without such schedules there can be much confusion and differences of opinion in the plant as to what should be used or even what is actually being used in the different equipment, especially when there has been a change in either supervision or the operating personnel in the plant. The mere issuing of the lubrication schedule, of course, does not ensure compliance. Field followup is usually required.

Lubrication Notification

Monthly lubrication work lists, as shown in Figure 12-8, are printed by the computer. These lists are compiled from the Master Lubrication Schedule for each worker assigned lubrication responsibilities. The notification system is essential for effective utilization of the lubrication schedule. These monthly work lists are simply memory joggers for the people with lubrication responsibilities. The work lists are also an effective communication link between the worker and his supervisor. The worker is required to note any unusual condition of the machine or lubricants in the machine observed when the lubrication job was being performed.

Many modern plants are now using portable data terminals to capture lubrication data. These data monitoring or data logging methods are explained in technical and sales literature which can be obtained from the major manufacturers of machinery vibration monitoring equipment.

Lubricant Handling and Storage

The large quantities of lubricants used in the operation of plants make the buying and handling of these materials within the plants an important

1. Printed and Distributed by 1st Working Day of Month
2. Worker Name and Area
3. Points of Lubrication
4. Lubricant
5. Signed and Dated
7. Returned to Supervisor When Complete

Figure 12-8. What monthly lubrication notification, or lubrication worklists contain.

item from the standpoint of housekeeping, safety, and costs. Major considerations should include storage methods, whether returnable containers should be used, etc.

To obtain the lowest prices, plants purchase their lubricants in the largest quantities consistent with the rate of usage and the capacity of their storage facilities. In the case of fluid lubricants purchased in quantities greater than 6,000 gallons per year, tank truck delivery is generally preferred.

In addition to the price advantage, purchase in bulk eliminates the storage and handling of large numbers of drums within the plant. Bulk storage is even more attractive if the oil is delivered directly to a tank at the point of consumption in the plant.

Good housekeeping can be maintained only if empty drums are returned from the consuming units to the supplier promptly. Most lubricant suppliers charge deposit fees of $15 or more for each drum of oil delivered to the consumer. Full credit is given for the return of each undamaged empty drum. Careful management of returnable empty drums can result in large annual savings.

For identification of lubricants in handling and applying them throughout the plant, the type designation or name must be stenciled on the lid and the side of each drum. The same name appears on the drums as on the lubrication schedules, and the chance of a mistake in application is small.

Oil Sampling and Testing Program

The technique of used oil analysis as a preventive and predictive maintenance tool has been used by industry for many years. The prevention of unexpected outages of critical machinery can be avoided or at least kept to a minimum when a program is properly initiated and maintained.

Oils from critical machinery are periodically tested to determine their suitability for continued use. Volume 2 of this series of books on Machinery Management contains additional information on this subject.

For conventional analysis, a sample of the oil is withdrawn from the system and several of its characteristics measured in the laboratory. The usual tests conducted to determine the condition of used oils include viscosity, pH and neutralization number, precipitation, color, and odor. There are however, additional tests that may be conducted for a more detailed evaluation of the used oil. These might include metals analysis by atomic absorption or spectographic methods, infrared, X-ray diffraction, gel permeation chromatography, etc.

The test results are subsequently reviewed and compared with the new oil specifications. Depending upon the nature or degree of departure from new oil, the laboratory will either approve the lubricant for continued use or recommend the necessary corrective action.

Evaluation of New Lubricants

Since most lubricant manufacturers conduct continuous research to develop lubricants with more desirable combinations of characteristics, it is difficult to state that any particular lubricant is the "best" lubricant. A lubricant that may appear to be the best this year may be surpassed by a newly developed lubricant next year.

For this reason, it is advisable to give consideration from time to time to new lubricants offered by suppliers. Whenever new lubricants are evaluated, detailed laboratory analyses, field testing, and cost should influence your ultimate decision.

First, the specifications of the lubricant may be compared with those of the lubricant you are now using and other available similar lubricants. Specifications furnished by the supplier are usually limited to some of the basic physical properties of the lubricant and are sufficient to indicate its type and general composition.

The specification and test data serve as good means for screening lubricants, but they do not give an absolute measure of the all-around performance of a lubricant in a given application. This can be determined only by observing its performance in actual service as closely as is practical.

Price, of course, is also a factor in the purchase of lubricants. The end user must be certain, however, that satisfactory performance has been established before price becomes the deciding factor in procurement. A sacrifice in performance can often make a seemingly low-priced lubricant very expensive. Also, purchase of the highest priced lubricant gives no assurance of superior performance.

Lubrication Training

Lubrication training is as important as any other training program for plant personnel. The increased skills and knowledge of the people responsible for the actual application of lubricants can substantially increase reliability of costly production equipment. In addition, trained operating and maintenance personnel will be more alert to equipment malfunction and report conditions before the equipment actually fails.

The prevention of unscheduled outage can reduce production losses and minimize maintenance costs, especially when catastrophic failures are prevented.

Lubrication training of plant personnel includes instruction in basic principles of lubrication, the computer-assisted scheduling and notification program, lubrication procedures, and centralized lubrication systems.

Lubrication Methods and Procedures

A highly rewarding phase of any lubrication program in most plants today is the adoption of improved methods for applying lubricants.

Many oil mist and centralized systems have been installed. In almost every case outstanding improvement in lubrication has resulted. Machinery life has increased and appreciable overall savings realized. In initiating such a program in a plant, it is important to keep records showing maintenance costs before and after installation of the new system to convince management of the value of continuing with this type of improvement.

An essential factor in a program of applying centralized lubrication systems is the training of personnel who will operate this equipment. These same persons should have the responsibility of inspecting and maintaining the systems in proper operating conditions.

We can now consider some of the more important methods and procedures that have been developed into very effective means of preventing failures and increasing improved machinery reliability: constant level oilers, oil-lubricated motors, grease-lubricated motors, bearing labyrinth purge, reservoir purge and vent system, gear coupling lubrication, and oil mist lubrication.

Constant Level Oilers

The oil in machinery equipped with constant level oilers as shown in Figures 12-9 and 12-10 must be routinely inspected visually to determine the condition of the working oil. The frequency of the visual inspection

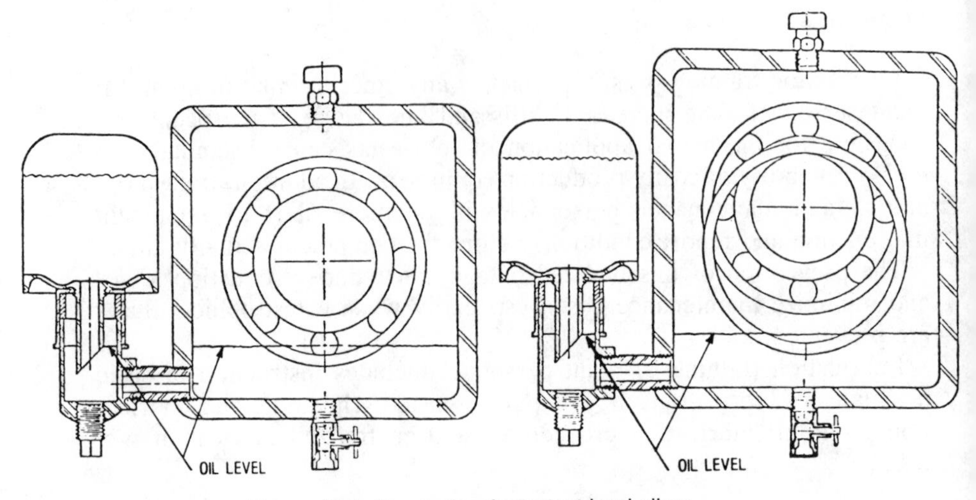

Figure 12-9. Conventional constant level oilers.

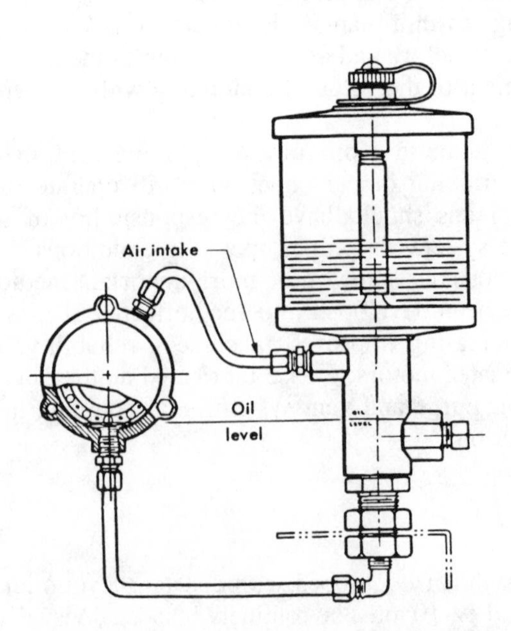

Figure 12-10. Constant level oil with equalizing tube (Oil-Rite Corporation).

can vary from one month to three months, depending on the severity of service.

The operators must be continually aware that all oiler bowls contain oil. The frequency for checking these may vary from each shift to once weekly, depending on experience with each machine.

The definition of "visual" is to drain approximately two ounces oil from the bottom of the bearing housing and to observe:

1. Oil appearance—it should be normal color. Dark color change normally indicates the oil has oxidized. Change the oil if it is discolored.
2. Check for contaminants. Look for water, product, or dirt. Change the oil if it is contaminated.
3. Verify that oil will flow into bearing after the two ounce bottom sample has been drawn. Air bubbles should rise from the bottom of the oiler bowl as oil flows into the bearing. If no air bubbles are observed, this indicates:
   a. The machine may be overfilled. Drain excess oil until air bubbles begin to rise in bowl.
   b. The connecting pipe nipple between the oiler and bearing may be plugged with sludge. Remove the pipe nipple and oiler lower bowl and clean.
   c. The bearing housing vent may be plugged, creating an unequal pressure between the bearing and the oiler.

Figure 12-10 illustrates a constant oiler with a balance line between the bearing housing and the oiler lower bowl. This type of installation is required for bearing housings having an excessive back pressure or vacuum. A constant level can be maintained in spite of pressure or vacuum in the housing as the equalizing type provides static balance of pressure between the bearing housing and oiler lower bowl.

Oil-Lubricated Electric Motors

Oil-lubricated motors are generally large frame designs. These motors may have oil supplied from a console common with the driven equipment. Alternatively, there could be an independent console just for the motor or each bearing housing could have its own oil supply. The first and second types are preferred since the oil condition in all consoles is monitored by oil analysis. The third type requires regular attention to obtain maximum bearing life.

Motors with each bearing having a captive oil supply are checked weekly for oil level. At three month intervals the oil in each bearing is checked visually. A sample is drawn from the bottom drain into a glass container and checked for appearance including moisture, dirt, or other discoloration. The oil should be changed if contaminated. This type of motor may be equipped with constant level oilers and in that case the procedure for checking constant level oilers is applied.

Grease Lubrication of Electric Motors

Electric motors are relubricated at regular intervals based on size, speed, duty, and environment.

The lubrication schedule for specific motors is shown on the computerized Programmed Lubrication Schedule.

Grease-lubricated electric motor types are:

1. Motors with grease inlet and outlet ports on the same side of the bearings, as shown in Figure 12-11. The bearings are commonly referred to as conventional grease flow design.
2. Motors with grease inlet and outlet ports on opposite sides of the bearings, Figure 12-12. The bearings are commonly referred to as cross flow lube design.

Each of these motor types requires a different lubrication procedure. The motor with inlet and outlet grease ports on the same side of the bearings must be lubricated with the motor stopped whereas the motor with

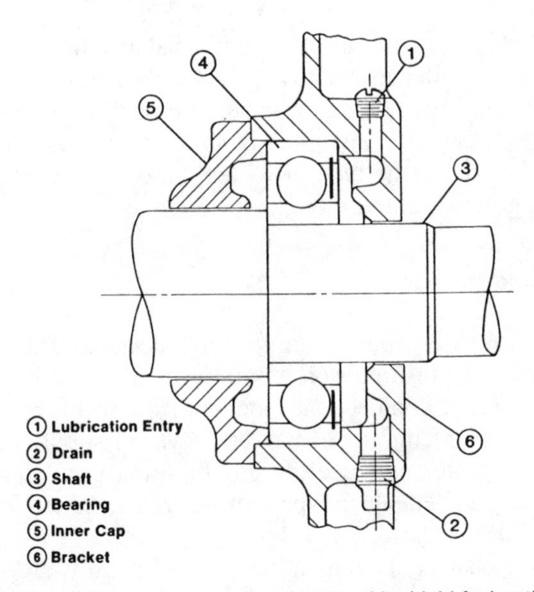

Figure 12-11. Single-shield motor bearing, with shield facing the grease cavity.

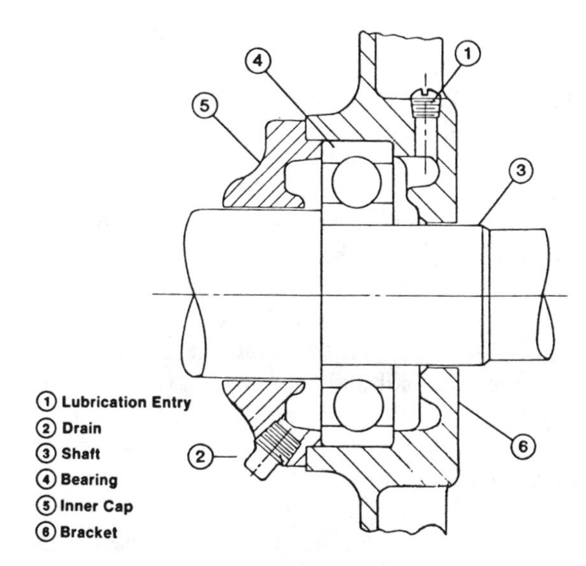

Figure 12-12. Open bearing with cross-flow grease lubrication.

inlet and outlet grease ports on the opposite sides of the bearings must be lubricated with the motor running.

Electric motor bearings should be regreased with a grease compatible with the original charge. It should be noted that the polyurea greases often used by the motor manufacturers are *not* compatible with lithium-base greases.

Single-Shielded Bearings

To take advantage of single-shield arrangements in electric motors, the Phillips Petroleum Company developed three simple recommendations:

1. Install a single-shield ball bearing with the shield *facing* the grease supply in motors having the grease fill-and-drain ports on that same side of the bearing. Add a finger full of grease to the ball track of the back side of the bearing during assembly.
2. After assembly, the balance of the *initial* lubrication of this single-shielded bearing should be done with the motor idle. Remove the drain plug and pipe. With a grease gun or high volume grease pump, fill the grease reservoir until fresh grease emerges from the drain. The fill and drain plugs should then be reinstalled and the motor is ready for service. It is essential that the initial lubrication *not* be attempted while the motor is running. It was observed that to do so will cause, by pumping action, a continuing flow of grease

through the shield annulus until the overflow space in the inner cartridge cap is full. Grease will then flow down the shaft and into the winding of the motor where it is not wanted. This will take place before the grease can emerge at the drain.

3. Relubrication may be done while the motor is *either* running or idle. (It should be limited in quantity to a volume approximately 1/4 the bearing bore volume.) Test results showed that fresh grease takes a wedge-like path straight through the old grease, around the shaft, and into the ball track. Thus, the overflow of grease into the inner reservoir space is quite small even after several relubrications. Potentially damaging grease is thus kept from the stator winding. Further, since the ball and cage assembly of this arrangement does not have to force its way through a solid fill of grease, bearing heating is kept to a minimum. In fact, it was observed that a maximum temperature rise of only 20°F occurred 20 minutes after the grease reservoir was filled. It returned to 5°F rise two hours later. In contrast, the double-shield arrangement caused a temperature rise of over 100°F (at 90°F ambient temperature the resulting temperature was 190°F) and maintained this 100°F rise for over a week.

Regreasing Double-Shielded Bearings

1. Ball Bearings

   a. Pack (completely fill) the cavity adjacent to the bearing. Use necessary precautions to prevent contaminating this grease before motor is assembled.
   b. After assembly, lubricate *stationary motor* until a full ring of grease appears around the shaft at the relief opening in the bracket.

2. Cylindrical Roller Bearings

   a. Hand pack bearing before assembly.
   b. Proceed as outlined in (a) and (b) for double-shielded ball bearings.

If under-lubricated after installation, the double-shielded bearing is thought to last longer than an open (nonshielded) bearing given the same treatment because of grease retained within the shields (plus grease remaining in the housing from its initial filling).

If over-greased after installation, the double-shielded bearing can be expected to operate satisfactorily without overheating as long as the excess grease is allowed to escape through the clearance between the shield and inner race, and the grease in the housing adjacent to the bearing is not churned, agitated and caused to overheat.

It is not necessary to disassemble motors at the end of fixed periods to grease bearings. Bearing shields do not require replacement.

Double-shielded ball bearings should not be flushed for cleaning. If water and dirt are known to be present inside the shields of a bearing because of water intrusion or other circumstances, the bearing should be removed from service. All leading ball-bearing manufacturers are providing reconditioning service at a nominal cost when bearings are returned to their factories. As an aside, reconditioned ball bearings are generally *less prone* to fail than are brand new bearings. This is because grinding marks and other asperities are now burnished to the point where smoother running and less heat generation are likely.

Regreasing Open Bearings

Motors with open, conventionally greased bearings are generally lubricated with slightly different procedures for drive-end and opposite end bearings.

Open bearings located at the motor drive-end should be regreased as follows:

1. Relubrication with the *shaft stationary* is recommended. If possible, the motor should be warm.
2. Remove plug and replace with grease fitting.
3. Remove large drain plug when furnished with motor.
4. Using a low pressure, hand operated grease gun, pump in the recommended amount of grease, or use 1/4 of bore volume.
5. If purging of system is desired continue pumping until new grease appears either around the shaft or out the drain plug. Stop after new grease appears.
6. On large motors provisions have usually been made to remove the outer cap for inspection and cleaning. Remove both rows of cap bolts. Remove, inspect and clean cap. Replace cap, being careful to keep dirt from bearing cavity.
7. After lubrication, allow motor to run for fifteen minutes before replacing plugs.
8. If the motor has a special grease relief fitting, pump in the recommended volume of grease or until a one inch long string of grease appears in any one of the relief holes. Replace plugs.
9. Wipe away any excess grease which has appeared at the grease relief port.

Open bearings located at the outboard end of the motor should be regreased as follows:

1. If bearing hub is accessible, as in drip-proof motors, follow the same procedure as for the drive-end bearing.
2. For fan-cooled motors note the amount of grease used to lubricate shaft end bearing and use the same amount for commutator-end bearing.

Open bearings arranged with housings provisions as shown in Figure 12-12, with grease inlet and outlet ports on opposite sides, are called cross-flow lubricated. Regreasing is accomplished with the motor running. The following procedure should be observed:

1. Start motor and allow to operate until normal motor temperature is obtained.
2. Inboard bearing (coupling end):

   a. Remove grease inlet plug or fitting.
   b. Remove outlet plug. Some motor designs are equipped with excess grease cups located directly below the bearing. Remove the cups and clean out the old grease.
   c. Remove hardened grease from the inlet and outlet ports with a clean probe.
   d. Inspect the grease removed from the inlet port. If rust or other abrasives are observed, *do not grease the bearing*. Tag motor for overhaul.
   e. Bearing housings with outlet ports:

      - Insert probe in the outlet port to a depth equivalent to the bottom balls of the bearing.
      - Replace grease fitting and add grease slowly with a hand gun. *Count strokes of gun* as grease is added.
      - Stop pumping when the probe in the outlet port begins to move. This is an indication the grease cavity contains an adequate quantity of grease.

   f. Bearing housings with excess grease cups:

      - Replace grease fitting and add grease slowly with a hand gun. *Count strokes of gun* as grease is added.
      - Stop pumping when grease appears in the excess grease cup. This indicates the grease cavity contains an adequate quantity of grease.

3. Outboard bearing (fan end):

   a. Follow inboard bearing procedure provided the outlet grease ports or excess grease cups are accessible.
   b. If grease outlet port or excess grease cup is not accessible, add 2/3 of the amount of grease required for the inboard bearing.
4. Leave grease outlet ports open—*do not replace* the plugs. Excess grease will be expelled through the port.
5. If bearings are equipped with excess grease cups, replace the cups. Excess grease will expel into the cups.

Dry Sump Oil Mist Lubrication for Electric Motor Bearings

Oil mist lubrication is a centralized system which utilizes the energy of compressed air to supply a continuous feed of atomized lubricating oil to multiple points through a low pressure distribution system of approximately 20 inches water column. Oil mist then passes through a reclassifier nozzle before entering the point to be lubricated. This reclassifier nozzle establishes the oil mist stream as either a mist, spray, or condensate, depending on the application of the system. (See Volume 1.)

For well over a decade, the oil mist lubrication concept has been accepted as a proven and cost-effective means of providing positive lubrication for centrifugal pumps. Centralized oil mist systems have also been used on gears, chains, and horizontal shaft bearings such as on steam turbines and steel rolling mill equipment. Although clearly demonstrated to be ideally suited for lubricating antifriction bearings in electric motors ranging from fractional horsepower to well over 1,000 horsepower, relatively few petrochemical installations have extended this superior lubrication concept to electric motors on a plant-wide basis. However, many experienced motor manufacturers are offering oil mist options as shown in Figure 12-13.

The actual method of applying oil mist to a given piece of equipment is governed to a large extent by the type of bearing used. For sleeve bearings, oil mist alone is not considered an effective means of lubrication because relatively large quantities of oil are required. In this case, oil mist is used effectively as a purge of the oil reservoir and, to a limited extent, as fresh oil make-up to the reservoir. Rolling element bearings, on the other hand, are ideally suited for dry-sump lubrication. With dry sump oil mist, the need for a lubricating oil sump is eliminated. If the equipment shaft is arranged horizontally, the lower portion of the bearing outer race serves as a mini-oil sump. The bearing is lubricated directly by a continuous supply of fresh oil condensation. Turbulence generated by bearing rotation causes oil particles suspended in the air stream to con-

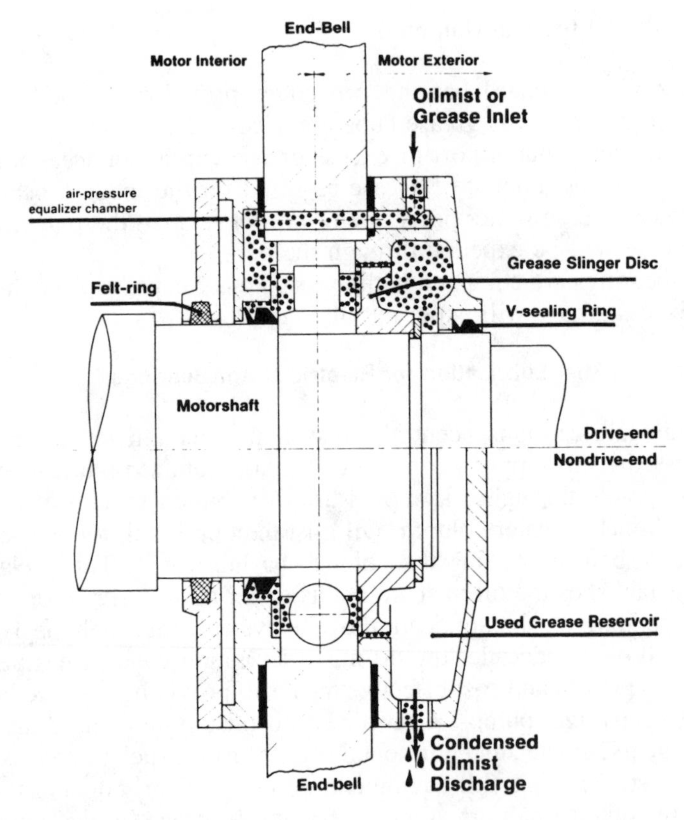

Figure 12-13. Oil mist lubricated bearings furnished with Siemens TEFC Motors from 18 to about 3000 kW.

dense on the rolling elements as the mist passes through the bearings and exits to atmosphere. This technique offers four principal advantages:

- Bearing wear particles are not recycled back through the bearing, but are washed off instead.
- The need for periodic oil changes is eliminated.
- Long-term oil breakdown, oil sludge formation, and oil contamination are no longer factors affecting bearing life.
- The ingress of atmospheric moisture into the motor bearing is no longer possible and even the bearings of standby equipment are properly preserved. Without oil mist application, daily solar heating and nightly cooling cause air in the bearing housing to expand and contract. This allows humid, often dusty air to be drawn into the housing with each thermal cycle. The effect of moisture condensation on anti-

friction bearings is extremely detrimental and is chiefly responsible for few bearings ever seeing their design life expectancy in a conventionally lubricated petrochemical environment.

It has been established that loss of mist to a pump or motor is not likely to cause an immediate and catastrophic bearing failure. Tests by various oil mist users have proven that bearings operating within their load and temperature limits can continue to operate without problems for periods in excess of eight hours. Furthermore, experience with properly maintained oil mist systems has demonstrated outstanding service factors.

Also, when oil mist lubrication of motors is discussed, the question of oil intrusion into the motor windings is often raised. The concern is voiced that lube oil would enter the motor housing and cause damage to winding insulation or overheating until winding failure. Initial efforts were, therefore, directed toward developing lip seals or other barriers confining oil mist to only the bearing areas.

When occasional seal failures were experienced on operating motors, oil mist entered the motor housings and coated the windings with lube oil. The potential explosion hazard was again investigated on this occasion and it was confirmed that the oil/air mixture was substantially below the sustainable burning point. The fire or explosion hazard of oil mist lubricated motors is thus no different than that of NEMA-II motors. No signs of overheating were found and winding resistance readings conformed fully to the initial, as-installed values.

Today's epoxy motor winding materials will not deteriorate in an oil mist atmosphere. This has been conclusively proven in tests by several manufacturers as well as occasional incidents of severe lube oil intrusion. In one such case, a conventional oil-lubricated, 3,000 hp, 13.8 kV motor ran well even after oil was literally drained from its interior. The incident caused significant dirt collection but did not adversely affect winding quality.

The failure experience of hundreds of motor bearings with conventional grease or oil ring lubrication was compared with that of oil mist lubricated motors. Feedback from petrochemical units using dry sump oil mist lubrication showed them to experience far fewer bearing problems than similar units adhering to conventional lubrication methods. Failure reductions of 75 percent seem to be the rule and have been documented in many technical papers. Larger reductions have sometimes been achieved. Installations which already employ oil mist on pumps will often find it possible to extend this lubrication method to the motor driver at a cost of approximately $50 per bearing. Needless to say, this is a fraction of the cost of replacing bearings or repairing consequential damage to motor windings.

Conversion to dry sump oil mist lubrication does not necessarily require that the motor be removed and sent to the shop. Motors with regreasable bearing lubrication are easiest to convert because they generally incorporate neither oil rings nor bearing shields. However, oil lubricated bearings are also easily modified for dry sump lubrication by providing only the piped oil mist inlet, vent, and overflow drain passages. Oil rings must be removed because there is, of course, no longer an oil sump from which oil is to be fed to the bearing. Figure 12-14 shows the bearing shields removed in order to establish unimpeded passage from the oil mist inlet pipe through the bearing rotating elements and finally the vent pipe to atmosphere. However, recent experience shows that the inboard bearing shield does not have to be removed to ensure a successful installation in horizontal motors. Only in vertical motors with thrust-loaded and/or multiple rows of bearings is it prudent to remove all shields and route the oil mist through the rotating elements. This is shown in Figure 12-15 which, incidentally, illustrates a typical dry sump configuration for vertical motors. One such motor, rated 125 hp, 3,560 rpm, experi-

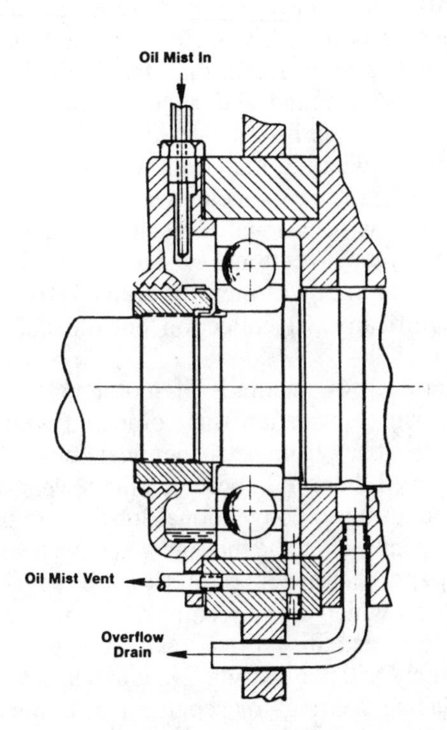

Figure 12-14. Dry-sump oil mist lubrication applied to horizontal electric motor.

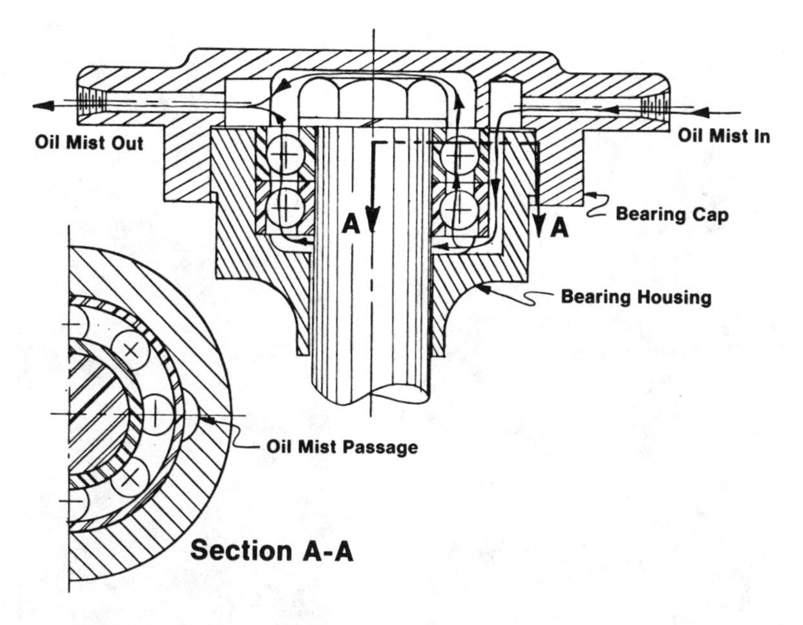

Figure 12-15. Oil mist lubricated thrust bearings on vertical motor.

enced frequent thrust bearing failures with conventional oil lubrication. Installation of dry sump mist apparently solved the chronic lubrication problem. Bearing housing temperature was lowered from 160°F to 110°F after the conversion to dry sump lubrication.

The simplicity of applying dry sump oil mist lubrication to a wide variety of electric motors is evident from Figures 12-16 and 12-17. Sloped stainless steel inlet tubing is used from the distribution block to the bearings. The vent tubing is terminated in a transparent plastic container.

It may be anticipated that a properly installed and maintained oil mist lubrication system will result in a high percentage reduction in bearing failure rates. It must be noted, however, that such bearing failure reductions will not be achieved if the basic bearing failure problem is not lubrication related. Oil mist cannot eliminate problems caused by defective bearings, incorrect bearing installation, excessive misalignment or incorrect mounting clearances.

Additional information on oil mist lubrication principles is given later in this chapter.

Figure 12-16. Dry sump oil mist applied on a vertical motor.

Figure 12-17. Dry sump oil mist applied on a horizontal electric motor.

Lubrication by Single-Point Automatic Grease Injectors

Single-point automatic grease injectors as shown in Figures 12-18 through 12-20 are a refinement of the age-old grease cup. Grease cups are, of course, small containers fitted to the bearing. The cup or container is filled with grease, which is forced into the bearing by manually forcing down the cap or piston covering the grease charge.

Single-point automatic grease injectors differ from the traditional grease cup by employing either a spring or an expanding gas force to exert pressure on the cap, piston, or diaphragm in contact with the grease volume. These continuous force grease injection devices are screwed into the thread which would typically accommodate a grease fitting. They range in size from 0.9 oz (28 grams) to 3.8 oz (112 grams), or approximately 26 cu. cm (1.6 cu. in.) to 120 cu. cm. (7.3 cu. in.) grease capacity.

The device shown in Figure 12-18 is spring-loaded. It claims to "adjust the flow of grease to a bearing automatically, by the use of a new, more effective metering control principle. This is accomplished by a special piston O-ring seal that creates a changing level of friction as it moves

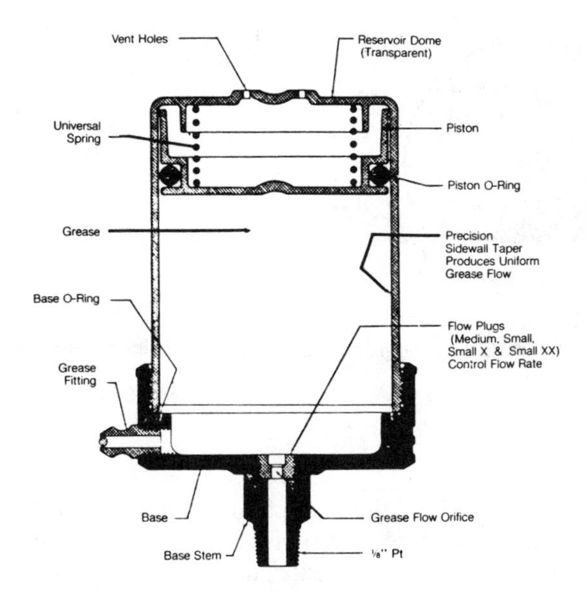

Figure 12-18. Spring-loaded single point grease injector.

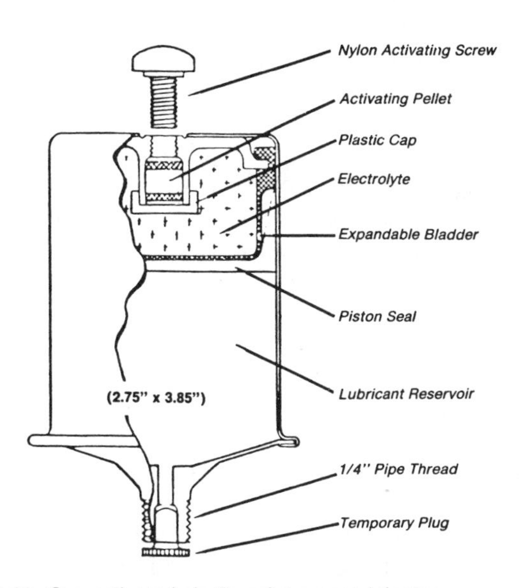

Figure 12-19. Gas-activated single point grease injector.

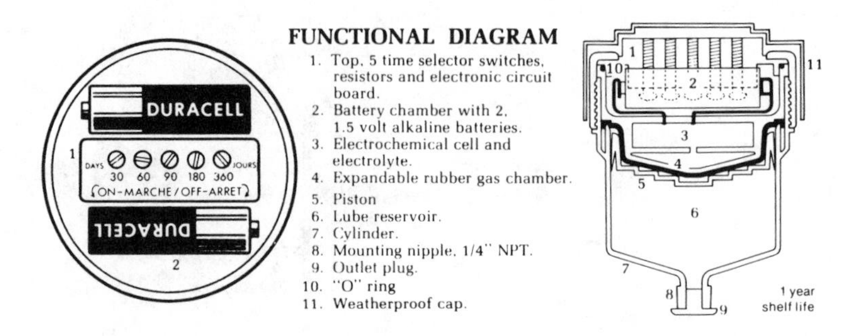

Figure 12-20. Electrochemically activated single point grease lubricator.

along the tapered wall of the transparent reservoir dome. The changing resistance is designed to counterbalance the changing force of the compression spring as it gradually expands. Because the lubricators operate with a single universal spring at the lowest reliable pressure *(under 2 psi)*, no grease is moved into the bearing until it is needed."

Variations in discharge flow rate are achieved by inserting different size orifices into the discharge nipple of this field-refillable lubricator.

The device shown in Figure 12-19 consists of a cylinder containing a pressure generator and a piston which, in response to the pressure generator, pushes the prepacked lubricant into the bearing.

The pressure generator is a rubber bladder containing an electrolytic solution and a sealed plastic tube containing a galvanic strip of specially treated metal. When the injector is installed, the plastic tube is broken by the activating screw exposing the galvanic strip to the electrolytic solution. This results in an electrochemical reaction within the bladder which produces a gas. As the bladder expands with the production of the gas, it pushes against the piston which in turn pushes the lubricant out of the injector and into the bearing. When all of the lubricant has been expelled into the bearing over the life of the particular unit installed, this nonrefillable unit is thrown away and a similar unit installed and activated.

The rate of lubricant ejection is a function of the gas production which in turn is a function of time and rate of the electrochemical reaction. Consequently the rate of lubricant discharge can be predesigned into this device so as to accommodate the user's discharge rate specifications.

The manufacturer states that "should the lubricant discharge flow from the unit be restricted due to an increase in lubricant viscosity, hardening of the lubricant, or mechanical restriction in the lubricant flow line, the flow would normally be reduced or stopped. It is a feature of the injector that under these conditions, the gas pressure will increase to a maximum of *136 psi* until normal flow is restored. If, for some reason, resistance to lubricant flow is reduced, the lubricant flow will temporarily increase until an equilibrium condition between the amount of gas generated and the amount of lubricant discharged is again reached.

"Discharge rates are affected by ambient temperature variations because of the increase or decrease in the speed of the electrolytic action resulting from temperature changes within the bladder. As the temperature rises, the discharge rate increases and as the temperature drops, the rate decreases. A sudden large rise in temperature also causes the lubricant to expand within the unit which will cause a temporary increase in discharge rate and conversely, a sudden drop in temperature will cause the lubricant within the unit to contract and result in a temporary decrease in discharge rate until the gas production within the bladder compensates for the reduced volume within the unit resulting from the sudden temperature drop."

For lubrication of electric motor bearings ranging from 25 to 400 horsepower, this injector manufacturer recommends a unit which, at an ambient temperature of 77°F (25°C), would discharge approximately 0.166 cu cm per day and would be in service for 24 months. Elevation of the ambient temperature to 113°F (45°C) would increase the grease discharge rate by a factor of 4, to 0.66 cu cm per day, resulting in 6 months of service life for the device.

The most intriguing working principle is found in the battery-powered grease dispenser shown in Figure 12-20. When a selector switch is closed both the electrochemical cell and the indicating 'Blip' light are activated. The choice of selector switch fixes the current and rate of gas generation in the electrochemical cell. Nitrogen gas is produced in the cell, and passes into the rubber gas chamber. The rubber chamber expands against the piston and pushes the lubricant into the bearing.

The electrochemical reaction is able to produce several times more gas than is needed to dispense all the lubricant in normal use. Normal use means against no external resistance to lubricant flow. The dispenser is adjusted to deliver lubricant at the specified rate against atmospheric pressure (14.7 psi absolute). Added back pressure will reduce the discharge rate.

Bearing Labyrinth Purge

Atmospheric water condensation in turbomachinery oil systems can be a serious detriment. Every reasonable effort must be made to prevent this source of contamination.

Experience has proven that circulating oil systems can be kept free of water by modifying the bearing labyrinth to provide an inert gas or dry air purge on each atmospheric shaft seal on each bearing. Figure 12-21 illustrates a typical bearing labyrinth purge system that has been successfully used to prevent water contamination.

Figure 12-22 illustrates the modification procedure. A 1/8-in. diameter hole is drilled through the bearing cap end wall to intersect the labyrinth. A 1/4-in. metal tubing from the purge hole in the bearing cap is connected to a 0–50 SCFH flow meter. The flow rate of the purge gas is adjusted on each seal to 10 SCFH with the flow meter. Increased flow may be required depending on the condition of the seal to keep the oil water-free.

Reservoir Purge and Vent System

The vapor space in the oil reservoir should also be purged to keep the system water free. Experience has proven that 25–50 SCFH purge on the reservoir combined with the bearing labyrinth purge is required to maintain a water free system.

Check the reservoir vent. It must be located in the top or in the end or side near the top of the reservoir. It must be free of baffles that would collect and return condensate to the reservoir. Keep the length as short as possible to provide a minimum surface on which water vapors could condense. When it is necessary to run the vent up and away from the reservoir, a water trap should be provided as close to the reservoir as possible to prevent condensed water formed in the vent stack from entering the oil system.

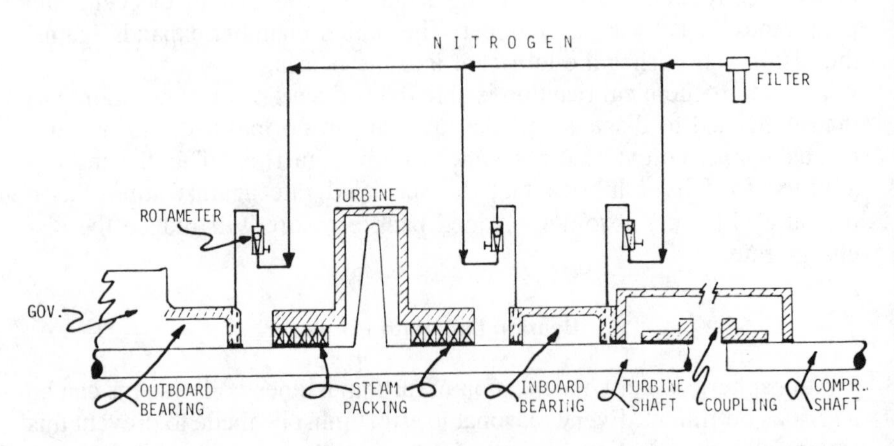

Figure 12-21. Bearing labyrinth purge system.

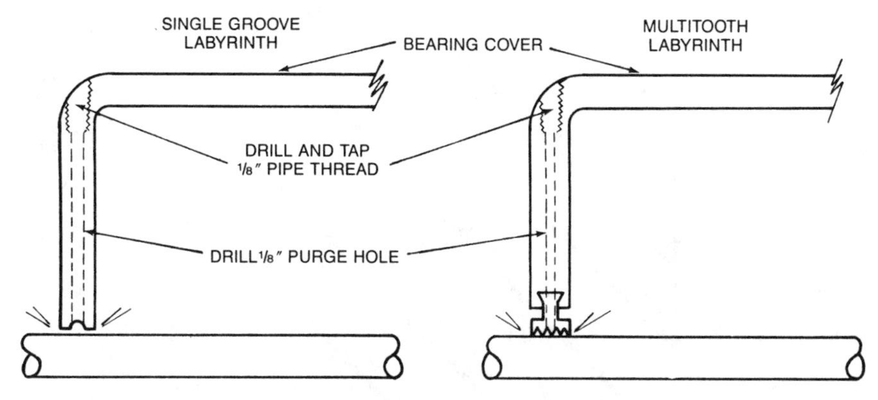

CAUTION: Do not allow drill to break into bearing housing. Purge holes must intersect center of labyrinth.

Figure 12-22. Typical bearing seal purge.

Gear Coupling Lubrication

If a user elects to use gear couplings instead of nonlubricated coupling types, he should be made aware of their vulnerability. The gear coupling is one of the most critical components in a turbomachine and requires special consideration from the standpoint of lubrication.

There are two basic methods of gear coupling lubrication: batch and continuous flow. In the batch method the coupling is either filled with grease or oil; the continuous flow type uses only oil, generally light oil from the circulating oil system.

The grease-filled coupling requires special quality grease. The importance of selecting the best quality grease cannot be overemphasized. A good coupling grease must prevent wear of the mating teeth in a sliding load environment and resist separation at high speeds. It is not uncommon (Figure 12-23) for centrifugal forces on the grease in the coupling to exceed 8,000 Gs.

Testing of many greases in high speed laboratory centrifuges proved a decided difference existed between good quality grease and inferior quality grease for coupling service. Testing also showed separation of oil and soap to be a function of G level and time. In other words, oil separation can occur at a lower centrifugal force if given enough time. The charac-

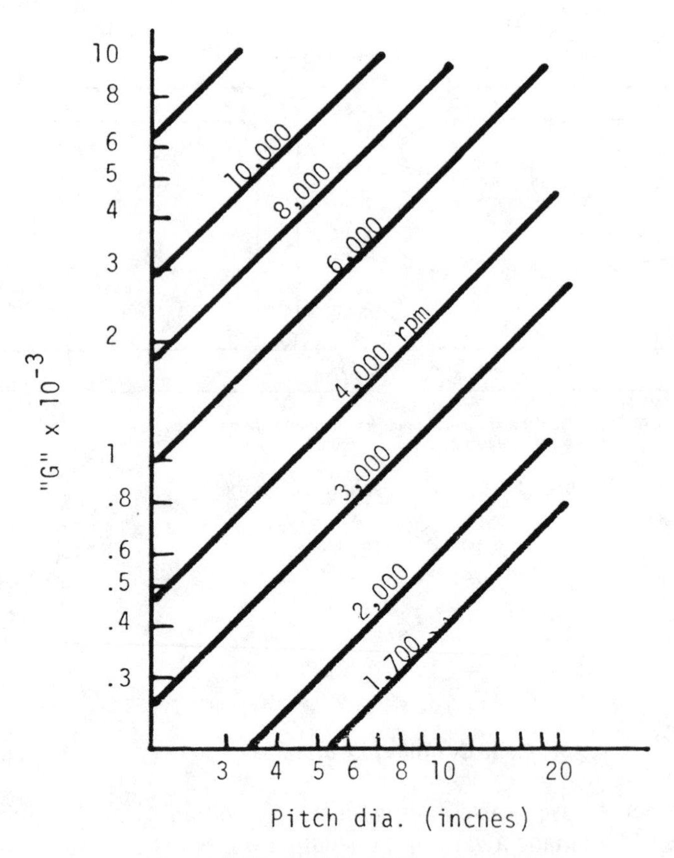

Figure 12-23. Centrifugal forces in couplings.

teristics of grease that allow the grease to resist separation are high viscosity oil (Figure 12-24), low soap content, and soap thickener and base oil as near the same density as possible.

In the late 1970's, a number of greases were tested for separation characteristics in a Sharples high speed centrifuge and for wear resistance on a Shell 4 Ball Extreme Pressure Tester. It was found (Table 12-2), that Grease B exceeded all other greases tested in separation characteristics. Zero separation was recorded at all speeds up to and including 60,000 Gs. Greases A, C, and D were rated poor in separation characteristics at all speeds tested.

Table 12-3 illustrates how these four greases performed on the Shell 4 Ball Extreme Pressure Tester in comparison with a typical Extreme Pressure gear oil. Based on these data, Greases A and B should provide excel-

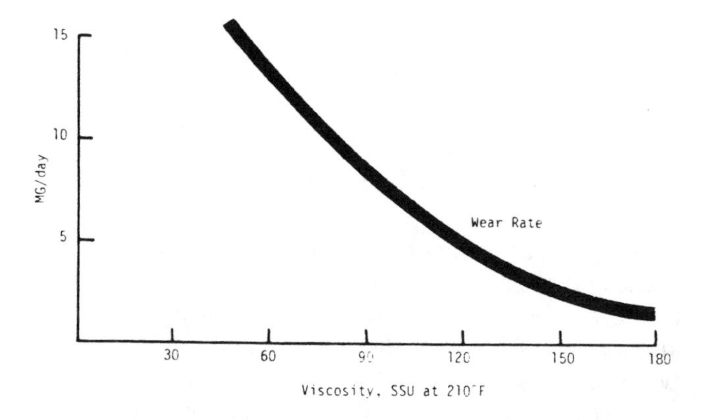

Figure 12-24. Effects of lubricant viscosity on coupling wear.

Table 12-2
Centrifugal Separation—% Oil Extracted

	"G" LEVEL		
GREASE	**60,000 2.5 hr**	**32,000 6 hr**	**9,700 6 hr**
GREASE "A"	53	35	—
GREASE "B"	0	0	0
GREASE "C"	79	75	40
GREASE "D"	51	—	—

Table 12-3
Shell 4 Ball Test—One Minute Wear Load

GREASE	KILOGRAM LOAD PASSED
GREASE "A"	80
GREASE "B"	90
GREASE "C"	50
GREASE "D"	20
TYPICAL E.P. 140 OIL	90

lent wear protection in severely loaded service. Grease B is in fact used at a very large petrochemical company for all grease lubricated couplings regardless of speed and load ranges.

Principles of Oil Mist Lubrication*

The application of lubricants has always posed problems. Manual lubrication of each point has disadvantages. Circulating systems pose problems. Centralized grease systems satisfy some of these problem areas, but there are applications where they are limited, also.

It is the purpose of this segment of our text to describe what mist lubrication is, how mist lubricants are used, and what constitutes a mist lubricant. The discussion will define mist lubrication, trace some of its history, and describe the equipment required. Then some of the properties of the lubricants will be outlined and how they are attained; and finally, the advantages of mist lubrication will be summarized.

How Oil Mist is Applied

Manual lubrication is time consuming and lacks dependability. Circulating systems entail large volumes of oil and pose the problem of disposal of the spent oil. Centralized grease dispensing systems have limitations, also. Speeds are often high enough that the grease is flung out of the bearing, or the heat created by the grease becomes excessive. All of these problems are overcome with mist lubrication. For a schematic representation, see Figure 12-25.

Mist lubrication may be defined as a centralized lubrication system in which the energy of compressed gas is used to disperse the oil which is then conveyed by the gas in a low pressure distribution system to multiple points of application. Several key words in this definition, by one manufacturer, require further emphasis. Centralized indicates one source of lubricant. The compressed gas is usually air. The dispersion is of micron-sized particles of oil, usually averaging around 1.5 microns. The low pressure is in the order of 1 to 2 psig, so it poses no confinement problems. Multiple points can be lubricated by appropriately sizing the application fittings, as will be seen later.

The idea of mist lubrication began in the late 1930's. A bearing manufacturer couldn't lubricate a high speed spindle. The speed was high

* Source: Products Training Course, compiled by P. E. Knoeller, Exxon Company, USA. Reprinted and adapted by permission of Exxon Company, USA.

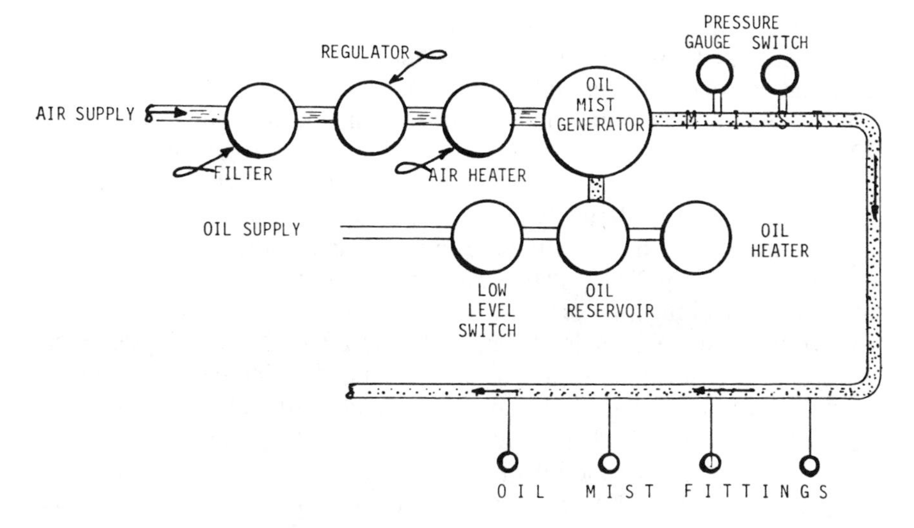

Figure 12-25. Oil mist schematic.

enough that grease was flung out. A liquid oil circulating system created too much frictional heat. The continuous film of lubricant provided by mist solved the problem. However, mechanical problems delayed the progress of mist systems. Air consumption was high and most of the oil coalesced in the piping. As little as 3 percent of the oil misted reached the point of application.

But these problems were overcome in the 1950's and the use of mist lubrication has spread to many industries. Recent refinements and improvements have increased present and potential applications on a wide variety of machine elements. Automotive use is limited to off-highway vehicles due to the current environmental trends. Initially, trailer truck fleets were a target, but with mist there is a problem of disposition of the oil, as it is fed continuously.

Now let us examine the basic components of a mist system and how they affect its operation. An air supply of 60–100 psig is required. The incoming air should be filtered as the internal part of the generator contains minute passageways. Also, the oil itself will tend to remove impurities and they will build up in the reservoir. The next component is an air pressure regulator. This controls the inlet pressure, and hence the air flow rate, and ultimately the mist output. The mist generator is where the liquid oil is transformed into a mist. Then the airborne mist is conveyed into the manifold where it is distributed to the reclassifiers. These reclassifiers transform the oil into the desired form for the application. Electrical power is required for auxiliaries as will be discussed later. The point is that mist systems are simple and contain no moving parts. Next, we examine each component and its function in detail.

The generator operates on a venturi principle as shown in Figure 12-26. The low pressure at the maximum velocity draws the oil up to the air flow where it is broken up into small droplets. "Atomized" is the term used, but it is technically inaccurate for this operation. The required air pressure varies, but usually 60 psig is adequate. A solenoid valve and alarm system to indicate status may be incorporated. A pressure regulator to control the inlet pressure and a pressure indicator will normally follow. For systems intended to handle oils of over 700 SUS @ 100°F, an oil heater and/or an air heater may be included to increase the output. The generator is sized to fit the application and is usually in the neighborhood of 4 to 20 standard cubic feet per minute. The oil output is a function of the venturi size and inlet air pressure. However, the oil output can also be affected by the characteristics of the oil as we will see later. The generator creates particles of 1.5 microns average size with a size distribution of approximately 0.5 to 2.5 microns. Those smaller than approximately one micron are difficult to coalesce again and those larger than two microns are removed by the outlet baffle and remain in the reservoir. Thus, particle size of the mist generated is an important factor.

Now let's turn our attention to the outlet side of the generator. The baffle, as mentioned previously, removes the larger particles that would subsequently coalesce in the distribution piping; hence only particles

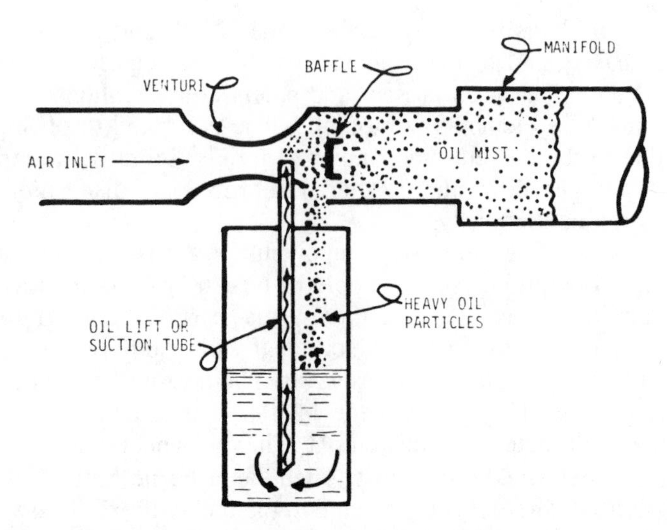

Figure 12-26. Principle of oil mist generation.

smaller than approximately two microns enter the manifold. These droplets are pneumatically conveyed through the manifold to the application fittings. The velocity of the oil in the manifold is a factor that must be considered. Design velocity should be less than 20 ft/sec in order to minimize the amount of oil coalesced in the distribution system. However, even under ideal circumstances some oil coalesces in the manifold, but this is usually only 10–20 percent of the generator output. For this reason, the manifold is usually sloped back toward the generator to avoid formation of low spots where liquid oil can accumulate and prevent the passage of the mist.

Manifold pressures vary from 5–40 inches of water, or approximately 1/4 to 1 1/2 psig. It is through the manifold pressure that the air flow and hence oil flow are controlled. Some mist units are designed around a generator output of .4 cubic inches of oil per cubic foot per minute of air throughput. This translates to .00023 parts of oil per part of air, or 1 in 240,000, to give some idea of the mist density. The manifold pressure must be correlated with the applications and the various fittings involved. The mist system oil flow will depend on the manifold pressure, the viscosity of the oil, and the temperature of operation.

Another interesting concept via which a mist is generated is the Vortex Mist Generator. This unit, originally developed by Farval, relies on a swirling action of the air to create the vacuum. As the air approaches the small end of the funnel, it reaches near supersonic speed, and as well as drawing the oil in, it tears it into small particles. Other principles and accessories for this unit are the same as the other mist units.

The four most common types of application fittings used with mist lubrication are pure mist, spray, condensing, and purge mist. These four types are shown in Figure 12-27. Each particular type has its own application. The fitting is sized to control the air flow rate and, consequently, the oil flow rate. However, the ultimate throughput is dependent on the manifold pressure.

The first type of fitting we will address is the spray or wetting spray fitting. This type is usually used on slower moving rolling motion applications. Examples of these are gears, chains, and antifriction bearings under certain conditions.

The second type of fitting is the pure mist fitting. This type is generally for the same applications as spray fittings, particularly where odd geometric designs are encountered and antifriction bearings are in enclosed housings. This type will also be found in high speed applications. The fitting merely meters the amount of oil to the machine element and relies on the turbulence of the air to coalesce the oil within the application. The relative velocity of the machine element versus that of the mist will cause the wetting out or condensation.

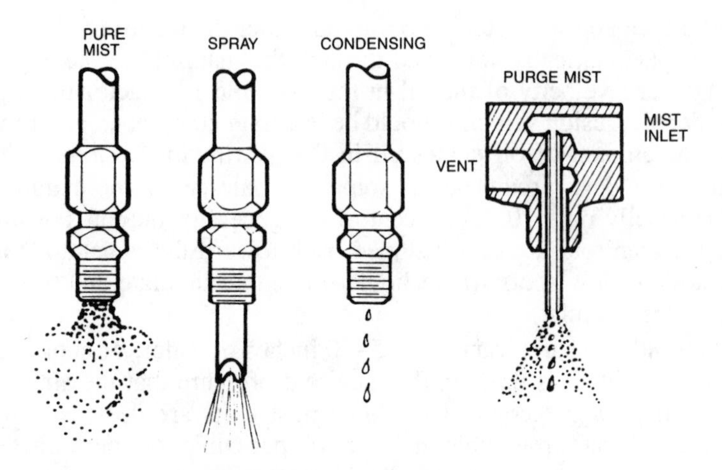

Figure 12-27. Application fittings.

The third common type of mist fitting is the condensing fitting. This type is generally used for sliding motion elements, such as plain bearings, slides and ways. There is no turbulence at the point of lubrication so the coalescing must be done by the fitting. This is done by creating the turbulence within it.

The fourth, or purge mist fitting, introduces oil mist into a bearing which is lubricated by a conventional "wet" sump.

Mist fittings are usually located as near to the objects they are lubricating as possible. This will often provide a positive pressure on the application. This in turn gives some protection from airborne dirt and corrosive air. It also provides protection from casual moisture, but not from high pressure water jets, etc. The proximity of the fitting also offers a slight cooling effect, bearing in mind that there are 240,000 parts of air per part of oil passing through.

There are several manufacturers of mist equipment such as Alemite, Lubrication Systems Company, Norgren and Trabon. Most are basically identical in principle with the venturi above the oil supply as we have seen. The vortex principle is relatively new and should be added to this list.

A host of auxiliary equipment is available to aid in the operation of mist systems. The most common are oil and/or air heaters that heat these fluids. The purpose of this is to reduce the oil viscosity to increase the ease with which the mist is formed. Level indicators may be installed on the reservoir and the reservoir supply, and these may be equipped with alarms that sound if the level is abnormal. Also, alarms may be installed on the air inlet and manifold to indicate anomalies in these places. Some vendors are offering a "mist monitor" on their system that monitors the density of the mist entering the manifold and warns if it becomes abnormally low.

Oil Mist Properties

ASTM has devised a procedure for evaluating misting properties of lubricants which is designated as D-3705. This procedure utilizes an Alemite mist unit and all the components of an industrial installation. One major point to be made here is the simplicity of the system. The generator is a 5-quart reservoir with a nominal 4 CFM mist head. Commercial units are available from less than 1 to approximately 40 CFM. The test unit has air and oil heaters which for comparative work are set at 40°C. It incorporates an air flow meter on the air inlet as well as a filter, pressure controller, and pressure gauge. The generator feeds one oil spray nozzle through the angled pipe simulating the distribution manifold. The oil spray nozzle has 11 holes in it, each .067 inch diameter. The unit, after being cleaned, is filled with a known amount of oil. After running for 19 hours, the oil reclassified and the oil coalesced in the manifold are weighed and that escaping as stray is calculated by difference. This method is being used, although with different equipment, to formulate the various oil mist lubricant grades. The general intent is to maximize the amount of oil delivered to the point of application per unit of air throughput.

With this brief introduction to the equipment involved with mist lubrication, let us now turn to the properties of the lubricant itself.

Oils designed for use in systems such as this usually have good mist generation characteristics and desirable reclassification characteristics. Both are economic advantages for the consumer with the latter reducing housekeeping and ecological problems. These two properties are probably those unique to mist lubricants, but other industrial lubricant properties such as extreme pressure and antiwear characteristics cannot be ignored. We now want to outline the properties of the lubricants and see how they are attained and how they affect performance.

Mistability, to use a coined word, will be defined as the ease with which the lubricant can be transformed into a mist. However, the mist must be of such a nature that it can be controlled. If it is too fine, the reclassification fittings will not coalesce enough oil out of the air and excessive amounts of oil will escape as stray mist. If the mist droplets are too large, they will not escape from the generator and the delivery rate will suffer radically.

The conventional mist lubricator utilizes compressed air directed through a venturi or vortex generator to create a vacuum. This vacuum draws oil up from the reservoir into the air stream where it is broken into small droplets.

The size distribution of the droplets controls the proportion of oil picked up to that carried into the manifold.

If the particles generated are too large, they are knocked out of the air in the mist generator and fall back into the reservoir. This merely reduces the ratio of oil to air exiting and increases the operating cost. Conversely, if all the droplets are of a very small size, they will be readily carried into the manifold, but are difficult to coalesce at the point of application. This results in higher stray mist losses. Oil *B* would have a lower generation rate than Oil *A* in Figure 12-28, but Oil *A* would be more difficult to recover at the reclassifiers.

Under ideal circumstances, the oil droplets would all be of the same size but to date this has not been achieved, probably due to the random chemical composition of mineral oils.

An examination of the factors that affect the mistability indicates that there are three major variables:

1. Base Oil
2. Additives
3. Operating Conditions

Base oil type has little effect on mistability as will be seen later. The viscosity of the base oil, however, can have a more noted effect under

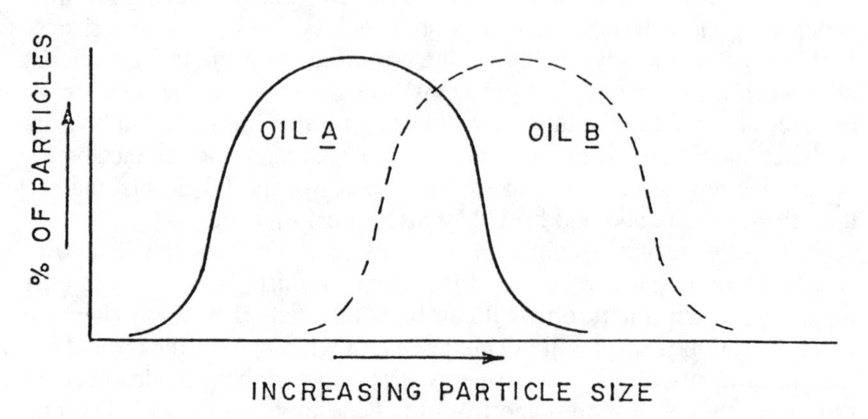

Figure 12-28. Effect of increasing particle size.

given operating conditions. The higher the oil viscosity, the lower the mistability. This is shown in Figure 12-29.

It must be pointed out that this is a general pattern under specific operating conditions and can be upset by radically varying the operating conditions of the mist system.

The second major factor that affects mistability is the additive system of the oil. Certain polymeric materials when added to the mineral oil will drastically affect the mistability. This is displayed by the data accumulated in Figure 12-30.

This indicates that the motor oil contains an additive package that changes the particle size distribution such that more oil is recirculated within the generator and less is permitted to enter the distribution manifold. This detrimental effect is probably due to the viscosity index improver which is a high molecular weight polymer. However, the more optimistic side of this situation is that there are polymers available that will not reduce mistability greatly, but do improve the reclassification characteristics of a mist lubricant as will be seen later. These additives affect the mistability as shown in Figure 12-31.

This displays that the mistability is reduced but the greater reclassification at the point of application represents a substantial reduction in stray mist, so that the loss to the atmosphere at the application becomes very small. The net result is that almost as much liquid oil reaches the point of application.

The third variable affecting mistability is the operating conditions of the mist unit. The air volume passing through the generator dictates the amount of oil picked up. Increasing the air flow will generally increase

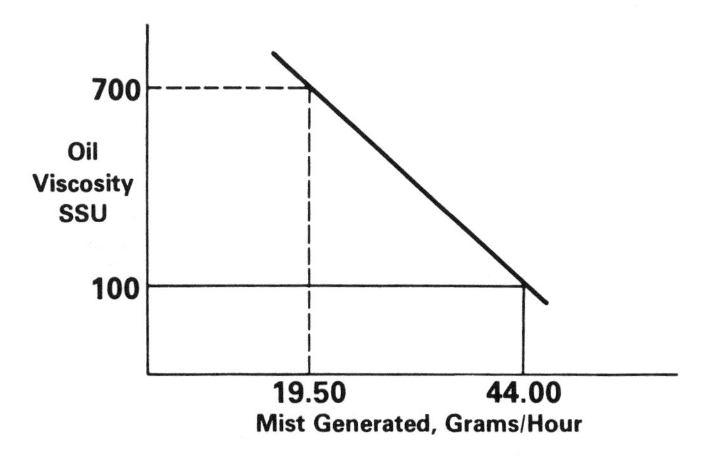

Figure 12-29. The higher the oil viscosity, the lower the mistability.

Effect Of Additive

Product	460 VIS SE Motor Oil	465 VIS Mist Oil
Temperature, °F	75	75
Delivery Rate, Grams/Hour	2.95	19.96

Figure 12-30. Certain polymeric materials added to the mineral oil will dramatically affect the mistability.

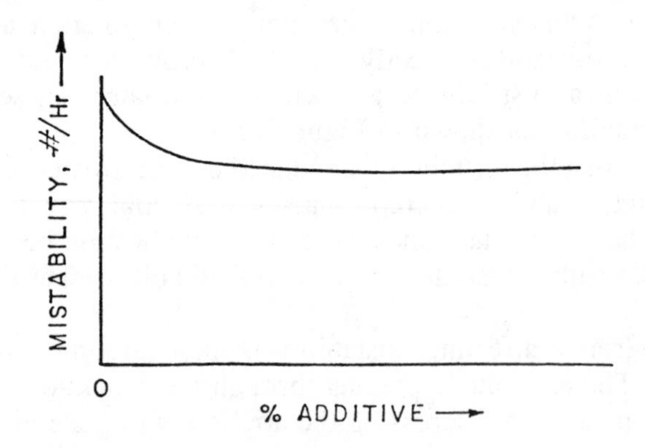

Figure 12-31. Some polymers improve the reclassification characteristics of a mist lubricant.

the amount of mist generated. Another change that can be instituted is that the operating air temperature can be increased which will increase the amount of mist generated. This is shown in Table 12-4.

The data indicate, as would be expected, that increased temperatures can significantly increase mistability. This phenomenon is due simply to the fact that elevated temperatures decrease oil viscosity and it has been shown that lower viscosity oils mist better than those with higher viscosities.

The most reliable method of determining the mistability of a lubricant is to quantitatively measure the amount of oil removed from the generator in a known period of time with a known air flow. Then it can be determined if the oil is being misted at a rate near that of the mist system designer's intent and the requirement of the application.

Table 12-4
Effect of Temperature—Delivery Rate, Grams/Hour

Air temperature, °F	700 VIS Oil	2100 VIS Oil
75	19.50	—
150	49.90	39.92
175	—	46.72
200	—	48.54

After the mist has been created and delivered to its point of application, in order for it to be useful it must be coalesced to provide a continuous oil film to the machine element being lubricated. While a rapidly moving application can coalesce the oil itself, slow moving parts must rely solely on the mist system reclassifiers to do this.

Two other considerations from the lubricant standpoint are that reclassification in the manifold is undesirable and also that stray mist is undesirable. Again, the objective is to get the most oil to the application since what is reclassified in the manifold usually drains back to the reservoir and recycles. That which escapes as stray mist is costly and could possibly create environmental problems.

The reclassification characteristics of a mist oil are also largely dependent upon the particle size created in the generator. Referring back to our bell-shaped curve of particle size distribution, it can be deduced that the smaller particles are more difficult to coalesce and most likely to escape as stray mist while the larger ones will be more easily coalesced. This is also an important consideration involved in the selection of a mist lubricant additive system.

It is the objective of the mist lubricant to provide a product that gives the maximum amount of oil per unit volume of air to the application. However, with the Occupational Safety and Health Act of 1970 limit of 5 mg. of oil mist per cubic meter of air in inhabited areas, it is now necessary to control stray mist as well. Previous to this innovation stray mist was only undesirable due to its unsightliness and housekeeping detriments as well as its wasted cost.

The reclassification characteristics of a mist lubricant are dependent upon certain oil additives and the mist system's operating conditions. Stray mist suppressants do, in fact, reduce the amount of stray mist dissipated but, as mentioned previously, affect the delivery rate and manifold reclassification. However, they do help to keep the stray mist to a tolerable level.

Table 12-5 represents a laboratory situation, but it does display the advantage of a mist lubricant over a mineral oil used in mist applications. While the amount reclassified is reduced by only 15 percent, the stray mist is reduced by 77 percent.

The air flow rate or manifold pressure of the mist system can also affect the quantity of stray mist. Usually an increase of these two variables will tend to reduce the stray mist strictly because as the air flow is increased through the reclassifier, its efficiency improves. However, at increased air flows, the quantity of oil coalesced in the manifold usually also increases. This is because the maximum design air velocity has been approached or exceeded. The design air velocity in the manifold should be in the neighborhood of 15 to 20 ft/sec, and if this is exceeded, it will cause more impingement on the sides and at turns which increases the amount coalesced enroute to the reclassifiers.

The most accurate measure of lube oil reclassification characteristics is obtained by actually misting it and preparing a material balance on the operation. Small test rigs are available for this purpose.

The mist lubricant is, for the most part, mineral base oil blended with additives required for the application. The choice of base oil type is dependent upon the general conditions under which the lubricant is to be used. Where the application is outside, the oil is light in viscosity, and there is no external heat, a naphthenic base oil is probably a better choice as it inherently has a lower pour point. Where heavier equipment is involved and the oil is heated due to its higher viscosity, a paraffinic base oil would be better due to its greater inherent oxidation stability. As mentioned previously, both behave nearly the same under a given set of operating conditions. However, the effect of the viscosity index will be seen to some extent as shown in Figure 12-32.

This shows that the viscosity of the naphthenic oil, although the same as the paraffinic at 100°F, will be slightly lower at higher temperatures.

Table 12-5
Distribution of Oil Generated

	Generation Rate Grams/Hr	Distribution, %		
		Reclassified	Coalesced In The Manifold	Stray Mist
Uninhibited Oil	40.8	64.5 (26.3)	15.2 (6.2)	20.3 (8.3)
Mist Oil	30.6	72.8 (22.3)	20.9 (6.4)	6.3 (1.9)

(Figures in parentheses are grams)

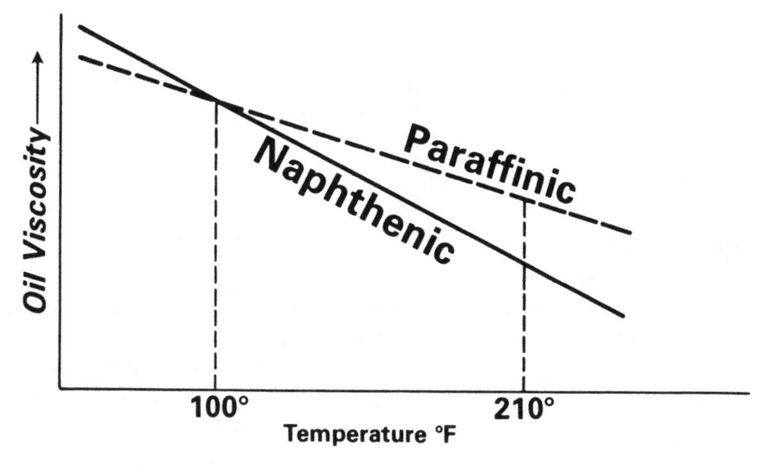

Figure 12-32. Effect of the viscosity index.

Consequently, it will mist slightly more readily, but this phenomenon is usually overcome by other factors such as the mist unit operating conditions.

The viscosity will usually be determined by the application involved. Often a host of applications are being lubricated by a given system, so the viscosity is a compromise.

The lubricant supplier usually has several viscosity grades of each type of base oil available and can blend similar types to the required viscosity before treatment with additives.

The option to blend the two types in order to attain intermediate properties also remains, but this has little advantage except economics in certain instances. This can also result in fictitious viscosity indices.

EP and antiwear additives are not always needed in mist supplied applications, but oil mist is often applied to such a variety of machine elements that even from one generator there are probably several points of lubrication that benefit by these additives. Consequently, most mist lubricants contain extreme pressure additives. These systems are often currently composed of sulfur or sulfur-phosphorus components. This is partially due to the recent ecological resistance to lead and certain other heavy metals. Chlorine is another source of EP properties, but is a bit more difficult to inhibit against corrosion of ferrous elements. The EP and antiwear properties of mist lubricants are measured by conventional laboratory tests.

Mist lubricants are most often utilized to lubricate ferrous machine elements so corrosion must be considered. In operating plants, moisture and corrosive air are often in contact with the equipment. Some applications are in contact with water which invites rust, especially during shutdown periods. Lubricants can be inhibited against ferrous corrosion with such inhibitors as sulfonates or naphthenates with little or no effect on other properties.

Consideration must also be given to corrosion of copper-containing alloys as these metals are occasionally used in the construction of the mist generators and in such applications as brass bearing cages that may be lubricated with mist. Since copper corrosion is usually caused by the choice of EP additives, careful selection of these additives is necessary. Another method of minimizing copper corrosion is through the use of metal deactivators. These materials render vulnerable surfaces less likely to react with the active compounds and hence prevent any accumulation on them. The result is that the lubricant will display very good corrosion test results.

Corrosion inhibition is measured through a host of different laboratory tests, with the D665 corrosion test being the most common.

Mist systems are generally a once-through operation where the oil is put in the mist generator, misted to the point of application, and not returned to the system. Its life in the generator is short, as it is in the final use, but both places may be relatively warm so oxidation stability cannot be overlooked. The elevated temperatures encountered in the generators, coupled with the constant aeration that is created, tend to invite oxidation of the oil or the additives. However, the highest recommended operating temperature for mist systems is in the neighborhood of 175°F which, when coupled with the short life of the oil in the generator, is not very detrimental to the mist lubricant of this vintage. Oxidation is overcome by careful selection of the base stock and the use of an oxidation-resistant additive system and, if necessary, the use of an oxidation inhibitor.

Oxidation resistance is measured again through a variety of laboratory tests designed for this purpose.

Mist lubricants, while not utilized where the final use requires an antifoaming oil, are aerated in the reservoir to the point that an excessively foaming oil can create problems. Again, for this reason, the foam tendencies of mist lubricants must be considered. Most base stocks seldom have the tendency to foam, but foaming can be caused by miscellaneous additives. Therefore, it is necessary to ensure that the EP, antiwear, corrosion inhibitors, or other additives incorporated in the mist lubricant do not impart foaming tendencies to the finished oil. Foaming tendencies of mist lubricants can be readily measured by the ASTM test method designed for this purpose.

The foregoing are the more obvious properties to be incorporated into mist lubricants, but there are other more subtle properties that may be built in. One such property that is not easily detected is the tendency of an additive system to deplete while in service. Some EP additives tend to deplete as their chemical constituents react or adhere to the surfaces that they are protecting. This may be particularly prevalent in mist generators where the oil is warm, constantly recirculated, and highly aerated. This property has historically been overcome in one of two ways. One is by the choice of an additive system that will not behave in this way at an appreciable rate due to its chemical composition. The other is by the addition of a depletion inhibitor to the compounded lubricant. Additive depletion has diminished as a problem with the current additive systems that are available.

Another such property relative to stability is that it is intolerable to have a mist lubricant that does not deliver exactly the same oil through the reclassifiers as that which is put in the generator. This situation indicates that some component of the lubricant is not misting properly and the result will be the buildup of this component in the mist unit reservoir.

These two properties are occasionally related to the base oil by the phenomenon that paraffinic oils tend to have less desirable solubility characteristics than the lower viscosity index oils. However, as previously stated, the paraffinic oils offer better thermal stability and thus provide an advantage in use at elevated temperatures.

The stability of mist lubricants can only be measured by chemical analysis of the fresh oil and that of an aged portion. The ideal circumstances for aging are actual field service, but accelerated laboratory tests are often utilized.

We have outlined some of the properties of the industrial mist lubricants and how they are attained. It should be kept in mind that the user wishes to have the maximum oil delivery to the application per unit volume of air with a minimum of loss or coalescing in the lines. These characteristics can be incorporated by careful selection of required additives in appropriate base oils. ASTM has recently standardized a misting property test unit and a procedure for determining the misting characteristics of mist lubricants. This is another step in the sophistication of the lubricant testing area, but the real proof of a mist oil, as any lubricant, is its field performance.

Finally, we must mention the highly favorable results reported by industrial users of dibasic ester synthetic lubricants in oil mist systems serving pump and electric motor bearings. The reader would be well advised to consider these superior lubricants for oil mist applications in a typical petrochemical plant.

Principles of Grease Lubrication*

Handling and Dispensing Properties

The first thing that has to be considered in any application is, can a grease be transferred to the part being lubricated? This, of course, depends on how the grease is to be transferred—by the simplest method of hand packing or swabbing it onto bearings and gears, by the somewhat more sophisticated pressure-gun, or by the more modern means, a centralized system. Regardless of the method of application, the suitability of the grease for that method is determined by two properties:

1. Consistency of the grease in its container.
2. Viscosity when it is flowing in piping, pumps, etc.

Consistency

Consistency is the degree of stiffness or hardness of a grease and is defined as "the resistance of a plastic substance, such as grease, to deformation under the application of force."

Obviously, therefore, the grease in the container—*at the temperature of application*—must not be so hard that it cannot be removed from the container, and yet be stiff enough at the operating temperature to be retained where it is needed.

Consistency is measured by the ASTM Cone Penetrometer. This is simply a scientific way of poking your finger into the grease to see how soft or hard it may be. The method consists of allowing a weighted metal cone to sink into the grease for five seconds and measuring, in tenths of a millimeter, the depth to which the point penetrated below the surface. This depth is known as the "penetration" of the grease. Thus the higher the penetration, the softer the grease.

In order to separate the wide range of available grease consistencies into smaller, more practicable classifications, the National Lubricating Grease Institute (NLGI) has established nine penetration ranges or NLGI grades as shown in Figure 12-33. Grade 000 is the softest with the highest penetration numbers, and Grade 6 is the hardest with the lowest range of penetration.

---

* Source: Products Training Course, compiled by P. E. Knoeller, Exxon Company, USA. Reprinted and adapted by permission of Exxon Company, USA.

NLGI
(NATIONAL LUBRICATING GREASE INSTITUTE)
GREASE CONSISTENCY GRADES

GRADE FLUID AND SEMI-FLUID GRADES	WORKED PENETRATION AT 77° F
000	445-475
00	400-430
0	355-385
1	310-340
2	265-295
3	220-250
4	175-205
5	130-160
6	85-115
BLOCK GREASES	

1-47-1131

Figure 12-33. Grease consistency grades.

Viscosity

Viscosity is defined as a measure of the resistance of a fluid to flow. Hence, for greases, it is the "yardstick" that is applied when flow in pipes and bearings is under consideration.

Thickened greases offer complications as compared to simple fluids such as oils, water, or even asphalt, which are known as "true" or Newtonian fluids. With these, the viscosity is constant regardless of how fast they are being moved. Viscosity can be expressed mathematically as the ratio of shear stress (the force causing the fluid to flow) to the shear rate. For Newtonian fluids this ratio remains constant, since as the rate of shear increases, a proportionately greater force is required to move the fluid.

With greases, this ratio does not remain constant, that is, the force required to move or shear a grease does not increase proportionately with the shear rate. Consequently, we cannot correctly use the term "viscosity" for greases but must refer to their "apparent viscosity." This is what the viscosity appears to be at any given shear rate. Since the apparent viscosity will be different for each different shear rate, the shear rate must always be stipulated whenever an apparent viscosity is quoted for a grease. Also, as in the case of Newtonian or true fluids, the temperature at which the viscosity or apparent viscosity determination is conducted must be stated.

Shear rate is defined as the rate of slip within a substance engaging in flow. If the substance is between parallel surfaces moving relative to each other, shear rate is equal to the relative velocity of the two surfaces divided by the distance between them and is expressed as the reciprocal of time, usually "reciprocal seconds."

It can be shown that the apparent viscosity of a grease decreases from a very high value at low shear rates to a much lower value at high shear rates and will eventually approach the true viscosity of the base oil in the grease.

Apparent viscosity is of particular concern to manufacturers of dispensing equipment, since they must design their equipment to handle many different types of greases. It is a matter of concern to the user primarily when dispensing equipment is to be operated at low ambient temperatures.

Working Properties

The properties of a grease that may show it to be suitable for the method of application are not sufficient to indicate if it will work in the unit being lubricated. The latter can be referred to as the "Working Properties."

Consistency Stability

Effect of mechanical working on consistency. The most fundamental working property of a grease is its ability to return to the consistency needed for proper performance after prolonged shearing or mechanical working. While shearing is taking place, the grease in the immediate area of the moving surfaces assumes semi-fluid to fluid characteristics depending on the rate of shear, but as soon as the process of working stops (shear rate becomes zero), the grease should, ideally, resume its desired consistency. For most applications, this desired consistency is approximately the original consistency of the grease in the container, but in some instances it is desirable for the grease to become materially stiffer after being subjected to the action of the working part. For example, greases used in high-speed ball bearings should harden to a controlled degree so that they will channel and stay out of the way of the rapidly moving parts, but at the same time allow small quantities to migrate to the contacting surfaces to provide lubrication.

There are a number of ways of measuring mechanical or shear stability, the most common being by prolonged working (at least 10,000 strokes) in a Grease Worker (ASTM or Federal Test Method) and by the ASTM Roll Stability Test. In both methods, the change in consistency of the grease is measured after being subjected to several hours of working in the test apparatus.

Both of these methods can be considered only as screening tests to detect radical changes that may take place at low shear rates. Most machine elements subject greases to much higher shear rates and, of course, for much longer periods of time. The ultimate criterion is to determine the performance of the grease by trial in the machine itself.

Effect of temperature on consistency. It is necessary that a grease not become excessively soft or hard with changes in temperature. Although most soap thickeners become softer with increases in temperature to the point where they are essentially a fluid, some become progressively harder upon exposure to high temperatures. Nonsoap thickeners as a whole show very little change in consistency with a temperature increase.

It is quite difficult to obtain a valid measurement of the effect of heat over prolonged periods since time of exposure to higher temperature levels has a profound effect on consistency characteristics. The ASTM Dropping Point Test indicates the temperature at which a grease becomes sufficiently soft to flow under the force of gravity but does not relate to high temperature service capability. However, since different thickeners have fairly characteristic dropping points, this test can be used as a means of identification.

The ASTM Trident Probe Method provides a curve of apparent viscosity versus temperature over a wide temperature range (R.T. to +315°C [+600°F]). This is some indication of its consistency, or flow properties, but it must be remembered that these results are for low shear rates, far below what is normally experienced in operating bearings.

Change in consistency or apparent viscosity as temperatures go below room temperature is primarily a function of the viscosity of the oil and is much less time-dependent. It can be determined by measuring the penetration and apparent viscosity at the desired temperature level.

Effect of water on consistency. A grease that comes into contact with relatively large quantities of water—as opposed to the small amount that may be introduced by condensation from a high humidity environment—must obviously contain a thickener that is not water soluble. Otherwise, it will emulsify, fluidize and not be capable of staying where it is needed.

One of the few generalizations that can be made is that calcium, lithium, and aluminum soaps are highly water resistant, and hence greases made from them should also be water resistant. Conversely, sodium soap greases have a reputation for poor water resistance, although they frequently display good resistance to cold water.

Water resistance can be measured by several methods. A common and very practical method is to work increasing increments of water into the grease until it fluidizes or until it rejects any further addition of water without becoming fluidized. A modification of this is to work a given quantity of water, such as 10 percent of the grease weight, into the grease and measure the change in consistency (see Specification MIL-G-10924C). Another method is by the ASTM Water Washout Test—this simulates the action of water on grease in a shielded or incompletely sealed ball bearing. A fourth method is to measure the grease retention on a steel plate after exposure to a specified water spray. The test method chosen should be reasonably analogous to the expected exposure to water. The Water Washout Test would not be suitable for an open bearing operating under water. Conversely, basing the choice of grease on its ability to withstand large volumes of water when only exposure to condensation is expected, may restrict the choice to greases that are unsuitable in other respects.

Ability to resist the effect of water on consistency change is not synonymous with rust protection. Many highly water-resistant greases provide very poor rust protection. Conversely, greases that combine readily with water generally offer good rust protection up to the point where complete washout occurs, this being particularly true of sodium soap greases.

Rust protective properties of greases are now measured almost universally by the ASTM Corrosion Prevention Test. In this test a tapered roller bearing coated with the test grease is dipped into distilled water, removed, and stored for 48 hours at 52°C (125°F) and 100 percent relative humidity. The presence of rust spots just visible to the naked eye determines if the grease fails the test.

Oxidation Stability

Greases will deteriorate from oxidation of both the petroleum oil and the thickener, particularly if the latter is a soap. Although the effects of oxidation can be quite varied, it is generally manifested as hardening throughout the grease, formation of varnish-like films, and eventual carbonizing. This is caused primarily by polymerization of the oil or soap, or both.

Rate of oxidation is profoundly affected by temperature, being approximately doubled for each 10°C (18°F) increase in temperature. Thus a grease that could provide 18 months of useful service at 100°C (212°F) before relubrication is necessary should, theoretically, require replacement every two months if the temperature were increased to 130°C (266°F).

Depending on their intended use, greases can be made to range in oxidation stability from those suitable for periods of service of only a few months at moderate temperatures to those that will operate for several thousands of hours of continuous operation at 149°C (300°F). In the latter case, this is accomplished by the selection of highly stable components and the use of potent inhibitors which prevent the oxidation reaction from proceeding.

The ability of a grease to withstand oxidation under dynamic conditions such as encountered in service can, at present, only be determined by running the grease in test rigs which simulate actual service conditions. Manufacturers of sealed-for-life anti-friction bearings, and of other sealed machinery, wherein the grease is intended to last the life of the machine, must rely heavily on functional testing such as the ASTM Lubrication Life Test (D 3336).

A popular bench test for oxidation stability is the ASTM Bomb Oxidation Test. Here, the grease is placed in a steel container or "bomb" containing pure oxygen at 758 kPa (110 psi) pressure and 99°C (210°F). As the grease reacts with the oxygen, the pressure of the oxygen decreases. This pressure drop is recorded versus time.

The bomb oxidation test should not be used as an indication of oxidation stability under dynamic service conditions. It is a static test intended only to show whether a grease will resist oxidation when it is placed in machine parts, such as ball bearings, and stored for long periods of time. Use of the test in any other way must be done with caution. A grease that may be sufficiently inhibited to provide a long induction period in this test could, under dynamic conditions in bearings, have a very limited life due to weaknesses in other respects.

More and more of the complaints that are received by the grease industry concerning the performance of so-called "high temperature greases" are directly the result of a lack of understanding of the effect of temperature on grease life. Since many of the newer greases are of the "nonmelting" or nonfluidizing type, they will stay in the bearings at high temperatures and are consequently allowed to remain far beyond their useful life. When this happens, they oxidize to form hard, carbonized residues and are then unjustly blamed for not being sufficiently oxidation stable.

Lubricating Ability

So far we have made no mention of the primary function of a grease—its lubricating capabilities. The ability of a grease to reduce friction and wear can be derived from three sources:

1. The oil in the grease
2. The thickener
3. Additives

The lubricating capability of the oil, if it is a petroleum product, is primarily a function of its viscosity and viscosity index. Hence, in the broadest terms, the same criteria are used in selecting a grease containing an oil of a certain viscosity as would be used if the working parts were lubricated with oil alone—low viscosity for high speeds, low loads, and low temperatures, and vice versa. A complete range of oil viscosities is available in modern greases—from 10 cSt (60 SUS at 100°F) for extreme low temperature service to 500 cSt (2300 SUS at 100°F) for steel mill requirements. However, the great majority of greases use oils in the 60 to 190 cSt (300 to 900 SUS at 100°F) range.

The lubricating abilities of thickeners vary widely. Most soaps are good lubricants, some being so good as to provide load-carrying capabilities far beyond those of the oil alone. Inorganic nonsoap thickeners generally do not contribute to the lubricating properties of a grease, but there is no evidence that they detract in any way from the lubricating ability of the oil.

The most common method of enhancing the extreme pressure and anti-wear properties of greases is, of course, to use additives. These will be discussed later.

Because of the excellent lubricating qualities of petroleum oils, greases made from them without the addition of load carrying additives are entirely adequate for the majority of applications. However, where shock loads or overloading of gears, plain bearings, and certain types of roller bearings (as opposed to ball bearings) are encountered, greases may be needed which have load-carrying properties beyond those of the oil alone. Ball bearings do not require such properties.

There are a number of methods, utilizing a variety of test instruments, for determining the lubricating capabilities of greases under severe load conditions. These instruments include the 4-Ball Wear Test, 4-Ball EP Test, Falex Wear Test, Timken Test machines, and others. In addition, various test rigs have been designed to simulate the action of gears and bearings under service conditions. A discussion of the details of these tests is beyond the scope of our text but can be obtained from NGLI texts dealing with grease tests. Interpretation of the relationship of results obtained in these tests to performance in actual service is largely a matter of experience with the particular application involved.

Extreme pressure and anti-wear characteristics are not necessarily synonymous. Greases which provide a high degree of protection against seizure, i.e., which have "extreme pressure" properties, may not be equally effective in reducing wear at more moderate loads. This is due to differences in the chemical nature of various load-carrying additives. The ability of many additives to prevent seizure and scoring depends primarily on their reacting chemically with the metal surfaces. Films are formed which are sacrificed under the high rubbing loads and thus protect the underlying metal. If these films are too readily formed and removed, high wear rates might result after long periods of operation under moderately severe loads.

Therefore, in comparing the load-carrying capabilities of various greases, it is always desirable to consider the results obtained from both types of procedures. This is particularly true if the greases and additives are of quite different types or their types are unknown.

The wide differences in response of certain greases of different types to the various load-carrying tests are shown in Table 12-6. With the excep-

Table 12-6
Response of Various Load-Carrying Tests

Grease Types	Timken Load	Four-Ball EP		Four-Ball Wear D 2266
		Weld Load	Load Wear Index	
Conventional Lithium - Non-EP	P	P	P	G
Conventional Lithium - with EP Additive	G	G	G	G
Lithium with 3% Moly Disulfide	P	G	G	G
Lithium-Complex	P	G	G	E
Lithium-Complex with EP Additive	G	G	G	E
Calcium-Complex (Type A-No EP Additive)	E	E	E	E
Calcium-Complex (Type B- With EP Additive)	G	E	E	G
Clay Thickener with EP Additive	E	G	G	P
Aluminum Complex with EP Additive	G	E	E	G

KEY: E - Excellent, G - Good, P - Poor

tion of the Type A calcium-complex grease, these variations are predominantly a function of the type of additive used rather than the thickener type. In the case of the Type A calcium-complex grease, the exceptional load-carrying properties are due wholly to the nature of the soap thickener.

It is obvious from this that reliance on a single test method could lead to quite false, and possibly unfortunate, conclusions as to the ability of a lubricant to provide the desired anti-wear or extreme-pressure properties.

Characteristics of Thickeners

In the early years of the grease-making "art," much could be learned about the performance properties of a grease simply by knowing its type of thickener and the viscosity of the oil component. The number of grease-making ingredients being utilized was very limited, and consequently, it was quite simple to adequately define the characteristics of the grease in terms of these ingredients. However, the many advances in grease technology have by now introduced an almost infinite number of variables, both in ingredients and methods of manufacture. Consequently, a knowledge of the type of thickener alone provides only very limited information as to the overall physical properties and performance capabilities of the grease. The final properties of the grease are drastically affected by the manufacturing process, the characteristics of the oil component and, of course, by the additives which may be used.

In the following discussion, a cursory review is made of the characteristics which the thickeners now in common use *generally* impart to a grease. In the case of soap thickeners, only the types of alkali (metal hydroxide) are covered. No attempt has been made to discuss the effect of the various fats or fatty acids or manufacturing techniques on the properties of the grease. The three major classes of soap thickeners (simple, mixed, and complex) are first reviewed. This is followed by a summary of the principal soap and nonsoap thickeners and their general grease-making characteristics.

Simple Soaps

A simple conventional soap is made by reacting a single alkali (calcium, sodium, lithium hydroxide, etc.) with a high molecular weight fat or fatty acid (oleic, stearic, palmitic, 12-hydroxystearic, or "12-OH stearic," etc.).

Mixed Soaps

A mixed soap is generally a simple mixture of the soaps formed by reacting two different alkalies (sodium-calcium, lithium-calcium, etc.) with a high molecular weight fat or fatty acid.

Complex Soaps

Complex soaps are the latest additions to the science of making soap-thickened greases. They have generally come to be considered as consisting of a soap-salt complex of the same alkali. That is, they would be formed by the concurrent reaction of an alkali with a high molecular weight fat or fatty acid to form a soap and the same alkali with a low molecular weight acid (usually a weak acid such as acetic, formic, boric, lactic, etc.) to form a salt.

Conventional Calcium Soap

- Since the hydrated form of the soap is usually required, the small amount of water that is present can be driven off if the grease is heated to above 82°C (180°F). When this occurs, the soap structure disintegrates, the soap loses its thickening ability, and the grease fluidizes. Hence, these greases are unsuitable for service above 82°C (180°F).
- Highly water-resistant
- Inexpensive
- Anhydrous calcium soap greases contain no water, have higher dropping points and, therefore, wider operating temperature ranges.

Calcium-Complex Soaps

- Do not require the soap to be in the hydrated state to form a stable grease structure, so high temperatures do not destroy its thickening ability. The low molecular salt (calcium acetate) has a very high melting point which imparts a high dropping point [260+ °C (500+ °F)] to the grease.
- May provide exceptional load-carrying properties
- Highly water-resistant if properly made
- Tend to harden in storage or on exposure to high temperatures
- Inexpensive

Sodium Soaps

- High melting point, thus imparting a high dropping point [over 177°C (350°F)] to greases. Melting point dependent on fat or fatty acid used. Sodium-complex soaps have dropping points above 260°C (500°F).
- Generally water-soluble and act as an emulsifier which causes the grease to emulsify, fluidize, and wash out from bearings if exposed to large amounts of water. Emulsification (water washout) greatly increased if water temperature above 38°C (100°F).
- Provide rust protection under high humidity since soap tends to preferentially absorb water and prevent contact with iron or steel surfaces.
- Simple soaps generally fibrous. Complex soaps generally smooth and buttery.
- Complex soaps may provide excellent anti-wear characteristics.

Lithium Soaps

- Water resistant
- High melting point—highly dependent on type of fat or fatty acid used. Dropping points of simple lithium soap greases 150°–190°C (300°–375°F). Complex-lithium soap greases over 260°C (500°F).

Mixed Soaps

- Properties generally intermediate between those of the individual components.

Other Soaps

- Although many other types of soaps such as barium, aluminum, strontium have been made, their use is so limited that their characteristics will not be discussed here.

Nonsoaps

Inorganic

- Treated clays (Bentonite) and silica gel are "nonmelting," hence greases have dropping points above 260°C (500°F). Clay greases, in particular, have low oil bleeding tendencies.

- Water resistance highly variable. If thickener or grease not made properly, water will be absorbed on surface of thickener and thus prevent thickening action, i.e., grease will fluidize.
- May be prone to rusting.
- Difficult to impart good load-carrying properties to greases by use of additives.
- May be incompatible with soap-thickened greases. Soaps can cause clays to lose thickening ability, so fluidization is possible.

Synthetic Organic

This category encompasses a wide variety of synthetic organic and metallo-organic compounds, including such diverse materials as dye pigments (copper phthalocyanine and indanthrene), aryl substituted ureas (ASU), polyureas, Teflon® or PTFE (polytetrafluorethylene), carbon black, and sodium octadecyl terephthalamate. These materials are generally characterized by their ability to withstand high temperatures. As a class they are relatively expensive and have found their principal use as thickeners for various synthetic fluids to make greases for extremely wide temperature range aerospace applications. However, some are now being used in combination with mineral oils for premium, high-performance industrial greases.

Application Limits for Greases

Bearings and bearing lubricants are subject to four prime operating influences: speed, load, temperature, and environmental factors. The optimal operating speeds for ball and roller type bearings—as related to lubrication—are functions of what is termed the DN factor. To establish the DN factor for a particular bearing, the bore of the bearing (in millimeters) is multiplied by the revolutions per minute, i.e.:

$$75 \text{ mm} \times 1000 \text{ rpm} = 75{,}000 \text{ DN value}$$

Speed limits for conventional greases have been established to range from 100,000 to 150,000 DN for most spherical roller type bearings and 200,000 to 300,000 DN values for most conventional ball bearings. Higher DN limits can sometimes be achieved for both ball and roller type bearings, but require close consultation with the bearing manufacturer. When operating at DN values higher than those indicated above, use either special greases incorporating good channeling characteristics or circulating oil.

Relubrication Frequency Recommended by Bearing Manfuacturers

Correct seal design is the prime factor in preventing contaminants from entering a bearing, but relubrication at proper prescheduled intervals offers the advantage of purging out any extraneous material from the seals before they have had an opportunity to gain access to the bearings or the housing cavity. Adherence to proper scheduled regreasing intervals will also ensure that the bearing has a sufficient amount of grease at all times, and will aid in protecting the bearing component parts against any damaging effects from corrosion.

The frequency of relubrication to avoid corrosion and to aid in purging out any solid or liquid contaminants is difficult to establish since relubrication requirements vary with different types of applications.

Anticipating a not-quite-clean to moderately dirty environment as can be assumed to be present in refineries and petrochemical plants, one authority suggests greasing intervals ranging from 1 to 8 weeks. Noting that the period during which a grease lubricated bearing will function satisfactorily without relubrication is dependent on the bearing type, size, speed, operating temperature and the grease used, a major bearing manufacturer suggests use of the graph shown in Figure 12-34. However, Figure 12-34 was developed for an age-resistant, average quality grease and for bearing operating temperatures up to +70°C (+158°F) measured at the outer ring. Its authors suggest that the intervals should be halved for every 15°C (27°F) increase in temperature above +70°C (158°F), but the maximum permissible operating temperature for the grease must not be exceeded. On the other hand, SKF has published lube interval data for motor bearings in very clean locations which exceed those shown in Figure 12-38 by a factor of 3.

SKF believes that if there is a definite risk of the grease becoming contaminated the above relubrication intervals should be reduced. This reduction also applies to applications where the grease is required to seal against moisture, e.g., bearings in paper making machines (where water runs over the bearing housing) should be relubricated once a week.

The FAG Bearing Company also opted for a graphical representation showing recommended relubrication intervals, Figure 12-35. Here, the horizontal scale depicts the ratio of running speed over the maximum allowable running speed for grease lubrication of a given bearing. This is basically similar to actual DN over the limiting DN of, say, 200,000. Most ball bearing-equipped motors are supplied with bearings operating at n/n_g approximately equal to 0.5.

Figures 12-34 and 12-35 can now be compared with our own experience value which has been published in nonproprietary data sheets by Exxon Company USA for use by its customers. These data sheets advocate

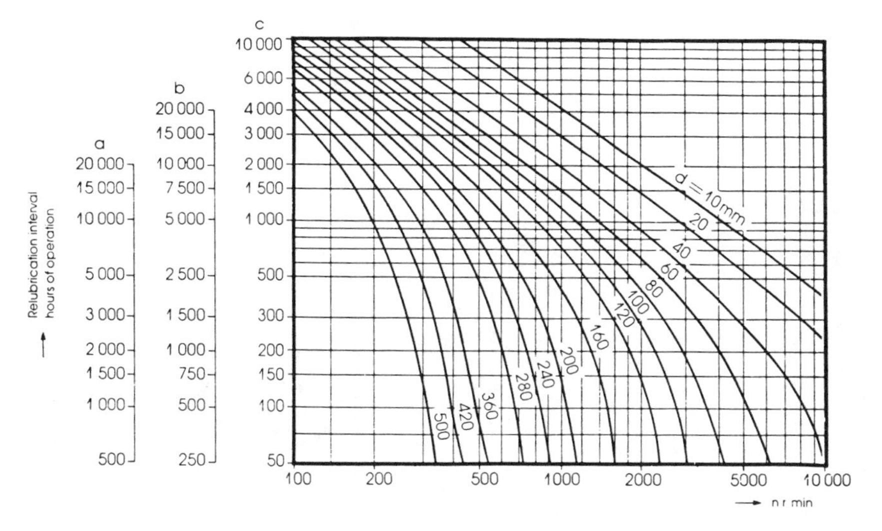

a Radial ball bearings
b Cylindrical roller bearings, needle roller bearings
c Spherical roller bearings, taper roller bearings, thrust ball bearings
d Bearing bore diameter

Figure 12-34. Relubrication intervals recommended by SKF.

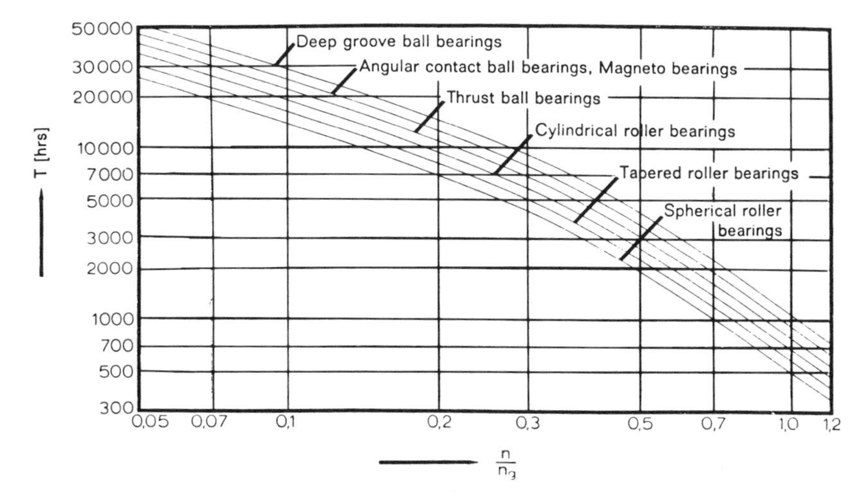

Figure 12-35. Relubrication intervals recommended by FAG Bearing Company.

Table 12-7
Maximum Relubrication Intervals for Motors Based on Petrochemical Plant Experience in the USA

Motor Size, Horsepower	1/4	7 1/2	10	40	50	150	Over 150	
Type of Grease	Ronex	Unirex	Ronex	Unirex	Ronex	Unirex	Ronex	Unirex
Type of Service								
I. Easy, infrequent operation (1hr/day). Valves, door openers, and portable tools.	60	120	60	84	48	48	12	12
II. Standard, 1 or 2-shift operation, Machine tools, air-conditioners, conveyors, refrigeration equipment, laundry and textile machinery, woodowrking machinery, light-duty compressors and pumps.	60	84	48	48	12	18	6	6
III. Severe, continuous running (24 hr/day). Motors, fans, motor-generator sets, coal and other mining machinery, steel-mill machinery and processing equipment.	36	48	12	18	6	9	2	3
IV. Very severe. Dirty, wet, or corrosive environment, vibrating applications, high ambient temperatures (over 40°C, 100°F), hot pumps and fans.	DO NOT USE	9	DO NOT USE	4	DO NOT USE	3	DO NOT	2

(1) Relubrication interval for Class F motors in Service Types III and IV should not exceed 12 months.

the intervals shown in Table 12-7 for a superior-quality multi-purpose grease (Ronex® MP) and a premium rolling-contact bearing grease (Unirex® N). Both are highly recommended for motor bearing lubrication.

References

1. McDivitt, N. E., "Inactive-Equipment Preservation Pays," in *Surveillance and Maintenance of Rotating Equipment,* Petroleum Publishing Co., Tulsa, Ok., 1978, Pages 122–125.
2. Correspondence and conference with Cooper-Bessemer Co. field service personnel, 1975.

Chapter 13

Vibration and Condition Monitoring

Vibration Measurement—Basic Parameters For Predictive Maintenance on Rotating Machinery*

Preventive maintenance programs for rotating machines, based on periodic sampling and inspections, have been greatly enhanced by the addition of vibration monitoring instrumentation. Although preventive maintenance has proven to be valuable for increasing production and decreasing unscheduled downtimes, a maintenance philosophy has evolved by which direct measurement of machine condition is used to provide "predictive maintenance."

The concept is explored here and the value of the concept is demonstrated that, in plants using predictive maintenance, machine turnarounds are based on machine condition rather than on elapsed time.

It has become well recognized that vibration monitoring can help to prevent major machinery failures and reduce costly downtime. "Preventive maintenance" type programs have proven their value in cost savings and increased production for many of the major refining, petrochemical, power generation, and industrial plants around the world. Most preventive maintenance programs are established with periodic inspection of machinery and permanent monitoring of critical, unspared machines, and scheduled periodic monitoring of less expensive, less critical, spared machinery. Generally, the parameters being monitored have been amplitude of vibration in peak-to-peak mils displacement, peak inches per second velocity, or peak Gs acceleration. Periodic checks of frequency of vibration are often included as part of this preventive maintenance type program.

* Pages 610 through 639 adapted from material compiled by Bently-Nevada Company, Minden, Nevada 89423. Adapted by permission.

In recent years, the introduction of new measuring and monitoring techniques has greatly helped the maintenance engineer in his monitoring and analysis of rotating machinery. In particular, the ability to monitor the direct shaft motion with proximity probes has increased the amount of information available to determine machine condition. Utilization of all available information allows for the implementation of "predictive maintenance" programs.

Many plants are in the process of changing their philosophy from that of shutting down on a time-scheduled basis to that of running the plant until the condition of the machinery indicates it is time to shut down. This condition is, of course, determined by instrumentation. This "predictive maintenance" philosophy is particularly applicable where the rotating machinery is the limiting factor in a plant maintenance schedule. Obviously this type of predictive maintenance program requires dependence upon instrumentation and the proper interpretation of the data it provides. In this respect, it is important that all available parameters of vibration and rotor position be measured and evaluated. A simple investigation of amplitude and frequency alone does not, and will not, provide sufficient information about machinery performance to provide a strong, accurate predictive maintenance program.

The following is a discussion of basic dynamic motion (vibration) and rotor position parameters that should be measured and analyzed in the diagnosis of rotating machinery in predictive maintenance programs.

Dynamic Motion (Vibration) Parameters

Amplitude—Amplitude, whether expressed in displacement, velocity, or acceleration, is generally an indicator of severity. It attempts to answer the question: "Is this machine running smoothly or roughly?" The ability to measure the shaft with proximity probes has helped greatly in providing more accurate information with regard to the amplitude of vibration. In the past, when only casing measurements were available, amplitude of the casing vibration was the only available parameter for severity. Whereas the casing measurements were able to indicate the presence of some machinery malfunction conditions, by and large, the casing measurements proved inadequate for proper machinery protection. This was primarily due to the variable transfer impedance between the shaft motion and the casing motion, depending upon the particular machine design, assembly, operating condition, and case pickup location.

Casing measurements have been utilized recently in an attempt to determine the presence of "high frequency" vibrations. In this manner, blade passage frequencies, blade resonance frequencies, gear mesh frequencies, etc., are observed to hopefully lead to the early discovery of possible malfunctions. These measurements are generally made on a periodic basis and are not usually continuously monitored. Since these are case measurements, it is somewhat difficult to evaluate the relative amplitudes and corresponding significance of the high frequency vibrations. Important is the fact that the vast majority of machine malfunctions manifest themselves at lower frequencies, usually less than 4 × running speed. High frequency measurements are useful only a small percentage of the time in machinery evaluation.

Amplitude of vibration on most machinery is expressed in peak-to-peak mils displacement. With proximity probes mounted at or near the bearings, vibration tolerances can be established which provide for the maximum excursion that the shaft makes with respect to the bearing. Today, most continuous monitoring of critical machinery is provided with a peak-to-peak displacement measurement either in mils or micrometers. A normal operating machine will generally have a stable amplitude reading of an acceptable low level. Any change in this amplitude reading indicates a change of the machine condition. Increases or decreases in amplitude should be considered justification for further investigation of the machine condition.

Frequency—The frequency of vibration (cycles-per-minute) is most commonly expressed in multiples of rotative speed of the machine. This is primarily due to the tendency of machine vibration frequencies to occur at direct multiples or submultiples of the rotative speed of the machine. It also provides an easy means to express the frequency of vibration. It is necessary only to refer to the frequency of vibration in such terms as one times rpm, two times rpm, 43 percent of rpm, etc., rather than having to express all vibrations in cycles-per-minute or cycles-per-second.

The emphasis on frequency analysis developed primarily from casing measurements where amplitude and frequency were the only major parameters available for measurement and evaluation. Also, the tendency of certain malfunctions to occur at certain frequencies has helped to segregate certain classes of malfunctions from others. It is extremely important to note, however, that the frequency/malfunction relationship is not mutually exclusive. That is to say, a vibration at one particular frequency often has more than one malfunction associated with it. There is no one-to-one vibration. One must not be easily swayed into attempting to directly correlate certain frequencies with particular *specific* malfunctions. Frequency is an important piece of information with regard to analyzing rotating machinery, and can help to classify malfunctions, but it is only one piece of data. It is necessary to evaluate all data before arriving at a conclusion.

The typical means of expressing frequency is as follows:

1 × rpm: The frequency of the vibration is the same as the rotative speed of the machine.

2 × rpm: The frequency of the vibration is twice the rotative speed of the machine.

½ × rpm: The frequency of the vibration is one-half the rotative speed of the machine.

.43 × rpm: The frequency of the vibration is 43 percent of the rotative speed of the machine, etc.

It is important to note a means of differentiating two types of vibration: Synchronous and nonsynchronous. Synchronous vibration occurs at a frequency which is some direct multiple or integer fraction of rotative speed of the machine. For example:

1 × rpm; 2 × rpm; ½ × rpm; ⅓ × rpm. In these examples, the vibration frequency is "locked-in" with the rotative speed of the machine. Nonsynchronous vibration occurs at some frequency other than a "locked-in" frequency with running speed. When viewed on an oscilloscope in time base presentation, a nonsynchronous waveform will appear to be continuously moving across the screen. If a once-per-turn event (Keyphasor®) is super-imposed on the waveform, it will appear to move along the waveform instead of remaining constant in one fixed position as it would with a synchronous vibration. Basic frequency measurements can be made with the use of a Keyphasor® and an oscilloscope. It is possible, with some minimal practice, to be able to pick out the major components of vibration present in a vibration waveform. For more discrete frequency analysis, it is necessary to employ some additional instrumentation, such as a tunable filter, swept frequency display, or digital spectrum display.

Phase Angle—Phase angle of vibration has long been ignored as an important criterion for analysis of rotating machinery by people in many areas of rotating machine use. The power generation people in the utility industry have, however, long recognized its value.

The phase angle measurement is a means of describing the location of the rotor at a particular instant in time. A good phase angle measuring system will define the location of the high spot of the rotor at each transducer location relative to some fixed point on the machine train. By determining these high spot locations on the rotor, it is possible to determine the balance condition and locations of the residual unbalances on a rotor. Changes in the balance condition of a rotor changing this high spot will be shown as a change in phase angle. Accurate phase angle measurements are extremely important in balancing of rotors, and can be extremely important in analysis of particular machine malfunctions. The phase angles of the rotor as determined by various transducers along a machine train can provide valuable information as to the performance of that machine train. It is this phase angle that provides timing information which helps answer the questions: What is happening, where, when, and how?

Phase angle is also valuable in determining the rpm location of the natural rotor balance resonances ("criticals").

The most accurate and reliable means of measuring phase angle is with the use of a Keyphasor® (shaft reference). The Keyphasor® can be accomplished with a proximity probe or optical pickup. This probe provides a synchronous, once-per-turn output signal that gives a direct shaft reference for phase angle measurements. In balancing, this direct shaft reference eliminates the need for a trial run to determine the high spots on the rotor. If the particular rotor response curve is known, the heavy spot can thus be located. Below the first balance resonance (first critical), the phase angle, as determined by the Keyphasor®, defines the heavy spot on the rotor. In a classical rotor, above the first balance resonance, the phase angle, as determined by the Keyphasor®, defines the location where a weight must be added. That is, it defines the light spot on a rotor or the position 180° from the heavy spot. The phase lag that occurs through this resonance region provides the number of degrees the high spot differs from the heavy spot.

In using the Keyphasor® as a phase reference mark, the phase angle is defined as the number of degrees from the Keyphasor® pulse to the first positive peak of the vibration. This first positive peak corresponds to the high spot on the rotor as it passes by the vibration input transducer.

In order to read phase angle accurately, instrumentation is required which filters the input signal to 1 × rpm, and then accurately measures and displays the number of degrees from the Keyphasor® to the first positive peak on the vibration input waveform. This filter must eliminate any possible error introduced by noise or harmonics on the imput waveform and must disallow any error introduced by the filter. This error is inherent in mistuned band pass filters of older strobe-light type vibration analyzers. Digital vector filters (DVFs), virtually eliminate the phase errors inherent in the instrument.

Phase angle is rapidly gaining acceptance as a very important parameter for diagnosis of rotating machinery as well as for balancing. Instrumentation for measuring vibration phase angle is generally included as part of the instrumentation package for large steam turbines in power generation service. For mechanical drive steam turbines and gas turbines, as well as compressors, pumps, fans, etc., vibration phase angle is usually provided by portable instrumentation. This methodology is a continuation of the balance work initiated by Theale 45 years ago and being employed in modern balance practice by some major electric power generation equipment manufacturers.

Vibration Form—The form of the vibration is perhaps the most important means of presenting vibration data for analysis. It is through this type of presentation that an understanding of particular machine behavior can be realized. The previously discussed three parameters have all been measurable quantities that can be displayed on an indicating meter or digital display, while "vibration form" is the raw waveform itself, displayed on an oscilloscope.

The oscilloscope is one of the most valuable, if not the most valuable, data-presenting instrument for machinery diagnosis.

Vibration form can be separated into two separate categories: *Time-base presentation* and *Orbital presentation*. Time base presentation is provided by displaying transducer inputs on the oscilloscope in the time base mode. In this mode, the oscilloscope displays the sinusoidal type waveform representing the shaft motion (utilizing a proximity probe input). This mode of the oscilloscope displays the position of the shaft relative to the input transducer versus time horizontally across the cathode ray tube of the oscilloscope. The orbit presentation is provided by displaying the output from two separate proximity probes at 90° angles to one another in the X-Y mode of the oscilloscope. In this mode, the oscilloscope displays the centerline motion of the shaft at that horizontal location along the rotor. If the probes are mounted at the bearing, the orbit is a presentation of the motion of the shaft centerline with relationship to the bearing.

These two presentations give the maintenance engineer the most data in one presentation. Basic amplitude, frequency, and phase angle can be determined by viewing the vibration form. The vibration form presentations inherently help the individual to understand "what the machine is doing" by observing the actual motion of the observed part. This is an important concept. The vibration form allows for the transition from determining what the amplitudes and frequencies are to determining *"What the Machine is Doing."* This is the ultimate parameter that we are attempting to understand in any preventive or predictive maintenance program.

Vibration Mode Shape—A recommended practice for capturing information on a rotating machine is to provide an extra set of X-Y (90°) probes some distance away from the bearing. This extra horizontal set of probes would not normally be monitored but would be available for diagnostics. The extra horizontal probes provide a third dimension to the machinery data and allow for an estimate of the mode shape of the machine rotor for the determination of nodal points. It is important to recognize that any set of X-Y probes along the machine train will describe the motion of the rotor at that horizontal location along the machine. By utilizing the extra set of X-Y probes at a different horizontal location along the machine train, we can attempt to determine the basic mode shape of the rotor itself. This mode shape can help to give closer estimates of the internal clearances between the rotor and stator elements and to give an estimate of the nodal points along the rotor shaft.

The parameters have all been a means of analyzing the vibration (dynamic motion) of a particular machine. They are all means of looking at *what the machine is doing* on a dynamic basis. The presentation of the amplitude, frequency, phase angle, form, and mode shape are all applicable to casing measurements as well as to shaft or rotor measurements. It is important from an overall system analysis to know what the casing is doing dynamically as well as what the rotor is doing. Such things as structural or piping resonance, loose or cracked foundations, external vibration input sources, etc., can be determined from measurements on the nonrotating machine parts. In the overall system analysis of a machine's mechanical performance, casing measurements can indeed be important.

The comparison of shaft or rotor vibrations with casing vibrations also can be an important parameter in determining the overall condition of a machine. As was mentioned before, the transfer impedance between the shaft and casing can vary widely due to various machine parameters. A comparison of both the amplitude and phase relationships of the casing and shaft vibrations may indeed prove valuable in solving a particular machine malfunction problem.

No discussion of dynamic motion (vibration) would be complete without the discussion of "relative" versus "absolute" measurements. A proximity probe, by its very nature, when mounted rigidly to the bearing cap or casing of a machine, provides a vibration measurement of the relative motion between the shaft and the mounting of the proximity probe. This relative measurement has proven a satisfactory parameter for continuous monitoring of most machines. However, on some machines the absolute motion of the rotor becomes an important parameter for continuous monitoring. This absolute motion can be provided with a "dual probe" which utilizes a relative proximity probe providing shaft motion relative to the casing and an absolute seismic-type transducer mounted on the casing of the machine in the same plane as the relative proximity probe. A vector summation of these two transducer inputs provided in the monitoring circuitry for this dual probe gives the "absolute" shaft motion. In this manner, four separate pieces of information are available.

1. The shaft motion relative to the casing.
2. The casing absolute (seismic) motion relative to free space.
3. The shaft absolute (seismic) motion relative to free space.
4. The shaft position measurement as supplied by the DC output of the proximity probe transducer.

The absolute measurement is most important on machines with flexible support structures or machines subject to high casing vibrations, as compared to the relative shaft vibrations.

The importance of X-Y (2 plane) monitoring has been well established for most types of machines. It is possible to have totally different vibrations in the vertical and horizontal directions at one particular bearing. It is entirely possible, for example, to have different amplitudes and different frequencies (and normally corresponding different phase angles) in two different planes at one bearing. This has been documented on many machines and the importance of X-Y mounted probes at radial bearings should not be underestimated.

Position Measurements

Other parameters which should be measured and evaluated for total machinery performance fall in a category of static and quasi-static position measurements. Depending upon the particular machine design and machine malfunction, these measurements can be important in evaluation and analysis. The following provides a discussion of these position type measurements.

Eccentricity Position—Eccentricity position is the measurement of the steady state position of the shaft in the journal bearings. Under normal operation with no internal or external preloads on the shaft, the shafts of most machines will ride where the oil pressure dam planes it. However, as soon as the machine gets some external or internal type preload (steady state force), the eccentricity position of the shaft in the journal bearing can be anywhere. This eccentricity position measurement can be an excellent indicator of bearing wear and heavy preload conditions such as misalignment. In installations where only single plane monitoring is present, it is imperative that eccentricity position be measured on a periodic basis. This is due primarily to the possibility of heavy preload condition precluding an increase in amplitude of vibration in the plane of sensitivity of the probe. This could possibly allow for a machine malfunction to occur without significant amplitude warning. By observation of the eccentricity, an early warning is possible.

Eccentricity position should also be closely watched during machine startup. During a machine startup with a vertically mounted proximity probe, one would normally expect the shaft to rise from the bottom of the bearing to someplace toward the center of the bearing. This is due fundamentally to the oil flowing under the shaft making the shaft rise in the bearing. It is generally believed that the oil film is about one mil in thickness. Observations on many bearings show that it is more often about 1/3 of the bearing clearance in the preloaded direction of the shaft.

The measurement of eccentricity position is accomplished by monitoring the DC output of the proximitor associated with the radial vibration probe at the bearings.

Because of this ability of the eccentricity position to change under varying conditions of machinery load, alignment, etc., it is important that the proximity probe transducer system have a long linear range sufficient to allow for these eccentricity position changes to occur without having the shaft move outside of the linear range of the proximity probe. This is especially true in large machines where large bearing clearances are normally present. Whereas eccentricity position is monitored continuously on some machines, on most machines a periodic check of eccentricity position appears to be acceptable for predictive maintenance. Especially where misalignment or other preload conditions may be considered as a possible malfunction condition, eccentricity position should be monitored very closely. It is important to document the cold eccentricity position and the hot eccentricity position so that a frame of reference is established for comparisons of eccentricity position at later dates.

Axial Thrust Position—Axial thrust position is the measurement of the relative position of the thrust collar to the thrust bearing. This measurement is perhaps one of the single most important monitored parameters on a centrifugal compressor and/or steam turbine. The primary purpose of an axial thrust position monitor is to ensure against an axial rub between the rotor and the stator. Axial thrust bearing failures can be catastrophic, and every attempt should be used to protect against this possible machine failure mode.

At least one, preferably two, axial thrust position probes should be mounted to provide axial thrust position protection.

Care must be taken in selection of probe mounting locations to ensure minimum effect of thermal growth of the rotor and minimum effect of springiness of the thrust bearing assembly on the accuracy of the reading. In early applications of proximity probes for axial thrust position measurements, it was very often found that the alarm and danger set points were established too close to the initial cold float zone of the machine. It was found that deflections occur in the thrust bearing assemblies and thermal growth of the rotor occurs such that, under normal operating conditions, the position of the rotor can indeed appear to be wider than the normal cold float zone of the rotor within its thrust bearing. It is important to note that most machines have sufficient axial clearance between the rotor and the stator that wide set points can be established allowing the thrust collar to severely wipe the babbitt of the thrust bearing shoes without having rotor to stator rubs. Under normal operating conditions of a centrifugal compressor or steam turbine, thrust position can vary with load of the machine, so varying thrust position measurements under differential loads and conditions of a machine are not uncommon. The thrust position measurement may also be important in the determination of surge or incipient surge conditions.

An added benefit received when applying thrust position probes for axial thrust position measurements is that axial vibration measurements can be read from the same proximity probe. Whereas axial vibration is not normally monitored continuously on centrifugal equipment, it has proven valuable in diagnosing some particular machinery malfunction conditions. If axial vibration is to be monitored or used for diagnosis of a particular machine, it is necessary for the observed surface to be smooth and to be perpendicular to the centerline of the rotor. This will minimize any effect of mechanical runout on the dynamic output of the probe, thus providing accurate axial vibration readings.

Eccentricity Slow-Roll (Peak-to-Peak Eccentricity)—In large steam turbines for power generation service and in some industrial gas turbines, it is very often desirable to provide an indication of eccentricity slow-roll, also called peak-to-peak eccentricity. Eccentricity slow-roll is the amount of bow the rotor takes while it is at rest. This bow can be indicated by the slowly changing DC peak-to-peak measurement from the proximitor as the rotor turns on turning gear. When the peak-to-peak amplitude is at an acceptable low level, the machine can be started without fear of damage to seals and/or rotor rubs caused by the residual bow and its corresponding unbalance. Eccentricity slow-roll is best measured with a probe mounted away from the bearing so that maximum bow deflections can be measured.

In large steam turbine applications, eccentricity slow-roll is often recorded on a strip chart recorder where the operators can visually see the amount of bow in the shaft and its rate of decreasing amplitude while the machine is on turning gear and is allowed to warm up. Eccentricity peak-to-peak can also be displayed on a multi-point type recorder or fed to a computer through sample-and-hold techniques providing a low frequency peak-to-peak measurement with an update of once-per-turn of the shaft.

Differential Expansion—In very large machines, such as large steam turbines in power generation service, it is extremely important that during startup the casing and the rotor both grow thermally at the same rate. If the rotor or the casing grow at different rates, there exists the possibility of damage to the machine caused by axial rubs. In order to measure this differential expansion, a proximity probe is mounted at the end of the machine opposite the thrust bearing where the relative growth between the case and the rotor can be observed. The typical range for this proximity probe is one inch. In very large machines, this range required of the proximity probe system can be as large as two or three inches.

Case Expansion—On very large machines it is also very common to provide, in addition to differential expansion, a case expansion measurement. This case expansion measurement is usually provided by a contacting linear variable differential transformer (LVDT) mounted externally to the machine case and referenced to the foundation.

This case expansion measurement helps to provide information with regard to the relative growth of the rotor to the case as just described in the paragraph on differential expansion. By knowing the amount of case growth and the amount of differential growth, it is possible to determine which is growing at a more rapid rate—the rotor or the case. If the case is not growing properly, the "sliding feet" of the case may be stuck.

Alignment—The need for proper alignment between different machine cases of a machine train has been well established. Misalignment has been determined to be one of the more frequently occurring malfunction conditions, especially in installation of process compressor trains and gas turbine driven machinery trains. The normal alignment procedures generally allow for the determination of an alignment drawing through calculations, using the anticipated growth of the various different machines involved. This has often been shown to be a "guesstimate" at best, and finding means of determining the final, hot running condition of the machine has been proven desirable and necessary.

Considerable work has been done in the area of hot alignment measurements. The more popular methods include optical measuring techniques, utilizing optical instruments similar to those used for surveying land, and the proximity reference system. The proximity probe has been used to determine the relative position changes between shafts of different machine cases. Popular techniques for utilizing proximity probes in this respect have been devised by Charles Jackson and Ray Dodd. These methods are described in Volume 3 of this series.

Other Parameters

Speed (RPM)—The measurement of speed (rpm) of the rotor has long been standard procedure. Most major centrifugal machine trains have continuous indication of the rpm of the machine. General practice in the past has been to provide this indication by means of some analog type meter. Most of these analog meters simply provide a general indication of rpm of the machine.

With the advent of new, reliable digital circuitry, digital tachometers have become more popular for speed indication. Besides being more accurate and easier to read, the digital tachometer lends itself very well to providing redundant overspeed trip protection. The digital indication of rpm eliminates many of the older problems associated with rpm measurement. Accurate, easily readable indication of rpm is generally provided in the operator control room. Many users also prefer a remote digital indicator at the machine train for use during startup operations. For machines where digital tachometers are not normally supplied, portable digital indicators have become popular for use during start-up for monitoring of rpm and for checking of mechanical overspeed trip mechanisms.

Transducer inputs for the digital tachometers can be from a variety of inputs. Among the more popular are the Keyphasor® proximity probe input, the photoelectric pickup, and the magnetic pickup. These transducers are all designed to observe a number of events per revolution of the shaft. This basically digital input is then translated into a direct readout of rpm by means of the digital tachometer circuitry.

The correlation of vibration measurement with rpm can be important for the final analysis of the mechanical performance of a particular machine. Centrifugal equipment is designed to operate in a speed range which will not coincide with the balance resonances of that particular machine, and at speeds which will not excite these particular resonances. A startup piece of information which is important for determining the balance resonances is an X-Y plot of amplitude and phase angle of the vibration vs. rpm of the machine. In plotting and correlating these parameters, it is possible to easily determine the machine balance resonances (criticals).

Temperature Measurement—In the final analysis of the condition of a particular piece of rotating machinery, other parameters also become important. One of the most popular and important parameters not yet discussed is that of *temperature measurements*. The temperature of the bearing babbitt in both radial and axial bearings is becoming more and more popular. A correlation of this temperature information with vibration and/or position measurements helps to give a better indication of possible machinery malfunctions.

Correlation—Correlation of temperatures, pressures, flows, and other external parameters which could affect the operation of a piece of machinery is extremely important for the overall system analysis of the machine in service. It is through this correlation that a good predictive maintenance program can be established. The ability of the engineer to utilize all of the available information in determining the mechanical running condition of the piece of machinery is extremely important in the overall objective of maintaining proper operation and continuous on-line service of critical machinery.

The engineer who has a thorough understanding of the parameters discussed in this text will better understand the mechanical performance of centrifugal equipment. It is through understanding these parameters that he will ultimately be able to determine *what a particular piece of machinery is doing*.

Transducer Types

Before we investigate the various monitoring approaches recommended for certain machinery categories, we will summarize the characteristics of the different transducer types available today.

Proximity Probe

Advantages

- Measures directly the motion of the shaft (the origin of most machine vibrations).
- Measures in terms of displacement (the most meaningful engineering unit for rotating machinery measurements).
- Measurement is via noncontact medium (will not influence the measured vibratory motion because of contact).
- Solid-state with no moving parts.

- One sensor simultaneously measures both dynamic motion and (average) position. System is modular with the most inexpensive part, the probe, requiring only occasional replacement (because of abuse).
- One extra transducer can be used as a rotor speed sensor and a phase reference.
- Excellent frequency response.
- Small size.
- Well-suited to most machinery environments.
- Ease of calibration.
- Accurate low frequency amplitude and phase angle information.
- High level low impedance output.

Disadvantages

- Control of observed shaft surfaces desirable to avoid excessive sensitivity to shaft mechanical and electrical runout.
- Somewhat sensitive to various shaft materials.
- Requires an external power source.
- Sometimes difficult to install.

Velocity Pickups

Advantages

- Ease of installation due to external machine mounting.
- Strong signal in the mid-frequency ranges.
- Some are suitable for relatively high temperature environments.
- No external power required.

Disadvantages

- Relatively large and heavy.
- Manufactured as a unit so that a transducer fault means replacement of the entire unit.
- Sensitive to input frequency (tendency to emphasize high frequencies).
- Relatively narrow frequency response with amplitude and phase errors at low frequencies.
- Device is mechanical, has moving parts, and is expected to degrade under extended normal use.
- Difficult to calibrate.
- Measures dynamic motion only (not static position).
- Can respond with excessive cross-axis sensitivity at high amplitude levels.

Accelerometers

Advantages

- Ease of physical installation due to external machine mounting.
- Good frequency response (especially high frequencies, although this could cause a disadvantage by increasing the noise level from various external vibrations).
- Small and lightweight.
- Some are suitable for relatively high temperatures.
- Strong signal in the higher frequency ranges.

Disadvantages

- Most sensitive to input frequencies (although this can be an advantage when measuring very high frequencies).
- Difficult to calibrate.
- Expensive.
- Requires external power source.
- Most sensitive to spurious vibrations (confusing the acquired data and making exact mounting location difficult).
- Impedance matching is needed.
- Normally requires some filtering for monitoring applications.

Note: Although ease of installation is listed as an advantage, this is only because of its physical location on the externals of a machine; an accelerometer is actually very sensitive to the *method* of attachment to the machine case; in addition, the *proper* location on the machine case for a meaningful measurement of vibration can be difficult to determine.

Signal Conditioning

Once a transducer type has been selected, it is usually best to use a readout in the same engineering units as the transducer, e.g., proximity probes should be used with a readout which will display mils peak-to-peak displacement. It is certainly possible to change the units of the readout through *electronic integration* or *differentiation;* however, this process should be carefully evaluated. One overriding concern is the reliability of the monitor system. Whenever additional electronics are added to the circuitry between the transducer and the alarm/readout circuits, the overall system reliability decreases. We therefore recommend, when possible, that monitors have a minimum of extra circuitry (integrators, differentiators, filters, etc.) between the transducer and the protection circuits in order to ensure maximum reliability and minimize confusion.

There *are* some applications where the benefits of electronic signal massaging outweigh the decrease in system reliability. Since electronic *differentiation* is difficult (expensive and less reliable) and *double integration* can introduce unreasonable errors and noise susceptibility problems through the use of high gain amplifiers, single integration is the most usable conversion for practical purposes. One common use of single integration involves changing a velocity signal to displacement. We have discussed why *displacement* is usually a more significant parameter to use for machine protection and diagnostics, but, at the same time, we realize that there are some situations where a proximity probe cannot be installed due to time available, mechanical constraints, environmental considerations, and the like. In some applications (usually machines with rolling element bearings), there may be significant housing motion and not necessarily as much shaft motion relative to the housing. In these cases, a velocity transducer could be used and the velocity signal integrated to displacement. Although the ideal transducer may not be used, through a minimum of additional electronics, the ideal readout parameter, displacement, can be evaluated.

There is one additional tool which can be employed to help optimize the use of various transducers. That tool is the electronic filter circuit. We would generally discourage the use of filters in monitor circuits because, as stated earlier, the addition of any extra electronics in the monitor tends to lower the overall reliability of the instrument. Since a filter, by definition, *discards* information, important *machinery data* may be discarded by the filter. Also, filters can introduce phase and amplitude errors. However, in some situations the advantages of using a filter will compensate for the disadvantage of somewhat decreased reliability. In some cases a filter is necessary to help overcome the limitations or undesirable characteristics of a given transducer. One valid application for the use of a filter is in monitoring a speed increaser or reducer (gearbox). Generally, high speed gear units use hydrodynamic sleeve or tilt-pad journal bearings, and as such these machines exhibit a significant amount of shaft motion relative to the bearing support structures. Thus, the transducer for basic machine protection against the common rotor-related malfunctions should be the proximity probe. However, gear units can also exhibit some very high frequency vibrations as a result of normal gear contact forces. These vibrations are generally characterized by high acceleration and low displacement amplitudes. The displacements are often so low as to be barely measurable by proximity probes. Therefore, a common application is to use a case-mounted accelerometer in addition to the proximity probes to measure/monitor these high frequency vibrations. If the unfiltered acceleration signal contains "noise" from the other sources of vibration, then a band-pass filter covering the range of expected gear mesh frequencies would be useful.

Another application of filters and signal integration involves monitoring of various special machine categories including rolling element bearing machines. On those machines which show very little shaft relative motion or in situations where proximity probes cannot be installed, a velocity pickup or accelerometer may be used. One possibility is to use a Dual Path Monitor® with a velocity transducer. The Dual Path Monitor® divides the transducer input signal into two paths, A and B. Path A retains the velocity units, and a filter in the path is optional for isolating those frequencies associated with rolling element bearing failures. Path B has an integration circuit to convert velocity to displacement and an optional filter to isolate rotor-related frequencies. Each path has individual alert and danger set points and one common meter which will display both velocity and displacement amplitudes. The Dual Path concept is not limited to rolling element bearing machines; it can be used in any situation where discrete vibration frequencies, representative of discrete machine malfunctions, must be separated and evaluated in different terms in order to provide an optimum machine protection system.

Shaft Rider

One last transducer consideration involves a device which we have not discussed up to this point, the *shaft rider*. The shaft rider includes a seismic transducer (velocity pickup or accelerometer) which is connected to a small rod, the end of which actually makes contact with the shaft, through or adjacent to a journal bearing. The rod is held in place against the shaft by means of a spring loaded mechanism and this, coupled with the seismic transducer, makes the entire transducer very much mechanical in nature. As such, it is susceptible to the wide variety of mechanical failures—wear of parts, change in spring and/or damping, improper contact between the rod and the shaft, etc. Since a seismic transducer produces the output signal, special equipment is required to verify calibration of the unit. In addition, the shaft rider provides a measurement of shaft absolute motion *only,* while the alternative transducer, the dual probe, provides four different measurements including shaft absolute motion. A dual probe is a relative proximity probe mounted at the same location as a velocity transducer. The dual probe provides (1) shaft relative dynamic motion and (2) shaft average relative position (the measurements made by the proximity probe), (3) case absolute motion (the measurement made by the velocity transducer, and (4) shaft absolute motion (the measurement provided by the electronic vector sum of both transducer dynamic signals). The dual probe provides *four* pieces of information whereas the shaft rider provides only *one*, and even that measurement, shaft absolute motion, is made more accurately with the dual probe than with the shaft rider due to the mechanical nature of the shaft rider. The dual probe also enables one to study the respective amplitudes and phase relationships of vibrations of the shaft and case.

Generalized Monitoring Recommendations for Specific Machine Types

The extent to which it is recommended to measure (and monitor) various parameters is based on the most common applications of the various machine types listed. That is, some machines such as centrifugal compressors, steam turbines, and generators are usually employed in critical paths of many plant processes. Other types such as small motors, pumps, and fans are more commonly employed as noncritical or semicritical machines. Therefore, our recommendations for monitoring compressors will generally be more extensive than those for fans. Special applications of some machine types are listed separately as necessary. As stated earlier, these machine measurements must be correlated with process variable measurements in order to obtain a complete machine protection system. The recommendations given here are not absolute. They apply to the most common design configurations and process applications of the machine types listed, but as with any general set of rules, there will be exceptions.

It should be noted that the selection of any transducer/monitor system involves many compromises. Let us assume that for a hypothetical machine, transducer system *A* is required for basic protection against the most frequently occurring malfunctions. If transducer system *A* is used with a monitor system, the machine may be protected against, for example, 70 percent of its potential malfunctions. The installation of transducer/monitor system *B* may offer protection against an additional 20 percent of potential malfunctions while transducer/monitor system *C* will protect against 5 percent of all malfunctions (in reality, 100 percent cannot actually be achieved). Of course it is nearly impossible to determine these percentages on a *real world* machine, but let us assume that they form the basis of our technical evaluation of protection for systems for this machine's application. Then the technical priorities are weighed against the economic considerations and a decision is made. For some semicritical and noncritical machines, the use of system *A* may be sufficient. For critical machines, it may be justified to monitor all three systems. For some semicritical machines we could monitor the basic system *A* and have system *B* and *C* transducers permanently installed (without monitors) for future complete machine analysis. In many applications, permanently mounted transducers (in addition to primary protection transducers), can be very valuable and economic indeed. In most cases, a permanently installed transducer will provide a much more accurate measurement than a hand-held one, or even one mounted with a magnetic base or vise grips. Also the transducer usually represents a small part of the cost of a complete monitor channel. When selecting a monitor/transducer system, considering the many varied applications, the choices are many.

In general, it is important to recognize that in order to determine the optimum protection system for machinery, each piece of machinery must be evaluated individually. Available data are often insufficient for a detailed analysis of a particular machine's expected behavior under normal and malfunction conditions. It then becomes necessary to use best engineering judgment and experience in determining what should be monitored. Often the user company has a machinery specialist group to provide the function of monitoring system specification. However, the user can also rely on the machinery manufacturer, the engineering consultant/contractor, and/or the machinery protection system manufacturer to accomplish this function. Also, specifications for the purchase of various types of machines and monitor systems have been published by national organizations.

If you are reading this segment of our text to determine the proper monitoring system for a particular type of machine, we suggest you first read the section on *Large Steam Turbines* and then the section pertaining to your particular machine. Large steam turbines typically require the most complete monitoring system when compared to other types of machines. By first reading our description of this complete system, it may help you understand our recommendations for the other machines covered in subsequent sections.

Steam Turbines

Large Steam Turbines usually consist of two or more steam turbine casings, or bodies. We generally call each body a case. Typically, large steam turbines consist of a high pressure case, an intermediate pressure case, and a low pressure case. The rotors are rigidly coupled together, and the entire turbine shaft is rigidly coupled (usually to a generator).

The fundamental monitored points for large steam turbines are vibration and thrust (axial) position. The general recommendation for vibration is to monitor X-Y proximity probes mounted at or near each radial bearing of the turbine and generator. These machines utilize fluid film bearings with a reasonable amount of damping so that shaft relative motion is more significant than casing motion. Traditionally, these machines have been monitored with shaft rider transducers mounted in the vertical plane only. In the first place, X-Y measurement is *always* recommended because it is impossible to determine at what radial angle the maximum vibration will occur. Even on machines which have a *normal* axis of greater freedom, such as with elliptical bearings (for shaft measurement) or cases which are more flexible in one direction (for case measurement), changing machine parameters such as alignment can restrict the motion of the machine in the normal direction. Secondly, if shaft absolute monitoring is desirable, the dual probe is a much better transducer than the shaft rider. The *ideal* installation would be X-Y dual probes at each radial bearing.

Actually, experience has shown that most large steam turbines do not require shaft absolute measurement, as the majority of the motion is shaft relative to the bearing. However, on some machines, the dual probe is indeed necessary. Those machines can be categorized as follows:

1. Those which exhibit a significant amount of case absolute motion in addition to shaft relative motion (case motion is generally considered significant when the case vibration amplitude is greater than 1/4 of the shaft relative amplitude).
2. Those which exhibit significant casing vibrations during start-up of the machine.
3. Those which were originally supplied with shaft riders and it is desirable to measure shaft absolute motion so that a better comparison can be made with past historical data on the machine. For these machines, X-Y dual probes are recommended.

For large steam turbine units, the leading manufacturer of machinery protection systems recommends a convenient *economic compromise* between X-Y relative proximity probes and X-Y dual probes. The compromise includes a relative probe in one plane (Y, for example) and a dual probe in the perpendicular plane (X, for example). This arrangement maintains the desired proximity configuration and, in addition, it allows for *some* measurement of casing vibrations. In this situation, the seismic transducer of the dual probe need not be continuously monitored if casing motion is not considered to be continuously significant. For example, the seismic transducer may only be measured during startup or during load changes or possibly during balancing runs (when coupled with the relative probe to read shaft absolute motion). If one of the two X-Y planes is predominantly horizontal, it would be desirable to choose this location for the dual probe and the predominantly vertical plane for the proximity probe. This manufacturer has found that, in general, the structures of these machines are *softer* in the horizontal plane.

Because of economic or other considerations, a machine may only be monitored by a limited number of transducers; however, other (unmonitored) transducers can be installed permanently in order to be available to provide the additional information necessary for complete analysis of the machine. Most diagnostic instruments are capable of accepting inputs directly from the various transducers without signal conditioning (as provided by a monitor) being required. Therefore, unmonitored transducers can still be used with diagnostic instruments, while the overall expense of monitoring may be kept to a reasonable level. We must caution that a monitoring system should not be compromised to an extent which will also compromise basic machine protection. Additional radial probes, mounted laterally away from from the bearings, will help determine the shaft mode shape when compared with the data provided by the monitored (radial vibration) probes. Auxiliary measurements can also be employed to determine foundation, turbine housing, piping and other structural vibrations.

Thrust position measurements (and monitors) are necessary to indicate impending thrust bearing failure and/or abnormal axial shaft action. The probe should be mounted near the thrust collar, within 12 inches (300 mm) for meaningful data, observing the thrust collar itself, the end of the shaft, and auxiliary collar, or a *shoulder* or *step* in the shaft diameter. When the thrust position monitor is connected to a machine shutdown circuit, it is recommended that a two-probe voting arrangement be used to minimize false shutdowns.

Another important application of proximity probes to steam turbines is the measurement of *shaft radial eccentricity position* in the journal bearings. This measurement is available from the same probes which are installed to monitor radial vibration. The measurement can provide meaningful data regarding preloads, electrostatic erosion, and other disturbances which affect the *average* shaft position relative to the bearing clearance. Measuring *X-Y* radial shaft position can result in a calculation of the *attitude angle* of the shaft average position, the position of the shaft centerline with respect to the bearing centerline.

Another type of eccentricity measurement is usually necessary on these large machines. This is called *eccentricity peak-to-peak* and is a measurement of the bow or sag in the rotor caused by gravity or uneven cooling of the rotor. An eccentricity peak-to-peak monitor is used to indicate the amount of shaft bow, a parameter which must be considered before a turbine startup.

A *Keyphasor*® probe, providing a once-per-turn timing mark, should be included on the turbine shaft at some radical location to provide speed, phase angle and timing information for the entire machine train. The Keyphasor® can be used directly to measure speed and can also be used to measure shaft speed acceleration (revolutions/minute/minute) for those machines where shaft acceleration during startup may be critical. The Keyphasor® can also be used for the measurement of zero speed, so that operators will know exactly when to engage the turning gear after turbine shutdowns; and, of course, the Keyphasor® provides a means to measure phase angle.

Large steam turbines are usually supplied with a set of monitored functions grouped together called the *Turbine Supervisory Instrumentation (TSI) System.* These measurements include shaft/casing differential expansion, casing/foundation expansion, valve position, and phase angle. Phase angle measurement can be provided by a Digital Vector Filter. The DVF can display simultaneously shaft rpm, vibration amplitude (overall, or filtered at running speed) and phase angle for any vibration point on the entire machine train. The unit provides continuous analog outputs for making two types of plots: *Bodé plot*—an XYY plot of phase angle versus rpm and amplitude (filtered at 1 ×) versus rpm, and *polar plot*—a polar coordinate plot of the rotational speed vector as speed is varied. These plots can be used directly to determine many machine characteristics such as the location of balance resonances (critical speeds), structural resonances, and amplification factors over the operating speed range of the machine. The phase angle measurement also provides information for the locations of the *high-spot* to be used for balancing.

Temperature should be monitored at the journal bearings and thrust bearings of the turbine and generator. At least two sensors should be located in each radial bearing and on each side of the thrust bearing. Also, temperature of the generator rotor and stator windings should be monitored as well as the oil systems and machine ambient temperatures. At times an accelerometer may be useful for measuring high frequency blade-related vibrations.

Mechanical Drive Steam Turbines are best monitored using X-Y relative proximity probes at each journal bearing for vibration, and a minimum of one axial probe for monitoring thrust position. A Keyphasor® probe should be used to provide measurements of speed and phase angle. Temperature of radial and thrust bearings should be monitored as well as oil system and ambient temperatures. TSI measurements are sometimes applicable on various types of mechanical drive steam turbines; these include differential expansion, valve position, and eccentricity peak-to-peak (shaft bow). Eccentricity position should also be monitored, using the same probes as for radical vibration.

Marine Turbines are generally two separate machine casings (high pressure and low pressure); cross coupled with piping, but operating on two shafts to drive the gear reduction unit to the final drive (propeller) shaft. Each turbine shaft should be monitored as described earlier for mechanical drive turbines. In addition, because of the relatively flexible support structure on most marine turbines, casing vibration and alignment measurements are of great importance. Shaft radial eccentricity position should be monitored (using the same probes which monitor radial vibration) and consideration should be given to some type of cold/hot alignment measurement using external references. Temperatures of the bearings (radial and thrust), oil systems, and ambient conditions should be monitored.

Gas Turbines

Large And Medium Industrial Gas Turbines are usually equipped with fluid film bearings and thus should be monitored with X-Y relative proximity probes at each radial bearing. Some of these machines have relatively flexible support structures and, as such, bearing housing vibration measurements are recommended in addition to shaft relative measurements. A *dual probe* should be used at these bearing locations to make both measurements. An additional vibration measurement, using an accelerometer, can be used to measure high frequency blade excitations. At least one axial probe should be employed to monitor thrust position.

A Keyphasor® should be used to provide phase angle and speed measurement. Speed measurement is of particular importance since many gas turbines have a specific rpm-related startup sequence. On two shaft gas turbines, it is desirable to have one Keyphasor® for the air compressor shaft and a second Keyphasor® for the power turbine shaft. Temperature of the machine bearings, oil supply, and ambient environment should also be monitored.

Hot Gas Expander (Jet) Turbines utilize an aircraft derivative jet engine design. The hot gas exhaust of the engine is fed into an overhung power turbine coupled to the driven equipment (compressor, generator, pump, etc.). Jet engines are usually equipped with rolling element bearings, and the mechanical impedance between the shaft and the case may make seismic casing measurements more meaningful than on other turbine designs. However, some of these rolling element bearings are themselves mounted in flexible supports (squeeze film damper bearings), and therefore the appropriate measurement is shaft motion as measured by X-Y proximity probes mounted to the *bearing support* structure.

The power turbine is usually equipped with sleeve bearings and should have X-Y proximity probes mounted at each radial bearing. At least one axial probe should be installed at each thrust bearing, and temperature of machine components should be monitored. Since the jet engine and power turbine are not directly coupled, a Keyphasor® should be mounted on each separate shaft to provide speed and phase angle measurement.

Small Industrial Gas Turbines with rolling element bearings fall into the same category as jet engines. Proximity probes mounted some lateral distance away from the rolling element bearings would be desirable; however, such installations are difficult because of the small size of the machines. Therefore, casing velocity transducers (or accelerometers for very high speed applications) can be used with Dual Path Monitors® as a compromise. In similar service are small gas turbines with fluid film bearings. These should be monitored with X-Y proximity probes. Thrust position probes and a Keyphasor® probe should also be used in addition to bearing and other temperature measurements.

Hydro-Electric Turbines

Hydro-Electric or Hydraulic Turbines are equipped with fluid film bearings and should be monitored with X-Y proximity probes at each radial bearing. Since these vertical machines have large bearing clearances, the monitoring of shaft radical *position*, in addition to shaft radial *dynamic motion,* is important. Of course, both position and motion measurements are made by the same proximity probes. The machine should also have at least one thrust probe and a Keyphasor® probe to provide speed, phase angle, and timing information for the entire machine train. Because of the low operating speed of most of these units, low frequency signal conditioning circuitry may be necessary in the monitors.

Electric Motors

Electric Motors are often used as prime movers in process compressor and pump applications. Most of these are equipped with fluid film bearings and should be monitored with X-Y proximity probes. Monitors have a magnetic center that provides for the centering of the rotor, so the *monitoring* of axial shaft position is not as important here as with other machine types. However, *position* measurements should be made periodically on the radial probes as well as on permanently installed axial probe. Sometimes axial centering is a particular difficulty, so an axial probe should be *monitored* for *axial vibration.* Small electric motors with rolling element bearings should be monitored with casing velocity transducers and Dual Path Monitors.

A Keyphasor® probe should be installed to provide speed, phase angle and timing information for the entire machine train. For motors in special critical service (e.g., nuclear), an important measurement can be *locked rotor* condition during startup. Temperatures should also be monitored at each bearing, the rotor and stator windings, and the oil system.

Compressors

Packaged Centrifugal Compressors are integrally geared, skid-mounted centrifugal units consisting of a driver (motor or turbine), a bull gear, and up to four pinions. Each pinion shaft terminates in an impeller. These machines are often described as packaged *air machines,* since the most common service is general plant compressed air supply. Generally, the driver and bull gear speed is 3,600 rpm or less, and the pinion speeds can be as high as 60,000 rpm. These machines are produced as a *package* with the entire machine mounted on a common foundation which also includes a panel with control and monitoring instrumentation. Because of the large number of these machines manufactured, proper monitoring locations for proximity probes have been established by the various machine manufacturers. Nearly all of these machines are supplied by the OEM with one proximity probe per impeller (one or two per pinion) and sometimes one probe on the bull gear shaft. In some instances, manufacturers have been able to respond to user specifications by supplying X-Y proximity probes and Keyphasor® probes on the various shafts.

Process Centrifugal Compressors are usually larger than packaged units and have radial-flow or axial-flow stages usually mounted in the center span of the rotor between two fluid film bearings. These machines should be monitored with XY radial proximity probes at each journal bearing and at least one axial probe at the thrust bearing. If thrust position measurement is connected into automatic shutdown, then *two* axial probes should be installed in a *voting logic* configuration to reduce the possibility of false trips. If radial vibration channels are also wired into automatic shutdown, then extra *false trip protection* should be incorporated in the radial vibration monitors.

It is not necessary to mount a Keyphasor® probe on the compressor if the coupled driver operates at the same speed and is equipped with a Keyphasor.® However, if a gearbox is part of the system, then the compressor(s) on the opposite side of the gearbox from the driver should have a Keyphasor.® Temperatures of all bearings, oil supply, and ambient conditions should also be monitored. On axial compressors, an accelerometer can be a useful auxiliary measurement for determining blade disturbances.

Reciprocating Compressors of the horizontal type can usually be monitored by using X-Y proximity probes to observe the piston rod or plunger of the compressor in addition to X-Y probes at accessible main crankshaft bearings. Monitoring the *average position* of the plunger can determine rod misalignment, packing wear, and wear on the sliding element and cylinder liner. Monitoring the *dynamic motion* signal of the probe can determine rod vibration or flexure (deflection). This is most necessary on hypercompressors. A Keyphasor® could be installed on one of the drive rods to provide speed and timing information. An alternative location for the Keyphasor® may be found on the crankshaft or motor driver. The Keyphasor® should be installed so that the voltage pulse occurs when one cylinder is at top dead center to observe the relationship of all plunger positions as a function of stroke.

Screw Compressors have two rotors with interlocking lobes and act as positive displacement compressors. If the machine is equipped with fluid film bearings, the optimum installation should include X-Y proximity probes at each radial bearing of each rotor as well as at least one axial probe for thrust position measurement of each rotor. Thrust position monitoring of each rotor is very important because a thrust bearing failure means an axial rub will occur between the rotating elements. If rolling element bearings are used, monitoring with casing velocity transducers and Dual Path Monitors® is acceptable; however, due to the very close axial clearances between rotors, *axial shaft position* monitoring is still of importance. The case-mounted transducer should be a velocity pickup with a Dual Path Monitor® measuring both velocity and displacement vibration. Temperatures of machine components should also be monitored. Two Keyphasor® probes, one for each rotor, would be desirable.

Generators

Generators are usually supplied with fluid film bearings and should be monitored with proximity probes at each radial bearing. Some generators have relatively soft support structures and require casing vibration and *shaft absolute vibration* monitoring in addition to shaft relative monitoring. (Refer to the same discussion in the large steam turbine section.) If the generator is not *rigidly* coupled to the driver, an axial probe should be installed to measure thrust position and axial vibration. Temperatures should be measured at each bearing and also at the rotor and stator windings, oil system, and ambient environment.

Pumps

Large Pumps such as those used in boiler feed water service or reactor coolant feed service in nuclear plants are generally equipped with fluid film bearings and should be monitored with X-Y proximity probes at each radial bearing and at least one thrust probe. A Keyphasor® should also be installed if pump speed is different from driver speed. Temperature of each bearing should be monitored, along with ambient and oil system temperatures.

Medium Pumps are generally used in semicritical or noncritical services and can be equipped with either fluid film or rolling element bearings. If economically justified, sleeve bearing pumps in semicritical service could use transducers as previously discussed for large pumps, but all transducers need not be monitored continuously. Those in noncritical service should have the same transducers permanently installed with measurements being made on a periodic basis as a minimum or ideally with a simple scanning monitor system. Pumps with rolling element bearings (particularly multistage designs) will usually show significant shaft deflection away from the bearing, so proximity probes would seem appropriate. However, in many cases such installations are difficult due to the small machine size, so velocity transducers may be used with a Dual Path Monitor®. A Keyphasor® should be installed unless provided on the driver at the same rotative speed; and the machine casing, oil system, and ambient temperatures monitored if economically justified.

Small Pumps are usually noncritical, are generally spared machines, and usually employ rolling element bearings. The economics generally favor only periodic measurements with portable instrumentation. On these machines, periodic measurements with case-mounted transducers may occasionally be selected for preventive maintenance; or some type of simple scanning monitor would be even more ideal.

Note also that large-scale acoustic incipient failure detection (IFD) systems have performed this function in modern plants. Refer to Volume 1 of this series for details.

Some small pump designs have flexible rotors (particularly overhung models) and/or squeeze-film damper rolling element bearings. These could be monitored with proximity probes measuring shaft motion relative to the *bearing support* structure.

Reactor Coolant Pumps and other pumps in nuclear service inside the containment of a reactor should receive special consideration. The operation of all of these pumps in a nuclear reactor is *critical*, so the best monitoring systems available should be installed. Many times these pumps are vertical and have one radial bearing at the top with a bushing and overhung impeller at the bottom. Since it is impossible at this time to monitor the shaft at the lower bushing of the pump on many of these machines due to pressures, temperatures, and Nuclear Regulatory Commission (U.S.) regulations, the top bearing of the machine should be monitored with X-Y proximity probes. Experience has shown that the motor generates significant casing motion and shaft motion relative to the casing. Therefore, it is desirable to install X-Y seismic transducers (velocity at the top and bottom motor bearings) in addition to X-Y proximity probes at the same bearings. It may be difficult to install the proximity probes, but the principal manufacturers have had good experience and have gathered meaningful data from X-Y proximity probes installed below the bottom motor bearing and above the coupling hub. A Keyphasor® should be installed on the pump or driver, and bearing temperature measurements are recommended.

Gears

Gearboxes in the form of speed increasers or reducers are usually equipped with fluid film bearings and should be monitored with X-Y proximity probes at each bearing. Since this means eight probes for the common four-bearing gearbox, economic considerations may dictate that the monitoring system be compromised. In any event, on large gear units all of the recommended transducers should be installed permanently for future diagnostic capabilities, and the input and output shafts should be continuously monitored at the coupling-end bearings. One or both rotors should be monitored for thrust position with at least one probe per rotor. Single helical gears have thrust bearings on both shafts to absorb the helical-generated thrust loads; therefore, axial probes for each rotor are *mandatory.* The thrust, or positioning bearing of double helical gears has little or no normal thrust loading, but gear coupling lockup could cause severe damage to the thrust bearing. Double helical gears usually have only one thrust bearing, typically on the bull gear. Therefore, that thrust bearing rotor should be monitored with at least one axial probe. For the nonthrust bearing rotor, it is desirable to have at least the axial probe permanently installed for full diagnostic capability. The thrust probes can also be used to monitor *axial vibration,* an important indicator of gear condition. In addition, bearing temperatures, oil system and ambient temperatures should be monitored. An auxiliary accelerometer can be used to monitor the high frequency excitations produced by gear tooth interaction. In some applications, it may be desirable to monitor the auxiliary accelerometer(s) with a Dual Path Monitor®. Smaller gearboxes with rolling element bearing may be monitored with casing velocity transducers and Dual Path Monitors®.

Fans

Fans have many industrial applications. Some large fans such as those used for forced-draft and induced-draft applications have fluid film bearings and thus fall into the same category as large compressors and blowers. They should be monitored with X-Y proximity probes as recommended for those machines. On large fans, *shaft bows,* both thermal and mechanical, create large problems. Therefore, proximity probes are necessary to differentiate between shaft bow and actual rotor dynamic action. This is particularly true for *balancing* work. Many of these machines have relatively flexible bearing support structures and/or a history of foundation problems. For these machines, bearing housing vibrations should be included with the shaft measurements. If it is not economically feasible to use dual probes on a particular large fan, then casing velocity transducers with Dual Path Monitors® may provide a convenient compromise. If the bearing housing is sufficiently compliant to exhibit enough shaft-generated motion, then casing measurements may be appropriate. In any event, unmonitored proximity probes should be installed permanently for balancing and analysis, and should be measured periodically (particularly the average gap DC output). For smaller fans with rolling element bearings, if significant shaft-generated motion can be detected on the casing, then velocity transducers may be used with Dual Path Monitors®. When case-mounted transducers are used, the monitor may require special low-frequency compensation electronics depending on the speed of the fan.

Cooling Tower Fans represent a unique application in rotating machinery. The prime malfunction of most of these machines is fan *unbalance.* Therefore, it would be desirable to monitor shaft motion with proximity probes. However, there is no suitable location to mount proximity probes as the entire tower structure exhibits significant dynamic motion. In addition, bearing (rolling element) and gear problems account for many fan failures which further supports the need for case measurement. These fans should be monitored with a velocity transducer mounted on the right angle gearbox and possibly a Dual Path Monitor® with alarms and readings in terms of velocity and displacement units. The velocity transducer must have adequate environmental protection and special frequency compensation must be provided in the monitor to offset the limitations of the transducer.

Centrifuges

Centrifuges have traditionally been monitored with seismic pickups measuring casing motion. However, some centrifuge designs allow the installation of X-Y proximity probes mounted to the bearing supports or to the outside housing of the machine. On several applications, shaft measurement has proven to be an earlier indicator of potential problems. It is expected that shaft measurements will be used much more extensively in the future. A proximity probe is also *essential* to monitor axial position between the inner and outer drums of certain centrifuges.

Pulp Refiners

Pulp Refiners, as used in the paper manufacturing industry, have received increased attention in recent years from the standpoint of vibration monitoring. The initial approach to monitoring these machines used velocity pickups mounted radially at each bearing. This method has been somewhat successful, especially when used with the Dual Path Monitor®. Recently, the monitoring equipment manufacturers report seeing proximity probes used in place of the velocity pickups on some machine designs, and this has also been successful. In addition, at least one proximity probe, or two for double (counter rotating) disk refiners, should be installed, on each separate rotating element, to provide speed, phase angle and timing information for the machine.

Principles of Machinery Condition Monitoring

The preceding segment of this chapter dealt with the basic parameters for predictive maintenance on rotating machinery. Having acquired an understanding of these parameters allows us next to address their integration into an overall monitoring system.

Machinery engineers in the process industries are aware of the shift in approaches that has taken place in the last few decades. The advent of machinery monitoring technology has enabled process plants to shift from the preventive maintenance of a few decades ago to essentially predictive maintenance in the mid-1970's and 1980's. However, there has been an unfortunate tendency since the late 1970's to place much emphasis on machinery diagnostics where prudence and in-plant experience should have directed primary attention on operator-friendly surveillance instead. Although not necessarily within the scope of a text on machinery maintenance and repair procedures, the topic is of such great importance that an overview is, indeed, appropriate.

Our perception and field experience is shared by the authors of a 1980 technical paper on the subject.* They, too, noted that rapid advances in electronic technology, particularly the microcomputer, have suddenly made it feasible to perform tasks which heretofore have been either impossible or prohibitively expensive. Within the field of machinery condition monitoring so much can now be accomplished that it is easy to lose sight of the environment and results required, and allow ourselves to be led astray by elegant solutions to uncommon, minor or nonexistent problems. While added power and features may often appear highly desirable, each must be carefully evaluated on its merits: Is it necessary? Does it add value? At what effect on reliability? What is its cost compared to the benefit expected? Appropriately then, we will attempt to take a fresh, objective look at machinery condition monitoring as well as a number of areas in which the introduction of new concepts and/or technology holds promise for a far better condition monitoring system.

Definition and Objective of Machinery Condition Monitoring

For the purposes of this text, condition monitoring is defined as a method or methods of surveillance designed to recognize changes from a norm and warning when the changes exceed safe or limiting values. As defined, condition monitoring must differentiate between good and bad condition, and if bad, answer the question of how bad.

At this point, it is important to distinguish between surveillance and diagnostics. Whereas the former is necessary for protection, the latter is not generally required until it becomes necessary to identify the source of a known anomaly. The medical analogy has been overused; however, it bears repeating. Most people are not able to justify the high cost of a rigorous and detailed physical examination including a stress EKG, EEG, and possibly CAT scan, unless they have good reason to suspect a problem. In this case, surveillance is effectively provided by our physical senses, how well we judge our own health, augmented by an occasional routine examination. When a problem does occur, detailed diagnostic efforts are focused on the specific symptoms to ascertain their exact cause, and to determine the necessary corrective action. Thus, diagnostic efforts which are generally not justified in the absence of a problem are easily justified once a problem has been recognized.

* "Time Marches on—Changing Concepts in Machinery Condition Monitoring," by John S. Mitchell, Palomar Technology International, Carlsbad, CA. and John L. Frarey, Shaker Research Corp., Latham, NY. Adapted by permission.

To continue this line of thinking, let's attempt to characterize machinery. First of all, and fortunately for us, the vast majority of machines are well engineered, operate well and tolerate without difficulty the abuse to which all are subjected from time to time. Problems, if any, are few and far between and minor in nature. Changes in condition typically develop slowly, and one can almost always judge severity with a high degree of accuracy even though the exact cause may be unknown.

Based on the foregoing, we submit that condition monitoring technology may be concentrated on performing the surveillance task efficiently and reliably. Diagnostics, to define the nature of any anomaly, are secondary as they are not immediately required to judge severity, nor do they, in most cases, improve the ability of a well designed surveillance system to recognize problems. The two functions can, however, be supportive and work together in the same system, but only if extreme care is taken not to confuse the two.

Environment in Which Condition Monitoring Must Function

A condition monitoring system must be designed for use by those who have operating responsibility for the machinery. Operating personnel understand measurements and trends, but generally do not have the same appreciation for, and understanding of, measurements made to describe machinery condition that they have for measurements of operating variables. Temperature, pressure, flow and level, typically have more meaning to an operator than vibration amplitude.

In an operating environment, diagnosing the cause of a problem is typically of little importance compared to assessing its severity. The question that must always be answered is will the machine continue to run? If so, how long and when is it mandatory to shut down?

Objectives of an Optimum Condition Monitoring System

An ideal machinery condition monitoring system should:

1. Reliably assess mechanical condition and thermodynamic efficiency. Display the results in a form such as Figure 13-1, which is readily understood and can be acted upon by those having operating responsibility for the machinery.
2. Quickly and accurately recognize changing conditions and their significance.
3. In the event of changing conditions, or a problem, the system must have enough intelligence to assess and display severity in a form which does not require additional interpretation.

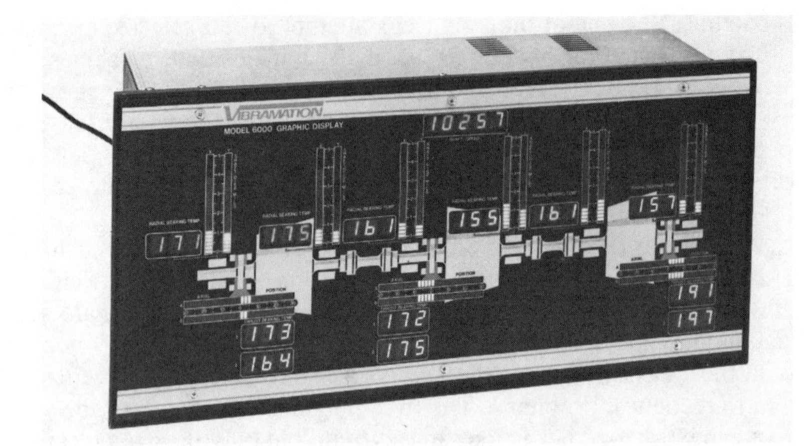

Figure 13-1. Graphic display of a condition monitoring system.

4. Provide a means to implement, if desired, an automatic shutdown which is accurate and reliable, protected against actuation by an instrument failure and, where possible, provide time for corrective action.

Secondary objectives of a machinery condition monitoring system are:

- Be capable of accepting and operating on both continuous and periodic information. Periodic information can be obtained with a portable terminal of the type illustrated in Figure 13-2.
- Be able to distinguish between valid and invalid information.
- Maintain annotated data files for long term trending and historical records.
- Maintain short term (10 to 15 minutes), detailed files of all continuous data which can be frozen by an event for later detailed analysis.
- Be capable of communicating with an external diagnostic system or central computer used for storage and/or management reports.
- Be ready to implement with data available at several points with minimum cost for interconnecting wiring.

Specific Recommendations for a Condition Monitoring System

In making the following recommendations, the authors have drawn heavily on experience with current condition monitoring systems, knowl-

Figure 13-2. Portable data acquisition terminal.

edge of the environment in which condition monitoring must function, as well as machinery and machinery problems. Recent developments in the philosophy and strategy of process control systems have been studied and incorporated where applicable and possible.

Before outlining specific recommendations for an advanced intelligent condition monitoring system designed to meet the objectives described in the previous section, we should briefly mention the criteria on which the concept is based.

First, and as we have pointed out earlier, any condition monitoring system must be primarily oriented toward providing accurate and reliable surveillance and protection. The addition of intelligence to enable continuously displaying *condition based on measurements* instead of measurements alone, is considered an important factor to maximize the utility and effectiveness of a condition monitoring system for operating personnel. Recognizing meaningful changes among related measurements and including rate of change in the assessment of current condition are both considered a necessary part of the intelligence that must be provided.

Second, the surveillance strategy to be described is directed toward the largest population of machines which require uncompromised protection, but generally operate with few problems. For this reason, features that are primarily diagnostic such as on-line spectrum analysis and automated response plotting on startup and shutdown are not included, but could be easily added when conditions warrant. As a by-product of this objective, the cost per channel of an optimized surveillance system in basic form will be less than that of current systems, thereby making it attractive for balance of plant applications.

Finally, any advance in surveillance technology must not be accomplished at the expense of reliability. We have thus added requirements such as redundant data busses with automatic transfer on failure and the ability to provide full protection with the data link between a remote alarm unit and the system controller completely failed.

A conceptual condition monitoring system meeting the objectives we have outlined is shown in Figure 13-3. The system consists of three major functional subsystems: a Data Acquisition Unit, Alarm Module, and System Controller. The three subsystems are connected together by redundant data busses with automatic transfer and alarm in the event of a failure. Once programmed and operating, the system must continue to provide protection even though both data links to the System Controller may be broken.

Within the concept it is envisioned that full protection at minimum cost can be provided by a combination of Data Acquisition Units, Alarm Modules and a simple System Controller for programming and display. Since all the information collected by the system is transmitted over the data bus, adding intelligence, operator oriented displays, fill capability and even diagnostics are easily accomplished by expanding the system controller.

Data Acquisition Unit

The Data Acquisition Unit provides the interface between the condition monitoring system and the various transducers used to measure machine condition. The Data Acquisition Unit provides power, where required, to the transducers, performs whatever conditioning is necessary to the input signals, and produces a normalized DC output proportional to each measured variable. It must be able to accommodate the signal from any conceivable transducer used in machinery surveillance in either analog or digital form in any mix. System inputs will be obtained from proximity (vibration and gap), acceleration or velocity transducers, and temperature and pressure sensors or may be a DC voltage or current (4-20mA) and contact closures.

The Data Acquisition Unit must have, where possible, the ability to determine the validity of an input signal. Providing a window detector to sense when each input is within its linear or expected maximum range and invalidate those which are outside should be adequate. Within the Data Acquisition Unit validated, conditioned DC outputs will be supplied to a high speed multiplexer where they, along with identification, are placed on a data bus.

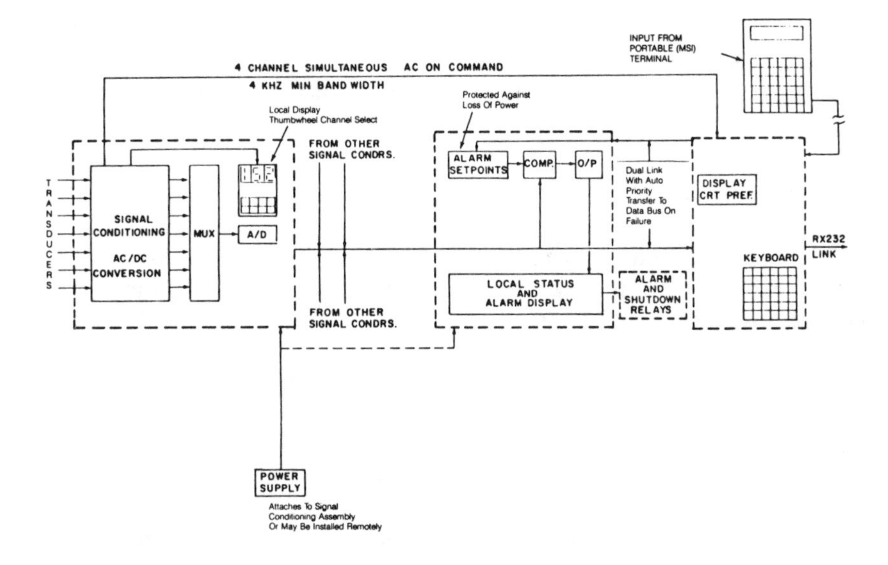

SIGNAL CONDITIONING

1. 32–64 channels per rack with addresses for rack and each channel.
2. Signal conditioners—
   a) analog with filtering where applicable;
   b) full scale range and engineering unit identifier
3. Designed for intrinsic safety (with remote power supply and barriers).
4. Capable of monitoring gap voltage.
5. Signal to indicate and identify failed channels.
6. X3 multiplier.

ALARM SYSTEM

1. Capable of comparing 100–200 channels at .5 sec. max interval between scans. More channels possible at slower rate.
2. Must be able to continue to function and provide undiminished protection with controller link broken.
3. Alarms validated by channel of signal from signal conditioning.
4. Auto restart following power outage.
5. Local alarm display could be programmable matrix over which a graphic representation could be overlaid.

CONTROLLER

1. Capable of controlling system by itself, allow remote programming and display.
2. Capable of controlling up to 1000 channels.
3. Display: preferably CRT, need flexibility to allow programming and display by alphanumeric reference designators.
4. Small keyboard for programming.
5. Capable of ordering and outputting up to 4 simultaneous AC outputs.
6. Maintains and displays historical files.
7. Program functions protected.
8. Preset programming such as ability to set all alarms simultaneously to a predetermined criteria.

Figure 13-3. Schematic of a condition monitoring system.

In addition to continuously supplying measurements, in digital form, to the data bus, the Data Acquisition Unit must also be capable of supplying AC signals, on command, for use elsewhere in the system.

Alarm Module

The Alarm Module may receive inputs from one or more Data Acquisition Units. Within the Alarm Module the incoming measurements are continuously compared to preprogrammed setpoints. Outputs, both local and remote, identifying the alarming channel must be produced for machine protection whenever a channel exceeds its setpoint. A scan rate in which successive comparisons of each channel occur at a maximum interval of .5 second is considered adequate to ensure protection. Within this interval, many options exist such as rapid sampling with .5 second averaging or voting techniques.

At this point, it should be emphasized that a Data Acquisition Unit and Alarm Module are intended as an optimized combination capable of providing full machine protection. Further, it is anticipated that this combination, together with a basic System Controller, will provide equivalent protection and reliability at lower cost compared to current analog systems.

System Controller

The final component of the monitoring system is a System Controller. Capable of controlling a number of Data Acquisition Units and Alarm Modules to a maximum of approximately 1,000 channels, the System Controller will provide control and display functions for the system, manage the information on the data bus, and maintain files of historical data. It can be located remotely from the Data Acquisition and Alarm Units, and should require only a single coaxial cable for full communication. Typically, two cables with an automatic transfer on failure should be provided for reliability. The System Controller is the heart and intelligence of the surveillance system, and as such, must be easily and economically adaptable to provide the specific functions required for a given application.

With this objective in mind, we will briefly describe all the functions of a System Controller considered necessary to realize the full benefits attainable with an advanced intelligent surveillance system. Recognize that functions not vital to system operation should be options that can be easily and economically added when required.

Functions to be performed by the System Controller are as follows:

1. Provide access to and control of the monitoring system. Most efficiently applied through a keyboard, an operator must be able to assign identification numbers to each channel, set and change alarm setpoints, call up for display the current value and/or status of any channel and perform all other control functions which might be required to operate the system. Access and control should have at least two levels of security with a higher level required to change control functions than would be required to change alarm setpoints.
2. Display any system measurement or alarm setpoint, in engineering units.
3. Provide a means for communicating with an external computer.
4. Provide a means to recover simultaneous AC signals from several channels for display, recording, or detailed analysis.
5. Accept periodic data from a portable terminal entered through the data bus.
6. Calculate and monitor thermodynamic efficiency from periodic and/or continuous measurements.
7. Recognize and call attention to bad or suspect measurements.
8. Maintain and display on demand, numerically or graphically, short and long term information files.

The short term file should be a circulating file containing the last 10 to 15 minutes of all continuous data recorded at approximately one second intervals. It must be capable of being frozen or transferred to permanent storage by an event such as an emergency trip so that data before, during and after the event are preserved for study and analysis. The definition of an event which freezes the file and the length of time after the event when the file is frozen should both be programmable. A means must also be provided to quickly recover the exact sequence at which the recorded data changed to identify the cause of the event.

The long term file should consist of approximately one year of data in a sequence such as the following:

a. One day file consisting of instantaneous values every five minutes, plus the minimum and maximum values recorded during each five minute period.
b. One week file consisting of an instantaneous value every hour or a one hour average of five minute measurements plus minimum, maximum.
c. One year file consisting of an instantaneous value every 12 hours or a 12 hour average of one hour measurements plus minimum, maximum.

In choosing between the use of an instantaneous or average value, we believe an instantaneous value, along with minimum and maximum values, provides much greater detail than average values. Trend logic could be utilized to blend data in short term files into longer term files.

The system should be able to accomplish the following from and on filed information:

a. Recognize changes in operating status; start, stop, and annotate files accordingly.
b. Recognize trends, identify measurements that are changing outside of predetermined limits.
c. Recognize and call attention to an increasing difference between maximum and minimum values.
d. Recognize and provide warning if, at some predetermined time after startup, input data are not within the minimum and maximum values recorded prior to shutdown or within some percentage of a "normal" value.

9. Provide a means to display overall condition in a form which is easily interpreted by operating personnel. The ability to group confirming measurements, assign weights and combine groups into a measure of overall condition is considered feasible, possible and highly advantageous.

As a suggested strategy to achieve this objective, a condition hierarchy might be considered as follows:

Level 0 All measurements within each group at or below approximately 75 percent of the low alarm (alert) value.
Level 1 Two or more confirming measurements in a group between 75 percent and 100 percent of the low alarm value.
Level 2 Single measurement between low and high alarm (danger) values.
Level 3 Two or more confirming measurements between low and high alarm values.
Level 4 One measurement above high alarm (danger) value.
Level 5 One measurement above high alarm value, one or more confirming measurements above low warning value.
Level 6 Two or more confirming measurements above high warning value, slow rate of increase.
Level 7 Two or more confirming measurements above high warning value, high rate of increase.

Within this hierarchy a slowly increasing measurement will advance the condition indicator one level, and a rapidly increasing measurement will advance the condition indicator two levels.

Determined as outlined, the condition indicator could be easily applied to drive an additional bar indicator on the graphic display illustrated in Figure 13-1, or change the color of a graphic representation of the machine on a CRT display. The numerical value should also be presented for an operator who might be color blind. In addition to displaying the overall condition of a machine, we anticipate a need for the ability to look at the condition indicator determined from a specific group of measurements that gauge the condition of an individual component. Along with the requirement for higher resolution to view individual components, there must also be provisions for displaying individual measurements, organized in groups, with measurements above warning values or changing highlighted for easy recognition.

What, then, is our overall conclusion? Based on observations of the environment in which machinery monitoring systems must function, the ability to consolidate and refine information into an easily understood condition indicator is the greatest single advance that can be made in monitoring technology. The strategy just described follows closely the thought process and logic used by an experienced person deciding whether to continue operating or to shut down and can be automated with a high degree of reliability if desired. Combined with a system capable of efficiently gathering and transmitting a wide range of condition related information, the final result will be far more effective and operator oriented than a system capable of displaying measurements alone.

Minimizing Electrical Runout During Rotor Manufacturing*

This vibration monitoring segment of our text would not be complete without comments on problems relating to eddy-current based transducer accuracy.

Noncontacting eddy-current proximity probes have proven to be the most successful transducers for vibration measuring and monitoring of high-speed rotating machinery. The proximity probe can prove superior to other vibration sensors, particularly velocity transducers and accelerometers, because of one major factor: The probe observes shaft motion directly while velocity and acceleration sensors observe bearing housing or machine casing motion. This can be a major advantage because the most frequently occurring machine malfunctions, unbalance, misalignment, rubs, whirls, etc., produce direct changes in shaft dynamic motion and/or average shaft centerline position within the bearing clearance.

---

* Source: Bently Nevada Company, Minden, Nevada 89423. Adapted by permission.

The bearing housings or machine casing do not directly show this change in shaft motion due to mechanical impedance, i.e., damping within the machine, phase lag, structural resonances and other factors.

Therefore, the proximity probe observes that part of the machine which is most indicative of the running condition of that machine. In addition, the casing mounted transducers have potential problems, electrical and mechanical noise and mechanical fatigue, for example, which are not associated with proximity probes. The probes, however, are not without disadvantages. Other than the isolated probe installation limitations on some machinery types, the only significant difficulty with proximity probes results from direct observation of the shaft. Proximity probe operation depends on a smooth, concentric shaft surface with consistent electrical properties of the shaft metallurgy.

"Runout" and "glitch" are terms associated with a noise source on the output signal of a proximity probe transducer. As has been documented in the literature, this runout is of two types—electrical and mechanical. The literature sufficiently explains how electrical and mechanical runout can be detected, analyzed, and corrected or compensated. Typically, electrical runout has not been a problem on the vast majority of new machinery shaft materials. Our data indicate that 90+ percent of all new machinery specified with proximity measurement probes has acceptable glitch levels. However, in the 10 percent of the cases where glitch is a problem, it can be significant, especially during machine acceptance testing at the manufacturer's shop when low levels of runout are a part of the purchase contract.

Obviously, if a shaft has a potential glitch problem, it should be discovered long before the machine reaches the test stand. So our first general recommendation to manufacturers is to check for glitch early in the shaft manufacturing procedure. Our further recommendations are in the areas of material selection, forging, material handling and finishing, and other processes in the manufacturing cycle.

Shaft Material Influence on Glitch

Electrical runout is generally caused by nonhomogeneous metallurgy in that portion of the shaft observed by the probe. It is recognized that many other factors, in addition to susceptibility to electrical runout, influence the selection of a particular shaft material. However, some general knowledge of shaft material and its susceptibility may help where material application flexibility is available.

Shafts machined from 1000/4000-series steels usually present minimal glitch problems as long as high-quality material is specified. Vacuum remelt or vacuum arc melt "aircraft" steels tend to be more uniform and thus yield lower glitch levels.

The 1000-series steels exhibit more electrical runout in an untreated state than the 4000-series but respond more readily to glitch removal techniques.

Forged Shafts

Forged shafts can present a potential electrical runout problem. Again, quality of the raw stock and the forging process are important. The handling of the hot forging can influence its glitch performance. Practices such as picking up the hot shaft in the probe area can selectively quench the shaft, thus inducing glitch. Forged shafts that are double tempered (such as used on steam turbines) appear to exhibit lower susceptibility to glitch. This is probably due to the more uniform tempering of the steel. Any specification or testing procedure, such as "ultrasonic inspection," which will ensure metallurgical uniformity of the shaft, will improve its glitch performance.

Heat Treating

Uniform heat treating of a shaft can improve glitch; however, nonuniform heat treating can induce glitch. In order to minimize glitch, tempering must be uniform. Rotating the shaft while treating may help in this respect. Certain case hardening techniques appear to cause glitch.

Precipitation Hardening

Steel of 17-4 pH nearly always presents an electrical runout problem. Some form of material replacement (shrink a collar, overspray a material) is normally required to eliminate glitch.

Other pH steels such as 15-5 pH seem less prone to glitch but any pH steel may cause difficulties.

Finishing Procedures

The finishing procedures used on the last 10 mils (250 micrometers) of shaft radius can significantly affect electrical runout.

Turning to a rough diameter, then wet grinding to ensure concentricity appears to cause little difficulty. Make sure that the finish feed rate and cooling is sufficiently controlled to avoid hot spots or chatter, especially during the last 5 mils (125 micrometers). The completion of this procedure generally results in a 15 to 35 microinch rms finish. At this stage, one of several final finishing procedures is often used; "micro" or "super" grinding, honing or lapping, or burnishing. Microgrinding may cause glitch. Honing is generally better. Burnishing, especially diamond burnishing, not only yields a bearing quality finish but in most cases substantially reduces glitch previously present. Demagnetizing after mechanical finishing has been known to reduce glitch, and is generally a necessity after grinding.

Handling

Care must be exercised to ensure that mechanical damage does not occur to an otherwise glitch-free surface. Nicks or scratches are obvious mechanical runout problems but an impact which does no deformation can work-harden the area and cause electrical runout.

Magnetic Particle Inspection

Magnetic particle inspection ("Magnafluxing") is a common practice applied to machine shafts. Residual magnetism from this or other operations must be removed by degaussing (demagnetizing), or glitch will result. The manufacturer of the particular testing system should be consulted as to degaussing equipment. A residual level of magnetism may be acceptable as long as it does not exceed 2 gauss.

Diamond Burnishing

Diamond burnishing is a very effective glitch reduction technique. To accomplish this technique: The shaft should be rotated in the lathe by hand or driven at 1 to 5 rpm for the purpose of observing the dial indicator readings. These readings should be recorded and the high and low levels noted as to their locations on the shaft surface. The dial indicator may now be removed from the rotor surface and the rotor operated to 20 to 50 rpm. The output of the proximitor should be observed on the oscilloscope during this initial run. A photograph of the scope trace should be taken at this time to document the initial runout pattern.

A diamond-tipped burnishing tool may now be applied per the instructions supplied with the tool. Upon completion of the burnishing operation, the oscilloscope trace of the transducer output can again be observed and compared to the original trace. If the electrical runout has not been reduced to an acceptable level, a second or third burnishing operation may have to be performed. In general, no further reductions will be achieved after a third pass of the burnishing tool. The final dynamic waveform on the oscilloscope should be photographed to document the reduced runout levels achieved with the burnishing operation.

Sample shafts which were burnished in an engineering lab consistently realized reductions in runout levels of 40 to 60 percent. In actual machine manufacturers' plants, runout level reductions of 80 to 90 percent have been achieved.

Your equipment procurement or repair specification could call for the integration of the preceding testing and burnishing operations into the normal manufacturing cycle. This should be done after the final machining or grinding operations have been performed on the shaft. If this procedure is regarded as a normal requirement, the diamond-tipped burnishing operation offers an efficient cost effective procedure for helping to provide glitch-free rotating elements.

Material Replacement

Shrink fit of a collar or sleeve, although cumbersome, will certainly control glitch. Some attempts at "flame spray" material replacement have worked but require careful operator technique. Current reevaluation of plating techniques shows promise but is preliminary and must be field tested for durability.

Bibliography

AGMA Standard 426.01, "Specification for Measurement of Lateral Vibration on High Speed Helical and Herringbone Gear Units," April 1972, American Gear Manufacturers Association, Arlington, Virginia.

API (American Petroleum Institute) Standard 612, "Special Purpose Steam Turbines for Refinery Services," Second Edition, June 1979.

API Standard 617, "Centrifugal Compressors for General Refinery Services," Fourth Edition, November 1979.

API Standard 670, "Noncontacting Vibration and Axial Position Monitoring System," First Edition, June 1976, American Petroleum Institute, Washington, D.C.

Arant, J. B. and Crawford, W. A., "Compressor Instrumentation Systems—Problems and Cures," presented at the 15th ISA Chemical and Petroleum Industries Instrumentation Symposium, San Francisco, California, 1974.

Bently, Donald E., "Need Advance Warning of Thrust Bearing Failure?", *Hydrocarbon Processing*, January 1970.

Bently, Donald E., "Shaft Motion and Position—Keys to Planned Machine Maintenance," presented at the Annual Meeting of the Technical

Association of the Pulp and Paper Industry, Miami Beach, Florida, 14-16 January 1974, *Tappi* Volume 7, Number 7, July 1974.

Bently, Donald E., "Monitor Machinery Condition for Safe Operation," presented at ASME Petroleum Mechanical Engineering Conference, Dallas, Texas, 15-18 September 1974, *Hydrocarbon Processing*, November 1974.

—Applications Note, "Standardized Rules for Measurements on Rotating Machinery," Bently Nevada Corporation, January 1978.

—Applications Note, "Polar and Bodé Plotting of Rotor Response," Bently Nevada Corporation, November 1978.

—Applications Note, "The Keyphasor—A Necessity for Machinery Diagnosis," Bently Nevada Corporation, February 1979.

—Applications Note, "Data Presentation Techniques for Trend Analysis and Malfunction Diagnosis," Bently Nevada Corporation, July 1979.

—Applications Note, "Vibration Measurement—Basic Parameters for Predictive Maintenance on Rotating Machinery," Bently Nevada Corporation.

Brozek, Bob, "How to Determine Motor Vibration Requirements," General Dynamics Corporation, Electro Dynamic Division, Avenel, New Jersey.

Clapis, A., Lapini, G., and Rossini, T., "Diagnosis in Operation of Bearing Misalignment in Turbogenerators," presented at the ASME Design Engineering Technical Conference, Chicago, Illinois, 26-30 September 1977, ASME publication 77-DET-14.

Diehl, G. M., "Machinery Vibration Analysis," *Chemical Engineering Progress*, May 1970.

Dodd, V. Ray, "Machinery Monitoring Update," presented at the Texas A & M Sixth Turbomachinery Symposium, Houston, Texas, December 1977.

Dwyer, John J., "Compressor Problems: Causes and Cures," *Hydrocarbon Processing*, January 1973.

Eisenmann, Robert C., "Shaft and Casing Motion of Large, Single Shaft, Industrial Gas Turbines," presented at the Machinery Vibration and Analysis Seminar, Vibration Institute, New Orleans, Louisiana, 9-13, April 1979.

Harker, Roger G., "A New Turbine Supervisory Instrumentation Package," Bently Nevada Corporation, August 1979.

Hibner, D. H., Kirk, R. G., and Buono, D. F., "Analytical and Experimental Investigation of the Stability of Intershaft Squeeze Film Dampers," presented at the ASME Gas Turbine Conference and Product Show, New Orleans, Louisiana, March 21-25, 1976, *Journal of Engineering for Power*, January 1977.

Hudacheck, R. J., and Dodd, V. R., "Synthesis—Gas Compressor Failure," *Oil and Gas Journal*, September 27, 1971.

Jackson, Charles, "Optimize Your Vibration Analysis Procedures," presented at the Petroleum Division Conference, Institute of Electrical and Electronic Engineers, Houston, Texas, September 1973, *Hydrocarbon Processing*, January 1974.

Jackson, Charles, "The Practical Vibration Primer," Gulf Publishing Company, Houston, Texas, 1979.

Jackson, Charles, "How to Prevent Turbomachinery Thrust Failures," *Hydrocarbon Processing*, June 1975.

Jackson, Charles, "Techniques for Alignment of Rotating Equipment," *Hydrocarbon Processing*, and "Alignment with Proximity Probes," edited by Bently Nevada Corporation.

Jackson, Charles, "Vibration Measurement on Turbomachinery," *Chemical Engineering Progress*, Volume 68, Number 3, March 1972.

Makay, Elmer, and Szamody, Olaf, "Survey of Feed Pump Outages," Energy Research and Consultants Corporation for Electrical Power Research Institute (EPRI), Palo Alto, California, April 1978.

Maxwell, A. S., "Vibration Monitoring—The Search for Optimum Protection," presented at the fourth Turbomechanics Seminar, Ottawa, Ontario, September 23, 1976.

McClain, L. R., "Vibration Monitoring—The Transition from Using Case and Bearing Cap Vibration Readings to Shaft Readings Obtained from Proximity Probes," Commercial Solvents Corporation, August 27, 1971.

Nimitz, Walter and Wachel, J. C., "Vibrations in Centrifugal Compressors and Turbines," presented at the ASME Petroleum Mechanical Engineering and Pressure Vessels & Piping Conference, Denver, Colorado, September 13-17, 1970.

Peters, G., "Vibration and Strain Measurement on Turbogenerators," *V.G.B. Kraftwerkstechnik*, Volume 53, Number 4, pages 224-233, April 1973.

Prentice, J. S., Smith, S. E., and Virtue, L. S., "Safety in Polyethylene Plant Compressor Areas," *Chemical Engineering Progress*, Volume 70, Number 9, September 1974.

Schollhammer, F., "New Experience with Shaft Vibration," *V.G.B. Kraftwerkstechnik*, Volume 53, Number 5, pages 320-332, May 1973.

Stuart, John W., "Retrofitting Gas Turbines and Centrifugal Compressors with Proximity Vibration Probes," Pacific Gas Transmission Company, San Francisco, California, 1974.

Wett, Ted, "Compressor Monitoring Protects Olefin Plant's Reliability," *Oil and Gas Journal*, September 10, 1973.

Chapter 14

Maintainability Considerations

We started this volume by pointing out how important a role maintainability plays within the "total picture" concept of machinery management. Throughout this text we have been stressing that if one has to do maintenance at all, if it cannot be easy, it should at least be possible. What, then, should we be watching for, if we want to address or even maximize maintainability? We define maintainability as everything that facilitates maintenance activities around a piece of process machinery. This includes the ease of supplying the necessary services to the machine and its availability, which should be reduced as little as possible by the maintenance activities. Our check list is as follows:

1. Safety. Maintenance personnel should not be subject to safety hazards while carrying out maintenance activities. Therefore:

- No dangerous elements must endanger personnel.
- Personnel must not be able to come in contact with components that carry electric tension.
- Maintenance personnel must not be exposed to moving components.
- Maintenance personnel must be protected from falling by having a safe work location at all times.

2. Availability. Maintenance activities should reduce the availability of the equipment as little as possible. Therefore:

- As many operations as possible should be carried out while the equipment is running normally.

- In the event that certain components or subsystems require shutdowns for maintenance activities, as few components as possible should be shut down.
- Machinery equipment must be accessible.
- If certain functions must be shut down for maintenance activities, the time required for shutting down the equipment should be as brief as possible.
- Should certain components require shutdowns for maintenance activities, these maintenance activities should be required as rarely as possible.
- Should certain maintenance activities require shutdown, the time needed for these activities should be as brief as possible.

3. Damage Potential. Very frequently, maintenance activities generate a substantial risk of damage to machinery equipment. In order to reduce this risk as much as possible:

- Should the application of certain media—for instance, lubricants and so forth—cause some risk of damage to other components, these media and their frequency of application should be clearly defined.
- Should certain media cause a risk of damage to other components, these components should be protected from the dangerous media—for instance, by a cover.
- Should certain maintenance activities cause a risk of damage to certain components, the endangered components should be designed in a way that will protect them as much as possible against that risk.
- Should maintenance activities on certain components require such operations as dismantling, assembling, or moving, they should be designed to minimize the risk of damage to these or other components.
- Should certain maintenance activities produce risk of damage because the components are difficult to see, the system should be illuminated as well as possible.

4. Serviceability. Process machinery maintenance activities require some unique specifications in addition to those previously mentioned. Serviceability covers the special attributes the equipment needs to have for efficient lubrication and sometimes cleaning. In order to improve serviceability the following specifications should be met:

- The media to be used for servicing—lubricants, impregnants, and detergents—should have the highest possible efficiency and endurance.
- Servicing media should be supplied automatically; waste media should be removed automatically.

5. Inspectability. Some components need to be inspected during the preventive maintenance activities to detect defects before they have passed the defect limit. To facilitate these inspections:

- Should rules or regulations require certain inspections of either the equipment or the component, the design must facilitate the carrying out of these inspections.
- Components that develop time dependent defects and components that develop random defects that can be controlled by periodic inspection should be accessible for inspection.

6. Repairability. Components that might need repair during the lifetime of the equipment should meet the following specifications:

- Should local and national laws force the operator to carry out certain repairs, those repairs must be possible.
- Should repairs be necessary, the design of the components must ensure that all attributes that are required for the proper performance of the system will be restored by the repair.

7. Materials Availability. Maintenance activities quite frequently require the application of certain materials, including spare parts, lubricants, and so forth. In order to facilitate the provision of this material:

- Materials that might be needed during the lifetime of the equipment should meet the in-plant, local and national standards.
- Materials required for maintenance but not stored in the plant should be designed so that they can be purchased from outside suppliers on short notice.
- Material that cannot readily be purchased from an outside supplier must be designed in such a way that it can be produced from prestored raw materials on short notice.

The foregoing "wish list" could be used to prioritize the different factors or objectives and then evaluate different alternative machines. This topic ties into the theme of Volume 1 of this series of books. Dealing with machinery selection criteria, Volume 1 ("Improving Process Machinery Reliability") conveys the importance of choosing machinery not on price alone.

Index

Practical Machinery Management for Process Plants

Other Volumes in the series:

Volume 1: Improving Machinery Reliability, *Heinz P. Bloch*
Volume 2: Machinery Failure Analysis and Troubleshooting, *Heinz P. Bloch and Fred K. Geitner*
Volume 3: Machinery Component Maintenance and Repair, *Heinz P. Bloch and Fred K. Geitner*